IN SITU REMEDIATION ENGINEERING

IN SITU REMEDIATION ENGINEERING

Suthan S. Suthersan
Fred C. Payne

CRC PRESS

Boca Raton   London   New York   Washington, D.C.

Library of Congress Cataloging-in-Publication Data

Catalog record is available from the Library of Congress

This book contains information obtained from authentic and highly regarded sources. Reprinted material is quoted with permission, and sources are indicated. A wide variety of references are listed. Reasonable efforts have been made to publish reliable data and information, but the author and the publisher cannot assume responsibility for the validity of all materials or for the consequences of their use.

Neither this book nor any part may be reproduced or transmitted in any form or by any means, electronic or mechanical, including photocopying, microfilming, and recording, or by any information storage or retrieval system, without prior permission in writing from the publisher.

The consent of CRC Press does not extend to copying for general distribution, for promotion, for creating new works, or for resale. Specific permission must be obtained in writing from CRC Press for such copying.

Direct all inquiries to CRC Press, 2000 N.W. Corporate Blvd., Boca Raton, Florida 33431.

Trademark Notice: Product or corporate names may be trademarks or registered trademarks, and are used only for identification and explanation, without intent to infringe.

Visit the CRC Press Web site at www.crcpress.com

© 2005 by CRC Press

No claim to original U.S. Government works
International Standard Book Number 1-56670-653-X
Printed in the United States of America 1 2 3 4 5 6 7 8 9 0
Printed on acid-free paper

*My utmost and sincere thanks to
Sumathy, Shauna, and Nealon
for their unending patience and endurance during the writing of this book.*

*This book was made possible by the enduring love and support of
Therese, Aaron, Nathan, and Meredith, and my parents,
Harold and Dorothy Payne. Thank you all for the encouragement and
guidance.*

Foreword

Dr. Suthan Suthersan's ground-breaking book, *Remediation Engineering*, published in 1996, has led the environmental industry in the practice of *in situ* remediation of contaminated soil and groundwater. While much has been accomplished, many complicated problems remain. *In situ* reactive zone (IRZ) technologies are commonly practiced now, but they still need to be better understood by many practicing professionals. Even after all these years and hundreds of successful IRZ projects, large-scale pump-and-treat systems are still routinely part of remediation strategies. These systems, while often much more expensive, also have been found to be marginally effective in their cleanup objectives.

While the continued protection of human health and the environment is still at the forefront of the industry, economical solutions that actually meet regulatory requirements and return properties to beneficial use are now being demanded by society. It is no longer acceptable to spend millions of dollars over decades with little or nothing to show for what is often taxpayer money. These economic and time-sensitive drivers are leading to an entirely new push for guaranteed site closures. Dr. Suthersan and Dr. Fred Payne, through their continued research and applications of science and engineering to complex problems, are again leading the industry with solutions to the challenges of the new millennium.

This book is truly the next generation in the science of groundwater cleanup. *In Situ Remediation Engineering* provides a comprehensive guide to the design and implementation of reactive zone methods for treatment of all major classes of groundwater contamination, including hydrophobic and miscible organics, metals, and other high-solubility inorganics. It teaches the fundamentals that underlie development of cost-effective reactive zone strategies, including contaminant distribution and transport patterns, chemical and biological mechanisms that can be used to achieve oxidation, reduction or precipitation of target compounds, and identification and management of by-products and secondary water quality impacts of treatment technologies. It guides the selection of cost-effective remedial strategies and provides environmental engineers and scientists with tools to achieve optimal deployment of source area, reactive barrier and site-wide treatments. The book provides extensive coverage of remedial system operation, with discussion of reagent injection strategies, interpretation of process monitoring results for both biological and chemical reactive zone systems and impacts of treatment processes on aquifer hydraulic characteristics. Special emphasis has been placed on describing the innovative and experimental treatment technologies for emerging contaminants such as radionuclides, explosives, pharmaceutical compounds, perchlorate, chlorophenols, and solvent stabilizers.

The authors bring to this book the collective experience of one of the world's leading companies in remediating contaminated soil and groundwater. They have personally led the design and implementation of more than 200 IRZ systems in North America, South America, Europe, and Asia. For more than 25 years, Dr. Suthersan and Dr. Payne have been pushing the envelope in search of new and better ways of remediating contaminated sites. Together they hold more than 15 patents on cleanup technologies. Never satisfied with the status quo, their tireless research, study, experimentation, collaboration, and application make the authors true leaders in the world of remediation engineering.

It has been my privilege and honor to work closely with Dr. Suthersan and Dr. Payne over the past several years as they have developed much of the material presented in this book. As the president of ARCADIS, I truly believe that the value of this book is to elevate the level of remediation engineering being practiced in the entire industry, which will help in further establishing the ARCADIS reputation within the environmental sector. Any practicing remediation engineer or scientist, regulator, student, or researcher will benefit from reading and studying this book.

Michael L. Myers
President, ARCADIS-US
Denver, Colorado

Preface

In Situ Remediation Engineering provides a comprehensive guide to the design and implementation of reactive zone methods for treatment of all major classes of groundwater contamination, including hydrophobic and miscible organics, metals and other high-solubility inorganics. This book teaches the fundamentals that underlie the development of cost-effective reactive zone strategies, including contaminant distribution and transport patterns, chemical and biological mechanisms that can be used to achieve oxidation, reduction or precipitation of target compounds, and identification and management of by-products and secondary water quality impacts. *In Situ Remediation Engineering* provides environmental engineers and scientists with tools to achieve optimal deployment of source area, reactive barrier, and site-wide treatments.

The book provides extensive coverage of remedial system operation, with discussion of reagent injection strategies, interpretation of process-monitoring results for both biological and chemical reactive zone systems and impacts of treatment processes on aquifer hydraulic characteristics. We have tried to develop the themes and discussions in this book to enable *In Situ Remediation Engineering* to be a next generation text in the science of groundwater cleanup. We have also tried to bring the collective experience of one of the world's foremost companies in the area of soil and groundwater cleanup. *In situ* reactive zone methods are proving to be the most cost-effective means of cleanup for compounds ranging from chlorinated solvents and metals, to the emerging contaminants such as perchlorate, 1,4-dioxane, and n-nitrosodimethylamine.

In Situ Remediation Engineering recognizes and tackles the difficult issues of sorption, adsorption, and the availability of contaminants for aqueous phase chemical and biological treatment methods. It also introduces the concept of linked reactive and desorption zones, which are sequenced to achieve remediation and compliance objectives in predictable time frames. The concept of remedy risk profiling, which provides a tool to examine the time and cost exceedance risks for various cleanup technologies has been introduced as a bridge between the technical and business aspects of remedial system design.

Throughout the roughly 30-year history of the environmental remediation industry, the appearance of new technologies in the market was often accompanied by excessive enthusiasm and unwarranted hopes for low-cost solutions to major soil and groundwater contamination problems. We have attempted to present *in situ* reactive zone technologies, not merely as magic bullets that can solve immense problems at low cost, but as continuously developing technologies that require careful design engineering and operational monitoring. We cannot expect these methods to solve intractable contamination problems at a small percentage of the cost of long-term containment.

Market pressures have driven site-characterization efforts to minimal levels and, in many cases, the data available at the completion of investigation is not sufficient to support *in situ* reactive zone designs. Cryptic approximations and large-scale extrapolations have displaced more careful analysis of conditions at many sites. We believe that the quality of site-characterization data available to *in situ* reactive zone designers, especially for aquifer matrix structure and hydraulics, must improve dramatically to assure the success of these methods. This is especially important as our focus shifts to source zone treatments, where the added complexities of nonaqueous-phase liquid mass is often a factor. This will require remedial system designers to be more assertive in expressing their information requirements.

Chapter 5 undertakes an extended discussion of injection zone design and management, a central element of any *in situ* reactive zone application. The aquifer matrix structure and hydraulics are defining elements in the design of remedial processes that rely on the distribution of injected fluids. In this discussion, we introduce the terms migratory fraction and static fraction to conceptually divide an aquifer's drainable porosity into a portion that carries most of the groundwater flux, the migratory fraction, and that which is drainable, but does not carry significant flow, the static fraction. The migratory fraction is distinct from the hydrogeology concept, effective porosity. We are focusing our attention on the migratory fraction and its interaction with the static fraction to better understand contaminant migration and injected reagent transport through aquifers.

We truly believe that the evolution of the understanding of injection zone hydraulics is the most important factor in determining the success of *in situ* remediation projects in the future. It is not the microbiology, or even biochemistry, but hydrogeology and injection zone hydraulics that will be considered as the primary design factor. Some of the concepts introduced in this book are groundbreaking as well as thought provoking and it is our hope that these concepts will have a significant impact in the direction our industry takes during the next 5 to 10 years.

Suthan S. Suthersan, Ph.D., P.E.
Fred C. Payne, Ph.D.

Acknowledgments

The writing of this book would not have been possible without the support of our like-minded colleagues throughout ARCADIS and its base of forward-looking clients, who share our enthusiasm for the development and evaluation of new methods for the restoration of contaminated sites. We would also like to thank the senior management at ARCADIS, specifically Steve Blake, Mike Myers, and Curt Cramer, for their unstinting support and enthusiasm that maintain ARCADIS at the frontiers of innovation.

The animated discussions and critical testing of ideas in brainstorming sessions with Dr. Scott Potter, Jeff Burdick, Frank Lenzo, John Horst, Mike Hansen, Carol Mowder, Jim Harrington, Jim Romer, Greg Page, Barry Molnaa, Scott Andrews, and Joe Quinnan were essential in developing and fine tuning these latest concepts. Dr. Scott Potter, John Horst, Jeff Burdick, Joe Quinnan, and Jim Romer were specifically helpful in the evolution of our understanding of injection zone hydraulics. This will be an area of continuing interest for all of us at ARCADIS and hopefully within our entire industry. These are ground-breaking and thought-provoking concepts, and it is our hope that these concepts will have a significant impact on the direction our industry takes during the next 5 to 10 years.

ARCADIS is blessed with young, highly capable staff who supported the development of this book. John Horst and Denice Nelson provided critical support as we worked to conceptually bridge our practical experience in enhanced reductive dechlorination and the developing literature on dechlorinating bacteria. They all have the energy for the pursuit of knowledge and a dedication to "getting it right" that can be an example to all of us. John Horst and Mark Klemmer found time to carefully read and edit chapters, and their critical thinking and willingness to question helped make this a better book.

We would also like to thank Amy Weinert for her patient typing of three chapters. Thanks are due also to Matt Wasilewski for his professional development of the figures for Chapters 1, 2, and 4. Song Wang should be commended for helping us on many fronts and thus maintaining the calm during the rush to meet deadlines.

About the Authors

Suthan S. Suthersan, Ph.D., P.E., is Senior Vice President and Director of Site Evaluation and Remediation Services at ARCADIS G&M, Inc., an international environmental and infrastructure services company. In his 14 years with the company Dr. Suthersan has helped make ARCADIS one of the most respected environmental engineering companies in the U.S., specifically in the field of *in situ* remediation of hazardous wastes. Many of the technologies he pioneered have become industry standards. His greatest contribution to the industry, beyond the technology development itself, has been to convince the regulatory community that these innovative technologies are better than traditional ones, not only from a cost viewpoint, but also for technical effectiveness. His experience is derived from working on more than 800 remediation projects in design, implementation, and technical oversight capacities during the past 20 years.

Dr. Suthersan has his B.S. in civil engineering from the University of Sri Lanka, his M.S. in environmental engineering from the Asian Institute of Technology, his Ph.D. from the University of Toronto, in addition to his consulting experience Dr. Suthersan has taught courses at several universities. He is the founding editor-in-chief of the *Journal of Strategic Environmental Management* and a member of the editorial board of the *International Journal of Phytoremediation*.

Dr. Suthersan is considered a pioneer in the development of *in situ* remediation technologies. He coined the phrase "*in situ* reactive zone" 10 years ago for the specific applications discussed in this book. His technology development efforts have been rewarded with seven patents awarded and more pending. His most important recent contributions are reflected by the following patents: Engineered *in situ* Anaerobic Reactive Zones, U.S. Patent 6,143,177; In Well Air Stripping, Oxidation, and Adsorption, U.S. Patent 6,102,623; *in situ* Anaerobic Reactive Zone for *in situ* Metals Precipitation and to Achieve Microbial De-Nitrification, U.S. Patent 5,554,290; and *in situ* Reactive Gate for Groundwater Remediation, U.S. Patent 6,116,816.

**Fred C. Payne, Ph.D.,** is a Vice President and Director of Remediation Services at ARCADIS G&M, Inc. He has been with ARCADIS for 5 years, and he provides technology development and design support for physical, chemical and biological *in situ* remedial systems throughout North and South America and Europe.

Dr. Payne received his B.S. in biology and botany and his M.S. and Ph.D. in limnology from Michigan State University. In 1982, Dr. Payne founded the company Midwest Water Resource Management (later MWR, Inc.), which became an international leader in physical *in situ* remedial technologies, including soil vapor extraction and aquifer sparging. His enhanced-mode soil vapor extraction process was responsible for many challenging site closures since its inception in 1985, including the first reported detection-limit closure of vadose zone soil for perchloroethene (Purdue Industrial Waste Conference, May, 1986). During his 17 years at MWR, he was the inventor of six and the co-inventor of a seventh U.S. patent for physical *in situ* remedial technologies and subsurface fluid delivery systems.

Since joining ARCADIS in 1999, Dr. Payne has supported the company's development of cost-effective chemical and biological *in situ* remedial technologies, including biostimulation for enhanced reductive dechlorination and chemical oxidation for destruction of miscible as well as hydrophobic organic compounds.

In addition to his remedial technology responsibilities, Dr. Payne leads the ARCADIS design team for sustainable stormwater treatment systems, which recently completed design and installation of the 67-acre Phase I Sustainable Stormwater Management system at the Ford Rouge Complex in Dearborn, Michigan.

Contents

Chapter 1 Introduction 1
  1.1 Legacy of Pollution 1
  1.2 Development of *in Situ* Approaches 2
  1.3 Types of *in Situ* Reactions 5
      1.3.1 Microbial Reactions 8
            1.3.1.1 Biochemical Energetics 10
            1.3.1.2 Microbial Oxidation Reactions 14
            1.3.1.3 Microbial Reduction Reactions 15
            1.3.1.4 Microbial Metals Precipitation 16
      1.3.2 Chemical Reactions 17
            1.3.2.1 Chemically Oxidizing Reactions 18
            1.3.2.2 Chemically Reducing Reactions 19
            1.3.2.3 Hydrolytic Reactions 19
            1.3.2.4 Acid–Base Reactions 22
  1.4 Reactants and Solvents.. 28
      1.4.1 Water–Solvent Reaction Medium 28
            1.4.1.1 Solubility in Water 30
      1.4.2 Reactants 31
            1.4.2.1 Molecular Structures of Compounds 32
            1.4.2.2 Functionalities and Environmental Behavior 45
  1.5 Reversible vs. Irreversible Processes 50
  1.6 Hydrogeology and Hydraulic Impacts 52
  1.7 System Poise and Flux of Competing Compounds 54
  1.8 Current Status 55
      1.8.1 Regulatory Acceptance and Permitting 55
      1.8.2 Biostimulation vs. Bioaugmentation 58
            1.8.2.1 Biostimulation 58
            1.8.2.2 Bioaugmentation 59
      1.8.3 Contaminant Types 60
      1.8.4 Emerging Contaminants 60
  References... 62

Chapter 2 Microbial Reactive Zones 65
  2.1 Types of Microbial Reactions 66
  2.2 Microbial Oxidizing Zones 68
      2.2.1 Engineered Aerobic Systems 68
      2.2.2 Direct Aerobic Oxidation 69
      2.2.3 Microbial Oxidation by Fe(III), Nitrate, and Sulfate 73

		2.2.4	Aerobic Cometabolic Oxidation 75

- 2.2.4 Aerobic Cometabolic Oxidation 75
- 2.2.5 Microbial Oxidation of MTBE 77
- 2.3 Microbial Reducing Zones 78
  - 2.3.1 Enhanced Reductive Dechlorination 78
    - 2.3.1.1 Microbial Ecology of Reductive Dechlorination ... 85
    - 2.3.1.2 The Aquifer as an Ecosystem 88
    - 2.3.1.3 Biochemistry of Reductive Dechlorination 102
    - 2.3.1.4 Dechlorinating Bacterial Species and Consortia ... 112
    - 2.3.1.5 Review of Field-Scale Bioaugmentation Studies ... 118
    - 2.3.1.6 Summary of Bioaugmentation 123
    - 2.3.1.7 Review of Field-Scale Biostimulation Studies ... 123
    - 2.3.1.8 Evaluating Degradation Signatures 149
    - 2.3.1.9 Summary and Conclusions 155
    - 2.3.1.10 Abiotic Reductive Dechlorination 158
  - 2.3.2 *In Situ* Precipitation of Heavy Metals 161
    - 2.3.2.1 Principles of Heavy-Metal Precipitation 165
    - 2.3.2.2 Chromium Precipitation 170
    - 2.3.2.3 Nickel Precipitation 174
    - 2.3.2.4 Arsenic Precipitation 175
    - 2.3.2.5 Cadmium 182
    - 2.3.2.6 Aquifer Parameters and Transport Mechanisms ... 182
    - 2.3.2.7 Contaminant Removal Mechanisms 184
  - 2.3.3 *In Situ* Denitrification 185
- 2.4 Emerging Contaminants 187
  - 2.4.1 Perchlorate Reduction 188
  - 2.4.2 1,4-Dioxane 195
    - 2.4.2.1 Prospects for Mass Transfer 196
    - 2.4.2.2 Biological Remedies 197
  - 2.4.3 Explosives 200
  - 2.4.4 Radionuclides 206
    - 2.4.4.1 Uranium 207
    - 2.4.4.2 Technetium-99 (^{99}Tc) 212
    - 2.4.4.3 Plutonium 214
  - 2.4.5 *N*-Nitrosodimethylamine (NDMA) 215
- References 216

Chapter 3 Chemical Reactive Zones 227
- 3.1 Introduction to Chemical Reactive Zones 227
- 3.2 Chemical Reactivity 228
  - 3.2.1 Oxidation States 229
  - 3.2.2 Thermodynamic Feasibility of Reactions 231
    - 3.2.2.1 Bond Energies 231
    - 3.2.2.2 Gibbs Free Energy 232
    - 3.2.2.3 Nonstandard Conditions —
      The Nernst Equation 233
  - 3.2.3 Kinetic Control of Reaction Rates 234

		3.2.3.1	Reaction Order and Rate Equations 234

              3.2.3.1   Reaction Order and Rate Equations 234
              3.2.3.2   Kinetic Feasibility 239
        3.2.4   Effect of Temperature on Reaction Rates 241
              3.2.4.1   Estimating Temperature Effect on Thiosulfate-Mediated Destruction of Methyl Bromide 242
        3.2.5   Effect of pH on Reactions 243
        3.2.6   Effect of Ionic Strength on Reactions 243
    3.3   Reaction Mechanisms 245
        3.3.1   Electron Transfer Reactions 245
              3.3.1.1   Electron Transfer (Oxidation–Reduction) Reactions 245
              3.3.1.2   Radical Reactions 248
              3.3.1.3   Carbonate Interference 250
        3.3.2   Nucleophilic Substitution and Elimination Reactions 255
              3.3.2.1   Nucleophilicity 258
        3.3.3   Precipitation Reactions 261
              3.3.3.1   Solubility in Aqueous Phase 262
              3.3.3.2   The Solubility Product 262
              3.3.3.3   Calculation of Nickel Concentrations from K_{sp} ... 263
              3.3.3.4   Aging of Precipitated Solids 265
        3.3.4   Heterogeneous Reactions 266
    3.4   Chemical Oxidation Zones 266
        3.4.1   Fenton's Reagent 266
              3.4.1.1   Hydroxyl Radical Production Strategies 268
              3.4.1.2   Conventional Fenton's Process 269
              3.4.1.3   Iron Chelation Strategies 270
              3.4.1.4   Mineral Iron Strategies 271
              3.4.1.5   Radical Consumption by Carbonates and Other Competitive Inhibitors 271
              3.4.1.6   Common Reaction By-Products 272
              3.4.1.7   Monitoring Fenton's Reagent Applications 273
        3.4.2   Permanganate Oxidation 274
              3.4.2.1   Permanganate Demand Estimation and Injection Loading Requirements 276
              3.4.2.2   Permanganate Reaction Mechanisms 277
              3.4.2.3   By-Product Formation 278
              3.4.2.4   Permanganate Reaction Kinetics 280
              3.4.2.5   Sorption Effects on Permanganate Oxidation 280
              3.4.2.6   Phase-Transfer Catalysts 284
        3.4.3   Ozone 286
              3.4.3.1   Ozone as the Sole Oxidant Source 286
              3.4.3.2   Ozone–Peroxide Systems 291
              3.4.3.3   Bench-Scale Ozonation of 1,4-Dioxane 292
              3.4.3.4   By-Products 295
        3.4.4   Persulfate 295
    3.5   Chemical Reducing Zones 297

		3.5.1	Hydrogen 297

- 3.5.1 Hydrogen 297
- 3.5.2 Dithionite 297
  - 3.5.2.1 Dithionite Chemistry 297
  - 3.5.2.2 Dithionite-Driven Reduction of Chromium 300
- 3.5.3 Zero-Valent Iron 303
  - 3.5.3.1 Zero-Valent Iron Reaction Mechanisms 304
  - 3.5.3.2 Zero-Valent Iron Applicability 305
- 3.5.4 Reduced Vitamin B_{12} 307
- 3.6 Chemical Precipitation Strategies 309
- References.. 310

Chapter 4 Components of an *in Situ* Reactive Zone 315
- 4.1 Site Screening 320
  - 4.1.1 Site Access 320
  - 4.1.2 Safety Considerations 326
  - 4.1.3 Vapor Migration 327
  - 4.1.4 Trends in Contaminant Concentration and Electron Acceptor Processes 329
- 4.2 Oxidation State of Contaminants 330
- 4.3 Site Conceptual Model 333
  - 4.3.1 Hydrogeologic Model 333
  - 4.3.2 Contaminant Distribution 337
  - 4.3.3 Biogeochemical Characterization 339
- 4.4 Performance Measures 341
  - 4.4.1 Process Monitoring 342
  - 4.4.2 Performance Monitoring 343
  - 4.4.3 Groundwater Sampling and Analysis 344
    - 4.4.3.1 Dissolved Gases 347
    - 4.4.3.2 Dissolved Oxygen Field Measurement 348
    - 4.4.3.3 Analysis of Biogeochemical Parameters 348
    - 4.4.3.4 Oxidation–Reduction Potential and E_H 351
    - 4.4.3.5 Microbial Assessments 353
- 4.5 System Design 357
  - 4.5.1 Design Considerations 357
    - 4.5.1.1 Geology 358
    - 4.5.1.2 Hydraulic Conductivity 359
    - 4.5.1.3 Groundwater Flow Characteristics 360
    - 4.5.1.4 Saturated Thickness and Depth to Water 360
    - 4.5.1.5 Geochemistry 361
  - 4.5.2 Groundwater Chemistry 363
    - 4.5.2.1 pH 364
    - 4.5.2.2 Role of Sulfur in Enhanced Reductive Dechlorination Systems 365
  - 4.5.3 IRZ Layout Options 366
    - 4.5.3.1 Injection Well/Point Placement 367
    - 4.5.3.2 Monitoring Well Placement 372

- 4.6 Reagents 372
  - 4.6.1 Microbial IRZ Systems 372
    - 4.6.1.1 Suitability of Different Electron Donors 374
  - 4.6.2 Chemical IRZ Systems 376
- 4.7 Delivery System Design 376
  - 4.7.1 Reagent Injection Strategy 381
- 4.8 Pilot Testing 383
  - 4.8.1 Pilot Test Wells — Number and Location 384
  - 4.8.2 Duration of Pilot Study 386
  - 4.8.3 Scaleup Issues 387
  - 4.8.4 Sustainability and Reliability 388
  - 4.8.5 Biofilm Developments 390
  - 4.8.6 Site Closure 391
- References... 392

Chapter 5 Building Reactive Zone Strategies 395
- 5.1 Introduction to Reactive Zone Strategies 395
- 5.2 A Conceptual Model for Contaminant Distribution 396
  - 5.2.1 Reactive Zone Hydrogeology 396
    - 5.2.1.1 Darcy's Law, Permeability, and Groundwater Velocity 397
    - 5.2.1.2 A Dual-Porosity Pore-Scale Model 400
  - 5.2.2 Contaminant Classes and Their Behaviors 402
    - 5.2.2.1 High-Solubility Inorganics (Salts) 402
    - 5.2.2.2 Transition Compounds 403
    - 5.2.2.3 Miscible Organics 403
    - 5.2.2.4 Hydrophobic Organics 403
  - 5.2.3 Sorption and Desorption Processes 404
    - 5.2.3.1 Adsorption 404
    - 5.2.3.2 Absorption 405
    - 5.2.3.3 Conventional Calculation of Sorbed-Phase and Total Contaminant Mass 406
    - 5.2.3.4 Multi-Compartment Sorption Models 408
    - 5.2.3.5 Cation Exchange Capacity 411
  - 5.2.4 Contaminant Distribution in Aquifer Matrices 412
    - 5.2.4.1 A Multi-Phase Conceptual Model for Contaminant Distribution 412
    - 5.2.4.2 Distribution of Hydrophobic Organics 415
    - 5.2.4.3 Distribution of Miscible Organics and Soluble Salts 418
    - 5.2.4.4 Distribution of Metals 419
  - 5.2.5 Steady-State and Non-Steady-State Distributions 420
  - 5.2.6 Propagation of a Clean Water Front 425
- 5.3 Reactive Zone Structure 427
  - 5.3.1 Injection Zone Management 429
    - 5.3.1.1 Injection Radius 430

		5.3.1.2	Injection Zone Tracer Test Example 431

        5.3.1.2 Injection Zone Tracer Test Example 431
        5.3.1.3 Injection Volume Limitations 433
        5.3.1.4 Balancing Injection Volumes and Frequencies ... 434
    5.3.2 The Reactive Zone 435
        5.3.2.1 The *in Situ* Bioreactor Concept 435
        5.3.2.2 Bioreactor Acclimation — Microbial Population Buildup Phase 437
        5.3.2.3 Chemical Reactive Zones 438
    5.3.3 The Desorption Zone 439
    5.3.4 The Recovery Zone 439
5.4 Reactive Zone Strategies 440
    5.4.1 Reactive Barriers 440
    5.4.2 Source Treatments 441
        5.4.2.1 Biological Reducing Zone Strategy for Nickel Precipitation 444
        5.4.2.2 *In Situ* Oxidation of *N*-Nitrosodimethylamine ... 449
    5.4.3 Reactive Zone Sequencing and Whole-Plume Treatments ... 451
5.5 Limitations of Aqueous-Phase Reactions in Contaminated Porous Media 452
    5.5.1 Achieving Contact between Reactants and Targets 452
    5.5.2 Contaminant Shielding by Organic Matter 454
    5.5.3 Phase Transfer Catalysts and Surfactants in Chemical Reactive Zones 454
    5.5.4 Partitioning Electron Donors, Surfactants, and Cosolvents in Biological Reactive Zones 455
5.6 Reaction Products and Consequences 455
    5.6.1 Type Curves for Enhanced Reductive Dechlorination 456
    5.6.2 The Taxonomy of Type Curves 456
    5.6.3 Results of Partial Oxidation 460
        5.6.3.1 Release of Sorbed 1,1,1-Trichloroethane Associated with Partial Oxidation by Fenton's Reagent 462
    5.6.4 Reactions with Nontarget Organics 463
    5.6.5 Redox Effects on Nontarget Inorganic Species 463
        5.6.5.1 Redox Reduction Effects 464
        5.6.5.2 Oxidation Effects 464
    5.6.6 Reaction Product Effects on Hydraulic Conductivity 464
        5.6.6.1 Chemical Oxidation Effects on Aquifer Conductivity 465
        5.6.6.2 Chemical Reduction Effects on Aquifer Conductivity 465
        5.6.6.3 Biological Reactive Zone Effects on Aquifer Conductivity 466
    References 466

Appendix 469

Index 487

1 Introduction

1.1 LEGACY OF POLLUTION

One hundred years ago, or even fifty, the disposal of industrial and household wastes was not a serious problem. Manufacturing plants were relatively small and there were only a few large cities. The population of the U.S. in 1900 was significantly less than half its present level. Disposal of several forms of waste by dumping it into streams, piling in isolated areas, or discharging into the atmosphere created no major recognized problems. The extent of pollution created by industries and cities, with only a few exceptions, was within the ability of nature to dilute or degrade to safer limits. Those placid days of existence have now receded into history.

The sources of pollution are many, and they differ from place to place. Manufacturing and industrial wastes have received the greatest notoriety, but other sources are of comparable importance. Agriculture- and mining-related contamination are extensive and likely to become increasingly serious. Defense-related activities have introduced certain types of contaminants specifically in the western U.S. Municipal and industrial landfills have and still let loose a complex collection of pollutants. It has become quite apparent that even "contained" applications and use of chemical compounds such as freons, chlorinated solvents, chromic acid, methyl-*tert*-butyl ether, polychlorinated biphenyls, and many others always result in a certain level of discharge into the environment. The total effect from all these different sources of pollutants is to contaminate only about 5% of the aquifers in this country. However, many of these contaminated aquifers are close to large population centers, and thus the environmental impact is disproportionately large. Since nearly 50% of the people of the U.S. depend on groundwater for their drinking water supplies, subsurface contamination in any form has significantly raised the level of public and regulatory concern during the last two to three decades.

The establishment of various regulations created a new industry for cleaning up environmental pollution and hazardous waste contamination. It has to be acknowledged that improved control of chemical usage in response to new scientific information and public concern over spills, releases, and environmental degradation has led to an improvement in the quality of the natural environment of our country. The public, government, and industry continue to rely on multidisciplinary environmental scientists to develop effective, practical, and cost-effective solutions. Technologies for remediating hazardous wastes, present in the subsurface, have evolved with time and have been influenced by various factors over the years.

Compliance with regulations was the primary driver during the early years and the type of contaminants influenced the technology evolution during the later years. Currently, economic factors influence technology development and application along with regulations and type of contaminants. This continuous evolution is a necessity due to the requirements of meeting stringent cleanup standards while finding the means to reduce cleanup costs and at the same time improving technical efficiencies and cleanup times.

Contaminated groundwater was the driving concern initially because it was mobile and, as a result, transported the contaminants and the resulting liability off site. The necessity to contain the contamination on site led to the universal application of groundwater pump and treat systems for both containment and mass removal during that time period. More than two decades of experience has taught us that pump and treat is not the solution; in fact, it is an inefficient technology for faster and cheaper site cleanups.

1.2 DEVELOPMENT OF *IN SITU* APPROACHES

The realization that mass removal efficiencies can be significantly enhanced using air as an extractive media instead of or in addition to water led to the development and application of *in situ* extractive technologies such as soil vapor extraction and *in situ* air sparging. While it can be argued that the initial motive for applying these technologies has been one of saving money, the end result is much quicker cleanup times to more acceptable cleanup levels. This win–win situation for the entire remediation industry fostered continuous innovation, which led to (1) faster, cheaper solutions, (2) less-invasive *in situ* technologies, and (3) technologies complementary to the natural environment, which took advantage of nature's capacity to degrade the pollutants. Thus, environmentally and economically sound and sustainable solutions were continuously developed and implemented.

The *in situ* extractive techniques subsequently gave way to *in situ* nonextractive techniques such as funnel and gate systems. *In situ* mass destruction techniques such as *in situ* reactive zones (IRZs) are the preferred remediation technologies today, due to their superior technical and cost efficiencies. Oxidation, reduction, and other biogeochemical processes play a major role in the mobility, transport, and fate of inorganic and organic contaminants in the subsurface, specifically in groundwater systems. Hence, engineering and manipulation of these conditions and processes to create different types of IRZs to remediate contaminated sites was a predictable evolution. Different processes to achieve various microbial reactions or chemical reactions can be designed to be implemented within these IRZs.

The concept of *in situ* reactive zones is based on the creation of a subsurface zone, where migrating contaminants are intercepted and permanently immobilized or degraded into harmless end-products. Figure 1.1(A) and (B) pictorially describe the concept of *in situ* reactive zones. The successful design of these reactive zones require the ability to engineer two sets of reactions: (1) between the injected reagents and the migrating contaminants, and (2) between the injected reagents and the subsurface

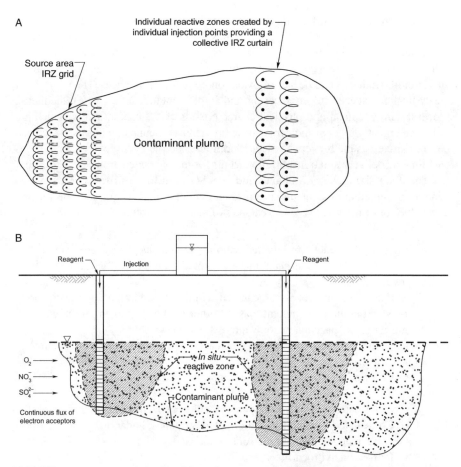

FIGURE 1.1 (A) Plan view of creation of a downgradient IRZ curtain and source area IRZ grid. (B) Cross-sectional view of IRZs created by individual injection wells.

environment to manipulate the biogeochemistry to optimize the required reactions, in order to effect complete remediation. These interactions will be different at each contaminated site and, in fact, may vary within a given plume and the site. Thus, the biggest challenge is to design an engineered system for the systematic control of these reactions under the naturally variable or heterogeneous conditions found in the field.

The effectiveness of the reactive zone is determined largely by the relationship between the kinetics of the target reactions and the rate at which the mass flux of contaminants and other reactants pass through it with the moving groundwater. Creation of a spatially fixed reactive zone in an aquifer requires not only the proper selection of the reagents, but also the proper delivery of the injected reagents uniformly within the reactive zone. Furthermore, such reagents must cause the fewest side reactions and be relatively nontoxic in both original and degraded forms.

Many processes that take place in groundwater systems have to be taken into consideration when designing an IRZ and thus the fate and transport of contaminants in the subsurface. Contaminants can sorb onto the soil, volatilize, react with water and hydrolyze, undergo chemical precipitation, be subjected to abiotic reactions and/or biodegradation and be part of oxidation–reduction sequences. The factors that are critical for successful engineering and implementation of *in situ* remediation processes are chemical and physical characteristics of the contaminants, which are influenced mostly by their molecular structure; environmental conditions, which can be summarized as the biogeochemical conditions; and the hydrogeologic conditions enabling the delivery and transport of reagents for the *in situ* reactions to take place.

There are significant advantages and benefits in utilizing *in situ* approaches in general for remediating contaminants present in contaminated aquifers. The specific advantages of *in situ* reactive zones can be summarized as follows:

- Most of the reactions targeted within the reactive zones are intended to destroy the contaminant mass except in the case of *in situ* metals precipitation.
- *In situ* treatment processes achieved within IRZs eliminate the need for transferring the contaminant mass to other media such as in groundwater pumping, air sparging, or soil vapor extraction systems.
- These processes have a potential application to a wide spectrum of contaminants such as:
  - Chlorinated aliphatic hydrocarbons — PCE, TCA, DCA, CT, MC, TCE, DCE, VC
  - Chlorinated cyclic hydrocarbons — chlorobenzenes and chlorophenols
  - Chlorinated pesticides — chlorinated propanes, lindane
  - Other halogenated organic contaminants
  - Petroleum hydrocarbons — BTEX
  - Polynuclear aromatic hydrocarbons — naphthalene and other multi-ring compounds
  - Nonchlorinated solvents — acetone, ketones such as MEK and MIBK, and alcohols
  - Ethers — such as MTBE and 1,4 dioxane
  - Dissolved metals — Cr^{6+}, Ni^{2+}, Pb^{2+}, Zn^{2+}, Cd^{2+}, Hg^{2+}, Cu^{2+}, As^{5+}, As^{3+}, and many other metals
  - Radionuclides — uranium, technetium, plutonium, neptunium
  - Explosives — RDX, HMX, TNT
  - Oxidants — such as perchlorate
- No *ex situ* waste is generated.
- The processes usually use reagents that are typically easily accepted by regulators and the public.
- Intended biologically mediated reactions can generally be driven by indigenous microflora.
- The technology is flexible in application, yielding a spectrum of contaminant mass treatment options from passive/containment barrier applications to aggressive source area applications.

- Promotes reduction of residual contaminant mass through enhanced desorption and disruption of the contaminant phase equilibrium.
- Enhances natural attenuation processes.
- Applicable to various geological settings and aquifer conditions.
- In most cases the reagents are highly soluble and can move through both diffusive and advective processes into difficult lithologies such as fractured bedrock.
- Systems can be designed with flexible operational approaches ranging from automated systems to manual bulk application.
- Can be used in tandem with existing remediation systems to optimize performance.
- Can be designed with minimal disturbance to site and facility operations, and also can be incorporated with ongoing site development activities.

1.3 TYPES OF *IN SITU* REACTIONS

Contaminants that are introduced into the subsurface environment are subjected to various physical, chemical, and biological processes. These processes act in an interconnected way in environmental systems that determine the overall fate of these compounds. Some of these processes leave the structure of the chemical unaffected and are generally termed as transport mechanisms. Other processes transform the chemical into one or several other products of different environmental behavior and effects. These types of processes alter the molecular structure of a compound and may include chemical, biochemical, and/or biological transformation reactions.

The transformation of contaminants most often occurs in several molecular events referred to as *elementary reactions*. An elementary reaction is defined as a process in which reacting chemical species pass through a single transition state without the intervention of an intermediate. A sequence of individual elementary reaction steps constitutes a *reaction mechanism*.

The types of *reactions* that can be designed and implemented within an IRZ are many. Chemical, physical, biochemical, and microbial reactions in water can be *homogeneous* — occurring entirely among dissolved species — or *heterogeneous* — occurring at the liquid–solid–gas interfaces. Most reactions in the subsurface environment are heterogeneous and purely homogeneous reactions are relatively rare in the groundwater environment.

Reactions can be defined as processes in which chemical bonds are broken, new bonds are formed, or one atom replaces another in the molecular structure of a compound. Reactions in general can be categorized as substitution, elimination, and configuration reactions. Within the IRZs we expect such transformation reactions lead to products that are harmless or less harmful compared to the parent compounds. The ideal case would be when the contaminants are mineralized within the reactive zone. Mineralization is defined as the complete conversion of an organic chemical to stable inorganic forms of C, H, N, P, and so on (i.e., conversion to CO_2, H_2O, NO_3^-, NH_4^+, PO_4^{3-}, etc.). There are also examples of "incomplete" transformation of contaminants that are of greater environmental concern due to the properties, reactivities, and toxicities of the intermediate products.

For our evaluation of the various transformation reactions that contaminants undergo naturally or to be engineered in the subsurface environment, it is convenient and common to divide these processes into two major categories: *chemical* and *microbial*. These reactions can be subdivided into those where there is no net electron transfer occurring between the contaminant and a reactant in the environment, and into oxidation–reduction (*redox*) reactions, where electrons are either transferred from (oxidation) or to (reduction) the contaminant.

Historically, geochemical processes were thought to be largely, if not entirely, driven by abiotic chemical reactions. But now we know that many important geochemical reactions are driven by bacteria. It should be pointed out that microorganisms have been found in even the harshest of environments — deep sea hydrothermal vents, toxic abandoned mine sites, and even deep under the Earth's surface — and hence, their influence is likely to be profound. Recent studies have demonstrated that the chemical reactions that occur close to microbial surfaces are very different from those of the bulk solution. The chemistry of this "microbial space" is complex and has a significant influence on the desirable reactions within an IRZ.

Consider a reaction that may give more than one product and if the main product is the one formed faster, the reaction is said to proceed under kinetic control. If the main product is the more stable one, then the reaction is said to be under thermodynamic control. Generally, the product formed faster is the one that is more stable and only reversible reactions can be subject to thermodynamic control. Both the extent to which a reaction can proceed and the rate at which it actually takes place determine which reactions actually occur within an engineered IRZ.

Reactions, chemical or biochemical, may be described from the standpoint of either *kinetics* or *equilibrium*. Kinetics describe the *rate* at which a reaction takes place and is significant when comparing a particular reaction rate to the rate at which a comparable process takes place. Equilibrium, by contrast, describes the final expected chemical composition in a control volume. A control volume with its biogeochemical components is often referred to as a *system*. Equilibrium is relevant in the case of reactions that are rapid, such as in the case of chemical oxidation processes.

When reactants first come together — before any products have been formed — their rate of reaction is determined in part by their initial concentrations. As the reaction products accumulate, the concentration of each reactant decreases and so does the reaction rate. Meanwhile, some of the product molecules begin to participate in the reverse reaction, which reforms the reactants. This reaction is slow at first but speeds up as the concentration of products increases. Eventually, the rates of the forward and reverse reactions become equal, so that the concentrations of reactants and products stop changing. The mixture of reactants and products is then said to be in chemical equilibrium.

Molecularity is often confused with the order of a reaction, which refers to the sum of the exponents of the reactants involved in a reaction. Determination of rate equations is developed from the proposed reaction mechanisms. A hypothetical first-order reaction, which is also known as a unimolecular reaction, is shown here. Reactant A can only undergo reaction if it acquires sufficient energy to overcome an energy barrier to the reaction. The differential rate equation for this reaction

$$A \rightarrow B \tag{1.1}$$

can be written as

$$-d[A]/dt = k_1[A] \tag{1.2}$$

$$[A] = [A]_o e^{-kt} \tag{1.3}$$

$$\ln A = \ln A_o - kt \tag{1.4}$$

When comparing the reactivity of compounds within a reactive zone, it is convenient to speak in terms of the half-life ($t_{1/2}$) of a reaction or the time for 50% of the chemical compound to disappear. For first-order reactions, if we set $[A] = 1/2[A_o]$, then the expression for $t_{1/2}$ becomes

$$t_{1/2} = 0.693/k \tag{1.5}$$

The half-life for a first-order reaction, therefore, is independent of the initial concentration of the chemical of interest.

For the more complex second-order reaction containing two reacting chemical species, A and B, also known as an association reaction,

$$A + B \rightarrow C \tag{1.6}$$

the rate of disappearance of A is given by

$$-d[A]/dt = k[A][B] \tag{1.7}$$

If one of the concentrations of a chemical species in a second-order reaction is present in excess, and as a result does not "effectively" change during the course of the reaction, the observed behavior of the system will be first order. Such a system is said to follow pseudo-first-order kinetics. If the concentration of the chemical species B above was in large excess of A, the disappearance of A could be written as

$$d[A]/dt = k_{obs}[A] \tag{1.8}$$

where $k_{obs} = k_2[B]$

Reaction mechanisms for the transformation of contaminants usually involve several elementary reactions. The kinetic expressions resulting from such mechanisms can become quite complex. Often, simplifying assumptions can be used to analyze these systems. One simplifying assumption is referred to as the *steady state approximation* and can be illustrated by the following hypothetical reaction scheme:[1]

$$A \underset{}{\overset{K_1}{\rightleftarrows}} B \tag{1.9}$$

$$B \xrightarrow{K_2} C \tag{1.10}$$

If B is a reactive, unstable species, its concentration will remain low. Applying the steady-state assumption, we assume that the rate of formation of the reactive species is equal to its rate of destruction:

$$K_1[A] = K_{-1}[B] + K_2[B] \qquad (1.11)$$

Solving for the steady-state concentration of B, we find that:

$$[B] = K_1[A]/(K_{-1} + K_2) \qquad (1.12)$$

If we assume that the conversion of B to C is the *rate limiting* or rate determining step, the overall rate of the reaction will be:

$$d[C]/dt = K_2[B] \qquad (1.13)$$

Substituting the term for the steady-state concentration of B (Equation (1.12)) in Equation (1.13) gives:

$$d[C]/dt = (K_2 K_1 [A])/(K_{-1} + K_2) \qquad (1.14)$$

If the experimental (observed) rate law was determined to be $d[C]/dt = k_{obs}[A]$, then k_{obs} would be related to the elementary reaction rate constants by:

$$k_{obs} = K_2 K_1/(K_{-1} + K_2) \qquad (1.15)$$

In general, establishing the identities of all products and the stoichiometry of the balanced chemical equation are necessary steps in determining a contaminant transformation reaction's mechanism and the kinetics. The structure of the reactant and the nature of the rate law may, of course, be reliable guides in characterizing the reaction, but there must be product analytical and structural evidence also to come to a reliable conclusion.

1.3.1 MICROBIAL REACTIONS

Microbial (biotic) reactions are biochemically mediated transformations and constitute a very important set of remediation processes that can be implemented within IRZs. A reaction is considered to be biotic if it directly involves the participation of metabolically active microorganisms. Sometimes abiotic reactions can also be influenced by microbial activity by controlling the availability of reactants via biochemical processes. Many important pathways by which contaminant molecules are microbially transformed within the reactive zones involve oxidative and reductive steps. The easiest way to recognize whether a contaminant has been oxidized or reduced during a reaction is to check whether there has been a net change in the oxidation state(s) of atoms like C, N, or S involved in the reaction.

Since most of the redox reactions that take place in the subsurface environment are microbially mediated, the evaluation of how much energy an organism may derive from a given reaction may provide very useful insight into the sequences of these reactions and the kinds of organisms expected under given conditions. To carry out their life functions, microorganisms require a carbon source, electron donors and electron acceptors, water, and mineral nutrients. Many organic compounds, including the frequently encountered contaminants like petroleum hydrocarbons and most chlorinated solvents, are degraded or transformed by microorganisms through use as a carbon source or as electron acceptors. It appears that those compounds that have been available for microbial metabolism in large quantities and for the longest period of time like the petroleum hydrocarbons are the most readily degradable.

Microorganisms, just as humans, obtain energy for growth and energy from physiologically coupling oxidation and reduction reactions. During these growth-promoting reactions, electrons are transferred from one group of compounds called *electron donors* to another group called *electron acceptors*, and energy is released. Electron donors are compounds that occur in relatively reduced states and include natural organic material, petroleum hydrocarbons, lightly chlorinated (less oxidized) ethenes, ethanes, methanes, sugars, organic acids, and dissolved hydrogen. Electron acceptors are elements or compounds that occur in relatively oxidized states and include dissolved oxygen, NO_3^-, Fe^{3+}, SO_4^{2-}, CO_2, and several highly chlorinated solvents.

In general, the more easily biotransformable contaminants have simple molecular structures (often similar to the structures of naturally occurring organic compounds), are water soluble and nontoxic, and can be transformed by aerobic metabolism. In contrast, contaminants that resist biotransformation may have complex molecular structures (especially structures not found in nature or of recent anthropogenic origin), low water solubility, or they may be toxic to the microorganisms. It should be noted that microbial transformations of contaminants via biochemical reactions can only take place when the contaminants are present and available in the *dissolved* phase (Figure 1.2).

In the past, and even today, the tendency is to categorize all microbially catalyzed reactions as *aerobic* or *anaerobic* reactions. Under aerobic conditions (in the

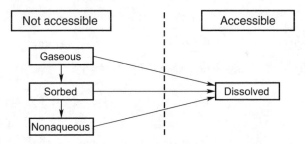

FIGURE 1.2 Schematic diagram describing the mechanisms by which a contaminant becomes available for microbial transformation via chemical reactions.

presence of oxygen), bacteria couple the oxidation of contaminants (electron donors) to the reduction of oxygen (electron acceptors). Under anaerobic conditions, most organic compounds are degraded by groups of interacting microorganisms referred to as a consortium. However, we would like to treat the microbial processes as *oxidation* and *reduction* reactions, which involve the microbial transfer of electrons from electron donors to electron acceptors. During these reactions, the electron donors and electron acceptors are both considered primary growth substrates because they promote microbial growth.

1.3.1.1 Biochemical Energetics

Microorganisms can cause major changes in the biogeochemistry of a groundwater system. Their small size and adaptability, as well as the diversity of nutritional requirements for different microbes, enable them to catalyze a wide range of reactions that often are the basis for biotransformations. Microorganisms use enzymes to accelerate the rates of biochemical reactions, which in turn help them to reproduce by creating daughter cells composed of various cellular components. The reactions are driven to completion by the expenditure of cellular energy in the form of a chemical known as adenosine triphosphate (ATP), which can be considered a cellular fuel (Figure 1.3).

The production of energy, its storage, and its use are central to the economy of the bacterial cell in a biogeochemical system. Energy may be defined as the ability to do work, a concept that is easy to grasp when it is applied to automobile engines and electric power plants. When we consider the energy associated with chemical bonds and biochemical reactions within cells, however, the concept of work becomes less intuitive.

There are two principal forms of energy: kinetic and potential. *Kinetic energy* is the energy of movement — the motion of molecules, for example. The second form

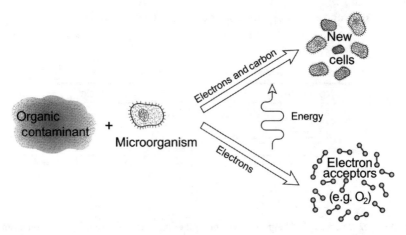

FIGURE 1.3 Conceptual description of microorganisms gaining energy and utilizing the contaminant for growth.

of energy, *potential energy*, or stored energy, is more important in the function of biological or chemical systems.

Several forms of potential energy are biologically significant.[2] Central to biology is the potential energy *stored in the bonds* connecting atoms in molecules. Indeed, most of the biochemical reactions involve the making or breaking of at least one covalent chemical bond. This energy is recognized when chemicals undergo energy-releasing reactions. Glucose, present in sugar for example, is high in potential energy. Microbial cells degrade glucose continuously, and the energy released when glucose is metabolized is harnessed to do many kinds of "microbial" work.

Another biologically important form of potential energy, to which we shall refer often, is the energy in a *concentration gradient*.[2] When the concentration of a substance on one side of a permeable barrier, such as a membrane, is different from that on the other side, the result is a concentration gradient. Microbes form concentration gradients between their interior and the external fluids by selectively exchanging substrate, nutrients, waste products, and ions with their surroundings. Also, compartments within bacterial cells frequently contain different concentrations of ions and other molecules.

According to the first law of thermodynamics, energy is neither created nor destroyed, but it can be converted from one form to another (conversion of mass to energy in nuclear reactions is an exception to this rule). In photosynthesis, for example, the radiant energy of light is transformed into the chemical potential energy of the covalent bonds between the atoms in a sucrose or starch molecule. In all cells, chemical potential energy, released by breakage of certain chemical bonds, is used to generate potential energy in the form of concentration and electric potential gradients. Similarly, energy stored in chemical concentration gradients or electric potential gradients is used to synthesize chemical bonds, or to transport other molecules "uphill" against a concentration gradient. This latter process occurs during the transport of nutrients such as glucose into microbial cells and transport of many waste products out of cells. Because all forms of energy are interconvertible, they can be expressed in the same units of measurement, such as the calorie or joule.[2]

1.3.1.1.1 Free Energy ΔG

Since biological systems are generally held at constant temperature and pressure, it is possible to predict the direction of a chemical reaction by using a measure of potential energy called free energy, or G, after the great American chemist Josiah Willard Gibbs (1839–1903), a founder of the science of thermodynamics. Gibbs showed that under conditions of constant pressure and temperature, as generally found in biological systems, all systems change in such a way that free energy is minimized. In general, we are interested in what happens to the free energy when one molecule or molecular configuration is changed into another. Thus our interest lies in knowing the relative, rather than absolute, values of free energy — with the difference between the values before and after the change. This free-energy change ΔG, is given by:

$$\Delta G = G_{\text{products}} - G_{\text{reactants}} \qquad (1.16)$$

In mathematical terms, Gibbs law can be described by the following set of statements about ΔG:

- If ΔG is negative for a biochemical reaction, the forward reaction or process (from left to right as written) will tend to occur spontaneously.
- If ΔG is positive, the reverse reaction (from right to left as written) will tend to occur.
- If ΔG is zero, both forward and reverse reactions occur at equal rates; the reaction is at equilibrium.

For example, if one mole of glucose is completely oxidized, 686 kcal of energy are produced along with carbon dioxide and water:

$$C_6H_{12}O_6 + 8O_2 \rightarrow 6CO_2 + 12H_2O + 686\,\text{kcal} \qquad (1.17)$$

Thus, microbial cells using one mole of glucose as the energy source cannot obtain more than 686 kcal of energy. This does not mean, however, that all of that 686 kcal of energy is available to the cells. The second law of thermodynamics states that some portion of the energy, the free energy, will be available to do work, whereas another portion, called entropy, will be unavailable to do work. Studies of energy utilization in living cells show that, of the 686 kcal released, a maximum of only about 280 kcal is actually available as free energy (ΔG), and can be utilized by the microbial cells. The rest (60%) of the energy released by glucose oxidation goes toward increasing entropy and is lost as heat.[2]

The value of ΔG, like the equilibrium constant, is independent of the reaction mechanism and rate. Reactions with negative values that have very slow rate constants may not occur, unless a catalyst is present, but the presence of a catalyst does not affect the value of ΔG.

At any constant temperature and pressure, two factors determine the ΔG of a reaction and thus whether the reaction will tend to occur: the change in bond energy between reactants and products and the change in the randomness of the system. Gibbs showed that free energy can be defined as

$$G = H - TS \qquad (1.18)$$

where H is the bond energy, or *enthalpy*, of the system; T is its temperature in Kelvin (K); and S is a measure of randomness, called *entropy*. If temperature remains constant, a reaction proceeds spontaneously only if the free energy change ΔG in the following equation is negative:

$$\Delta G = \Delta H - T\Delta S \qquad (1.19)$$

The enthalpy H of reactants or of products is equal to their total bond energies; the overall change in enthalpy ΔH is equal to the overall change in bond energies. In an *exothermic* reaction, the products contain less bond energy than the reactants, the liberated energy is usually converted to heat (the energy of molecular motion), and

ΔH is negative. In an *endothermic* reaction, the products contain more bond energy than the reactants, heat is absorbed, and ΔH is positive. Reactions tend to proceed if they liberate energy (if $\Delta H < 0$), but this is only one of two important parameters of free energy to consider; the other is entropy.

Entropy S is a measure of the degree of randomness or disorder of a system. Entropy increases as a system becomes more disordered and decreases as it becomes more structured. Consider, for example, the diffusion of solutes and nutrients from one solution into another one in which their concentration is lower. This important biological reaction is driven only by an increase in entropy; in such a process ΔH is near zero. The negative free energy of the reaction in which nutrient molecules are liberated to diffuse over a larger volume will be due solely to the positive value of ΔS in Equation (1.19).

Another example is when the formation of hydrophobic bonds is driven primarily by a change in entropy. When a long hydrophobic molecule is dissolved in water, the water molecules are forced to form a cage around it, restricting their free motion. This imposes a high degree of order on their arrangement and lowers the entropy of the system ($\Delta S < 0$). Because the entropy change is negative, hydrophobic molecules do not dissolve well in aqueous solutions and tend to stay associated with one another.

We can summarize the relationships between free energy, enthalpy, and entropy as follows:

- An exothermic reaction ($\Delta H < 0$) that increases entropy ($\Delta S > 0$) occurs spontaneously ($\Delta G < 0$).
- An endothermic reaction ($\Delta H < 0$) will occur spontaneously if ΔS increases enough so that the $T \Delta S$ term can overcome the positive ΔH.
- If the conversion of reactants into products results in no change in free energy ($\Delta G < 0$), then the system is at equilibrium, that is, any conversion of reactants to products is balanced by an equal conversion of products to reactants.

Thermodynamics determines which reactions are energetically favorable and which are not. However, the *rate* at which the reactions are actually carried out does not depend on free energy change but rather on the reaction mechanism itself. If one were to mix glucose with oxygenated water in a beaker, glucose and O_2 will react only at an immeasurable rate to form CO_2 and water, even though that reaction is energetically favorable. At ambient temperatures (25°C) collisions of oxygen and glucose molecules occur too infrequently and without enough force for the reaction to proceed. This reaction can only take place at extremely high temperatures, say 500°C or higher, without the presence of a catalyst.

Microbial organisms, which depend on thousands of biochemical reactions like the uptake of glucose in order to sustain life and grow, solve this problem through the use of enzymes. Enzymes are proteins that catalyze chemical reactions in living cells. In the previous example, if glucose oxidase, an enzyme that brings glucose and oxygen together, is added to the solution in the beaker, the reaction will be completed to CO_2 and water with some intermediate steps. During the biochemical reactions of microbial metabolism, glucose oxidase is produced by the bacteria.

Enzymatically catalyzed reactions can be summarized as:

$$E + S \rightleftarrows ES \tag{1.20}$$

$$ES \rightleftarrows E + Products \tag{1.21}$$

where E is the enzyme, S is the substrate, ES is the enzyme–substrate complex, and P is products.

1.3.1.1.2 Parameters That Affect the ΔG of a Reaction

The change in free energy of a reaction (ΔG) is influenced by temperature, pressure, and the initial concentrations of reactants and products. Most biological reactions — like others that take place in aqueous solutions — also are affected by the pH of the solution.

The standard free-energy change of a reaction $\Delta G°$ is the value of the change in free energy under the conditions of 298 K (25°C), 1 atm pressure, pH 7.0 (as in pure water), and initial concentrations of 1 M for all reactants and products except protons, which are kept at pH 7.0. The sign of $\Delta G°$ depends on the direction in which the reaction is written. If the reaction A → B has a $\Delta G°$ of $-x$ kcal mol^{-1}, then the reverse reaction B → A will have a $\Delta G°$ value of $+x$ kcal mol^{-1}.

Most biological reactions differ from standard conditions, particularly in the concentrations of reactants. However, we can estimate free-energy changes for different temperatures and initial concentrations:

$$\Delta G = \Delta G° + RT \ln Q = \Delta G° + RT \ln \frac{[\text{products}]}{[\text{reactants}]} \tag{1.22}$$

where R is the gas constant of 1.987 cal K^{-1} mol^{-1}, T is the temperature (Kelvin), and Q is the initial ratio of products to reactants.

1.3.1.2 Microbial Oxidation Reactions

Microbial oxidation of organic contaminants occurs when a microorganism uses the compound as an electron donor (primary growth substrate) in a coupled oxidation–reduction reaction. In other words the microbes transfer electrons from the contaminant to electron acceptors. Microbial oxidation can occur under either aerobic or anaerobic conditions. As long as there is molecular oxygen present, aerobic reactions take place. This includes the oxidation of organic compounds and inorganics like ammonium and sulfide ions.

Rapid depletion of dissolved oxygen caused by aerobic reactions results in anaerobic conditions. Biological oxidation reactions continue within contaminant plumes even after the oxygen is consumed. Microbes use "oxygen substitutes" to serve as electron acceptors. Denitrification is observed until nitrate is virtually absent. In the region where denitrification occurs, biological oxidation by oxidized manganese species also takes place. Under those conditions iron may still be present in oxidized forms. Then, a marked decrease in redox potential occurs when only electron acceptors that exhibit lower redox potentials are still left within the reactive zone (Figure 1.4). This redox sequence for biological oxidation is sometimes referred to as *oxic*

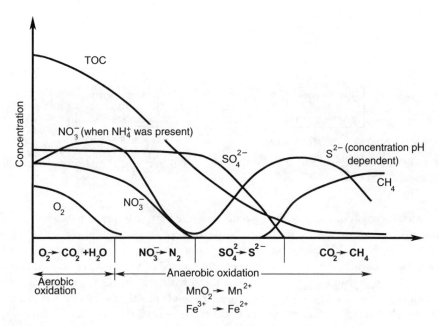

FIGURE 1.4 Sequential pathways of microbial oxidation reactions. (Adapted and modified from Schwarzenbach.[3])

(aerobic), *suboxic* (denitrification, manganese reduction), and *anoxic* (iron reduction, sulfate reduction, and methanogenesis) condition.[3]

The advantages of anaerobic microbial oxidation systems, implemented within an IRZ, are many. The reagents containing the electron acceptors have high water solubility and thus it is easier to deliver and distribute within the reactive zone. In addition, these reagents are resistant to abiotic reactions. All sites have some form of anaerobic bio-oxidizing and microbial populations and specific stimulation processes may vary from site to site. Anaerobic bio-oxidation systems can be designed to implement the following target reactions within an IRZ:

- Nitrate reduction
- Iron reduction — with or without catalyst
- Sulfate reduction
- Methanogenic systems

1.3.1.3 Microbial Reduction Reactions

Microbial reduction takes place when the compound is used as an electron acceptor (primary growth substrate) during reductive reactions (Figure 1.5). Most interest in reductive transformations of environmental contaminants involves dechlorination of aliphatic and aromatic organic compounds and the reduction of nitroaromatic compounds. Other examples of reductive transformations that may occur in the

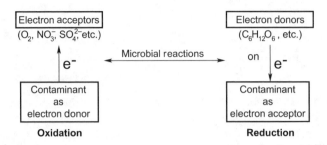

FIGURE 1.5 Depiction of contaminants(s) participating in microbial oxidation or reduction reactions.

subsurface environment include the reduction of ethers, azo compounds, disulfides, and sulfoxides.

The term dehalorespiration was recently coined to describe reductive dechlorination caused by microorganisms that utilize chlorinated compounds as electron acceptors. During dehalorespiration, the chlorinated hydrocarbon is used directly as an electron acceptor and dissolved hydrogen is used directly as an electron donor. Although there are many reports of a variety of electron donors (e.g., molasses, lactate, and methanol) being used to stimulate dehalorespiration of chlorinated solvents, it is now established that these substrates actually serve as indirect electron donors via fermentation of hydrogen. These substrates do not serve as direct electron donors for dehalorespiration, but rather, are fermented *in situ* to produce dissolved hydrogen, which in turn acts as the electron donor for the dehalorespiring microorganisms. Dechlorination can occur by several reductive pathways. The simplest results in replacement of a C-bonded chlorine atom with a hydrogen and is known as hydrogenolysis or reductive dechlorination.

1.3.1.4 Microbial Metals Precipitation

Dissolved metal contaminants can be caused to precipitate as solids within an IRZ by changes in pH or oxidation–reduction potential or by reacting with species in injected reagents. Precipitation often produces finely divided solids that get filtered out by the soil matrix, within the pore throats, by gravity or mechanical filtration processes. Breaking the precipitation process into separate conceptual steps helps to visualize them. Metal contaminants can be precipated as metallic sulfides, hydroxides or carbonates or as a combination of these species depending on the biogeochemical conditions within the reactive zone.

A typical *in situ* metals precipitation system, via an IRZ, is essentially a highly complex biogeochemical system. Complete and precise description of the full range of processes occurring in any one system over a given period of time is not feasible. Rather, we have to content ourselves with identifying the predominant processes that are influencing water quality improvement within a particular biochemical system. Most anaerobic IRZs may contain significant aerobic "pockets" within the system, yet the designation of a given biochemical strategy is generally an adequate simplification for most engineering purposes. If we add hydrogeologic fluctuations to

TABLE 1.1
Examples of Microbial Reactions That Can Be Applied with an IRZ

Microbial Oxidation	Microbial Reduction
Aerobic biostimulation	Enhanced reductive dechlorination
Denitrification	Biostimulation by hydrogenation
$Fe^{3+}-Mn^{4+}$ oxidation	*In situ* metal precipitation
Sulfate reduction	Perchlorate reduction
Methanogenesis	
In situ metal precipitation	

biogeochemical complexity, we might quickly despair of the likelihood of ever deriving sufficient system understanding that we could develop a predictive tool suitable for design purposes. The reality is not quite as daunting as these considerations might suggest, but it certainly always pays to bear in mind the inherent complexity of *in situ* microbial metal precipitation systems.

In purely chemical terms, there are two distinctions, to keep in mind, between *in situ* microbial precipitation systems and aboveground chemical metals precipitation systems. First, the principal alkaline reagent in *in situ* microbial systems is bicarbonate (HCO_3^-), as opposed to hydroxyl (OH^-) in aboveground chemical systems. Since bicarbonate tends to buffer the pH of groundwaters between about 6 and 8.5, successful *in situ* metal precipitation demands the utilization and optimization of reactions that operate briskly within this pH range. This differs substantially from the circumstances in aboveground systems, in which alkali dosing commonly results in pH ranging between 9 and 12, in which range most metals of interest will precipitate as hydroxides. In *in situ* precipitation systems, for most common metals present in the groundwater, their carbonate and sulfide minerals are the more favorable sinks.

The understanding of the second distinction between aboveground and *in situ* systems is critical for the complete removal of dissolved metals present in groundwater. In groundwater systems, metal removal and alkalinity generation processes are commonly separated in space and time, whereas they tend to be synchronous in aboveground treatment reactors. This has important implications for the layout of *in situ* metal precipitation systems.

In summary, the full spectrum of potential microbial reactions that can be implemented within an IRZ is shown in Table 1.1.

1.3.2 CHEMICAL REACTIONS

The term *in situ* chemical reactive zone, as used in this book, refers to the use of chemical reagents to destroy or chemically modify target contaminants by means of *in situ* chemical reactions. This section will address the basic principles of chemical oxidation, reduction, and precipitation reactions that can be implemented within engineered IRZs (Table 1.2). It addresses processes within these classes that are sufficiently advanced for full-scale application.

TABLE 1.2
Examples of Chemical Reactions That Can Be Applied within an IRZ

Chemical Reaction	Applications
Chemical oxidation	Organic contaminants
	Inorganic contaminants
Chemical reduction	Organic contaminants
	Inorganic contaminants
Precipitation	Metallic contaminants
Nucleophilic substitution	Organic contaminants

If we look at the multitude of varied and interesting chemical reactions, it is possible to classify a great many of them as either oxidation or reduction reactions. Oxidation and reduction is a theme that runs through the very core of environmental chemistry and it is difficult indeed to think of a remediation project that does not use such a reaction instituted by microbial metabolism or chemical reactants.

Oxidation of a contaminant may be defined as[4]
The addition of oxygen,
or
the removal of hydrogen,
or
the removal of electrons from that compound.

Similarly, *reduction* can be defined as
The removal of oxygen,
or
the addition of hydrogen,
or
the addition of electrons to the compound.

1.3.2.1 Chemically Oxidizing Reactions

As stated above, oxidation is defined as a loss of electrons. Oxidizing reagents gain electrons and are by definition *electrophiles. Electrophilic* species are defined to be electron liking and, hence, an electron-poor species. Oxidation reactions can either be associated with the introduction of higher oxygen or lower hydrogen content into a contaminant molecule or the conversion of a molecule to a higher oxidation state.[1] Almost all oxidations are kinetically second-order reactions in which the rate is proportional to the concentrations of both the oxidizing reagent [oxidant], and the contaminant, A:

$$-\frac{dA}{dt} = k \,[\text{oxidant}]\,[A] \qquad (1.23)$$

In *in situ* remediation, there are many oxidants available, whose importance is highly variable due to changes in reactivity. A potent oxidant such as the hydroxy radical, OH•, may not be equally efficient in every biogeochemical environment. This could be due to either inadequate production rates for the species or rapid side reactions that diminish its steady-state concentrations.

Chemical oxidation processes have been widely used for treatment of pollutants present in potable water and wastewater. These processes can be chemical oxidation alone, photodegradation/photolysis, or a combination of chemical oxidation and photolysis. Chemically oxidizing reactions within an IRZ are achieved by properly delivering the potent chemical oxidants. As the target contaminants involved tend to be refractory or as the concentration is high, the amount of oxidizing agent required becomes high.

The primary advantage of *in situ* chemically oxidizing reactions is the relatively high-speed destruction of contaminants. Since the reaction is nearly immediate, mass destruction of contaminants is far more rapid than what can be achieved by microbial reactions. The preferred oxidants typically used for chemical oxidation within an IRZ are

- Hydrogen peroxide and its decay products
- Fenton's reagent
- Ozone
- Permanganate (both potassium and sodium)
- Hypochlorite

Although all the above-mentioned reagents are strong oxidants, they are not completely interchangeable. The chemical reactions and pathways they follow in oxidizing the contaminants are somewhat different. Detailed discussions of chemically oxidizing reactions are presented in Chapter 3.

1.3.2.2 Chemically Reducing Reactions

Chemically reducing reactions take place when there is a transfer of electrons from an electron donor "or reductant" to the contaminant that functions as the electron acceptor. In reductive transformations the reductant will be a *nucleophilic* species, in other words nucleus-liking and, hence, an electron-rich species. Reductive transformations are most conveniently categorized according to the type of functional group that is reduced.

There are natural analogs to the possible *in situ* reductive chemical reactive zones due to the presence of Fe(II) in certain groundwater systems. Reduction of Cr^{6+} and certain chlorinated aliphatics, such as *cis*-dichloroethene (DCE), by Fe(II) via abiotic reactions are well known. Ferrous sulfate ($FeSO_4$), sodium dithionite (or hydrosulfite, $Na_2S_2O_4$), sodium bisulfite ($NaHSO_3$), and sulfides, sodium boro-hydride ($NaBH_4$) are known chemical reductants that can be effectively used to reduce specific target contaminants.

1.3.2.3 Hydrolytic Reactions

Because of its great abundance, water plays a pivotal role among nucleophiles present in the environment. A reaction in which a water molecule (or hydroxide ion) substitutes for another atom or group of atoms present in an organic molecule is commonly called a *hydrolysis* reaction. It should be noted that in a hydrolysis reaction, the compound is transformed into a more polar product that has quite different properties.[3] Therefore, the product of any hydrolysis reaction has different environmental behaviors than the starting chemical. It should also be noted that the products of hydrolysis are often of somewhat less environmental concern in comparison to the original contaminant. This is, however, not necessarily true for the products of reactions involving nucleophiles other than water or hydroxide ion.[3]

Hydrolytic reactions encompass several types of reaction mechanisms that can be defined by the type of reaction center — the atom bearing the leaving group, X — where hydrolysis occurs. In general terms, hydrolysis is a chemical transformation in which an organic molecule, RX, reacts with water, resulting in the formation of a new covalent bond with OH and cleavage of the covalent bond with X (the leaving group) in the original molecule.[1] The net reaction is the displacement of X by OH^-:

$$RX + H_2O \rightarrow ROH + X^- + H^+ \tag{1.24}$$

Two example reactions showing the hydrolysis of methyl bromide and ethyl acetate ester are presented in Equations (1.25) and (1.26), respectively.

$$CH_3Br + H_2O \rightarrow CH_3OH + H^+ + Br^- \tag{1.25}$$

$$CH_3COOC_2H_5 + H_2O \rightarrow CH_3COO^- + C_2H_5OH + H^+ \tag{1.26}$$

For the most part, it can be assumed that under ambient groundwater conditions of pH, temperature, and reactant concentrations, hydrolysis reaction proceeds spontaneously and to an extent that, for practical purposes, it can be considered to be irreversible.

Hydrolysis results in reaction products that may be more susceptible to biodegradation as well as more soluble. The likelihood that a halogenated solvent will undergo hydrolysis depends in part on the number of halogen substituents. More halogen substituents on a compound will decrease the chance for hydrolysis reactions to occur and will therefore decrease the rate of the reaction. Hydrolysis rates can generally be described using first-order kinetics, particularly in groundwater where water is the dominant nucleophile. Bromine substituents are more susceptible to hydrolysis than chlorine substituents. As the number of chlorine atoms in the molecule increases, *dehydrohalogenation* may become more important.[5,6]

Dehydrohalogenation is an elimination reaction involving halogenated alkanes in which a halogen is removed from one carbon atom, followed by subsequent removal of a hydrogen atom from an adjacent carbon atom. In this two-step reaction an alkene is produced. Although the oxidation state of the compound decreases due to the removal of a halogen, the loss of a hydrogen atom increases it. This results in no external electron transfer, and there is no net change in the oxidation state of the reacting molecule.[6] Contrary to the patterns observed for hydrolysis, the likelihood of dehydrohalogenation increases with the number of halogen constituents. Under normal environmental conditions, monohalogenated aliphatics apparently do not undergo dehydrohalogenation. The compounds 1,1,1-trichloroethane (TCA) and

1,1,2-TCA are known to undergo dehydrohalogenation and are transformed to 1,1-DCE, which is then reductively dechlorinated to vinyl chloride (VC) and ethene. Tetrachloroethanes and pentachloroethanes are transformed to trichlorothene (TCE) and pentachloroethene (PCE) via dehydrohalogenation pathways.[6]

Methods to predict the hydrolysis rates of organic compounds for use in the environmental assessment of pollutants have not advanced significantly since the first edition of the Lyman *Handbook*.[7] Two approaches have been used extensively to obtain estimates of hydrolytic rate constants for use in environmental systems.[8] The first and potentially more precise method is to apply quantitative structure/activity relationships (QSARs).[8] To develop such predictive methods, one needs a set of rate constants for a series of compounds that have systematic variations in structure and a database of molecular descriptors related to the substituents on the reactant molecule. The second and more widely used method is to compare the target compound with an analogous compound or compounds containing similar functional groups and structure, to obtain a less quantitative estimate of the rate constant.

Some preliminary examples of hydrolysis reactions illustrate the very wide range of reactivity of organic contaminants. For example, triesters of phosphoric acid hydrolyze in near-neutral solution at ambient temperatures with half-lives ranging from several days to several years,[2] whereas the halogenated alkanes such as tetrachloroethane, carbon tetrachloride, and hexachloroethane have half-lives of about 2 h, 50 years, and 1000 millennia (at pH = 7 and 25°C), respectively.[5,9] On the other hand, pure hydrocarbons from methane through the polycyclic aromatic hydro-carbons (PAHs) are not hydrolyzed under any circumstances that are environmentally relevant. Thus, hydrolysis can explain the disappearance of 1,1,1-TCA from a mixed chlorinated plume over time, when TCE and its daughter product *cis*-DCE persist.

Types of contaminants that are generally susceptible to hydrolysis include[10]

- Alkyl halides
- Amides
- Amines
- Carbamates
- Carboxylic acid esters
- Epoxides
- Nitriles
- Phosphoric acid esters
- Sulfonic acid esters
- Sulfuric acid esters

Compounds resistant to hydrolysis include

- Alkanes, alkenes, and alkynes
- Benzene
- Polycyclic aromatics
- Halogenated aromatics
- Nitro aromatics

- Alcohols and phenols
- Ethers
- Aldehydes and ketones
- Carboxylic acids

1.3.2.4 Acid–Base Reactions

It is important to understand that contaminants in groundwater systems react in an environment far more complicated than if they simply were surrounded by a large number of water molecules in a laboratory container. The various impurities in water interact in ways that can affect their chemical behavior markedly. The water quality parameters generally defined as controlling variables have an especially strong effect on water chemistry. For example, a pH change from pH 6 to pH 9 will lower the solubility of Cu^{2+} by four orders of magnitude. At pH 6 the solubility of Cu^{2+} is about 40 mg L^{-1} while at pH 9 it is about 4×10^{-3} mg L^{-1}.

The solvent inside microbial cells and in all extracellular fluids is water. An important characteristic of any aqueous solution is the concentration of positively charged hydrogen ions (H^+) and negatively charged hydroxyl ions (OH^-). These ions are the dissociation products of H_2O, they are constituents of all living systems, and they are liberated by many reactions that take place between contaminant molecules within the microbial cells.

When a water molecule dissociates, one of its polar H–O bonds breaks. The resulting hydrogen ion, often referred to as a proton, has a short lifetime as a free particle and quickly combines with a water molecule to form a hydronium ion (H_3O^+). For convenience, however, we refer to the concentration of hydrogen ions in a solution $[H^+]$, even though we really mean the concentration of hydronium ions $[H_3O^+]$. The dissociation of water is a reversible reaction:

$$H_2O \rightleftharpoons H^+ + OH^- \tag{1.27}$$

and at 25°C,

$$[H^+][OH^-] = 1.0 \times 10^{-14} \, (mol \, L^{-1})^2 = K_{w,25°C} \tag{1.28}$$

In pure water,

$$[H^+] = [OH^-] = 10^{-7} \, (mol \, L^{-1}) \tag{1.29}$$

When enclosed in square brackets the species concentration is in moles per liter. Because the degree of dissociation increases with temperature, K_w is temperature dependent. At 50°C,

$$K_{w,50°C} = [H^+][OH^-] = 1.83 \times 10^{-13} \, (mol \, L^{-1})^2 \tag{1.30}$$

If, for example, an acid is added to water at 25°C, the H^+ concentration increases but the product expressed by Equation (1.28) will always be equal to

TABLE 1.3
The pH Scale[11,12]

	Concentration of H⁺ Ions (mol L⁻¹)	pH	Example
	10^{-0}	0	
	10^{-1}	1	Gastric fluids
	10^{-2}	2	Lemon juice
	10^{-3}	3	Vinegar
	10^{-4}	4	Acid soil
	10^{-5}	5	Lysosomes
⇑ Increasing acidity	10^{-6}	6	Cytoplasm of contracting muscle
Neutral	10^{-7}	7	Pure water and cytoplasm
⇓ Increasing alkalinity	10^{-8}	8	Sea water
	10^{-9}	9	Very alkaline natural soil
	10^{-10}	10	Alkaline lakes
	10^{-11}	11	Household ammonia
	10^{-12}	12	Lime (saturated solution)
	10^{-13}	13	
	10^{-14}	14	

1.0×10^{-14} (mol L⁻¹)². This means that if [H⁺] increases, [OH⁻] must decrease. Adding a base causes [OH⁻] to increase and [H⁺] to decrease correspondingly.

The concentration of hydrogen ions in a solution is expressed conventionally as its pH:

$$\text{pH} = -\log [\text{H}^+] \tag{1.31}$$

In pure water at 25°C,

$$[\text{H}^+] = 10^{-7}\,\text{M}$$

and hence

$$\text{pH} = -\log 10^{-7} = 7.0 \tag{1.32}$$

On the pH scale, 7.0 is considered neutral, pH values below 7.0 indicate acidic solutions, and values above 7.0 indicate basic (alkaline) solutions (Table 1.3). In a 0.1 M solution of hydrochloric acid (hydrogen chloride (HCl)) in water [H⁺] is equal to 0.1 M, because virtually all the HCl has dissociated into H⁺ and Cl⁻ ions. For this solution pH = $-\log 0.1 = 1.0$. In fact, pH values can be less than zero, since a 10 M solution of HCl will have a pH of -1.

One of the most important properties of a biological fluid is its pH. The fluid inside the cells normally has a pH of about 7.2. In certain microbial cells, the pH is much lower, about 5; this corresponds to a H⁺ concentration more than 100 times higher than normal and such cells contain many degradative enzymes that function optimally in an acidic environment, whereas their action is inhibited in near-neutral environment.[11] Maintenance of a specific pH is imperative for some cellular structures to function properly, such as in an engineered *in situ* microbial reactive zone.

In general, any molecule or ion that tends to release a hydrogen ion is called an *acid*, and any molecule or ion that readily combines with a hydrogen ion is called a *base*. Thus, hydrogen chloride is an acid. The hydroxyl ion is a base, as is ammonia (NH_3), which readily picks up a hydrogen ion to become an ammonium ion (NH_4^+). Some organic molecules are acidic because they have functional groups like a carboxyl group ($-COOH$ (and other groups)), which tends to dissociate to form the negatively charged carboxylate ion ($-COO^-$):

1.3.2.4.1 Solubility of CO_2 in Water

Carbon dioxide plays a fundamental role in determining the pH of natural waters. Although CO_2 itself is not acidic, it reacts in water (reversibly) to make an acidic solution by forming carbonic acid (H_2CO_3), as shown in Equation (1.33). Carbonic acid can subsequently dissociate in two steps to release hydrogen ions, as shown in Equations (1.34) and (1.35).

$$CO_2 + H_2O \leftrightarrow H_2CO_3 \tag{1.33}$$

$$H_2CO_3 \leftrightarrow H^+ + HCO_3^- \rightarrow H_2CO_3 \leftrightarrow H^+ + HCO_3^- \tag{1.34}$$

$$HCO_3^- \leftrightarrow H^+ + CO_3^{2-} \tag{1.35}$$

As a result, pure water exposed to air is not acid–base neutral with a pH near 7.0 because dissolved CO_2 makes it acidic, with a pH around 5.7. The pH dependence of Equations (1.33) through (1.35) is shown in Figure 1.6. Between pH = 6.3 and 10.3, a range common to most environmental waters, bicarbonate ion (HCO_3^-) is the dominant species.

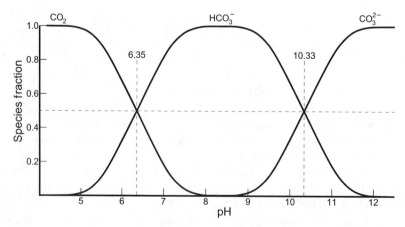

FIGURE 1.6 Distribution diagram showing pH dependence of carbonate species in water.

Natural processes such as biodegradation of organic matter and respiration of plants and organisms, which commonly occur in the subsurface, consume O_2 and produce CO_2. In the soil subsurface, air in the pore spaces cannot readily equilibrate with the atmosphere, and therefore pore space air becomes lower in O_2 and higher in CO_2 concentrations.[12] Oxygen may decrease from about 21% in the atmosphere to between 15% and 0% in the soil depending on the amount of easily degradable organic carbon present in the soil. Carbon dioxide may increase from about 0.04% in the atmosphere to between 0.1% and 10% in the soil.

When water moves through the subsurface, it equilibrates with soil gases and may become more acidic because of a higher concentration of dissolved CO_2. Acidic groundwater has an increased capacity for dissolving minerals. The higher the CO_2 concentration in soil air, the lower the pH of groundwater. Acidic groundwater may become buffered, minimizing pH changes, by dissolution of soil minerals, particularly calcium carbonate. Limestone (calcium carbonate, $CaCO_3$) is particularly susceptible to dissolution by low-pH waters. Limestone caves are formed when low-pH groundwaters move through limestone deposits and dissolve the limestone minerals.[12]

1.3.2.4.2 Acidity and Alkalinity

The alkalinity of water is its acid-neutralizing capacity. The acidity of water is its base-neutralizing capacity. Both parameters are related to the buffering capacity of water (the ability to resist changes in pH when an acid or base is added). On the one hand, water with high alkalinity can neutralize a large quantity of acid without large changes in pH; on the other hand, water with high acidity can neutralize a large quantity of base without large changes in pH.[12]

1.3.2.4.2.1 Acidity

Acidity is determined by measuring how much standard base must be added to raise the pH to a specified value. Acidity can be a net effect of the presence of a single constituent or several constituents, including dissolved carbon dioxide, dissolved multivalent metal ions, strong mineral acids such as sulfuric, nitric, and hydrochloric acids, and weak organic acids such as acetic acid. Dissolved carbon dioxide (CO_2) is the main source of acidity in groundwater systems. Acidity from sources other than dissolved CO_2 is not commonly encountered in unpolluted groundwater and is often an indicator of contamination.[12]

1.3.2.4.2.2 Alkalinity

In groundwater systems that are not contaminated, alkalinity is more commonly found than acidity. Alkalinity is often a good indicator of the total dissolved inorganic carbon (bicarbonate and carbonate anions) present. Since groundwater contains dissolved carbon dioxide, it will have some degree of alkalinity contributed by carbonate species — unless acidic pollutants would have consumed the alkalinity. It is not unusual for alkalinity to range from 0 to 750 mg L^{-1} as $CaCO_3$. For surface waters, alkalinity levels less than 30 mg L^{-1} are considered low, and levels greater than 250 mg L^{-1} are considered high. Average values for rivers are around 100 to 150 mg L^{-1}. Alkalinity in groundwater systems is beneficial because it minimizes pH changes, reduces the toxicity of many metals by forming complexes with them, and provides nutrient carbon for aquatic plants.[12]

Because alkalinity is a property caused by several constituents, some convention must be used for reporting it quantitatively as a concentration. The usual convention is to express alkalinity as ppm or mg L^{-1} of calcium carbonate (CaCO$_3$). This is done by calculating how much CaCO$_3$ would be neutralized by the same amount of acid as was used in titrating the water sample when measuring either phenolphthalein or methyl orange alkalinity. Whether it is present or not, CaCO$_3$ is used as a proxy for all the base species that are actually present in the water. The alkalinity value is equivalent to the mg L^{-1} of CaCO$_3$ that would neutralize the same amount of acid, as does the actual water sample.

1.3.2.4.3 Relationship between pH and K_{eq} of an Acid–Base System

Many molecules used by microbial cells have multiple acidic or basic groups, each of which can release or take up a proton. In the laboratory, it is often essential to know the precise state of dissociation of each of these groups at various pH values. The dissociation of an acid group HA, such as acetic acid (CH$_3$COOH), is described by

$$HA \rightleftharpoons H^+ + A^- \tag{1.36}$$

The equilibrium constant K_a for this reaction is

$$K_a = \frac{[H^+][A^-]}{[HA]} \tag{1.37}$$

By taking the logarithm of both sides and rearranging the result, we can derive a very useful relation between the equilibrium constant and pH as follows:

$$\log K_a = \log \frac{[H^+][A^-]}{[HA]} = \log[H^+] + \log \frac{[A^-]}{[HA]} \tag{1.38}$$

or

$$-\log[H^+] = -\log K_a + \log \frac{[A^-]}{[HA]} \tag{1.39}$$

Substituting pH for $-\log[H^+]$ and pK_a for $-\log K_a$, we have

$$pH = pK_a + \log \frac{[A^-]}{[HA]} \tag{1.40}$$

From this expression, commonly known as the *Henderson–Hasselbalch* equation, it can be seen that the pK_a of any acid is equal to the pH at which half the molecules are dissociated and half are neutral (undissociated). This is because when pK_a = pH, then log ([A$^-$]/[HA]) = 0, and therefore [A$^-$] = [HA]. The Henderson–Hasselbalch equation allows us to calculate the degree of dissociation of an acid if both the pH of the solution and the pK_a of the acid are known. Experimentally, by measuring the concentration of A$^-$ and HA as a function of the pH of the solution, we can calculate the pK_a of the acid and thus the equilibrium constant for the dissociation reaction.[11]

1.3.2.4.4 Buffering

A growing microbial cell must maintain a constant pH in the cytoplasm of about 7.2 to 7.4 despite the production, by metabolism, of many acids, such as lactic acid and CO_2, which reacts with water to form carbonic acid (H_2CO_3).[11] Cells have a reservoir of weak bases and weak acids, called buffers, which ensure that the cell's pH remains relatively constant. Buffers do this by "soaking up" H^+ or OH^- when these ions are added to the cell or are produced by metabolism.

If additional acid (or base) is added to a solution of an acid (or a base) at its pK_a value (a 1:1 mixture of HA and A^-), the pH of the solution changes, but it changes less than it would if the original acid (or base) had not been present. This is because protons released by the added acid are taken up by the original A^- form of the acid; likewise, hydroxyl ions generated by the added base are neutralized by protons released by the original HA.

This ability of a buffer to minimize changes in pH, its buffering capacity, depends on the relationship between its pK_a value and the pH. To understand this point, we need to recognize the effect of pH on the fraction of molecules in the undissociated form (HA). The *titration curve* for acetic acid shown in Figure 1.7 illustrates these relationships: at one pH unit below the pK_a of an acid, 91% of the molecules are in the HA form; at one pH unit above the pK_a, 91% are in the A^- form. Thus the buffering capacity of weak acids and bases declines rapidly at more than one pH unit from their pK_a unit, from their pK_a values. In other words, the addition of the same number of moles of acid to a solution containing a mixture of HA and A^- that is at a pH near the pK_a of the acid will cause less of a pH change than it would if the HA and A^- were not present or if the pH were far from the pK_a value.

All biological systems contain one or more buffers. Phosphoric acid (H_3PO_4) is a physiologically important buffer: phosphate ions are present in considerable quantities in microbial cells and are an important factor in maintaining, or buffering, the pH of the cytoplasm.[11]

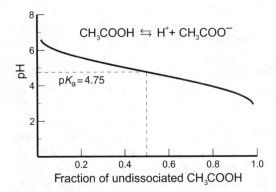

FIGURE 1.7 The pK_a for the dissociation of acetic acid to hydrogen and acetate ions is 4.75. At this pH, half the acid molecules are dissociated. Because pH is measured on a logarithmic scale, the solution changes from 91% CH_3COOH at pH 3.75 to 9% CH_3COOH at pH 5.75. The acid has maximum buffering capacity in this pH range.

1.4 REACTANTS AND SOLVENTS

When confronted with the task of designing an IRZ system to remediate specific contaminants, we obviously need to be able to quantify each of the individual processes occurring within the system designed. We have to be able to quantify all relevant compound-specific and system-specific parameters that are required to define the reactions and processes. To this end, we need to develop a feeling of how chemical structures cause the molecular interactions that govern the various reactions and processes. It should be stressed that without an understanding at the molecular level, a sound assessment of the environmental behavior of contaminants, organic or inorganic, is not possible. Finally, in order to be able to understand the dynamic nature of these reactions we need to understand both the transport and fate processes of contaminants together in the subsurface environment.

Chemistry as a science helps us understand molecular changes that take place during reactions, whether chemical, physical, biochemical, or biogeochemical. How do molecules or ions interact with each other and evolve into new harmless end-products? Why are some reactions slow and others fast, and why do the reaction rates depend on environmental conditions such as temperature, pH, redox potential, presence or absence of catalysts, and the concentrations of the contaminants themselves? Recent advances in the science of *in situ* remediation were possible with the eventual intersection of the disciplines of geology, hydrogeology, microbiology, chemistry, and biochemistry. The science of biogeochemistry is attempting to uncover the processes that occur at microscopic and macroscopic scales.

If it were possible to measure or, ideally, to predict the rates and products of the fundamental processes, then the more complex degradation or transformation reactions, involving many elementary steps, can be easily understood. Indeed, without the detailed knowledge of elementary reactions it would be impossible to provide a quantitative understanding of complex reactions necessary to remediate contaminants *in situ*.

1.4.1 WATER–SOLVENT REACTION MEDIUM

It is important to understand that chemical constituents in environmental water bodies, particularly in groundwater systems, react in an environment far more complicated than if they simply were surrounded just by a large number of water molecules. The various constituents in water, other than the contaminants themselves, interact in ways that can affect their chemical behavior markedly. In addition to the naturally present groundwater constituents, these variables include pH, oxidation–reduction potential, alkalinity, acidity, temperature, and dissolved solids.

It is also important to make a distinction between a solution and a mixture. When we talk of a solution, it is implied that the solute is not a major component of the bulk solvent. Therefore, the presence of a dissolved compound does not have a significant impact on the properties of the bulk liquid. In contrast, in a mixture it is recognized that the major components contribute substantially to the overall nature of the "dissolved" medium. This is reflected in both macroscopic properties and molecular scale phenomena.

Whether a contaminant "likes" or "dislikes" being surrounded by liquid water, or alternatively whether water "likes" or "dislikes" accommodating a given solute compound, is of utmost importance to the environmental behavior and the reactions that the compound will undergo. Environmentally relevant compounds have aqueous solubilities ranging over more than ten orders of magnitude — from completely soluble compounds (miscible) to levels of saturation that are so low that special methods are required to detect their concentrations.

Solvents are used to bring reactants together at suitable and required concentrations. Solids do not react together well. Even with finely ground powders complete mixing of reactants is difficult. Gas-phase reactions are limited to those that involve volatile reactants, and high pressures are necessary to increase reactant concentrations. Hence, to get reactants to mix well at a molecular level requires the use of a solution. When solutions are used, the concentrations of reactants can be readily adjusted over some ten orders of magnitude. Compounds too reactive in one solvent can be used in another solvent.

The transfer of a reaction from one medium to another need not be just from one solvent to another. Subtle changes to the bulk properties of a solvent, such as water, can be brought about by adding a low concentration of a nonreacting cosolvent or an electrolyte. They alter the solvent's intermolecular structure — especially, in the case of water, the nature of its hydrogen bonding — and thereby change its relative "permitivity." Reactions that involve ions as reactants, products, or reactive intermediates are affected most.

Solutions are a convenient way of delivering reagents at required concentrations and quantities within an IRZ. Solutes that are more soluble in a particular solvent than in water can be concentrated into the nonaqueous phase. Nonpolar molecular compounds dissolve in nonpolar solvents. Polar molecular compounds and ionic compounds may dissolve in polar solvents. The word "polar" is used frequently when referring to solvents. We need to examine carefully the concept of polarity, as it can apply to individual bonds, individual molecules, or to a bulk solvent.

Solubility is a process that results from the interplay between intermolecular forces. All molecules have attractive forces acting between them. The attractive forces are electrostatic in nature, created by nonuniform distribution of valence electrons around the positively charged nuclei of a molecule. When electrons are not uniformly distributed, the molecule is attracted to oppositely charged regions on adjacent molecules, resulting in the so-called *polar attractive forces*.[12]

In discussing solvents the idea of *polarity* is central. The idea of solvent polarity refers not to bonds or to molecules, but to the solvent as an assembly of molecules. Qualitatively, polar solvents promote the separation of solute moieties with unlike charges and they make it possible for solute moieties with like charges to approach each other more closely. Like the concept of electronegativity, the more precisely you want to define polarity, the more difficult it is to define. Polarity is thus a rather ill-defined term. It covers the solvent's overall solvation capability (solvation power) for solutes. The polarity depends on the action of all possible, nonspecific and specific, intermolecular interactions between solute ions or molecules and solvent molecules. It covers electrostatic, directional, inductive, dispersion, and charge-transfer forces, as well as hydrogen bonding forces, but excludes interactions leading to definite chemical alterations of the ions or molecules of the solute.

For a solute A to be soluble in a solvent S, the S–S intermolecular forces must be broken to create a cavity in the solvent. Also A–A intermolecular (or inter-ionic) forces must be broken to split the solute into its constituents. Both these processes require energy. When the solute is in the solvent S–A interactions can take place. If these are large enough they compensate for breaking the S–S and A–A links. On average, molecular arrangements will favor the lower energy attractive positions, and the attractive forces always prevail over the repulsive forces.[11]

1.4.1.1 Solubility in Water

Solubility in water is one of the most important physical–chemical properties of a chemical compound, with numerous applications to the prediction of its fate and its effects in the environment. In one sense it can be seen as a direct measurement of hydrophobicity, that is, the tendency of water to *exclude* the substance from solution. It can be viewed as the maximum concentration that an aqueous solution will tolerate before the onset of phase separation. In another sense it is the situation in which the molecules of a pure compound (gas, liquid, or solid) are partitioned so that its concentration reflects equilibrium between the pure material and aqueous solution. In this case, we refer to the equilibrium (or saturation) concentration in the aqueous phase as the *water solubility* or *aqueous solubility* of the compound.

Random motion, resulting from attractive and repulsive intermolecular forces, is the primary force causing substances to dissolve in water or any other solvent. Gases and liquids dissolve more quickly than solids due to the increased level of these random motions. Contaminants are more soluble in water when intermolecular attractions between the contaminant and water are similar in magnitude to the attractions between the molecules of the pure substances.

Aqueous solutions can be thought of as rearrangements of the structure of liquid water in order to accommodate "foreign" solute molecules, which can interact with the "solvent water" in more or less consequential ways. Two extremes can be imagined: a virtual noninteraction in which the water molecules merely surround the solute, arranging themselves around it, or an almost complete incorporation in which the functional groups of the solute molecule placidly combine in the hydrogen-bonded lattice of the "water solvent."

Substances that are readily soluble in water, such as lower-molecular-weight alcohols, will dissolve freely in water if accidentally spilled and will tend to remain in aqueous solution due to the strong affinity of the alcohol molecules to the water molecules. On the contrary, sparingly soluble substances dissolve more slowly and, when in solution, have a stronger tendency to partition out of aqueous solution into other phases. Oil and water do not mix because water molecules are attracted strongly to one another, and oil molecules are attracted strongly to one another, but water molecules and oil molecules are attracted only weakly to one another.

The combined effects of molecular weight and molecular structure on solubility (in water) can be described by the following examples. Because of their –OH group, alcohols have high water solubilities due to their ability to form hydrogen bonds, as described later. However, the alcohol solubility decreases as the number of carbons in the molecular structure increases. For example, methanol (CH_3OH, molecular weight — 32) is miscible (infinite solubility) in water while 1-dodecanol ($C_{12}H_{25}OH$, molecular weight — 186) has a water solubility of 0.000019 mol L^{-1} at 25°C. Even though the –OH group in alcohols is hydrophilic (attracted to water), the hydrocarbon skeleton is hydrophobic (repelled from water). Since the hydrocarbon skeleton of 1-dodecanol is very large, the hydrophobic behavior overcomes the hydrophilic behavior of the –OH group. On the other hand, the double alcohol 1,5-pentanediol ($C_5H_{10}(OH)_2$, molecular weight — 104) is miscible in water as a result of its two –OH groups, and 1-pentanol ($C_5H_{11}OH$, molecular weight — 88) has a water solubility of only 0.25 mol L^{-1}. Double alcohols (diols) or triple alcohols (triols) have much higher water solubility than single alcohols of comparable molecular weight due to their affinity to form more hydrogen compounds.[12]

Solubility normally is measured by bringing an excess amount of a pure chemical phase into contact with water at a specified temperature so that equilibrium is achieved and the aqueous phase concentration reaches a maximum value. It is rare to encounter a single compound as the contaminant present in the groundwater at a contaminant site.

Under natural conditions, dissolved organic matter such as humic and fulvic acids frequently increase the apparent solubility. This is the result of sorption of the chemical to organic matter which is sufficiently low in molecular mass to be retained permanently in solution. The *true* solubility or concentration in the pure aqueous phase probably is not increased. The apparent solubility is the sum of the *true* or dissolved concentration and the quantity that is sorbed.

So far we have considered only situations in which a given organic contaminant is present as the *sole* solute in an aqueous solution. In reality, in any relevant aquatic environmental system there may be other natural and/or anthropogenic organic chemicals present that may or may not impact the solubility of the compound of concern. The effect of highly water-soluble organic compounds (organic cosolvents) is to significantly change the solution properties of an "aqueous" phase. The concept of the flooding with a cosolvent (such as methanol) to remove separate phase, nonaqueous phase liquids (NAPLs) such as PCE and TCE, is a good example of utilization of the cosolvency effect for remediation within an engineered IRZ. The beneficial effects of formed microbial surfactants acting as cosolvents within microbial IRZs will be discussed in Chapter 2. Most of the systematic experimental research on cosolvency effects in the aqueous phase have focused only on the effects of completely water-miscible organic solvents such as methanol, ethanol, acetone, methyl–ethyl ketone, glycerol, etc.[3]

1.4.2 REACTANTS

Any contaminant, when spilled in the subsurface environment, will partially dissolve into the water, partially volatilize into the soil gas and air, and partially sorb to the soil matrix. The relevant mechanisms are described in Figure 1.8. The relative amounts of the contaminant that are found in each phase with which it is in contact depends on intermolecular forces existing between the contaminant, water, and the soil matrix. The most important factor for predicting the partitioning behavior, transport, and the fate of contaminants in the subsurface environment is an understanding of the relevant intermolecular forces in play.[12]

Each chemical's structure, which defines that compound's "personality," provides a systematic basis to understand and predict the physical and chemical

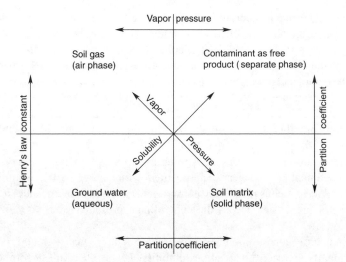

FIGURE 1.8 Partitioning of a contaminant among air, water, soil and free product phases when spilled in a subsurface environment. (Adapted and modified from Weiner.[12])

behavior in the environment. Thus in order to quantify the dynamics of the behavior of a chemical and design a remediation process within a reactive zone, we need to understand and visualize the molecular structures from a microscopic perspective.

The factors that contribute to the reactivity of a contaminant molecule are all interrelated and include the energy possessed by the species (which may help overcome activation barriers), the intrinsic reactivity of the specific electronic arrangement within the molecule, and the relative efficiencies of the different competing pathways for the change of the particular electronic state. There are a number of related issues, all closely linked to the effect of altering the electronic configuration, that explain why excitation from one state to another can lead to such marked differences in chemical characteristics. These include the effects of changes in geometry, dipole moment, redox characteristics (i.e., electron donating and accepting ability), and the related acid–base properties. Both the sizes and the shapes of molecules may be affected by excitation and the concurrent redistribution of electrons between bonding, nonbonding, and anti-bonding orbitals.[3]

1.4.2.1 Molecular Structures of Compounds

To understand the nature and reactivity of contaminant molecules, we first have to consider the composition of the molecules. Molecules are made of the various atoms and the chemical bonds linking them. It should be noted that most of the contaminants, organic or inorganic, are combinations of only a few elements: carbon (C), hydrogen (H), oxygen (O), nitrogen (N), sulfur (S), phosphorus (P), and the halogens (Cl, F, Br, or I). Fortunately, the knowledge of a few governing rules about the nature of these elements and chemical bonds enables us to understand important relationships between the structure of a compound and its properties and reactivities and thus the compound's behavior in the environment.[3,11] If the relative attractive forces between the molecules within a biogeochemical system can be predicted, then we can also predict the potential reactions and behavior within this system.

When describing the chemistry of a compound, the elemental composition has to be specified. For example, a chlorinated hydrocarbon, as the name implies, will contain chlorine, hydrogen, and carbon atoms in its molecular structure. The molecular formula describes how many atoms of each of these elements are present in the structure: CH_4 for methane, C_2HCl_3 for TCE. The molecular formula allows us to calculate the molecular weight of each compound. The exact connection of each atom within the molecule is referred to as the molecular structure.

Electrons move around the nucleus of an atom in clouds called *orbitals* which lie in a series of concentric *shells*, or energy levels; electrons in outer shells have more energy than those in inner shells. Each shell has a maximum number of electrons that it can hold. Electrons fill the innermost shells of an atom first, then the outer shells. The energy level of an atom is lowest when all of its orbitals are filled, and an atom's reactivity depends on how many electrons it needs to complete its outermost orbital. The noble gases, helium (He), neon (Ne), argon (Ar), and radon (Rn), are nonreactive since all their orbitals are "filled."

Core electrons are those electrons that are never utilized in chemical bonding. Their high ionization energies and contracted nature mean that they are not perturbed by the orbitals of neighboring atoms. In general, core electrons may be defined as all those electrons in orbitals which are associated with the noble gas before that element in the periodic table (Figure 1.9). The first orbital holds only 2 electrons (He structure), and the second shell holds 8. The third can hold 18, but a stable configuration is reached when the shell is filled with 8 (Ar structure).

It is important to know that the number of electrons present in a particular atom in its outer shell (called the valence electrons) determine the chemical characteristics of an element. All valence electrons do not necessarily become involved in bonding in all the possible compounds containing that element. For example, chromium, arsenic, and manganese do not form any compounds in the oxidation state corresponding to the utilization of all electrons in their outer shells. They do, however, form compounds in only a couple of oxidation states. Examples include chromium (+3) and (+6), arsenic (+3) and (+5), and manganese (+7) and (+4).

Intermolecular forces are electrostatic in nature. Molecules and atoms are composed of electrically charged particles (electrons (positive) and protons (negative)), and it is common for them to have regions that are predominantly charged positive or negative. Attractive forces between molecules arise when electrostatic forces attract positive regions on one molecule to negative regions on another. The strength of the attractions between molecules depends on the *polarities* of chemical bonds within the molecules and the *geometrical shapes* of the molecules.[12]

The number and nature of bonds that each of the various elements present in organic or inorganic molecular structures are again dependent on the electronic characteristics of the constituent elements. At the simplest level, the chemical bonds that hold atoms together in a molecule are of two types, namely, covalent bonds and ionic bonds.

1 1A	2 2A	3 3B	4 4B	5 5B	6 6B	7 7B	8 8B	9 8B	10 8B	11 1B	12 2B	13 3A	14 4A	15 5A	16 6A	17 7A	18 8A
1 **H** 1.008																	2 **He** 4.003
3 **Li** 6.941	4 **Be** 9.012											5 **B** 10.81	6 **C** 12.01	7 **N** 14.01	8 **O** 16.00	9 **F** 19.00	10 **Ne** 20.18
11 **Na** 22.99	12 **Mg** 24.31											13 **Al** 26.98	14 **Si** 28.09	15 **P** 30.97	16 **S** 32.07	17 **Cl** 35.45	18 **Ar** 39.95
19 **K** 39.10	20 **Ca** 40.08	21 **Sc** 44.96	22 **Ti** 47.87	23 **V** 50.94	24 **Cr** 52.00	25 **Mn** 54.94	26 **Fe** 55.85	27 **Co** 58.93	28 **Ni** 58.69	29 **Cu** 63.55	30 **Zn** 65.41	31 **Ga** 69.72	32 **Ge** 72.64	33 **As** 74.92	34 **Se** 78.96	35 **Br** 79.90	36 **Kr** 83.80
37 **Rb** 85.47	38 **Sr** 87.62	39 **Y** 88.91	40 **Zr** 91.22	41 **Nb** 92.91	42 **Mo** 95.94	43 **Tc** (98)	44 **Ru** 101.1	45 **Rh** 102.9	46 **Pd** 106.4	47 **Ag** 107.9	48 **Cd** 112.4	49 **In** 114.8	50 **Sn** 118.7	51 **Sb** 121.8	52 **Te** 127.6	53 **I** 126.9	54 **Xe** 131.3
55 **Cs** 132.9	56 **Ba** 137.3	57 **La** 138.9	72 **Hf** 178.5	73 **Ta** 180.9	74 **W** 183.8	75 **Re** 186.2	76 **Os** 190.2	77 **Ir** 192.2	78 **Pt** 195.1	79 **Au** 197.0	80 **Hg** 200.6	81 **Tl** 204.4	82 **Pb** 207.2	83 **Bi** 209.0	84 **Po** (209)	85 **At** (210)	86 **Rn** (222)
87 **Fr** (223)	88 **Ra** (226)	89 **Ac** (227)	104 **Rf** (261)	105 **Db** (262)	106 **Sg** (266)	107 **Bh** (264)	108 **Hs** (277)	109 **Mt** (268)	110 **Ds** (281)								

Lanthanide Series	58 **Ce** 140.1	59 **Pr** 140.9	60 **Nd** 144.2	61 **Pm** (145)	62 **Sm** 150.4	63 **Eu** 152.0	64 **Gd** 157.3	65 **Tb** 158.9	66 **Dy** 162.5	67 **Ho** 164.9	68 **Er** 167.3	69 **Tm** 168.9	70 **Yb** 173.0	71 **Lu** 175.0
Actinide Series	90 **Th** 232.0	91 **Pa** 231.0	92 **U** 238.0	93 **Np** (237)	94 **Pu** (244)	95 **Am** (243)	96 **Cm** (247)	97 **Bk** (247)	98 **Cf** (251)	99 **Es** (252)	100 **Fm** (257)	101 **Md** (258)	102 **No** (259)	103 **Lr** (262)

FIGURE 1.9 Periodic table of the elements.

Covalent bonds: Covalent bonds, which hold the atoms within an individual molecule together, are formed by the sharing of electrons called bonding electrons, in the outer atomic orbitals. The electron-attracting properties of covalent-bonded atoms are not different enough to allow one atom to pull an electron entirely away from the other. However, unless both atoms attract bonding electrons equally, the average position of the bonding electrons will be closer to one of the atoms. The atoms are held together because their positive nuclei are attracted to the negative charge of the shared electrons in the space between them. The distribution of shared as well as unshared electrons in outer orbitals is a major determinant of the three-dimensional shape and chemical reactivity of molecules. For instance, the shape of larger organic molecules is crucial to their function and their interactions with small molecules.

Ionic bonds: Ionic bonds occur when one atom attracts an electron away from another atom to form a positive and a negative ion. The ions are then bound together by electrostatic attraction. The electron transfer occurs because the electron-receiving atom has a much stronger attraction for electrons in its vicinity than does the electron-losing atom. Ionic bonds result from the interactions that result from attraction of a positively charged *cation* for a negatively charged *anion*. Unlike covalent bonds, ionic bonds do not have fixed or specific geometric orientations because the electrostatic field around an ion — its attraction for an opposite charge — is uniform in all directions.

1.4.2.1.1 Covalent Bonds

The outermost orbital of each atom has a characteristic number of electrons. Some atoms readily form covalent bonds with other atoms and rarely exist as isolated entities. As a rule, each type of atom forms a characteristic number of covalent bonds with other atoms. For example, a hydrogen atom, with one electron in its outer shell, forms only one bond, such that its outermost orbital becomes filled with two electrons. A carbon atom has four electrons in its outermost orbital; it usually forms four bonds, as in methane (CH_4), in order to fill its outermost orbital with eight electrons. The single bonds in methane that connect the carbon atom with each hydrogen atom contain two shared electrons, one donated from the C and the other from the H, and the outer (s) orbital of each H atom is filled by the two shared electrons (Figure 1.10(A)).

Nitrogen and phosphorus each have five electrons in their outer shells, which can hold up to eight electrons. Nitrogen atoms can form up to four covalent bonds. In ammonia (NH_3), the nitrogen atom forms three covalent bonds; one pair of electrons around the atom (the two dots on the right) is in an orbital not involved in a covalent bond. In the ammonium ion (NH_4^+), the nitrogen atom forms four covalent bonds, again filling the outermost orbital with eight electrons (Figure 1.10(B)). Phosphorus can form up to five covalent bonds, as in phosphoric acid (H_3PO_4). The H_3PO_4 molecule is actually a "resonance hybrid," a structure between the two forms shown below in which nonbonding electrons are shown as pairs of dots (Figure 1.10(C)).

The difference between the bonding patterns of nitrogen and phosphorus is primarily due to the relative sizes of the two atoms: the smaller nitrogen atom has only enough space to accommodate four bonding pairs of electrons around it without creating destructive repulsions between them, whereas the larger sphere of the phosphorus atom allows more electron pairs to be arranged around it without the pairs being too close together.

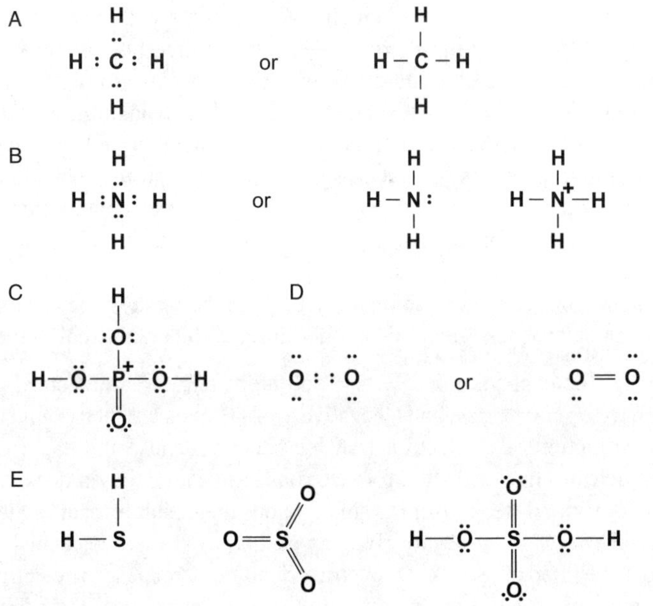

FIGURE 1.10 (A) Structure of CH_4 molecule with its four covalent bonds. (B) Molecular structure of NH_3 and NH_4^+. (C) Molecular structure of phosphoric acid (H_3PO_4). (D) Structure of the O_2 molecule. (E) Structures of hydrogen sulfide (H_2S), sulfur trioxide (SO_3), and sulfuric acid (H_2SO_4).

Both oxygen and sulfur contain six electrons in their outermost orbitals. However, an atom of oxygen usually forms only two covalent bonds, as in molecular oxygen, O_2 (Figure 1.10(D)). Primarily because its outermost orbital is larger than that of oxygen, sulfur can form as few as two covalent bonds, as in hydrogen sulfide (H_2S), or as many as six, as in sulfur trioxide (SO_3) or sulfuric acid (H_2SO_4) (Figure 1.10(E)).

Covalent bonds tend to be very stable because the energies required to break or rearrange them are much greater than the thermal energy available at room temperature (25°C) or the ambient temperatures under which we encounter these compounds. For example, the thermal energy at 25°C is less than 4 kJ mol^{-1} (kcal mol^{-1}), whereas the energy required to break a C–C bond in ethane is about 348 kJ mol^{-1} (Table 1.4).

When two or more atoms form covalent bonds with another central atom, these bonds are oriented at precise angles to one another. The angles are determined by the mutual repulsion of the outer electron orbitals of the central atom. These bond angles give each molecule its characteristic shape. In methane, for example, the central carbon atom is bonded to four hydrogen atoms, whose positions define the four points of a tetrahedron, so that the angle between any two bonds is 109.5° (Figure 1.11(A)). Like methane, the ammonium ion also has a tetrahedral shape. In these molecules, each bond is a *single bond*, a single pair of electrons shared between two atoms.

When two atoms share two pairs of electrons — for example, when a carbon atom is linked to only three other atoms — the bond is a *double bond* (Figure 1.11(B)). In this case, the carbon atom and all three atoms linked to it lie in the same plane. Atoms

TABLE 1.4
The Energy Required To Break Some Important Covalent Bonds Found in Compound Molecules[a]

Type of Bond	Energy (kJ mol^{-1})	Type of Bond	Energy (kJ mol^{-1})
Single bond		*Double bond*	
C–F	485		
O–H	465	C=O	737
H–H	436	C=N	620
P–O	422	C=C	612
C–H	415	C=S	536
C–O	360	P=O	508
C–C	348	O=O	498
S–H	342		
C–Cl	339	*Triple bond*	
C–N	306	C≡N	838
C–Br	284	C≡C	888
C–S	270	N≡N	946
N–O	224		
C–I	209		

[a] Note that double and triple bonds are stronger than single bonds.
Source: Adapted and modified from Lodish et al.[11] and Schwarzenbach.[3]

connected by a double bond cannot rotate freely about the bond axis, while those in a single bond generally can. The rigid planarity imposed by double bonds has enormous significance for the shape of large biological molecules such as proteins and nucleic acids. In *triple bonds*, two atoms share six electrons, such as ethyne, and are rare in biological molecules (Figure 1.11(C)).

All outer electron orbitals, whether or not they are involved in covalent bond formation, contribute to the properties of a molecule, in particular to its shape. For example, the outer shell of the oxygen atom in a water molecule has two pairs of nonbonding electrons; the two pairs of electrons in the H–O bonds and the two pairs of nonbonding electrons form an almost perfect tetrahedron. However, the orbitals of the nonbonding electrons have a high electron density and thus tend to repel each other, compressing the angle between the covalent H–O–H bonds to 104.5° rather than the 109.5° in a tetrahedron (Figure 1.11(D)).

1.4.2.1.2 Unequal Electron Sharing in Polar Covalent Bonds
In certain cases, the bonded atoms in a covalent bond exert different attractions for the electrons of the bond, resulting in unequal sharing of the electrons. The power of an atom in a molecule to attract electrons to itself, called *electronegativity*, is measured on a scale from 4.0 (for fluorine, the most electronegative atom) to a hypothetical zero (Table 1.5). Knowing the electronegativity of two atoms allows us to predict whether a covalent bond can form between them; if the differences in electronegativity are considerable — as in sodium and chloride — an ionic bond, rather than a covalent

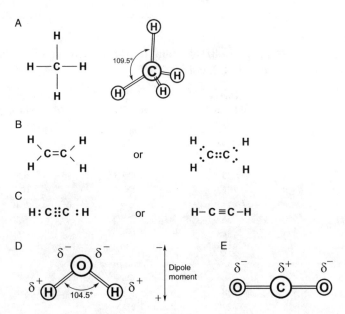

FIGURE 1.11 (A) Molecular shape of methane. (B) Molecular structure of ethene with a C = C double bond. (C) Molecular structure of ethyne with a –C≡C– triple bond. (D) The water molecule with two polar O–H bonds and a net dipole moment. (E) Molecular structure of CO_2 without a net dipole moment.

TABLE 1.5
Electronegativity Values, Shown in Parentheses, of Some Atoms

			H (2.2)				He
Li (1.0)	Be (1.6)	B (2.0)	C (2.6)	N (3.0)	O (3.5)	F (4.0)	Ne
Na (0.43)	Mg (1.3)	Al (1.6)	Si (1.4)	P (2.2)	S (2.6)	Cl (3.2)	Ar
K (0.82)	Ca (1.3)	Ga (1.6)	Ge (2.0)	As (2.2)	Se (2.6)	Br (3.0)	Kr
Rb (0.82)	Sr (0.95)	In (1.8)	Sn (2.0)	Sb (2.1)	Te (2.1)	I (2.7)	Xe
Cs (0.79)	Ba (0.89)	Te (2.0)	Pb (2.3)	Bi (2.0)	Po (2.0)	At	Rn

Source: Adapted and modified from Lodish et al.[11]

bond, will form. Atoms located to the upper right tend to have high electronegativity, fluorine being the most electronegative. Elements with low electronegativity values, such as the metals lithium, sodium, and potassium, are often called electropositive. Since the inert gases (He, Ne, etc.) have complete outer shells of electrons, they neither attract nor donate electrons, rarely form covalent bonds, and have no electronegativity values.

In a covalent bond in which the atoms either are identical or have the same electronegativity, the bonding electrons are shared equally. Such a bond is said to be *nonpolar*. This is the case for C–C and C–H bonds. However, if two atoms differ in electronegativity, the bond is said to be *polar*. The electronegativities of several atoms abundant in chemical compound molecules differ enough that they form polar covalent bonds (e.g., O–H, N–H) or ionic bonds (e.g., Na^+Cl^-). One end of a polar bond has a partial negative charge (δ^-), and the other end has a partial positive charge (δ^+). The symbol δ represents a partial charge, a weaker charge than the one on an electron or a proton. In an O–H bond, for example, the oxygen atom, with an electronegativity of 3.5, attracts the bonded electrons more than does the hydrogen atom, with an electronegativity of 2.2. As a result, the bonding electrons spend more time around the oxygen atom than around the hydrogen. Thus the O–H bond possesses an *electric dipole*, a positive charge separated from an equal but opposite negative charge. We can think of the oxygen atom of the O–H bond as having, on average, a charge of 25% of an electron, with the H atom having an equivalent positive charge. The dipole moment of the O–H bond, defined as a measure of the nonuniform charge separation, is a function of the size of the positive or negative charge and the distance separating the charges. The dipole moment is equal to the magnitude of positive and negative charges at each end of the dipole multiplied by the distance between the charges.

In general, if the electronegativity difference between two bonded atoms is zero, they will form a nonpolar covalent bond. Examples are the molecules of O_2, H_2, and N_2. If the electronegativity difference between two atoms is between 0 and 1.7, they will form a polar covalent bond. Examples are NO and CO. If the electronegativity difference between two atoms is greater than 1.7, they will form an ionic bond. Examples of such a bond are NaCl, HF, and KBr.[12] Because electronegativity differences can vary continuously between 0 and 4, bond character also can vary continuously between nonpolar, covalent, and ionic.

Polarizability is a measure of how easily the electron distribution can be distorted by an electric field — that is, how easily a dipole moment can be induced in an atom or a molecule. Large atoms and molecules have more electrons and larger electron clouds than small ones. In large atoms and molecules, the outer shell electrons are farther from the nuclei and, consequently, are more loosely bound. The electron distributions can be more easily distorted by external charges. In small atoms and molecules, the outer electrons are closer to the nuclei and are more tightly held. Electron charge distributions in small atoms and molecules are less easily distorted.[11]

Therefore, large atoms and molecules are more polarizable than small ones. Since atomic and molecular sizes are closely related to atomic and molecular weights, we can generalize that polarizability increases with increasing atomic and molecular weights. The greater the polarizability of atoms and molecules, the stronger is the intermolecular dispersion forces between them. Molecular shape also affects polarizability. Elongated molecules are more polarizable than compact molecules. Thus, a linear alkane is more polarizable than a branched alkane of the same molecular weight.

In a water molecule both hydrogen atoms are on the same side of the oxygen atom. As a result, that side of the molecule compared with the other side has a slight net positive charge and the other side has a slight net negative charge. Because of this separation of positive and negative charges, the entire molecule has a net dipole moment (Figure 1.11(D)). Some molecules, such as the linear molecule O=C=O, have two polar bonds. Because the dipole moments of the two C=O bonds point in opposite directions, they cancel each other out, resulting in a molecule without a net

FIGURE 1.12 Water molecule showing a high net dipole moment as a result of the two O–H bonds.

dipole moment (Figure 1.11(E)). Similarly, if any molecule is symmetrical in a way that the bond polarity vectors add to zero, then the molecule is nonpolar although it contains polar bonds.

Knowing whether a molecule is polar or not helps predict its water solubility and other properties. Nonpolar molecules invariably have low water solubility. A molecule with no polar bonds cannot be a polar molecule. Thus, all diatomic molecules where both atoms are the same, such as H_2, O_2, N_2, and Cl_2, are nonpolar because there is no electronegativity difference across the bond. Carbon dioxide, carbon tetrachloride, and hexachlorobenzene are all symmetrical and nonpolar, although all contain polar bonds.

Water is a particularly important polar molecule. Its bond polarity vectors add to give the water molecule a high polarity (i.e., dipole moment) (Figure 1.12). The dipole–dipole forces between water molecules are greatly strengthened by hydrogen bonding, discussed below, which contributes to many of water's unique characteristics. These characteristics include relatively high boiling point and viscosity, low vapor pressure, and high heat capacity.[12]

All molecules are attracted to one another because of electrostatic forces. Polar molecules are attracted to one another because the negative end of one molecule is attracted to the positive ends of other molecules, and vice versa. Attractions between polar molecules are called *dipole–dipole forces*. Similarly, positive ions are attracted to negative ions. Attractions between ions are called *ion–ion forces*. If ions and polar molecules are present together, as when sodium chloride is dissolved in water, there can be *ion–dipole forces*, where positive and negative ions (e.g., Na^+ and Cl^-) are attracted to the oppositely charged ends of polar molecules (e.g., H_2O).

However, nonpolar molecules also are attracted to one another although they do not have permanent charges or dipole moments. Evidence of attractions between nonpolar molecules is demonstrated by the fact that nonpolar gases such as methane (CH_4), oxygen (O_2), nitrogen (N_2), ethane (CH_3CH_3), and carbon tetrachloride (CCl_4) condense to liquids and solids when the temperature is lowered sufficiently. This attraction is caused by the transitory dipole movements due to the interaction between the electron clouds caused by the collision of the molecules.[12]

1.4.2.1.3 Hydrogen Bond

Normally, a hydrogen atom forms a covalent bond with only one other atom. However, a hydrogen atom covalently bonded to a donor atom, D, may form an additional weak association, the *hydrogen bond*, with an acceptor atom, as shown in Figure 1.13. In order for a hydrogen bond to form, the donor atom must be highly

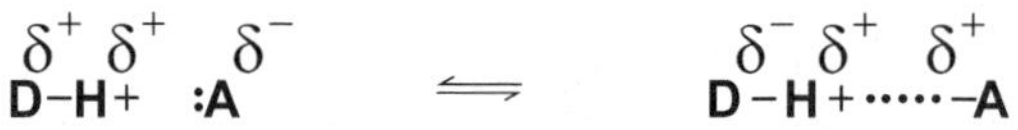

$$\overset{\delta^+}{D}\text{-}\overset{\delta^+}{H}+ \quad \overset{\delta^-}{:A} \quad \rightleftarrows \quad \overset{\delta^-}{D}\text{-}\overset{\delta^+}{H}+\cdots\cdots\overset{\delta^+}{-A}$$

FIGURE 1.13 Formation of a hydrogen bond.

electronegative, so that the covalent D–H bond is polar. The acceptor atom also must be electronegative, and its outer shell must have at least one nonbonding pair of electrons in its valence shell that attracts the δ^+ charge of the hydrogen atom. Fluorine, oxygen, and nitrogen are the smallest and most electronegative elements that contain nonbonding valence electron pairs. Although chlorine and sulfur have similarly high electronegativities and contain nonbonding valence electron pairs, they are too large to consistently form hydrogen bonds (H-bonds). Because hydrogen bonds are both strong and common, they influence many substances in important ways.

Hydrogen bonds are very strong (10 to 40 kJ mol^{-1}) compared to other dipole–dipole forces (from less than 1 to 5 kJ mol^{-1}). The hydrogen atom's very small size makes hydrogen bonding so uniquely strong. Hydrogen has only one electron. When hydrogen is covalently bonded to a small, highly electronegative atom, the shift of bonding electrons toward the more electronegative atom leaves the hydrogen nucleus nearly bare. With no inner core electrons to shield it, the partially positive hydrogen can approach very closely to a nonbonding electron pair on nearby small polar molecules. The very close approach results in stronger attractions than with other dipole–dipole forces.

Because all covalent N–H and O–H bonds are polar, their H atoms can participate in hydrogen bonds. By contrast, C–H bonds are nonpolar, and hence these H atoms are almost never involved in a hydrogen bond. Water molecules provide a classic example of hydrogen bonding. The hydrogen atom in one water molecule is attracted to a pair of electrons in the outer shell of an oxygen atom in an adjacent molecule. Not only do water molecules hydrogen bond with one another, they also form hydrogen bonds with other kinds of molecules, as shown in Figure 1.14(A). The presence of hydroxyl (–OH) or amino (–NH$_2$) groups makes many molecules soluble in water at high concentrations due to the ability to form several hydrogen bonds (methanol (CH$_3$OH) and methylamine (CH$_3$NH$_2$)). In general, molecules with polar bonds that easily form hydrogen bonds with water can dissolve at high concentrations in water and are said to be *hydrophilic*. In addition to the hydroxyl and amino groups, peptide and ester bonds are other examples (Figure 1.14(B)).

Because of the strong intermolecular attractions, hydrogen bonds have a strong effect on the properties of the substances in which they occur. Compared with non-hydrogen-bonded compounds of similar size, hydrogen-bonded substances have relatively high boiling and melting points, low volatilities, high heats of vaporization, and high specific heats. Molecules that can H-bond with water are highly soluble in water. In liquid water, each water molecule apparently forms transient hydrogen bonds with several others, creating a fluid network of hydrogen-bonded molecules.

The mutual attraction of its molecules causes water to have melting and boiling points at least 100°C higher than they would be if water were nonpolar; in the absence of these intermolecular attractions, water on Earth would exist primarily as a gas. The

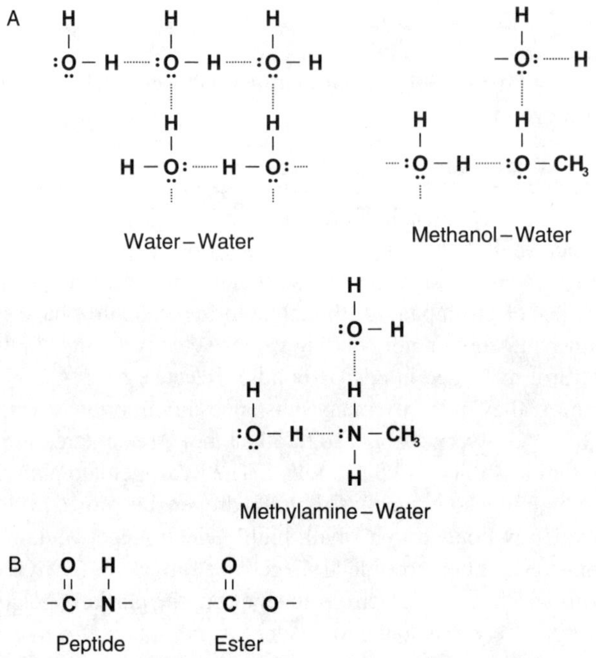

FIGURE 1.14 (A) Water readily forms hydrogen bonds. (B) Other examples of molecular structures that can form hydrogen bonds.

exact structure of liquid water is still unknown.[12] It is believed to contain many transient, maximally hydrogen-bonded networks. Most likely, water molecules are in rapid motion, constantly making and breaking hydrogen bonds with adjacent molecules. As the temperature of water increases toward 100°C, the kinetic energy of its molecules becomes greater than the energy of the hydrogen bonds connecting them, and the gaseous form of water appears.

1.4.2.1.4 Ionic Bonds
In some compounds, the bonded atoms are so different in electronegativity that the bonding electrons are never shared: these electrons are always found around the more electronegative atom (Table 1.5). In sodium chloride (NaCl), for example, the bonding electron contributed by the sodium atom is completely transferred to the chlorine atom. Even in solid crystals of NaCl, the sodium and chlorine atoms are ionized, so it is more accurate to write the formula for the compound as Na^+Cl^-. Because the electrons are not shared, the bonds in such compounds cannot be considered covalent. They are ionic bonds (or interactions) that result from the attraction of a positively charged ion — a cation — for a negatively charged ion — an anion.

Unlike covalent or hydrogen bonds, ionic bonds do not have fixed or specific geometric orientations because the electrostatic field around an ion — its attraction for an opposite charge — is uniform in all directions. However, crystals of salts such as

Na$^+$Cl$^-$ do have very regular structures because that is the energetically most favorable way of packing together positive and negative ions. The force that stabilizes ionic crystals is called the *lattice energy*. The lattice energy is the energy required to convert one mole of a solid ionic compound in its usual lattice structure at absolute zero into its gaseous ions.

In aqueous solutions, simple ions of biogeochemical significance, such as Na$^+$, K$^+$, Ca^{2+}, Mg^{2+}, and Cl$^-$, do not exist as free, isolated entities. Instead, each is surrounded by a stable, tightly held shell of water molecules. An ionic interaction occurs between the ion and the oppositely charged end of the water dipole, as shown below for the K$^+$ ion (Figure 1.15(A)). In the case of a magnesium ion (Mg^{2+}), six water molecules are held tightly in place by electrostatic interactions between the two positive charges on the ion and the partial negative charge on the oxygen of each water molecule (Figure 1.15(B)).

Most ionic compounds are quite soluble in water because a large amount of energy is released when ions tightly bind water molecules. This is known as the *energy of hydration*. Oppositely charged ions are shielded from one another by the water and tend not to recombine. Salts like Na$^+$Cl$^-$ dissolve in water because the energy of hydration is greater than the lattice energy that stabilizes the crystal structure. In contrast, certain salts, such as Ca$_3$(PO$_4$)$_2$, are virtually insoluble in water; the large charges on the Ca^{2+} and PO$_4^{3-}$ ions generate a formidable lattice energy that is greater than the energy of hydration.

Most dissolved inorganic compounds are present in groundwater in an ionic form. These include dissolved metals such as Fe^{2+}, Mn^{2+}, Ni^{2+}, Cr^{6+}, Pb^{2+} and nonmetal species such as NH$_4^+$, Cl$^-$, CN$^-$, F$^-$, NO$_3^-$, SO$_4^{2-}$, S^{2-}, CO$_3^{2-}$, HCO$_3^-$, etc. Some of these ions are considered to be contaminants while others play important and significant roles in the biogeochemical systems within groundwater systems. Important parameters such as pH (concentrations of H$^+$ ions), electron acceptors (concentration of NO$_3^-$, SO$_4^{2-}$, CO$_3^{2-}$, etc.), alkalinity, and acidity (concentration of HCO$_3^-$, CO$_3^{2-}$, OH$^-$, and H$^+$), which strongly influence the remediation processes

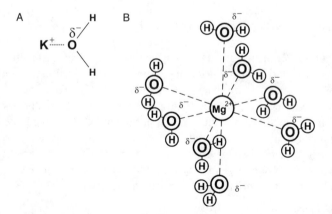

FIGURE 1.15 (A) Ionic interaction between a cation and water. (B) A shell of water surrounding a cation in aqueous solution.

to be implemented within an IRZ, depend on the presence or absence of these ions within the biogeochemical system.

1.4.2.1.5 van der Waals Interactions

When any two atoms approach each other closely, they create a weak, nonspecific attractive force that produces a van der Waals interaction, named for Dutch physicist Johanne Diderik van der Waals (1837–1923), who first described it. These nonspecific interactions result from the momentary random fluctuations in the distribution of the electrons of any atom, which give rise to a transient unequal distribution of electrons, that is, a transient electric dipole. If two noncovalently bonded atoms are close enough together the transient dipole in one atom will perturb the electron cloud of the other. This perturbation generates a transient dipole in the second atom, and the two dipoles will attract each other weakly. Similarly, a polar covalent bond in one molecule will attract an oppositely oriented dipole in another.

These van der Waals interactions, involving either transient induced or permanent electric dipoles, occur in all types of molecules, both polar and nonpolar. In particular, van der Waals interactions are responsible for the cohesion between molecules of nonpolar liquids and solids, such as heptane, $CH_3-(CH_2)_5-CH_3$, which cannot form hydrogen bonds or ionic interactions with other molecules. When these stronger interactions are present, they override most of the influence of van der Waals interactions. Heptane, however, would be a gas if van der Waals interactions could not form.

The strength of van der Waals interactions decreases rapidly with increasing distance. However, if atoms get too close together, they become repelled by the negative charges in their outer electron shells. When the van der Waals attraction between two atoms exactly balances the repulsion between their two electron clouds, the atoms are said to be in *van der Waals contact*. Each type of atom has a van der Waals radius at which it is in van der Waals contact with other atoms. The van der Waals radius of an H atom is 0.1 nm, and the radii of O, N, C, and S atoms are between 0.14 and 0.18 nm. Two covalently bonded atoms are closer together than two atoms that are merely in van der Waals contact. For a van der Waals interaction, the internuclear distance is approximately the sum of the corresponding radii for the two participating atoms. Thus the distance between a C atom and an H atom in van der Waals contact is 0.27 nm, and between two C atoms is 0.34 nm. In general, the van der Waals radius of an atom is about twice as long as its covalent radius. For example, a C–H covalent bond is about 0.107 nm long and a C–C covalent bond is about 0.154 nm long.[11]

The energy of the van der Waals interaction is about 1 kcal mol^{-1}, only slightly higher than the average thermal energy of molecules at 25°C. Thus, the van der Waals interaction is even weaker than the hydrogen bond, which typically has energy of 1 to 2 kcal mol^{-1} in aqueous solutions. The attraction between two large molecules can be appreciable. If they have precisely complementary shapes, they make many van der Waals contacts when they come into proximity. These van der Waals interactions, as well as other noncovalent bonds, mediate the binding of many enzymes with their specific substrates (the substances on which an enzyme acts).

1.4.2.1.6 Hydrophobic Bonds

Nonpolar molecules do not contain ions, possess a dipole moment, or become hydrated. Because such molecules are insoluble or almost insoluble in water, they are said to be *hydrophobic* ("water fearing"). The force that causes hydrophobic molecules or nonpolar portions of molecules to aggregate together rather than to dissolve in water is called the *hydrophobic bond*. This is not a separate bonding force; rather, it is the result of the energy required to insert a nonpolar molecule into water. A nonpolar molecule cannot form hydrogen bonds with water molecules, so it distorts the usual water structure, forcing the water into a rigid cage of hydrogen-bonded molecules around it. Water molecules are normally in constant motion, and the formation of such cages restricts the motion of a number of water molecules; the effect is to increase the structural organization of water. This situation is energetically unfavorable because it decreases the randomness (entropy) of the population of water molecules.

The opposition of water molecules to having their motion restricted by forming cages around hydrophobic molecules or portions thereof is the major reason molecules such as heptane are essentially insoluble in water and interact mainly with other hydrophobic molecules. Nonpolar molecules can also bond together, albeit weakly through van der Waals interactions. The net result of hydrophobic and van der Waals interactions is a very powerful tendency for hydrophobic molecules to interact with one another, and not with water.

Small hydrocarbons like butane (CH_3–CH_2–CH_2–CH_3) are somewhat soluble in water, because they can dissolve without disrupting the water lattice appreciably. However, 1-butanol (CH_3–CH_2–CH_2–CH_2OH) mixes completely with water in all proportions. The replacement of just one hydrogen atom with the polar –OH group allows the molecule to form hydrogen bonds with water and greatly increases its solubility. Simply put, *like dissolves like*. Polar molecules dissolve in polar solvents such as water, while nonpolar molecules dissolve in nonpolar solvents such as hexane.

1.4.2.1.7 Dispersion Forces

Molecules that have no permanent dipole still have their electrons in movement. Although the time-averaged distribution of electrons is symmetrical, at any instant the electrons are not uniformly distributed, so the molecule has a small instantaneous dipole, μ. This instantaneous dipole can polarize electrons in a neighboring molecule, giving a small dipole in the neighbor, and the dipole–dipole interaction results in attraction between the molecules. This is the dispersion attraction responsible for molecules sticking together. These dispersion forces (sometimes called London forces after the theoretician F. London) are the weakest of all intermolecular forces.[11] If the distance between two molecules is r, the dispersion interaction energy is proportional to $1/r^6$, so dispersion forces only operate over very short ranges. The van der Waals forces comprise the repulsive forces between electrons and nuclei on adjacent molecules, as well as the dispersion attractions.

All molecules, including nonpolar molecules, are attracted to one another by dispersion forces. The larger the molecule the stronger the dispersion force. Nonpolar molecules, large or small, have low solubilities in water because the small-sized water molecules have weak dispersion forces, and nonpolar molecules have no dipole moments. Hence, there are neither dispersion nor polar attractions to encourage solubility.

1.4.2.2 Functionalities and Environmental Behavior

In an attempt to completely understand the science of *in situ* remediation, special emphasis must be placed on the interrelationship between chemical structure and environmental behavior of the compounds. Structurally, organic molecules have a skeleton of carbon and hydrogen atoms and a group of heteroatoms (such as halogens, O, N, S) attached to that skeleton. Such "sites" are called *functional groups*. Functional groups, as the name implies, are commonly the site of reactivity or function. Classification of organic chemicals by the structural features will be dependent on the type of carbon skeleton, the type of functional group(s) present, or a combination of both. The carbon skeletons form the framework to which functional groups may be attached.

Chemical structure and thus the physical–chemical properties of contaminants have considerable impact on the rate and pathways of remediation mechanisms. Some of these structure/biodegradability relationships have some biochemical mechanistic underpinnings. For example, highly branched compounds frequently are resistant to biodegradation. Another structure-related trend is where functional groups commonly seen by microorganisms in natural products are usually degraded easily, because the microbes have had eons to develop the required enzyme systems for the metabolism.

In general, a *functional group* is a small set of atoms, held together by covalent bonds in a specific, characteristic arrangement that is responsible for the principal chemical and physical properties of a compound. It would be both foolhardy and futile to try to learn the chemistry and behavior of each of the thousands upon thousands of individual organic compounds introduced into the environment. Instead we focus our attention on the functional groups that form the most reactive portions of their molecular structures. Since all compounds that bear the same functional group undergo the same sort of biochemical reactions — some faster than others in the same class, some slower, some with interesting and unusual twists — learning the chemistry of each of these relatively few functional groups allows us to generalize our knowledge.

It has to be recognized that what a functional group is connected to can influence the course of its metabolism. For example, a chlorine atom bonded to a carbon atom differs substantially depending on whether the carbon atom is part of an alkane, alkene, or aromatic-ring skeleton. Moreover, the number of chlorine atoms connected to a single carbon atom will impact its metabolic fate. These features highlight the richness of biochemistry as a science. It is advantageous to focus on functional groups and describe their individual metabolisms as a way to begin to simplify the complex array of molecules and the reactions they undergo in the subsurface environment.

It is not a new idea to formally classify remediation reaction mechanisms by using organic functional groups as the elements. But this treatment has often been biased by the perceived idea that certain functional groups are xenophores and hence resist biodegradation. A common functional group, which was considered to be a xenophore for a long time is a chlorine atom, bonded to a carbon. However, the approach of calling a chlorine substituent a xenophore is too simplistic. It has to be noted that over 1000 halogenated natural products are known. It is true that the presence of many halogen substituents on a molecule may slow its biodegradation. However, the presence of multiple methyl groups on a single alkyl carbon or an aromatic ring will also slow its biodegradation, and a methyl group was never considered a xenophore. Today we know that a wide range of chlorinated organic compounds is biodegradable.

Some of the most commonly encountered functional groups, while addressing subsurface contamination, are briefly described below, as examples.

1.4.2.2.1 Organohalogens

Many of the organic contaminants contain one or several halogen atoms, especially chlorine (Cl). Because of their high electronegativities (Tables 1.4 and 1.5), halogens form rather strong bonds with carbon. Hence, in many cases, substitution of carbon-bound hydrogens with halogens enhances the inertness of a molecule (and thus its persistence in the environment).

In many respects, the chemical behavior of halogenated organic compounds is due to the unique physicochemical properties of their halogen substituent (F, Cl, Br, or I). At the start of the series, the carbon–fluorine (C–F) bond is very strong with high polarity, and with increasing molecular weight of the halogen, carbon–halogen bond energies decrease markedly (C–F > C–Cl > C–Br > C–I).

The presence of the strongly electronegative halogen atom — for example, chlorine — on the aromatic ring should cause electron withdrawal through the inductive effect. The inductive effect leads to a decreased affinity for electrophilic attack on the ring and the net effect is that the chlorobenzenes are less reactive than benzene.

Halogens present in organic compounds, particularly when bound to aromatic hydrocarbons, have a very weak tendency to form hydrogen bonds with water. Hence, the presence of larger halogens such as chlorine (Cl) and bromine (Br) renders an organic compound hydrophobic, thus decreasing its solubility significantly. Substitution of more than one carbon-bound hydrogen atoms by halogen atoms, specifically in low molecular weight aliphatics, also increases the oxidation state of the corresponding carbon atom. Hence, it should not be surprising that highly chlorinated hydrocarbons act as effective electron acceptors in biotic and abiotic redox reactions.

1.4.2.2.2 Oxygen-Containing Functional Groups

Due to its high electronegativity (Table 1.5), oxygen forms polar bonds with many atoms — H, C, N, P, and S. These "functional sites" have a significant impact on the physical–chemical properties as well as on the reactivity of that compound. Some of the oxygen-containing organic chemicals are alcohols, ethers, phenols, aldehydes, ketones, carboxylates, and furans.

1.4.2.2.2.1 Alcohols and Ethers

The simplest oxygen functional groups are the *alcohol* and *ether* groups. The hydroxyl group is indicated by a suffix -ol when it is the principal group attached to the carbon skeleton. Polyfunctional alcohols contain two or more hydroxyl groups — diol and triol. Alcohol group hydroxyl function has an (R–OH) structure and the ether function has an (R_1–O–R_2) structure. Both these groups have a significant impact on the partitioning behavior, and thus the solubilities, due to the ability of the oxygen atom to participate in hydrogen bonds. Alcohols have a greater ability because R–OH can act as both an H-donor and an H-acceptor, while R_1–O–R_2 is only an H-acceptor.

There are several ways in which the –OH group of an alcohol is affected by other reactants. One reaction is the rupture of the oxygen–hydrogen bond, and the other is the rupture of the carbon–oxygen bond. The oxidation of primary alcohols gives an aldehyde (Figure 1.16).

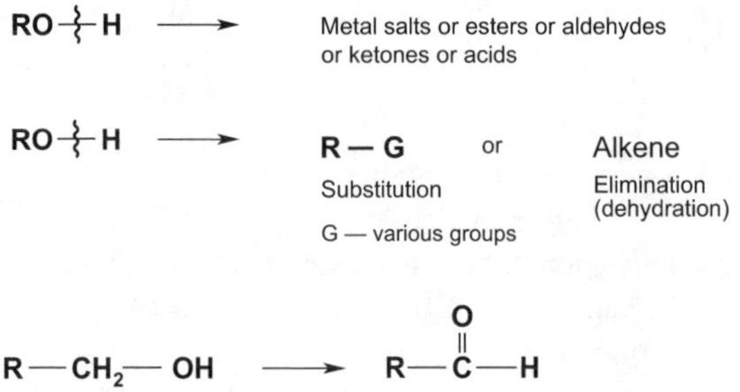

FIGURE 1.16 Various reactions of the –OH group in an alcohol.

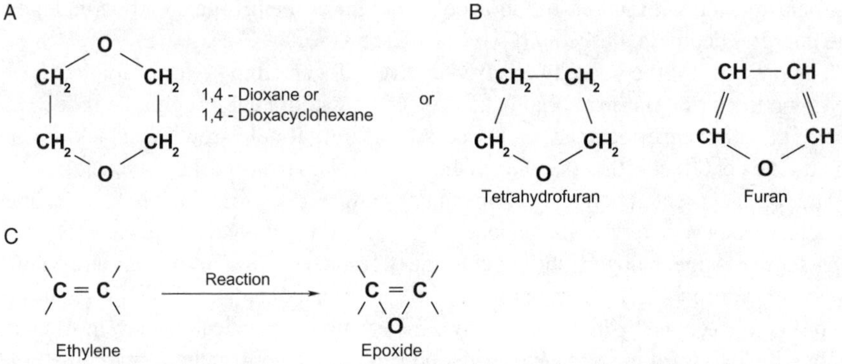

FIGURE 1.17 (A) Cyclic ethers. (B) Unsaturated cyclic ethers. (C) Epoxide.

The reactivity of alcohols toward various metals is closely related to their acidity. Primary alcohols are more acidic than secondary or tertiary alcohols. For example, methyl alcohol reacts rapidly (almost explosively) with potassium metal, whereas tertiary butyl alcohol reacts very slowly — about 24 h is required for complete reaction.[13]

All ethers contain the C–O–C linkage, and several subclassifications of them depend on whether alkyl or aryl groups are present: R–O–R, dialkyl ether; Ar–O–R, alkyl aryl ether; Ar–O–Ar, diaryl ether. Methyl-*tert*-butyl ether (MTBE), for example, is an alkyl–alkyl ether. When oxygen is part of a cyclic ring system, the result is a cyclic ether. A contaminant that is a cyclic ether is 1,4-dioxane (Figure 1.17(A)) (the name is a contraction of 1,4-dioxacyclohexane). Cyclic ethers can also be unsaturated, as is the case with furan (Figure 1.17(B)). Another cyclic ether, for example, ethylene oxide, is the simplest member in the family of *epoxides*. Epoxides are unusually reactive toward many electrophilic and nucleophilic reactants[13] (Figure 1.17(C)).

1.4.2.2.2.2 Aldehydes and Ketones

There are a number of carbon–oxygen functions in which oxygen forms a double bond with the carbon atom. If there is only one oxygen involved and if the carbon is bound to a hydrogen, the function is called an aldehyde (Figure 1.18(A)). If the carbon is bound to another carbon, and not to a hydrogen, the function is called a ketone

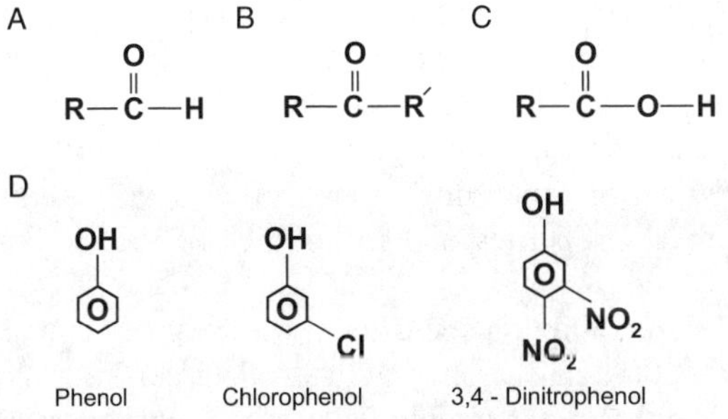

Introduction

FIGURE 1.18 (A) Aldehyde. (B) Ketone. (C) Carboxylic acid. (D) Phenolic structures.

(Figure 1.18(B)). As in the case of ethers, the aldehyde and ketones are H-acceptors, which makes some aldehydes and ketones suitable solvents.

Both aldehydes and ketones are similar in their molecular structures and their reactions with many reactants. The planar and polar carbonyl group in aldehydes and ketones permits them to associate with neighboring molecules in the liquid phase through dipole–dipole interactions. On the other hand, ether contains polar carbon–oxygen bonds that are "hidden" beneath adjacent hydrogen atoms or substituents. Lower molecular weight aldehydes and ketones show appreciable solubility in water because of the hydrogen bonding between water and the carbonyl group.

1.4.2.2.2.3 Carboxyl Groups

If the aldehyde (carbonyl) group is oxidized, that is, the hydrogen is replaced by hydroxyl, we obtain an acidic function that is referred to as carboxylic acid (Figure 1.18(C)). As the name implies, carboxylic acids may dissociate in aqueous solution. Furthermore, carboxylic acid functions are both strong H-donors and H-acceptors. Hence, the presence of a carboxylic function increases the water solubility of a compound significantly.

The name *carboxyl* describes this functional group in two ways. First, the *carboxylic* is a contraction of the words *carbonyl* and *hydroxyl*, which are the two structural units in the –COOH group. Second, the word *acid* is appended because of the acidic properties of this family of compounds, such as acetic acid (CH_3COOH), etc.

Carboxylic acids have considerable acidic nature and are the "acids" among organic compounds. There are other organic compounds that are acidic to varying extents due to rupture of the –O–H bond: sulfonic acids, Ar–SO_2–H; alcohols, R–O–H; and phenols, Ar–O–H.

1.4.2.2.2.4 Phenols

Phenols are aromatic alcohols where the hydroxyl group is attached to the aromatic ring. Phenols are more acidic than alcohols, but most importantly, the –OH group bonded to an aromatic ring is not susceptible to substitution as it is bonded to an aryl group. There are some similarities between alcohols and phenols, but for the most part these two families are quite different (Figure 1.18(D)).

As a family, phenols are active bactericides. The parent compound, phenol (Figure 1.18(D)), is more water soluble than the substituted phenols. This is attributed to hydrogen bonding between water and the –OH group.

1.4.2.2.3 Nitrogen-Containing Functional Groups

Among the atoms present in organic chemical molecules nitrogen is a special case in that, like carbon, it can assume many different oxidation states. Hence, there are many nitrogen-containing functional groups exhibiting very different properties and reactivities. Nitrogen-containing groups often have a significant impact on the physical–chemical properties and reactivities of the contaminants in which they occur.

Nitrogen is trivalent, and this opens up a whole range of functional groups and structural variants that are inaccessible to its divalent neighbor oxygen (in the periodic table). Organic nitrogen-containing functional groups are many and it is impossible to imagine the evolution of life in which organonitrogen chemistry does not play a crucial role. Organonitrogen functional groups include amines, ammonium compounds, imines, amides, nitriles, urethanes, ureas, nitro compounds, nitroso compounds, etc.[3]

1.4.2.2.4 Amino Group

Amino group is an important group present in numerous natural and man-made compounds (amino acids, aniline (aromatic amino compound)). Amino groups may engage in hydrogen bonding, both as H-acceptors and to a somewhat lesser extent as H-donors. Their reactivity is dependent on the nitrogen lone electron pair, which facilitates reactions in which the nitrogen acts either as a nucleophile or as a base.[3]

1.4.2.2.5 Nitro Group

A nitro group affects the properties and reactivity of a compound due to its strong electron-withdrawing character. Nitro compounds play a central role in organonitrogen chemistry. The synthesis and properties of aromatic vs. aliphatic nitrocompounds are very different.[3]

1.4.2.2.6 Sulfur-Containing Functional Groups

Sulfur, like carbon and nitrogen, is also a special case since it can also assume many different oxidation states. Sulfur has a smaller electronegativity (Table 1.5) and has a much weaker tendency to be engaged in H–bonding. Furthermore, as compared to oxygen, sulfur forms weaker bonds with carbon and hydrogen. In addition, sulfur in organic molecules may assume various different oxidation states due to its capability of valence shell expansion.[3]

Various functional groups that are found in an environment are shown in Figure 1.19.

1.5 REVERSIBLE VS. IRREVERSIBLE PROCESSES

Reversible processes are characterized by a mixture of products and reactants remaining at chemical equilibrium. Instead of the forward reaction going to essential completion, the reaction only proceeds spontaneously to a mixture that results in the largest change in free energy. At chemical equilibrium, the rate of the forward reaction is equal to the rate of the reverse reaction. Most acid–base and complexation reactions in groundwater systems are reversible reactions.[10]

Functional group	General formula	General name	Functional group	General formula	General name
None	C_nH_{2n+2}	Alkane	$-C\equiv N$	$R-C\equiv N$	Nitrile
$\diagup C=C\diagdown$	C_nH_{2n}	Alkene	$-NO_2$	$R-NO_2$	Nitro
$-C\equiv C-$	C_nH_{2n-2}	Alkyne	$-SH$	$R-SH$	Thiol
$-Cl$	$R-Cl$	Chloride	$-S-$	$R-S-R$	Thioether (sulfide)
$-Br$	$R-Br$	Bromide	$-S-S-$	$R-S-S-R$	Disulfide
$-OH$	$R-OH$	Alcohol	$\underset{-C=OR}{\overset{O}{\parallel}}$	$\underset{R-C-OR}{\overset{O}{\parallel}}$	Ester
$-O-$	$R-O-R$	Ether	$\underset{-C=NH_2}{\overset{O}{\parallel}}$	$\underset{R-C-NH_2}{\overset{O}{\parallel}}$	Amide
$-NH_2$	RNH	Amine	$\underset{-C=Cl}{\overset{O}{\parallel}}$	$\underset{R-C=Cl}{\overset{O}{\parallel}}$	Acid chloride
$-NR_2^+X^-$	$R_4\overset{+}{N}H^-$	Quaternary ammonium	$\underset{-C=O-C-}{\overset{O\quad O}{\parallel\quad\parallel}}$	$\underset{R-C=O-C-R}{\overset{O\quad O}{\parallel\quad\parallel}}$	Acid anhydride
$\underset{-C=O}{\overset{}{\mid}}H$	$R-C=O$ $\quad\mid$ $\quad H$	Aldehyde	$\underset{-S=OH}{\overset{O}{\parallel}}_{\overset{}{\parallel O}}$	$\underset{R-S=OH}{\overset{O}{\parallel}}_{\overset{}{\parallel O}}$	Sulfonic acid
$-C=O$ $\mid$	R $\mid$ $R-C=O$	Ketone	$\underset{-S-}{\overset{O}{\parallel}}$	$\underset{R-S-R}{\overset{O}{\parallel}}$	Sulfoxide
$\underset{-C-OH}{\overset{O}{\parallel}}$	$\underset{R-C=OH}{\overset{O}{\parallel}}$	Carboxylic acid	$\underset{-S=O}{\overset{O}{\parallel}}$	$\underset{R-S=O}{\overset{O}{\parallel}}$	Sulfone

FIGURE 1.19 Examples of various functional groups found in an enviro–chemical system.

Some examples of reversible reactions and/or processes that take place within an engineered IRZ are

- Adsorption–desorption from the source mass in and out the dissolved phase
- Gas transfer — volatilization of dissolved contaminants in and out of the soil gas
- Acid–base reactions
- Precipitation reactions

However, the whole concept of engineered IRZs is based on the creation of a subsurface zone, where migrating contaminants are intercepted and permanently immobilized or destroyed into harmless end-products. When the contaminant mass is *destroyed*, then the reaction is completely irreversible and, hence, the treatment that took place is permanent. There will be no rebound effect at all if the total amount of contaminant mass is destroyed. Achieving this condition may not be a technically feasible, and also not an economically viable option. However, enough mass can be destroyed/transformed within the contaminated zone within a reasonable time frame in an economic manner so that at the rate of dissolution the contaminant concentration will not exceed the regulatory standards.

It should be noted that the end-products of the transformation reaction(s) will result in dissolved and gaseous species and that the impact of these end-products on the natural REDOX equilibrium will be short term. If the impact is expected to be significant, it can be controlled by limiting the reaction kinetics and transport of the end-products from the reaction zone.

Immobilization mechanisms, which include heavy metal(s) precipitation reactions, in reality transform the contaminant into a form (precipitate) which is much less soluble. The longevity and permanence of the treatment depends on the groundwater-accessible redox-buffering capacity of the aquifer soil matrix and on the concentration of the reactive species entering the "remediated" reactive zone. Metallic precipitates in the form of sulfide (MeS) have such a low K_{sp} value that resolubilization will be extremely small and the concentration will not exceed the regulatory standards.

1.6 HYDROGEOLOGY AND HYDRAULIC IMPACTS

It is important to deliver the required reagents and other additives (such as buffering solutions, tracers, etc.) at the desired concentrations and distribution to implement the target reactions within the IRZs. While a complicated lithology can place constraints on the implementation of an IRZ, in most cases it does not have to completely eliminate the *in situ* remediation option. By properly placing injection well screen zones or other delivery mechanisms such as fracturing to target-specific impacted groundwater-bearing layers where the contamination resides, the desired remediation reactions can be achieved in most environments. Delivery in complex settings will more often be dictated by remediation goals, desired time frames, economic considerations, or nontechnical factors such as regulatory and public perception.

A contaminated groundwater formation's hydraulic conductivity is critical, along with hydraulic gradient to determine the design rates of reagent injections. These two

Introduction

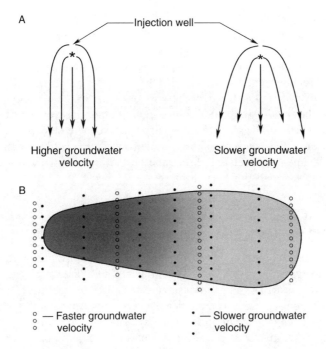

FIGURE 1.20 (A) Variation in lateral distribution of reagent from a single injection point under varying hydraulic conductives. (B) Differing number of injection curtains required to achieve the same cleanup goals under the same time frame, but under varying groundwater velocities.

variables also play a critical role in determining injection well spacing and distances between injection well arrays (Figure 1.20(A) and (B)). The higher the hydraulic conductivity of the formation the easier it is to deliver the reagent into the subsurface and the more effective a single delivery point and event can be. As the hydraulic conductivity increases — all other factors remaining equal — the distribution of reagent from a single injection point along the direction of advective flow increases but the radius of influence (or the lateral distribution) perpendicular to the flow direction decreases. Lower permeability will generally contribute to lower groundwater velocities and advective transport of the reagent. This, in turn, must be carefully considered when evaluating the design of full-scale IRZs for implementing a remediation system. It is critical to pick the spacing and locations of injection points to meet the remediation goals within the acceptable time frame and cost during a full-scale application. This factor is also important during pilot scale testing so that the performance monitoring wells can be placed at the right locations to see desirable results within the time frame of the study.

Groundwater flow characteristics are another important consideration in the design of reactive zones. Groundwater velocity, flow direction, and horizontal and vertical gradients have a great influence on the effectiveness of reagent injections, delivery, and the speed with which the reagent will spread and mix with the flowing groundwater within the IRZ. Low-velocity systems typically require lower reagent mass feed rates since the groundwater flux is reduced — all other conditions being equal. While the composition of interstitial water is the most sensitive and reliable indicator of the types and extent of reactions that will take place between the contaminants and the injected reagents in the aqueous phase, it is critical to understand the dynamics of groundwater flow to assure that the injected reagents will form a reactive zone in the targeted area. Advective transport accounts for the majority of the flow dynamics for the creation of the IRZ. However, the authors have observed in the field that some influences of the reactive zone will propagate more quickly than would be predicted by advective transport, although this contribution is likely immaterial for design considerations. A detailed discussion on the impacts of hydrogeology and hydraulics will be provided in later chapters.

1.7 SYSTEM POISE AND FLUX OF COMPETING COMPOUNDS

Oxidation–reduction processes play a major role in the mobility, transport, and fate of inorganic and organic compounds and contaminants in groundwater systems. Hence, the manipulation of REDOX conditions to create an IRZ to meet the cleanup objectives at a contaminated site was a predictable evolution. Any and all processes that influence the charge balance of groundwater systems — either through proton consumption or release by redox processes or uptake and release of either cations or anions, will generally influence the mechanisms of the intended remediation reactions. It is generally agreed that (bio)chemical reactions in nature take place because of electrical and chemical gradients.

Redox is one of the catchy phrases invented by someone unhampered by commitment to the use of scientifically correct terminology. The name is reversed (RED–OX instead of OX–RED) for the sake of easy pronunciation. The RED stands for reduction and it signifies gain of electrons by the chemical species called electron acceptors; the OX connotes oxidation, or electron loss by a chemical species called electron acceptors. Oxidation–reduction (REDOX) reactions, along with hydrolysis and acid–base reactions, account for the vast majority of (bio)chemical reactions that occur in groundwater systems.

IRZs are simple engineered systems consisting of a submerged zone where the contaminants are mixed with the injected reagents simulating a plug flow reactor. The flow of the contaminated groundwater is intercepted within the reactive zone created. Injected reagents create the optimum biogeochemical regime and generate the specific reactants to target the specific contaminants, in addition to other by-products. Improperly designed injection frequencies and dosages, not matching the hydrogeologic and hydraulic requirements, will result in the buildup of unintended conditions and products. This happens as a result of the development of reaction conditions mimicking a "closed system" rather than a continuous flow reactor — for example, the buildup of organic acids in a microbial IRZ resulting in a lower pH.

The effectiveness of an IRZ is determined not only by the relationship between the kinetics of the target reactions and the rate at which the mass flux of contaminants passes through it, but also by the flux of electron acceptors and other parameters which are said to define the *system poise*. There appears to be a tendency for an engineered biogeochemical system to maintain a redox balance, that is, poise, by donating electrons to surplus electron acceptors or by accepting electrons from surplus electron donors. It is important to create a point of poise or buffered redox region, where the electron-donating and electron-accepting tendencies cross for the successful implementation of any IRZ.

Significant redox trends along groundwater flow paths led to the hypotheses of successive redox zones characterized by the activity of specific thermodynamically favored electron acceptors. These redox zones may be classified as oxic, sub-oxic, or anoxic, or based on the dominant electron acceptors, aerobic (O_2), denitrifying (NO_3^-), iron-reducing (Fe^{3+}), sulfate-reducing (SO_4^{2-}), and methanogenic (CO_2) zones.

Even in the absence of engineered or natural biotic or abiotic transformation, contaminants, electron acceptors, and other naturally occurring compounds are always subject to transport processes. The manner in which the remediation mechanisms are implemented with an IRZ will be greatly dependent on the means of overcoming the oxidative or reductive poise of the system. An influx of electron acceptors will tend to increase the oxidative poise of the system, while an influx of electron donors enhances the reductive poise of the system.

Recharge and the migration of uncontaminated groundwater from upgradient locations are the primary modes of supplying the redox poise to any biogeochemical system. Infiltration of precipitation through the vadose zone brings the water into contact with the soil and thus allows the introduction of electron acceptors (such as NO_3^- and SO_4^{2-}) in addition to the dissolved oxygen in the recharge water. Dependent on the location of a particular groundwater system, a significant amount of natural organic carbon (electron donors) also will be introduced. Infiltration and recharge, therefore, provides fluxes of water, electron acceptors, and electron donors into any target IRZ. Thus, the groundwater entering the IRZ system will not only aid in dilution, but it also adds oxidative and reductive poise to the system.

After a recharge event, the effects of the additional flux of these parameters, more specifically the electron acceptors, may be apparent from the differing patterns in the reaction mechanisms and by-product formation. However, the effects of short-term variations in such a system (which are likely due to the intermittent nature of the precipitation events in most climates) may not easily be quantified.

It is clear from this discussion that the flux of oxidative poise contributed by dissolved oxygen, NO_3^-, Fe^{3+}, Mn^{4+}, and SO_4^{2-} has to be depleted as quickly as possible to achieve successful implementation of reducing IRZs. Similarly, the reductive poise has to be overcome for the implementation of oxidizing IRZs (Figure 1.21(A) and (B)).

1.8 CURRENT STATUS

1.8.1 Regulatory Acceptance and Permitting

Potential regulations that affect IRZ projects are similar to those applicable to any other *in situ* remediation technologies. While this book does not attempt to cover the range of regulatory environments and issues that exist, it is important to consider the regulatory arena in which the remediation work is being conducted and during the

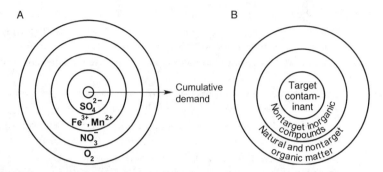

FIGURE 1.21 (A) Pictorial depiction of oxidative poise to be overcome during implementation of a reducing IRZ. (B) Oxidative poise that needs to be overcome before the chemical oxidant demand for the target contaminants can be met.

reagent selection process. Characteristics of the various reagents that are used in various IRZ systems can be characterized as

- Food grade
- Laboratory grade
- Commercial grade
- Industrial grade
- No chemical residuals
- Historic applications approved by other state or federal agencies

The degree of interaction with regulatory agencies required to execute IRZ projects is sometimes substantially greater than with traditional technologies, until a particular regulatory agency becomes familiar and comfortable with these technologies. However, the long record of successful applications discussed in later chapters attests to the fact that the technology has been successfully permitted in numerous jurisdictions and that the regulatory personnel's experience base is growing.

Many states regulate the injection of materials into the subsurface and may require a Safe Drinking Water Act–mandated Underground Injection Control (UIC) permit prior to implementing the demonstration. The UIC permit includes information regarding the chemical nature of the reagent solution, and addresses potential concerns with water quality resulting from the injection process. UIC permitting for injection of reagents is generally waived or is implemented with minimal paperwork (e.g., permit by rule). This issue is not considered to be a major impediment to implementation of IRZ projects.

Previous experience with state regulatory agencies where IRZ systems have been implemented indicates that an initial meeting to establish the proposed course of action for the project is the most effective process. The concerns of the UIC permit staff at state regulatory agencies must be addressed at the onset of the project to avoid delays. Usually, the information required to satisfy the requirements of the UIC permits is readily available, and should not represent a major regulatory hurdle. Continued close communications with the regulatory agencies during the planning and execution of IRZ projects will greatly increase the potential for a successful demonstration. A teaming relationship with the local environmental regulatory agencies is important for the successful implementation of the technology.

Public participation during the technology implementation process should be addressed on a site-specific basis. Inquiries on behalf of public entities should be addressed, in a timely manner, by the project management members. The IRZ technology, particularly the microbial zones, is a relatively straightforward nature-based and nonthreatening process, and thus it is anticipated that any public communications will be favorably received.

The production of intermediate products is a potential concern to regulatory agencies. For example, the microbial IRZ processes convert more highly chlorinated CAHs to less chlorinated and eventually nonchlorinated end-products. The cascading reactions can result in the production of VC as an intermediate product during the microbial reactions. This product is more harmful, than the parent compound. Complete reductive dechlorination of VC to ethene can be easily controlled within the IRZ system, and it is also quickly biodegraded by aerobic microorganisms. For these reasons, the production of VC or other intermediate products is considered a temporary situation and does not represent a major impediment to the technology but should be monitored during application of the technology.

Similarly, reaction of $KMnO_4$ with contaminants produces manganese dioxide (MnO_2), excess Mn^{2+} ions, carbon dioxide, by-products, and intermediate compounds. The kinetics of reaction mechanisms between permanganate and contaminants are obviously an important factor in the overall process implementation. Gas binding and excessive production of O_2 gas are operational factors that need to be monitored regularly during injection of Fenton's reagent.

Another regulatory issue can be the production of gases such as methane, hydrogen sulfide, and carbon dioxide, high concentrations of O_2, and the migration and potential accumulation of these gases in the vadose zone. Concentrations of these gases can accumulate in the subsurface when structures in the vicinity do not allow for passive diffusion of these gases. For this reason, vapor-phase concentrations of these compounds are monitored when a potential concern exists to ensure that safe conditions are maintained. If required, venting of subsurface gases or a modified electron donor injection routine will be used to protect against exposure or accumulation. This issue is not considered to be a major impediment to technology implementation, but it must be considered.

There are no unusual issues involving the transport, storage, or disposal of wastes and treatment residuals. The standard issues of drill cuttings produced during injection well installation and purge water generated during sampling apply.

Secondary water quality impacts within IRZ zones can occur due to the desorption and mobilization of contaminants, formation of by-products, and intermediate compounds. Secondary water quality impacts can also occur from mobilization of metals naturally occurring in the solid phase into the groundwater. Other parameters of interest with regard to secondary water quality impacts are COD, BOD, TDS, taste, odor, and sulfides.

In general, enhanced anaerobic *in situ* bioremediation processes will reduce the mobility of many metals (it has been successfully used for the treatment of many dissolved metals) but it will initially solubilize some other naturally occurring metals within the reactive zone (e.g., iron, manganese, and arsenic). However, even in solubilized form, under anaerobic conditions metals such as arsenic are substantially retarded by adsorption to the aquifer matrix. Furthermore, it is generally believed that arsenic will be reprecipitated/immobilized immediately downgradient of the reactive zone when the conditions return to their preexisting state (which, for the purposes of this discussion, is assumed to be aerobic). Similarly, reprecipitation/immobilization of solubilized iron and manganese will occur within the IRZ area sometime after the sulfate-reducing conditions are established.

COD, BOD, TDS, taste, and odor are necessarily elevated in the reactive zone because the injected reagents contribute to these parameters. However as the reagent is consumed downgradient these parameters typically return to near background values. Sulfides are typically produced under anaerobic conditions, but are again not typically found in aerobic zones further downgradient. Thus the potential for secondary water quality impacts needs to be fully identified and addressed during design and in consultation with all applicable regulatory agencies and the public.

1.8.2 Biostimulation vs. Bioaugmentation

There is a significant level of debate going on today among the designers of microbial IRZs for enhanced reductive dechlorination with respect to the efficacy of biostimulation vs. bioaugmentation. The debate is centered around whether the first level of the treatment hierarchy for microbial transformation of chlorinated ethenes should be biostimulation or bioaugmentation. Biostimulation or enhanced biodegradation involves stimulating the indigenous microbial populations and thus enhancing microbial activity to destroy the target contaminants at a rate that meets the cleanup objectives. Bioaugmentation involves injection of selected exogenous microorganisms with the desired metabolic capabilities directly into the contaminated zones along with any required nutrients to effect the rapid biodegradation of the target contaminants.

The scientific and technical applications knowledge base supporting the use of microbial reductive dechlorination is now very large and continuing to grow at a rapid pace. The themes of reductive dechlorination are now well developed.

1.8.2.1 Biostimulation

Naturally occurring aquifer bacteria can be managed within an engineered IRZ to very high rates of reductive dechlorination, through the injection of electron donors at rates that exceed the recharge of electron acceptors. This drives the aquifer microbial communities into sulfate reduction, acetogenesis, and methanogenesis, providing broad-spectrum cometabolic dechlorination. Direct utilization of chlorinated alkenes as alternative electron acceptors is also stimulated among bacterial species in several genera, including *Desulfuromonas*, *Dehalospirillum*, *Dehalococcoides*, *Dehalobacter*, and *Desulfomonile*.

Biostimulation is a robust approach that relies on coarse adjustments of the microbial ecology to induce the desired dechlorinating behaviors. In some cases, a lag phase is observed, during which dechlorination proceeds slowly. But in all cases observed to date by the authors, the dechlorinating community emerges from the lag phase within 1 to 6 months, and complete dechlorination to ethene is achieved.

1.8.2.2 Bioaugmentation

Aquifer microbial communities can be augmented through the injection of large masses of commercially grown bacteria intended to overwhelm the native communities. In laboratory microcosms, these bacteria can achieve dechlorination rates that exceed those achieved through stimulation of native communities at field scale. The challenge in application of bioaugmentation is to establish tight control over key microbial habitat variables required to elicit high-rate dechlorinating performance.

In field-scale applications that have been reported, effective habitat control could only be achieved through significant groundwater pumping, amendment, and reinjection. Adhesion properties of the injected bacteria limited their migration to small zones surrounding the injection points. In effect, bioaugmentation amounts to a very large scale-up of the highly controlled laboratory microcosms on which the technology is based. Treatment can only be achieved in the limited area around injection points, where populations of the effective, but hypersensitive, bacterial inoculum are established. There is no reliable information available regarding the life span of these pure cultures, even with the smaller area of influence.

The authors' working hypothesis, based on experience derived from involvement in more than 250 enhanced reductive dechlorination projects, is as follows:

- Distribution of the electron donor is critical.
- Electron donor loading rates must exceed electron acceptor recharge rates.
- More complex electron donors induce more complex microbial communities and involve a broader range of potential dechlorinators.
- Biostimulation is an ecosystem-level process — we manage for ecosystem-level, not species-level, effects.
- Aquifer microbial continuum — the aquifer microbial community structure forms a continuum, aligned along the groundwater flow axis, controlled by the rate of supply and quality of electron donors and electron acceptors available at any point in an aquifer.
- The community structure responds rapidly to increases in the rate of electron donor influx induced by biostimulation.
- A continuum of dominant metabolic strategies, from oxygen reducing in the upgradient reaches, to sulfate reducing and methanogenic in the lower reaches, is typically observed.

Based on the authors' experience, a practical approach to enhanced reductive dechlorination can be summarized as follows:

- Sites with ongoing transformation to VC or ethene only need enhancement through electron donor addition.
- Sites without VC or ethene should be biostimulated for 6 to 12 months, before any other action is considered.
  - Pre-biostimulation PCR analysis for *Dehalococcoides* has been shown to be unreliable.

- Biostimulation is a necessary precursor to bioaugmentation at any site where enhanced reductive dechlorination process is to be implemented.
- Lag times to VC and ethene production in these sites have generally been less than 1 year, normally around 3 to 9 months depending on pre-injection biogeochemical conditions.
- Bioaugmentation may be applicable at sites that are not fully dechlorinating after a suitable period of effective biostimulation.
  - Many sites that have been reported to be a failure to fully dechlorinate may be carbon limited, not microbially limited.
  - Bioaugmentation must be demonstrated successfully under natural-gradient conditions, before it can be adopted for full-scale applications.
- All peer-reviewed field applications (Dover AFB, Kelly AFB, Bachman Road site, Schoolcraft site) have been conducted with groundwater recirculation.
- Bacterial adhesion properties may significantly constrain distribution of injected bacteria under natural aquifer gradients.
- Closely spaced recirculation systems may render unacceptable the costs of bioaugmentation treatment at full-scale sites.

1.8.3 Contaminant Types

This section summarizes the various classes of contaminants and possible reactions that can be applied with IRZs to remediate those contaminants. Potential remediation mechanisms for classes of inorganic and organic contaminants are listed in Tables 1.6 and 1.7.

1.8.4 Emerging Contaminants

There are many compounds encountered in the subsurface environment — in soil, groundwater, and surface water systems—which have not fallen completely into regulatory control regimes. Some of these compounds were disposed into the environment a long time ago, in a legal and acceptable fashion, based on the regulatory controls of that time period and a lack of a regulatory standard for these compounds. However, with the advancement of analytical detection techniques and increased level of understanding of the epidemiological concerns, these compounds are being referred to as "emerging contaminants." Perchlorate, 1,4-dioxane, nitrosodimethylamine (NDMA), pharmaceutical compounds such as antibiotics and endocrine disruptors are some of the compounds that are considered to be *emerging contaminants*.

Widespread environmental detection of such contaminants (mainly due to the development of advanced analytical techniques) is currently receiving extensive media coverage and scientific notice. In the U.S., comprehensive screening of these contaminants from an epidemiological perspective is being performed to develop appropriate and acceptable regulatory standards.

Research and development activities to assess and improve the treatment efficiency of these emerging contaminants are being performed extensively in the

TABLE 1.6
Potential Remediation Mechanisms for Classes of Organic Contaminants When Present in Groundwater[15]

Contaminant Class	Example Compounds	Potential Remediation Mechanisms
Aromatic hydrocarbons		
Low molecular weight	Benzene, toluene, ethyl benzene, xylene (BTEX)	Aerobic and anaerobic microbial oxidation
Higher molecular weight	Polycyclic aromatic hydrocarbons (PAHs), creosotes	Aerobic microbial oxidation for the lighter ends but at a very, very slow rate
		Chemical oxidation for some
Chlorinated aromatic hydrocarbons		
Highly chlorinated	Pentachlorophenol, tetrachlorophenol, hexachlorobenzene, PCBs	Reductive dechlorination
Less chlorinated	Chlorobenzene, dichlorobenzene, chlorophenol, pesticides	Microbial oxidation
		Reductive dechlorination
		Chemical oxidation
Aliphatic hydrocarbons		
Oxygenated	Alcohols, ketones, esters, ethers, phenols (acetone, MEK, MTBE, phenol)	Aerobic microbial oxidation
		Anaerobic microbial oxidation
Highly chlorinated	PCE, TCE, 1,1,1-TCA, carbon tetrachloride	Microbial reductive dechlorination
		Abiotic dehydrodehalogenation for the ethanes
		Chemical oxidation
Less chlorinated	DCAs, DCE, vinyl chloride, methylene chloride, chloro ethane	Microbial reductive dechlorination
		Aerobic microbial oxidation
		Abiotic oxidation reactions
		Chemical oxidation for some
Nitro aromatics		
Explosives	TNT, RDX, HMX	Microbial reduction
		Chemical oxidation
Cyclic compounds		
Cyclic ethers	1,4-dioxane, tetrahydrafuran	Microbial oxidation
		Microbial reduction
		Chemical oxidation

U.S. Technology development activities include *in situ* and aboveground treatment systems for contaminated groundwater systems. In addition, significant focus and attention is being paid to improve the removal efficiencies of these compounds in conventional and advanced wastewater treatment systems as well as drinking water treatment systems.

TABLE 1.7
Potential Remediation Mechanisms for Classes of Inorganic Contaminants when Present in Groundwater[15]

Contaminant Class	Example Compounds	Potential Remediation Mechanisms
Metals	Cr^{6+}, Cu^{2+}, Cu^+, Ni^{2+}, Pb^{2+}, Hg^{2+}, Cd^{2+}, Zn^{2+}, etc.	*In situ* microbial precipitation — preferably as sulfides (MeS) except Cr^{6+} which precipitates out as $Cr(OH)_3$
		Chemical precipitation (with lime) as $Me(OH)_2$ or MeS with sodium sulfide or calcium polysulfide
Metalloids	As, Se, Sb	*In situ* microbial precipitation — coprecipitation with FeS
		Coprecipitation with Fe^{3+} as ferro-metallic complex
Radio nuclides	U, ^{99}Tc, Pu, Ne	*In situ* microbial precipitation
		Coprecipitation with FeS
Non-metals		
Oxyanions	Nitrate	Microbial denitrification
	Perchlorate	Microbial reduction in the presence of organic substrates
	Sulfate	Precipitation as $CaSO_4$ or reduction as sulfide and precipitation
	Sulfide	
Cyanide	CN^-	Aerobic and anaerobic microbial reactions

REFERENCES

1. Larson, R.A. and E.J. Weber, *Reaction Mechanisms in Environmental Organic Chemistry*, Lewis Publishers, Boca Raton, FL, 1994.
2. Chappelle, F.H., *Ground-Water Microbiology and Geochemistry*, John Wiley & Sons, New York, 2001.
3. Schwarzenbach, R.P., P.M. Gschwend, and D.M. Imboden, *Environmental Organic Chemistry*, 2nd ed., Wiley-Interscience, New York, 2003.
4. Donohoe, T.J., *Oxidation and Reduction in Organic Synthesis*, Oxford University Press, Oxford, U.K., 2000.
5. Jeffers, P.M. et al. Homogeneous hydrolysis rate constants for selected methanes, ethenes, and propanes, *Environ. Sci. Technology*, 23, 965–969, 1989.
6. Wiedemeier, T.H. et al., *Natural Attenuation of Fuels and Chlorinated Solvents in the Subsurface*, John Wiley & Sons, New York, 1999.
7. Lyman, W.J., W.F. Reehl, and D.H. Rosenblatt, *Handbook of Chemical Property Estimation Methods*, McGraw-Hill, New York, 1982.
8. Boethling, R.S. and D. MacKay, *Handbook of Property Estimation Methods for Chemicals*, Lewis Publishers, Boca Raton, FL, 2000.

9. Mabey, W.R. and T. Mill, Critical review of hydrolysis of organic compounds in water under environmental conditions, *J. Phys. Chem. Ref. Data*, 7, 383–415, 1978.
10. Schnoor, J.L., *Environmental Modeling; Fate and Transport of Pollutants in Water and Soil*, John Wiley & Sons, New York, 1996.
11. Lodish, H. et al., *Molecular Cell Biology*, W.H. Freeman & Company, New York, 2000.
12. Weiner, E.R., *Applications of Environmental Chemistry*, Lewis Publishers, Boca Raton, FL, 2000.
13. Wingrove, A.S. and R.L. Caret, *Organic Chemistry*, Harper & Row, New York, 1981.
14. Streitwieser, A., Jr. and C.H. Heathcock, *Introduction to Organic Chemistry*, Macmillan, New York, 1985.
15. Rittmann, B.E. and P.L. McCarty, *Environmental Biotechnology: Principles and Applications*, McGraw-Hill, New York, 2001.

2 Microbial Reactive Zones

Soil microbiology traditionally has been preoccupied with soil fertility related to agricultural practices. Prior to the 1970s, there were virtually no microbiologic studies that extended below the root zone of the plants. Beginning in the 1970s, there was an increasing awareness of subsurface contamination, caused by various waste-disposal or waste-management practices. At that time, considerable concern was also raised about the effects of contamination on groundwater quality, and this concern began to draw microbiologists toward studying subsurface environments. The microbiology of shallow aquifers became an active topic of investigation.

Throughout the 1980s, laboratory studies of microbial activity in shallow aquifer systems, and the ability of those microorganisms to degrade petroleum hydrocarbons, was a topic of intense inquiry.[1] In addition, numerous field studies showed that microbial oxidation of petroleum hydrocarbons was clearly reflected in groundwater chemistry. These field studies were important because they suggested, for the first time, that petroleum hydrocarbons were also being oxidized under anaerobic conditions. Throughout most of the 1970s and early 1980s, it was widely believed that molecular oxygen was necessary to biodegrade petroleum hydrocarbons. Studies have now shown that petroleum hydrocarbons can also be degraded under iron-reducing, sulfate-reducing, and methanogenic conditions.

Petroleum hydrocarbons were only part of the contamination problem. Industrial solvents like trichloroethene (TCE) and perchloroethene (PCE) had been used since the mid-1900s and huge amounts of these chemicals were simply spilled or dumped into seepage basins and unlined landfills. Because solvents like TCE and PCE are entirely synthetic compounds, with no natural analogs, it was widely believed in the 1970s that microorganisms could not degrade them. But in 1985, it was shown for the first time that aerobic methane-oxidizing bacteria were capable of cometabolically oxidizing TCE to carbon dioxide.[2] Later, it became evident that chlorinated ethenes were also subject to anaerobic reductive biodegradation, with chlorine atoms being sequentially stripped off of the ethene molecule to form dichloroethene (DCE), vinyl chloride (VC), and eventually ethane under anaerobic conditions. Later, it was discovered that VC and DCE were also subject to oxidation by microorganisms.

Starting from the initial assumption that chlorinated ethenes were not biodegradable at all, it is now evident that there are a number of microbial processes that biodegrade these compounds in groundwater systems, and these processes are very important in restricting contaminant transport. This experience has been repeated with a number of so-called "recalcitrant" contaminants. Polychlorinated biphenols, explosives such as TNT, RDX, HMX, 1,4-Dioxane, and gasoline additives such as methyl-*tert*-butyl ether (MTBE) are other examples of compounds that were initially thought to be completely inert (biologically recalcitrant), but which now are known to undergo microbial degradation in groundwater systems. This, of course, is the normal progression of scientific inquiry. Understanding the microbial transformation of man-made contaminants in groundwater systems requires both an observational approach (looking for evidence that degradation is occurring *in situ*) and confirmation with an experimental approach (showing that a particular microbial process occurs under controlled laboratory conditions). This takes time. From the time that TCE was first identified as an important contaminant of groundwater systems (in the mid-1970s) it took 15 years to identify a cometabolic microbial degradation process, and another 10 to fully appreciate the importance of reductive and then oxidative degradation processes. As new anthropogenic contaminants are introduced in the future, it is safe to assume that an understanding of their microbial degradation processes will lag in a similar fashion.[1]

The industrial expansion during and after World War II necessitated heavy use of metal plating operations at manufacturing plants across the U.S. As a result, many heavy metal containing liquids were spilled into the subsurface. The mechanisms that can be used to reduce the concentrations of heavy metals dissolved in groundwater are transformation and immobilization. These mechanisms can be induced by both abiotic and biotic pathways. Abiotic pathways include oxidation, reduction, sorption, and precipitation. Examples of microbially mediated processes include: reduction, oxidation, precipitation, biosorption, bioaccumulation, organo-metal complexation, and phytoremediation. Currently, *in situ* microbial precipitation of dissolved Cr^{6+}, Pb^{2+}, Ni^{2+}, Zn^{2+}, As^{5+}, Cu^{2+}, Cd^{2+}, uranium, and other radio-nuclides has been implemented successfully by the authors.

The recent advances in the science of *in situ* bioremediation were possible primarily as a result of the convergence between the different disciplines of geology and microbiology. Observations of groundwater chemistry had suggested, from at least the 1920s, the possible importance of microbial processes. However, an observational approach was not sufficient to demonstrate that this was, in fact, the case. Rather, rigorous sampling methods and experimental techniques were required to demonstrate adequately the role of microbial processes. On the other hand, experimental methods themselves had significant disadvantages, such as extreme cost, disturbance of sediments, and an inherent small scale that was inappropriate for large hydrogeologic systems. In the end, an integration of microbiology and the geosciences has proved to be the most effective means of studying subsurface microbiology for the advancement of bioremediation as a technology.[1]

2.1 TYPES OF MICROBIAL REACTIONS

Reactions mediated by microorganisms constitute a very important set of transformations affecting the fate of almost all organic compounds and most inorganic compounds in both natural and engineered environments. Biochemical processes resulting from these reactions change the structure of the contaminants of concern, thereby, removing the risk and harm caused by the presence of that contaminant in the subsurface environment. However, the reactivities, fate, and other properties of the final end-products that result from these biotransformations have to meet the desired objective of remediating the contamination present.

Biochemical transformations of contaminants to harmless end-products are important because many reactions, although thermodynamically feasible, occur at extremely slow rates due to kinetic limitations. Although examples of biodegradation and biocatalysis abound in nature, the pathways and reactions used by microorganisms have achieved serious recognition as a research area only in the last 40 years since the publication of Rachel Carson's *Silent Spring*.[3] The use of microorganisms and "Mother Nature" to remedy anthropogenic contamination of the environment has met with increasing success and continues to be an exciting area of research with the identification of many emerging contaminants.

Most often, the term microorganism is used without definition. The taxonomic division of living things depicts three branches, also known as kingdoms. Two of them are constituted by single-celled *prokaryotes*, Bacteria and Archaea. The third, Eukaryota, also contains many single-celled forms.

The terms *biotransformation*, *biodegradation*, and *biocatalysis* are used many times to explain the same process. While they have similar meanings, they should be used in different contexts. Biotransformation is more of a technical definition and should be used to describe a metabolic process. The term biocatalysis is used similarly to biotransformation, but it has the additional connotation of metabolism for the purpose of making a useful compound. The term biodegradation, by contrast, is most typically used when the objective is taking a compound away, a microbial process by which a potentially toxic contaminant is transformed into a nontoxic one or few. *Bioremediation* is a more recently coined term that refers to the application of biodegradation reactions for the practical cleanup of chemical contamination.[3]

Biodegradation can be categorized into three types that have importance in an ecosystem setting:

Primary biodegradation: biodegradation to the minimum extent necessary to change the identity of the compound.
Ultimate biodegradation: biodegradation to water, carbon dioxide, and inorganic compounds (if elements other than C, H, and O are present). This is also called mineralization. Under anaerobic conditions, methane may be formed in addition to carbon dioxide during fermentation reactions.
Acceptable biodegradation: biodegradation to the minimum extent necessary to remove some undesirable property of the compound, such as toxicity. Conversion of VC to ethene is an example and in many instances this can also be considered *biotransformation*.

In general, one cannot say that certain bacteria are predominant in metabolizing certain classes of compounds, such as aliphatic hydrocarbons, aromatic hydrocarbons, chlorinated aliphatic hydrocarbons, or heterocyclic ring compounds. The presence or absence of oxygen often dictates the type of biodegradative pathway and the types and number of bacteria involved in biodegradation. In the past, conventional thinking always led to the choice of fostering aerobic or anaerobic conditions for implementing biodegradation. However, recent thinking in the science of bioremediation encourages the dividing lines to be drawn around microbial oxidative vs. microbial reductive reactions.

In order to identify structural features of contaminants that enhance or inhibit their biodegradability, many studies have looked into the structure–biodegradability relationships. Most of these efforts have helped in quantifying a contaminant's biodegradability on a scale from "labile" to "recalcitrant." It should be noted that structural influences on biodegradability exhibit "fragment–fragment" interactions that complicate the simple models that treat each functional group in isolation. For example, the presence of a carboxylic acid moiety improves the degradability of an otherwise recalcitrant aromatic contaminant; conversely, the addition of a halogen to a chlorinated contaminant disproportionately decreases the biodegradability of that contaminant under aerobic conditions.

The oxidative approach used by microorganisms to initiate the degradation of chemical contaminants involves the use of electrophilic forms of oxygen to oxidize the contaminant. *Oxidation* is defined as a loss of electrons. Oxidizing agents, called electron acceptors, gain electrons and are by definition electrophiles. In biodegradation, oxidation can either be associated with the introduction of oxygen into a molecule or the conversion of a molecule to a higher oxidation state. Almost all oxidative reactions are kinetically second-order reactions in which the rate is proportional to the concentrations of both the oxidizing agent (electron acceptor) and the substrate.

The other set of biologically mediated reactions used for the transformations of chemical contaminants are reductions. Reduction, which is defined as a gain of electrons, occurs when there is a transfer of electrons from an electron donor to the contaminant, which acts as the electron acceptor in this instance. In other words, the contaminant functions as the "oxidant." Microbially mediated reductive transformations involve the same structural moieties that are susceptible to abiotic reductions.

2.2 MICROBIAL OXIDIZING ZONES

Microorganisms transform contaminants by means of electron-transfer reactions in order to extract free energy by synthesizing ATP. Within microbial oxidizing zones, contaminants (e.g., BTEX) act as reductants, while electron acceptors (such as O_2, NO_3^-, SO_4^{2-}) function as oxidants. Thermodynamics mainly determines the order in which microorganisms use electron acceptors, as microorganisms tend to perform redox reactions close to thermodynamic equilibrium. The Gibbs free energy for microbial oxidation decreases at neutral pH in the order O_2, NO_3^-, Mn(IV) oxide, Fe(III) oxide, SO_4^{2-}, and CO_2 (Figure 2.1). When there is a large content of metal oxides in the aquifer sediments within contaminated groundwater plumes, reductive dissolution of metal oxide becomes an important microbial oxidation process. Zones of iron reduction, sulfate reduction, and methanogenesis are often observed to overlap depending on pH, redox species concentrations, and solubility of iron-oxide minerals.

2.2.1 ENGINEERED AEROBIC SYSTEMS

Aerobic bioremediation utilizes the approach used by microorganisms to initiate the transformation of chemical contaminants using oxygen as the electron acceptor.

Electron Acceptors (oxidants)	Microorganisms ⟹	Change in ΔH⁺	Aqueous Chemistry Reduced Species
O_2	Aerobic respiration	+ve	CO_2, H_2O
NO_3^-	Nitrate reduction	+ve	N_2
Mn^{4+}	Manganese reduction	−ve	Mn^{2+}
Fe^{3+}	Iron reduction	−ve	Fe^{2+}
SO_4^{2-}	Sulfate reduction	+ve	S^{2-}, HS^-
CO_2	Methanogenesis	+ve	CH_4

FIGURE 2.1 Primary microbial oxidation processes.

Since O_2 in the subsurface environment is not very reactive with most contaminants, microorganisms must invest metabolic energy to convert this oxygen into a more effective oxidant. Generally, this involves enzymes with metals (e.g., oxygenases) and coenzymes like NAD(P)H. If both atoms of oxygen from O_2 are transferred, the enzymes are called *dioxygenases*. In contrast, *monooxygenases* deliver only a single oxygen atom. Whether one oxygen atom or two oxygen atoms are transferred to the organic contaminant, the key that permits these reactions to occur is the attraction of the biologically produced oxidant to the electrons of the organic contaminant. Microorganisms accomplish this by using an enzymatically prepared form of electrophilic oxygen.[4] Recognition of these "electrophilic attack" approaches of microorganisms for executing oxidation reactions enables us to make important predictions regarding such biotransformations.

Engineered *in situ* bioremediation systems are classified according to the means by which stimulatory reagents are added. Although technical details depend strongly on site-specific conditions and are advancing rapidly, these broad classifications are valuable for identifying what approaches have merit. Engineered aerobic *in situ* reactive zones have been applied successfully mostly for the remediation of petroleum hydrocarbons.

2.2.2 DIRECT AEROBIC OXIDATION

The majority of the compounds in petroleum products are biodegradable at significantly faster rates under aerobic conditions. The amount of oxygen required for complete aerobic mineralization of 1 g of hydrocarbon ranges from 3 to 3.5 g. In simplistic volumetric terms, 300,000 kg of oxygen-saturated water must be delivered and mixed in order to mineralize 1 kg of petroleum hydrocarbons. This illustrates the need to select the technically and economically most effective method of delivering O_2 into the groundwater and also to maximize the efficiency of O_2 utilization by the microorganisms in the subsurface. The total cost of a pound of dissolved O_2 delivered into the subsurface could range from $0.80 to $10.00 depending on the method selected and the geologic, hydrogeologic conditions encountered at a site. The cheapest method of delivering dissolved O_2, if hydrogeologic conditions are conducive, is by injecting compressed air into the contaminated plume. The next available method is to inject dilute hydrogen peroxide (at about 500 to 1000 ppm concentrations) into the contaminated zone. Other methods of oxygen delivery include various methods of air injection and expensive methods such as oxygen release compounds.

Originally developed in Europe, air sparging has become a popular means of engineered *in situ* bioremediation for strictly aerobic biodegradation. Injection of compressed air directly into the contaminated zone is an efficient way to deliver oxygen to achieve the target reactions. In addition, air sparging can strip volatile contaminants from the saturated zone. Nutrients and other amendments also can be added from injection wells. Some nutrients can be added in a gaseous state.

This method of introducing oxygen for enhanced aerobic biodegradation is one of the inherent advantages of air sparging. However, the oxygen transfer into the bulk water is a diffusion limited process. The diffusion path lengths for transport of oxygen through groundwater are defined by the distances between air channels. Where channel spacing is large, diffusion alone is not sufficient to transport adequate oxygen into all areas of the aquifer. The pore-scale channels formed and the induced mixing during air sparging enhances the rate of oxygen transfer.

During air sparging, 1 scfm of air contains approximately 0.0161 lb of oxygen (7.3 g). Mass transfer efficiencies for sparge systems range from a maximum of 1 to 2% to a more typical 0.1 to 0.01%. If delivery of oxygen for the aerobic microorganisms is the primary objective for air injection, the volume of airflow does not have to be at the same level required to achieve stripping and volatilization. Thus injection of air at very low flow rates (0.5 scfm to less than 2 to 3 scfm per injection point) into the water-saturated contaminated zones is sometimes called *biosparging* (Figure 2.2).

In some instances, hydrogen peroxide (H_2O_2) is used as a dissolved source of oxygen. Using this method, it is possible to achieve substantially greater dissolved oxygen concentrations than by air saturation. In most cases, H_2O_2 and other stimulants

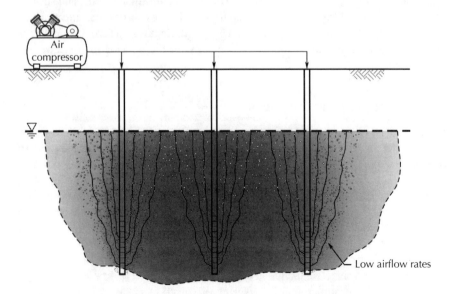

FIGURE 2.2 Conceptual description of a biosparging system.

Microbial Reactive Zones

are added by circulating water either vertically or along the path of flow of groundwater. One problem that can occur as part of water-circulation systems is clogging near well screens. Significantly enhanced localized growth of bacteria, sometimes coupled with chemical precipitation and/or gas evolution, can reduce the soil's hydraulic conductivity.

Providing an electron acceptor such as oxygen for direct aerobic bioremediation often becomes the critical limiting factor during system design. Continuous or intermittent oxygen delivery into the saturated zone is a challenging task, with field options primarily limited to sparging air, adding hydrogen peroxide or oxygen release compounds. Electrochemical methods to generate oxygen *in situ* also have been utilized in aerobic bioremediation systems.

An experimental technique to inject pure oxygen in the form of microbubbles has been reported in the literature.[5] A coarse soil matrix in the saturated zone is preferred for this technique to provide both a high permeability for flowing groundwater and a suitable saturated matrix for adhering and retaining microbubbles. The use of oxygen microbubbles for *in situ* bioremediation has the advantages of increased oxygen transfer rate to flowing groundwater (DO pickup), and increased oxygen utilization (percent of O_2 injected) compared to air injection.

The microbubbles are typically made from a low-surface-tension water containing 100 ppm or more of an appropriate surfactant. The microbubbles upon generation resemble a thick cream with much of the volume made up of microbubbles. The bubbles are generated by a colloidal gas apron. In a field study, it was demonstrated that approximately 15 to 20% of the oxygen injected can be dissolved into the flowing groundwater.[5] With 10% committed to biodegrading the surfactant, a minimum of 5 to 10% net utilization was available for biodegrading contaminated groundwater.

Formulations of very fine, insoluble magnesium peroxide (MgO_2) release oxygen at a slow, controlled rate when hydrated. Their use has been demonstrated to increase the dissolved oxygen concentrations within contaminated plumes, and thus enhance the rate of aerobic biodegradation. Magnesium peroxide releases oxygen when it comes in contact with water as shown by the following equation:

$$MgO_2 + H_2O \rightarrow \tfrac{1}{2}O_2 + Mg(OH)_2 \quad (2.1)$$

The by-products of the reaction are oxygen and magnesium hydroxide, which will also help in maintaining moderate pH levels within the contaminated plume. MgO_2 is normally placed in an inert matrix and is available in easily installable socks of various diameters. These socks can be stacked in wells screened across the entire thickness of the contaminated zone. In addition, these oxygen-release compounds are also available in injectable slurry form and can be injected via direct push points into the contaminated zone.

In addition to the petroleum hydrocarbons, other compounds that are more conducive for aerobic biodegradation are: nonchlorinated phenolic compounds, alcohols, ketones, carboxylic acids, amides, amines, esters, aldehydes, aliphatic and/or aromatic hydrocarbons with a single chlorine atom or two chlorine atoms. Among the chlorinated compounds chlorobenzene, methylene chloride (MC), chloromethane, chloroethane, and vinyl chloride are among the commonly encountered contaminants that are biodegradable faster under aerobic conditions.

The most significant biological mechanism for the aerobic degradation of chlorinated solvents is when they are used as a primary substrate (electron donor). In these direct oxidation reactions, the chlorinated compound acts as an electron donor and the microorganism uses molecular oxygen as an electron acceptor. The microorganism obtains energy and organic carbon from the degraded chlorinated compound. The more chlorinated compounds, PCE, carbon tetrachloride (CT), and hexachloroethane (HCA), due to their highly oxidized character, are neither susceptible to aerobic oxidation nor are they degraded under anaerobic oxidizing conditions when used as a primary substrate. TCE undergoes very slow aerobic degradation to trichloroethanol and then to acetic acid, but the reaction is not thermodynamically favorable. The tendency of chloroethene compounds to undergo oxidation increases with decreasing number of chlorine substituents. Therefore, discussion of aerobic oxidation and mineralization has always been focused on DCE and VC.

Rates of aerobic oxidation are more rapid for the less chlorinated organics, such as VC, when compared to their reductive dechlorination rates; it has been well documented in the literature that VC is oxidized directly to carbon dioxide and water.[6] Aerobic oxidation of cis-DCE has been speculated in the past and has been proven recently. However, it could not be ascertained, in certain cases, whether DCE was reduced to VC and then direct oxidation of VC produced carbon dioxide or direct oxidation of DCE occurred to produce carbon dioxide. Other reports suggest that DCE can be degraded with DCE acting as the primary substrate in aerobic microbial reactions.

Under aerobic conditions, chlorinated aliphatic compounds with one or two carbons per molecule can be transformed by three types of microbial enzymes:[4,6] dehalogenases, hydrolytic dehalogenases, and oxygenases (Figure 2.3). Dehalogenases, which require reduced glutathione as a cofactor, dehalogenate the substrates by means of nucleophilic substitution. The first product of this degradation pathway is an S-chloralkyl-glutathione, which is probably nonenzymatically converted to glutathione and an *aldehyde*. Hydrolytic dehalogenases hydrolyze their substrates, yielding *alcohols*. Oxygenases use molecular oxygen as a reactant for the attack on the halogenated compounds; the products could be alcohols, aldehydes, or epoxides, depending on the structure of the compound. Numerous chlorinated short-chain

$$
-C-Cl \begin{cases} \text{a. Oxygenolytic dechlorination} \\ \quad \frac{1}{2}O_2 \\ \text{b. Hydrolytic dechlorination} \\ \quad (H_2O) \\ \text{c. Reductive dechlorination} \\ \quad 2e^-, 2H^+ \\ \text{d. Elimination} \end{cases} \begin{array}{l} C-C-OH \longrightarrow -C=O \\ \quad | \\ \quad Cl \\ \\ -C-OH \\ \\ \\ -C-H \\ \\ =C \end{array} + Cl^-
$$

FIGURE 2.3 Four fundamental dechlorination mechanisms capable of cleaving the carbon–chlorine bond. (Adapted and modified from Wackett and Hershberger.[3])

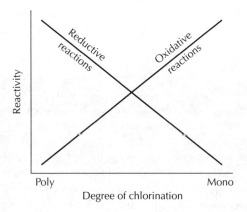

FIGURE 2.4 Relative trends of oxidative vs. reductive microbial reactions as a function of chlorination. (Adapted from Reference 17.)

aliphatic hydrocarbons have been demonstrated to undergo aerobic transformation. However, compounds that have all the available valences on their carbon atoms substituted by chlorine, such as PCE or CT, have never been shown to transform through any other but reductive pathways. Generally, as the degree of chlorination increases, the likelihood of aerobic transformation decreases (Figure 2.4); the opposite is true for anaerobic (reductive) transformations. Unfortunately, under typical field conditions, aerobic microbial oxidation of VC and DCE is often limited by the absence of oxygen within the chloroethene plumes.

Among the methane compounds, MC and chloromethane have been found to be amenable to aerobic microbial transformation. Pure cultures of the genera *Pseudomonas* and *Hyphomicrobium* have been isolated which can grow on MC as the sole carbon and energy source.[4,6,7] Alkylhalides (haloalkanes), such as 1,2-dichloroethane (1,2-DCA), are frequently hydrolytically dehalogenated. *Xanthobacter autotrophicus* utilizes 1,2-DCA as the sole carbon source. Complex communities consisting of methanotrophs and heterotrophs, which inhabit groundwater aquifers, mineralize 1,2-DCA. A *Pseudomonas fluorescens* strain isolated from water and soil contaminated by chlorinated aliphatic hydrocarbons was shown to utilize 1,2-DCA, 1,1,2-trichloroethane (1,2,1-TCA) and TCE, but not PCE or 1,1,1-TCA.[4,6,7]

2.2.3 MICROBIAL OXIDATION BY Fe(III), NITRATE, AND SULFATE

One promising alternative to the saturation limitations or high costs of the alternative forms of oxygen involves the use of nitrate as an electron acceptor. As stated earlier, oxygen is not very soluble and adding it to groundwater is an inefficient process, specifically within source zones. Dissimilatory nitrate reduction and denitrification are microbial oxidation processes that have been tested for the degradation of petroleum-related contaminants.[8]

BTEX compounds are readily degraded under aerobic conditions, but petroleum-contaminated aquifers often contain anaerobic zones. Therefore, a better understanding of the fate of these compounds, specifically benzene, under anaerobic conditions is necessary to design IRZ systems. Anaerobic benzene biodegradation has been shown to occur under iron-reducing, sulfate-reducing, perchlorate-reducing, methanogenic, and even under nitrate-reducing conditions.[1,6] However, despite these reports the microorganisms and pathways of anaerobic benzene degradation remain largely unknown.

Although anaerobic benzene biodegradation has been demonstrated, benzene has frequently been found recalcitrant under anaerobic conditions.[9] This is of particular concern because benzene is the most toxic of the BTEX compounds. Even though it has been suggested that introducing nitrate into contaminated groundwater as an electron acceptor can enhance partial or complete biodegradation of BTEX, the authors' experience indicates that nitrate alone is not sufficient to degrade benzene. Our experience in the field suggests that addition of NO_3^- as a single electron acceptor selectively stimulates the degradation of TEX contaminants leaving benzene untreated. If there is a significant amount of Fe(III) present in the aquifer sediments, total degradation of BTEX can be expected even if only nitrate is present.

A wide variety of organic contaminants can be microbially oxidized by Fe(III)-reducing microorganisms. Fe(III) is the most prevalent anaerobic electron acceptor present in many subsurface environments. In addition, there are several ways to increase the bioavailability of Fe(III) to further enhance microbial oxidation by Fe(III).[8]

Humic substances are natural organics that result from the incomplete breakdown of complex organic matter. Fe(III)-reducing bacteria can directly reduce humic substances and form reduced redox species. At the same time the oxidized Fe(III) can also *abiotically* accept electrons from the reduced humic substance species. The humic substances become re-oxidized and are again free to accept electrons in microbial metabolism. As such, the humic substances act as an *electron shuttle* between the microorganisms and Fe(III). This is significant because the prevalent form of iron in aquifer sediments is Fe(III) oxides. It is generally accepted that bacterial cells must physically contact the insoluble Fe(III) oxides in order to reduce them; however, because they are not freely soluble, the Fe(III) oxides are occluded from rapid reduction. Soluble humic substances stimulate Fe(III) reduction because of this electron shuttling phenomenon, and increased Fe(III) reduction hastens the degradation of many organic contaminants, including benzene.[8–10] Synthetic electron shuttles with quinone moieties, for example the synthetic humic substance anthraquinone-2,6-disulfonate (AQDS), can also stimulate Fe(III) reduction to the same extent as naturally occurring humic substances.[8,11]

The concern of whether NO_3^- alone can be injected as a single electron acceptor to achieve complete degradation of BTEX compounds led to the concept of combined injection of more than one electron acceptor. This combined injection provides the dual advantages of (1) increasing the total electron acceptor capacity and (2) increasing the potential for stimulating the degradation of all BTEX compounds. From an engineering design perspective it is easier to maintain higher concentrations of NO_3^- and SO_4^{2-} than in combination with O_2. In addition, sulfate accepts eight electrons per mole reduced, as opposed to oxygen which accepts two electrons per mole reduced. In the appropriate environment, this increased electron-accepting capacity compensates for the lower free energy yield of sulfate as an electron acceptor.

The injection of nitrate and sulfate was able to stimulate benzene degradation via microbial oxidation reactions within an *in situ* reactive zone.[9] However, it should be noted that benzene removal was achieved only after the removal of all the other hydrocarbons that limit benzene degradation. The authors of this study further recommended

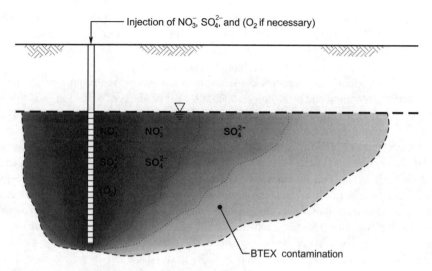

FIGURE 2.5 Expected biogeochemical zones during the injection of NO_3^-, SO_4^{2-}, and O_2 for microbial oxidation of BTEX compounds.

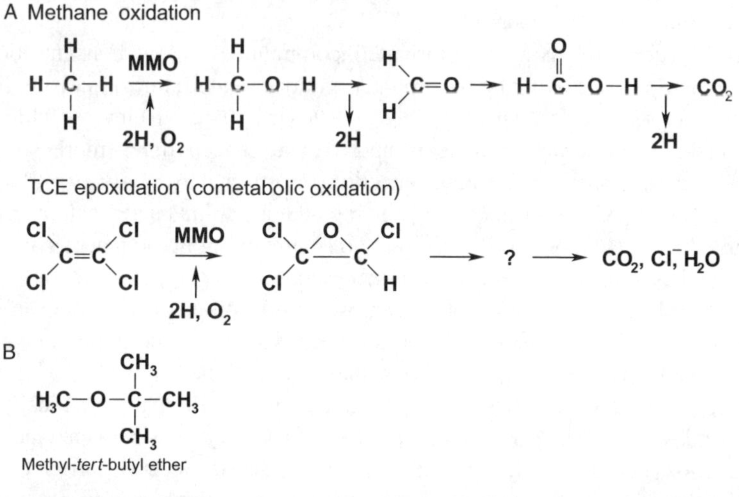

FIGURE 2.6 (A) Cometabolic oxidation of TCE. (B) Methyl-*tert*-butyl ether.

that one way to improve this concept as a full-scale technique for BTEX removal would be to include dissolved oxygen along with nitrate and sulfate in the injected water. Representation of the geochemical zones that are expected to be established upon the combined injection of O_2, NO_3^-, and SO_4^{2-} is shown in Figure 2.5.

2.2.4 Aerobic Cometabolic Oxidation

Several of the chlorinated aliphatic hydrocarbons (CAHs) can be microbially transformed through cometabolism. This was first demonstrated to be true for TCE in 1985.[2] The two steps are shown in Figure 2.6(A). Here, methane monooxygenase (MMO) initiates the oxidation of methane by forming methanol. This also requires molecular oxygen and a supply of reducing power noted as 2H. MMO also fortuitously oxidizes TCE, converting it into TCE epoxide, an unstable compound that degrades biochemically into many compounds that can be mineralized by many microorganisms. Thus TCE biodegradation is affected through cometabolism.

Methanotrophic communities consisting of methanotrophs that initiate the oxidative transformation and heterotrophs that utilize the products of oxidation and hydrolysis are very active, and can achieve complete degradation of chlorinated alkenes. The same communities fail to transform PCE, however, because this compound is too oxidized. Pure cultures of methanotrophs such as *Methylosinus trichosporium* OB3b or *Methylomonas* sp. MM2 have been shown to partially transform TCE, *trans*-DCE, and *cis*-DCE.[4,6,7] Other microorganisms capable of transforming chlorinated alkenes belong to the genera *Pseudomonas*, *Alcaligens*, *Mycobacterium*, and *Nitrosomonas*. All of these microorganisms, except the genus *Nitrosomonas*, are heterotrophs that grow on various organic substrates (e.g., toluene, cresol, phenols, propane, etc.); *Nitrosomonas* is a chemolithotroph that derives energy from oxidation of ammonia. All of them *cometabolize* chlorinated compounds such as TCE or 1,2-DCE while growing on their respective growth substrates; the chloroalkenes are only fortuitously transformed, not utilized for growth. However, VC seems to be an exception. It has been demonstrated that a *Mycobacterium* strain isolated from soil contaminated by VC could grow on VC as sole carbon and energy source.[12]

Aerobic cometabolism of chlorinated compounds at low concentrations by methane- and propane-utilizing bacteria is well documented. In comparison, *butane-utilizing bacteria* are less susceptible to the toxic effects of elevated chlorinated compound concentrations. Butane is approximately four times more soluble in groundwater than methane. Butane injection results in large radii of influence at injection wellheads. The difficulty of utilizing alkanotrophic bacteria stems from the low solubility of alkanes and the difficulty of maintaining homogeneous concentration of the dissolved alkane within the reactive zone.

As described earlier, methanotrophs grow on C1 compounds as sole carbon and energy sources. Their catabolic oxygenases are *MMOs* that incorporate one atom of oxygen from the oxygen molecule into methane to yield methanol.[4,6,7,12] This alcohol is further oxidized via a series of dehydrogenation steps through formaldehyde and formic acid to CO_2, which is the final product of catabolism. MMO enzymes utilize molecular oxygen as a reactant, and require a reduced electron carrier to reduce the remaining oxygen atom to water. MMO enzymes have relaxed substrate specificity and will oxygenate many compounds that are not growth substrates for methanotrophs Such compounds include various alkanes, alkenes, ethers, alicycles, aromatics, nitrogen heterocycles, and chlorinated alkanes, alkenes, and aromatics.[6,13]

Two types of MMOs have been suggested: a particulate (membrane bound) and a soluble enzyme.[6] The soluble MMO (purified from *M. trichosporium* OB3b and *Methylococcus capsulatus*), which is produced under the conditions of copper limitation and increased oxygen tension, has been considered to have broader substrate specificity. It has been stated that only the soluble MMO can transform TCE. However, recent findings indicate that the particulate MMO in some methanotrophs may be as effective in the transformation of chlorinated solvents as the soluble MMO. Since the soluble MMO is not constitutively expressed whereas the particulate MMO is, the latter methanotrophs (*Methylomonas* sp.) have a significant potential for *in situ* bioremediation.

Thus TCE can be transformed (upon the induction of the oxygenase enzyme by its substrate) in the presence of the microorganismal growth substrate (*cometabolism*), or in its absence (*resting cells transformation*). However, TCE is not utilized by the bacteria as a carbon, energy, or electron source; this transformation is only *fortuitous*. Based on the findings with methanotrophs, it can be concluded that TCE is most likely oxygenated to TCE-epoxide (Figure 2.6). The epoxide is unstable and is quickly nonenzymatically rearranged in aqueous solution to yield various products including carbon monoxide, formic acid, glyoxylic acid, and a range of chlorinated acids. Recent findings with purified MMO from *M. trichosporium* OB3b indicate that TCE-epoxide is indeed a product of TCE oxygenation. In nature, where cooperation between the TCE oxidizers and other bacteria (most prominently heterotrophs) occurs, TCE can be completely mineralized to carbon dioxide, water, and chloride.

Toluene, phenol, and cresol oxidizers, such as *Pseudomonas putida* or *P. cepacia*, express the TCE transformation activity upon induction by their aromatic substrates. These bacteria have a great potential for remediation of groundwater aquifers that are contaminated by mixtures of gasoline or jet fuel (or other petroleum derivatives), and chlorinated solvents, such as TCE, DCE, or VC. If the aromatic contaminants are not present, however, bacterial growth substrates need to be injected into the site in order to stimulate the transformation of chlorinated solvents. In this situation, methanotrophs become more attractive agents of bioremediation: methane, their preferred substrate, is a nontoxic and inexpensive chemical. Once methane and oxygen are injected into the site, methanotrophs (if present) will start cometabolizing chlorinated solvents, as well as a great number of other contaminants (see below), and the accompanying heterotrophs will mineralize their transformation products. As mentioned earlier it is important to maintain a reasonable high and uniform O_2 and CH_4 concentrations to achieve significant methanotrophic degradation. Cometabolic oxidation often is not considered a significant, long-term mechanism for nonengineered bioremediation of chloroethenes in groundwater, but has been successfully exploited within engineered IRZ systems.

2.2.5 MICROBIAL OXIDATION OF MTBE

The capacity to degrade MTBE (Figure 2.6(B)) is found in a wide variety of microorganisms. Some organisms can use MTBE as their sole substrate for growth and others need another substrate for cometabolism. MTBE-degrading organisms require higher concentrations of oxygen than the usual aerobic bacteria. Maintaining adequate concentrations of oxygen is an important design concentration for aerobic biodegradation of MTBE. There are many reports in the literature which state that the effective half saturation constant (K_s) for oxygen during MTBE degradation is significantly higher than general oxidative respiration.[14] In addition, inhibition of MTBE degradation has been reported at oxygen concentrations less than 1 mg L^{-1}.[15]

The prospects for aerobic biodegradation by native microorganisms may be related to the age of the spill and the time that has been available for acclimation of the native microorganisms to MTBE. In general, groundwater plumes impacted by

MTBE from petroleum spills are devoid of dissolved oxygen. Dissolved oxygen must be added to the groundwater to meet biochemical demands of MTBE, other contaminants, and other nontarget organic compounds associated with the aquifer matrix and the reduced minerals. There seems to be a consensus that at least 4 mg L^{-1} or higher of dissolved oxygen needs to be maintained within the contaminated plume to achieve MTBE degradation by native aerobic microorganisms.[14]

At some sites, adding dissolved oxygen to the groundwater is the only action that is necessary to stimulate aerobic biodegradation of MTBE. At other sites, it was necessary or advantageous to augment the aquifer with MTBE-degrading microorganisms. However, it is not easy to conclude whether the introduction of specific organisms or other factors were responsible for the disappearance of MTBE. Others have injected alkanes to cometabolically degrade MTBE by alkanotrophic microorganisms. Propane was injected at a site in New Jersey with an air sparging system and MTBE was degraded at significant rates. When the sparge system was turned off, MTBE concentrations rebounded in the monitoring wells. MTBE concentrations were reduced from 1.5 to 0.01 mg L^{-1} in an *in situ* bioreactor, simulating a funnel and gate system, at dissolved oxygen concentrations near 15 mg L^{-1}.[14]

Many site-specific factors influence the feasibility and selection of *in situ* oxygen amendment methods. An important factor for aerobic MTBE biodegradation is the need to maintain a high level of dissolved oxygen and the resulting magnitude of oxygen demand. Other contaminants, nontarget organics and inorganics, reduced metals, and the groundwater flow velocity will influence the amount of oxygen demand. Available methods for oxygen introduction include air or oxygen sparging, hydrogen peroxide injection, diffusive oxygen emitters, in-well oxygenation, electrolytic cells, and solid forms of oxygen.[11]

Early studies that focused on anaerobic biodegradation of MTBE were inconclusive, but led to a general speculation that MTBE will not biodegrade in the absence of oxygen. More recent studies, which included many field sites and laboratory research, suggest that biodegradation of MTBE and *tert*-butyl alcohol (TBA) under anaerobic subsurface conditions is possible. Some studies reported that there appeared to be a good correlation between strongly anaerobic biogeochemical conditions and biodegradation of MTBE.[16] Strongly methanogenic conditions seem to indicate favorable conditions for the degradation of both TBA and MTBE.

However, more recent research indicates that anaerobic microbial oxidation of MTBE is possible with each of the dominant anaerobic electron acceptors — nitrate, Mn(IV), Fe(III), sulfate — and under methanogenic conditions.[8] Finding that anaerobic degradation is possible was the first step in developing bioremediation strategies to be implemented within engineered anaerobic IRZs.

2.3 MICROBIAL REDUCING ZONES

2.3.1 ENHANCED REDUCTIVE DECHLORINATION

Industrial production of halogenated organic chemicals has been increasing since the mid-20th century. During this period, halogenated organic compounds have also been found increasingly as contaminants in the environment. CAHs are the top organohalogen compounds produced industrially. This class of compounds includes widely used solvents such as CT, methylene chloride (MC), TCA, trichloroethene (TCE), and tetrachloroethene (PCE). Other industrial compounds include a wide variety of chlorinated organics such as chloroflurocarbons, chlorinated phenolics, and chlorinated benzenes, representing products with diverse end uses as solvents, degreasers, fumigants, biocides, dielectric fluids, flame retardants, and chemical intermediates.

Polychlorinated biphenyls (PCBs) are another type of halogenated organics which were widely used from the mid-20th century to the 1980s for a great variety of industrial applications, including hydraulic fluid, dielectric fluid in capacitors and transformers, solvent extenders, and more. Chlorinated pesticides and insecticides were also in wide use by the mid-20th century. Until recently, polychlorinated phenols have been widely used for impregnation or short term preservation of wood and wooden products. Due to improper usage, work practices, and accidental releases, all these compounds have contaminated the subsurface environment during the last few decades. In the last 10 to 15 years, the anaerobic degradation of chlorinated contaminants has become a matter of special interest due to the almost ubiquitous presence of these contaminants in the subsurface environment. Despite efforts to replace chlorinated solvents and chemicals by compounds that are of lesser environmental concern, large amounts of chlorinated chemicals are still produced for a variety of applications.

Microorganisms have evolved different strategies to remove the chlorine substituent and utilize these chlorinated compounds for their own benefit. Commonly observed metabolic mechanisms for dechlorination are (1) oxygenolytic dechlorination, where under aerobic conditions the chlorine is replaced by a hydroxyl group derived from oxygen; (2) hydrolytic dechlorination, where the chlorine is replaced by the hydroxyl group derived from water; (3) reductive dechlorination, where mostly under anaerobic conditions the chlorine is replaced by hydrogen; and (4) elimination (Figure 2.3). However, reductive dechlorination is the most dominant and prevalent microbial reaction responsible for the complete transformation of chlorinated organics.[17]

Enhanced reductive dechlorination technology implemented within an IRZ is intended to facilitate and expedite the biological reductive dechlorination of chlorinated organics, specifically CAHs, through the well-documented mechanisms shown in Table 2.1 and Figures 2.7 and 2.8. The general mechanism behind enhanced reductive dechlorination relies on enhancing or inducing the biological transformation of chlorinated compounds through periodic injection of a soluble electron donor solution (a carbohydrate such as molasses or corn syrup, whey, lactate, butyrate, or benzoate) into the IRZ. Carefully designed periodic substrate injections alter existing aerobic or mildly anoxic contaminated groundwater plume(s) to anaerobic, microbiologically diverse, reactive treatment zone.

Reductive dechlorination is a biologically mediated reaction that entails transferring electrons to the chlorinated contaminant of interest from the electron donors. The more oxidized the chlorinated compound is, the more susceptible it is to reduction. The reduction of chlorinated solvent molecules that are used as electron acceptors cleaves their chlorine atoms and replaces them with hydrogen, leading to the sequential dechlorination pattern observed in many contaminated aquifers (Figure 2.9). Several microbially mediated reactions that may lead to reductive dechlorination are discussed in detail in later sections.

TABLE 2.1
Reductive Dechlorination Processes

Process	PCE	TCE	cis-DCE	VC	TCA	DCA	CT	CF	DCM
Direct aerobic	N	N	Y&N	Y	N	N	N	N	Y
Cometabolic (methanotrophic)	N	Y	Y	Y	Y&N	N*	N	Y	NR
Cometabolic (organic substance)	N	Y	Y	Y	N	N*	N	Y&N	NR
Cometabolic w/NH$_4$	N	Y	Y	Y	Y	N*	N	Y	NR
Direct anaerobic	N	N	N	Y	N	N	N	N	Y
Anaerobic/denitrification	Y&N	Y&N	N*	N*	N*	N*	Y	Y&N	NR
Anaerobic/sulfate reduction	Y	Y	Y	Y	Y	Y	Y	Y	NR
Anaerobic/methanogenic	Y	Y	Y	Y	Y	Y	Y	Y	NR

N, not documented in the literature.
Y, documented in the literature many times; consensus opinion.
Y&N, both occurrence and absence documented in the literature more than once.
N*, not documented in the literature to date, but not investigated significantly.
NR, process may occur but is not relevant because competing process occurs more rapidly.

Source: ITRC training session 2002, ITRC 1999.

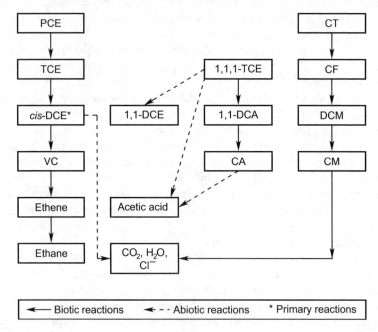

FIGURE 2.7 Biological and abiotic degradation pathways of the common chlorinated compounds encountered at contaminated sites.

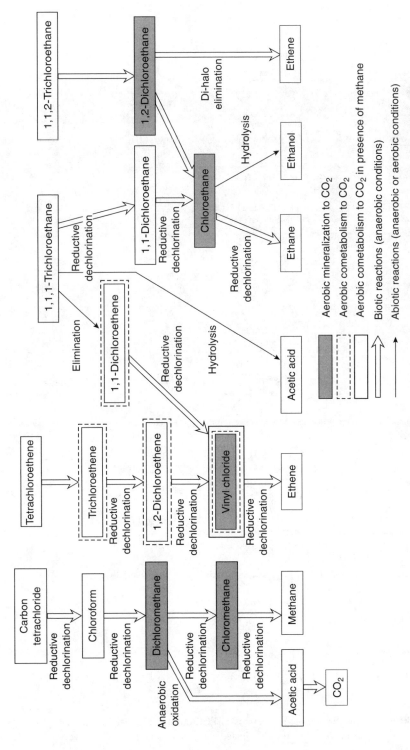

FIGURE 2.8 Common degradation pathways. (From Interstate Technology and Regulatory Cooperation work group (ITRC), May 1999.)

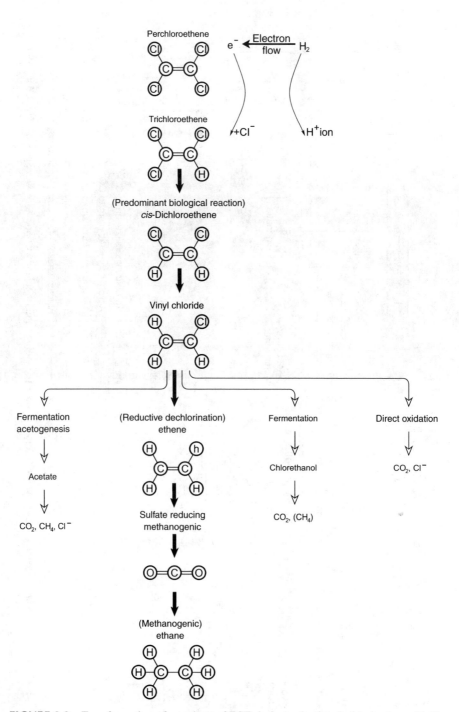

FIGURE 2.9 Transformation of reactions of PCE during reductive dechlorination with H_2 acting as the electron donor and the chlorinated compounds acting as electron acceptors.

The energy gained by bacteria in metabolic reactions is determined by the nature of the electron acceptor and electron donor compounds. Chlorinated solvent molecules yield very little energy to the bacteria that utilize them as electron acceptors (Figure 2.10). As a result, the populations of bacteria that can utilize chlorinated solvents as electron acceptors can be suppressed by competing species that utilize more beneficial electron acceptors such as oxygen, nitrate, and oxidized forms of iron and manganese. That is why reductive dechlorination is only observed in aquifers where oxygen and nitrate replenishment is minimal or cut off, or they have been consumed in the degradation of available carbon supplies. Bacterial consumption of the degradable carbon consumes matching quantities of electron acceptor compounds. When the rate of carbon consumption exceeds the rate of high-yield electron acceptor recharge, an anaerobic IRZ is created that provides full dechlorination of target alkenes and other chlorinated compounds.

In practice, enhanced reductive dechlorination can be implemented as an *in situ* bioreactor that forms downgradient from a line of degradable carbon substrate injection wells placed in a line perpendicular to groundwater flow (Figure 2.11). If sufficient carbon substrate is injected, oxygen and nitrate metabolism dominates near the injection line, while sulfate reduction, methanogenesis, and reductive dechlorination zones form farther downgradient. The technology operates most effectively when groundwater is passing through the sulfate-reducing zone, still bearing a degradable carbon load that will support methanogenesis and reductive dechlorination. Under these circumstances, *cis*-dichloroethene (*cis*-DCE) is degraded (in net effect) to ethene or ethane without a measurable buildup of VC.

A "high-performance" reductive dechlorination system can only be achieved when the rate of electron donor consumption substantially exceeds the rate of electron acceptor recharge. The carbon source must be highly mobile and highly degradable, and injected at rates commensurate with the overall flux of groundwater, electron acceptors, and

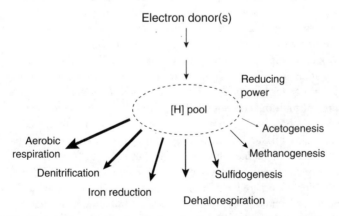

FIGURE 2.10 Energy flow and hydrogen transfer for dehalogenation and other respiration processes. The relative placement of the electron-accepting process and boldness of the arrows provide a comparative measure of energy yield as a function of electron (hydrogen) flow. (Adapted from Loeffler et al.[17])

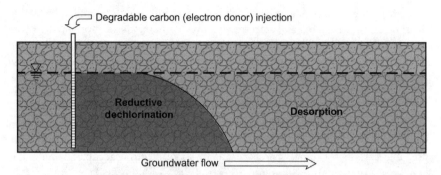

FIGURE 2.11 Simple reactive zone.

CAHs that move into the treatment zone. The organic acids that form must be buffered by aquifer carbonates or by the addition of carbonates and bicarbonates to the injection mix. Certain site hydrogeologic characteristics will necessitate modification of the high-performance approach, but most sites can still be treated by enhanced reductive dechlorination (ERD) systems that are more cost effective than many alternative approaches. Examples of sites requiring modification are those with very low and very high groundwater flow velocities and those with high electron acceptor recharge rates.

It is desirable at the majority of sites to introduce a relatively large mass of degradable carbon into the system, so as to consume influxes of the electron acceptors very rapidly and sustain optimal conditions for high-rate degradation. For this purpose soluble substrates have a distinct advantage over low solubility or sparingly soluble materials, and slow release carbohydrates. However, there are a few sites where the influx of electron acceptors and the groundwater velocity are so low that the use of slow release substrates is more appropriate.

A nearly uniform release rate of degradable volatile fatty acids and hydrogen, as a result of the breakdown of the injected substrates, over a period of many months creates optimal reducing conditions to develop a microbial consortium that is capable of ERD. Thus a substrate that is very rapidly bioavailable may require more frequent injections in order to develop and sustain this consortium, which has cost implications. Hydrogen gas is the extreme example of rapid bioavailability and utilization. Methanol is another example of rapid bioavailability. However, highly soluble and bioavailable substrates include both pure compounds (i.e., glucose, lactate) and complex mixtures of multiple compounds (i.e., molasses). The use of the complex food-grade mixtures (i.e., corn syrup and molasses) has two important advantages: the ability of complex mixtures to encourage the growth of a more diverse microbial community, and the moderate rate of biological utilization.

Microbially catalyzed processes that take place within an IRZ during enhanced reductive dechlorination are a result of the interaction between the biochemical conditions engineered and the complex, dynamic, and unique bacterial community. Ecological selection pressures can define the site-specific reductive dechlorination bacterial community. These pressures include changing availability of electron acceptors, pH, and electron donor availability. Within this chemically and microbially complex environment, there is no one biochemical mechanism or any one single bacterium that is completely "responsible" for the complete transformation process.

It is a consortium of microorganisms and a variety of mechanisms that bring about the desired transformation pathways of the target contaminant.

2.3.1.1 Microbial Ecology of Reductive Dechlorination

Though of unquestionable importance with regard to the function of subsurface ecosystems, we understand very little about the structure of soil microbial communities, how these communities respond to changes in their environment, or the consequences that alteration in microbial community structure have on ecosystem function. The dearth of information about soil microbial communities is a consequence of their enormous complexity and genetic diversity, and the fact that the microorganisms that can be isolated from soil and studied in isolation represent only a small portion of the microbial groups present *in situ*. Soil microorganisms do not behave as a homogeneous trophic level and the species composition of a soil microbial community can influence microbial processes in the soil both qualitatively and quantitatively. As a result, it is important to examine the *internal dynamics* of soil microbial communities not simply for the sake of characterizing these fascinatingly complex biotic systems but also to understand how bacterial populations evolve based on changes in the biogeochemical environment.

The great diversity of microorganisms and their versatile metabolism mirrors the wide range of substrates that are used for microbial growth and the natural environments in which they are found. Many types of microorganisms are able to adapt and metabolize different and often relatively new industrial compounds, which are often considered to be recalcitrant and persistent in the environment. Even though the conventional assumption is that chlorinated alkenes and alkanes were introduced into the environment only in the last 50 to 60 years, volcanic emissions have been shown to contain significant amounts of chlorinated alkanes and alkenes, and chlorinated benzoates.[18,19] Significant amounts of chloroorganic compounds which are of biogenic and geogenic origin have been identified in both marine and terrestrial habitats. The fact that these chloroorganic compounds did not accumulate in the environment indicates that many microorganisms have evolved strategies to benefit from the presence of these compounds, possibly from the earliest stages of microbial evolution. A number of bacteria that obtain energy for growth by metabolizing chlorinated compounds have been recovered over the last few decades from contaminated and uncontaminated sites. These findings suggest that nature has provided appropriate mechanisms for long-term microbial adaptation and for the evolution of enzyme systems specific for chloroorganic compounds.[17]

One of the earliest observations of reductive dehalogenation was reported by researchers at the University of California in 1966. They reported studies on biological degradation of 1,2-dibromoethane (EDB) and related soil fumigants. They later observed reductive dehalogenation in donor-fed cultures, and summarized the microbial flora as follows: "... although many bacteria could be obtained from the active cultures no single or mixed colony was effective when incubated with sterile soil-water or any other media."[20] Their mixed culture soil enrichment dehalogenated EDB, an exact analog of 1,2-dichloroethane, to ethene by a dihaloelimination reaction. But their attempt to isolate a species that could achieve the dehalogenation was unsuccessful.

Increasing awareness of surface water and sediment contamination by chlorinated compounds led to an expanding interest in the fate of these compounds in the environment, particularly their degradation mechanisms. Early studies were based on cultures of well-known bacterial species. For example, the biochemical feasibility for microbial degradation of insecticide residues was demonstrated for DDT using cell membranes isolated from *Escherichia coli* and by cell-free extracts of *Clostridium rectum* for lindane.[21–23] Dechlorination of hexachlorocyclohexanes (which include lindane) in intact cultures of *Clostridium sphenoides* was also demonstrated.[22]

In a parallel line of research, studies were conducted to isolate bacteria from river sediments and wastewater sludges in a search for dechlorinating species. The technical challenges in isolating anaerobic bacteria from natural habitats are very great due to the fact that anaerobic biodegradation studies often involve complex, undefined bacterial consortia. This is due partly to the slower cultivation cycles and also to the complexity of anaerobic metabolic interactions that make pure culture isolation sometimes difficult or impossible. Nonetheless, this work has provided numerous examples of dehalogenating species and consortia, even in uncontaminated river sediment environments.[6] By the late 1980s, it was clear that reductive dechlorination was occurring in many contaminated groundwater systems, where ethene and ethane solvents, chlorobenzenes, and wood preservatives such as the pentachlorophenol were present.

As reported extensively in the literature, noteworthy progress has been achieved over the past few years in our knowledge of the transformation of chlorinated compounds by anaerobic bacteria. Clearly, it must be admitted that the research and development in this area of bioremediation is still in its infancy. However, on the basis of the results published to date, we can anticipate some concrete trends and outline some interesting perspectives in this field. Taking into account their physiological characteristics and phylogenetic mapping, there is increasing evidence for the vast diversity of microorganisms involved in reductive dechlorination. Considering the minute fraction of microorganisms that has been cultured so far and identified, and the fact that the dechlorinating abilities of available cultures have not been comprehensively explored, it is very likely and almost certain that a large number of new dechlorinating species await discovery. It is very likely that only a fraction of the species capable of reductive dechlorination have been captured in cultures so far, and of those that have been cultured, only a small percentage have been isolated and identified. What has been clearly demonstrated is that reductive dechlorination can be a growth-supporting process, in which chlorinated alkenes are used as terminal electron acceptors in the oxidative metabolism of electron donors that include molecular hydrogen, C1 compounds (CO_2, CH_3OH, CH_2OO^-, and CH_2O), acetate (CH_3COO^-), and other low-molecular-weight organic compounds, all of which can be formed in the anaerobic metabolism of higher-molecular-weight carbohydrates.

Many reports have summarized the recent developments in the field of microbiology of reductive dechlorination, particularly during the last 5 years. Hence, we have made an attempt only to interpret the results and findings of these studies rather than delving too deeply into the scope of these microbiological studies themselves. Our focus is to develop a cohesive understanding of reductive dechlorination from the available studies and create a firm foundation for designing enhanced reductive dechlorination systems within an IRZ.

From the recent literature it can be summarized that there is increasing evidence for the vast diversity of anaerobic microorganisms involved in dechlorination (Table 2.1). Regardless of whether alkyl or aryl dechlorination is considered, the major known anaerobic bacterial groups are apparently involved as can be seen from the following nonexhaustive list: sulfate-reducing bacteria, methanogenic bacteria, clostridia, facultative anaerobic bacteria, and homoacetogenic bacteria.[24]

A variety of metabolic processes have been identified in anaerobic dechlorinating bacteria. Chlorinated compounds can serve three different metabolic functions in anaerobic bacteria: (1) as carbon or energy source or both, (2) as substrate of cometabolic activity, and (3) as terminal electron acceptor in an anaerobic process (dehalorespiration). The utilization of chlorinated compounds as growth substrates is restricted to very few compounds and organisms. In contrast, the main metabolic processes observed in reductive dechlorination are cometabolism and dehalorespiration.

Reductive dechlorination can also be a cometabolic process, in which enzymes and cofactors produced for other purposes perform dechlorination reactions. These cometabolic reactions are linked to bacteria that are surviving by means other than reductive dechlorination, but which produce enzymes and cofactors that reduce chlorinated solvents. Bacteria cannot gain energy for growth by cometabolic reactions, so it is necessary to provide electron donors, electron acceptors, and a carbon source to support populations of a density that can perform meaningful reductive dechlorination.

Dehalorespiration seems to be the important and promising biochemical process, especially in kinetic terms, for reductive dechlorination. It has been reported that dechlorination rates of dehalorespiring bacteria are considerably higher (often by one or more orders of magnitude) than those found for cometabolically dechlorinating anaerobes. Another important aspect of the dehalorespiring bacteria is the fact that the selective pressure favoring the growth of these microorganisms is imposed by the chlorinated substrate itself.[24] These observations imply that it may become practical to increase the standing biomass of indigenous reductive dechlorinating microorganisms, and consequently the *in situ* dechlorinating activity by providing appropriate chlororespiration substrates.

2.3.1.1.1 Principal Modes of Reductive Dechlorination

The scientific and technical applications knowledge base supporting the use of microbial reductive dechlorination, specifically for the purpose of designing an enhanced reductive dechlorination system, is now very large and continuing to grow at a rapid pace. Three major themes of reductive dechlorination are now well developed: natural attenuation, biostimulation, and bioaugmentation.

> *Natural attenuation.* Naturally occurring soil bacteria utilize chlorinated organics as terminal electron acceptors in aquifers that lack oxygen, nitrate, sulfate, or other more energy-efficient electron acceptors. Chlorinated contaminants are biotransformed via cometabolic reactions in the presence of natural and/or anthropogenic organic carbon. Natural attenuation does not occur in all aquifers, however, and it is often a decades-long process.
>
> *Biostimulation.* Naturally occurring aquifer bacteria can be managed to very high rates of dehalogenation through the injection of electron donors at rates

that exceed the recharge of electron acceptors. This drives the aquifer microbial communities into sulfate reduction, acetogenesis, and methanogenesis, providing broad-spectrum cometabolic dechlorination. Direct utilization of chlorinated alkenes as alternative electron acceptors is also stimulated among bacterial species in several genera, including *Desulfuromonas*, *Dehalospirillum*, *Dehalococcoides*, *Dehalobacter*, and *Desulfomonile*.[17]

Bioaugmentation. Aquifer microbial communities can be augmented through the injection of large masses of commercially grown bacteria intended to overwhelm the native communities. In laboratory microcosms, these bacteria can achieve dehalogenation rates that exceed those achieved through stimulation of native communities at field scale. The challenge in application of bioaugmentation is to establish tight control over key microbial habitat variables required to elicit high-rate dechlorinating performance.

2.3.1.2 The Aquifer as an Ecosystem

Until recently, the design and engineering of *in situ* bioremediation systems have been performed using methods that were more empirical than theoretical. It should be emphasized that engineered biological processes, whether implemented in an aboveground bioreactor or within an IRZ, are functionally complex ecosystems. As a result, design of wastewater treatment systems was, and still is, largely empirical, relying on equations developed over time based primarily on previous experience. These equations are algebraic and deterministic, whereas much evidence suggests that biological treatment processes are far too complex to permit such simple descriptions.[25] The design basis for engineered microbial IRZs, particularly to achieve complete reductive dechlorination, should account for the basic ecological interactions that influence the performance and successful operation of such systems.

Biological treatment processes can be grouped into three basic categories: (1) processes designed to treat complex, less-defined wastes, such as domestic wastewater; (2) processes that are designed to remove a narrower range of compounds present at comparatively high concentrations, such as industrial wastewaters; and (3) processes that must remove specific compounds at low to moderate concentrations, like most *in situ* bioremediation systems. The type of microbial community present, its relative stability, and how the community might be "engineered" differ for each case.

It is sometimes stated, largely without proof, that most biodegradation in nature occurs via consortial metabolism, that is, by sequential metabolism in which part of the pathway is found in one microorganism and part in others. This is a difficult conjecture to prove or disprove and impossible to evaluate adequately without knowing the enzymological details of biodegradation derived from the various studies conducted with pure cultures. When one wishes to consider the fate of specific compounds in the natural or engineered environments, it is wise to look at the aggregate potential of all of the bacteria and consider the entire microbial community itself as a "super-organism." The microbial community — considered as a super-organism — metabolizes collectively, shares biodegradative genes, and evolves collectively to biodegrade new compounds that enter an environmental niche.[3]

Microbial Reactive Zones

An aquifer is a very complex and dynamic system, teeming with life. Every cubic meter of an aquifer formation is composed of intermingled solids and liquids, populated by a diverse array of flora and, in some cases, fauna. In an unconsolidated aquifer formation, the habitable surface area of particles that comprise the porous medium may reach 1 million square meters for every cubic meter of aquifer. The specific surface of nonswelling clays such as kaolinite range from 10 to 30 $m^2 g^{-1}$ while specific surface for swelling clays such as montmorillonite and vermiculite may reach 800 $m^2 g^{-1}$.[26] These surfaces are populated by organisms that make a living from organic matter flowing past in the groundwater, by consuming organic matter trapped among the mineral particles, or by the oxidation and reduction of inorganic matter that may include the mineral particles, themselves. The soil, the groundwater, and the organisms populating that physical structure comprise the aquifer ecosystem.

The number of microbial species populating an aquifer ecosystem is typically very high. Studies showed up to 4000 distinct genomes in soil samples analyzed for DNA content.[27,28] This was more than 200-fold higher than results for enumeration of phenotypes in isolation plate counts.

Aquifer microbial populations may include representatives from all three of the biological domains: Bacteria, Archaea, and Eucarya. Bacteria include genera such as *Bacillus*, *Clostridium*, and *Pseudomonas*, while Archaea include the methanogens such as those in the genus *Methanosarcina*. Aquifer communities are dominated both in biomass and in numbers by members of the Bacteria and the Archaea. The Eucarya include all of the higher plants and animals, and may be represented in aquifer communities by members of the fungi, the ciliates, and the flagellates.

Bacterial populations in aquifer ecosystems are numerically immense. Groundwater samples often contain 10^9 or more colony-forming units (CFUs) per milliliter, and soil samples may contain more than 10^6 CFUs per gram. CFUs are spores and viable bacterial cells that form colonies when placed in favorable laboratory culture conditions. Most of the bacterial biomass in an aquifer is firmly attached to the soil particles and the groundwater fraction of the population is quite small in comparison. Generation times for bacteria can be very short, ranging from tens of minutes to tens of hours. This makes the aquifer ecosystem very responsive to changes in habitat conditions.

2.3.1.2.1 Aquifer Ecosystem Structure

Within every ecosystem, there is a network of interacting organisms competing for limited resources. Each of these organisms must fulfill four basic needs:

1. An *energy source* to support the endergonic reactions that are entailed in the building of proteins and other molecules. Organisms capture energy from reduction–oxidation reactions they organize between *electron donors* and *electron acceptors*, and store that energy in molecules such as adenosine-5'-triphosphate (ATP).
2. A *carbon source* to build proteins and other molecules. In many cases the electron donor and carbon source are the same. However, chemotrophic species gain energy from reduction–oxidation reactions arranged between

inorganic compounds, so they must gather carbon to build biomass from other sources.
3. *Mineral nutrients* such as nitrogen and phosphorus are required in the formation of many biochemical compounds critical to life. ATP, DNA, and RNA are a few examples.
4. Many organisms require *organic nutrients* that they cannot synthesize, but which are essential to their metabolism. Vitamins such as B_{12} and other cofactors are examples of organic nutrients.

These four components, combined with variables such as temperature, pH, and redox potential, comprise the backbone of the aquifer habitat.

The composition of aquifer microbial communities shifts continuously in response to changes in the physical and chemical habitat. Ecologists use the term succession to describe a predictable progression in the species composition of plant and animal communities in response to a major disturbance of habitat. Species early in the succession prepare the habitat for species that appear later. For terrestrial ecosystems, the succession occurs over time, on a particular piece of real estate. Over many generations, a dynamic equilibrium is achieved. In an aquifer, groundwater flow is a dominant feature of the habitat, and the concept of succession needs to be adjusted somewhat to support analysis of typical aquifer microbial ecology. To make those adjustments, we borrow from concepts of stream ecology.

2.3.1.2.2 The Aquifer Microbial Continuum

In 1980, Robin Vannote and his coworkers published a benchmark study on aquatic ecology titled "The River Continuum Concept."[181] They provided a framework for understanding how communities of organisms are structured in flowing ecosystems. From their analysis, it became clear that stream communities form a continuum of species associations that are structured in response to the carbon and energy sources arriving at any point in the continuum. In the uppermost reaches of a stream, carbon arrives largely unprocessed, as particulate organic matter (in the form of leaves and twigs, for example). Here, the stream communities are dominated by species that perform the first stages of carbon decomposition. In the lower reaches, carbon is highly processed, arriving mostly in the form of dissolved organic matter. At any point in the continuum, the composition of species is determined by the quality and quantity of arriving carbon and energy.

It is very useful to borrow from the River Continuum Concept in building an understanding of the aquifer ecosystem, because the structure and dynamics of an aquifer resemble those of a surface stream ecosystem in many ways: there is a unidirectional flow of water through the system; the biomass is predominantly affixed to solid surfaces and the organisms survive by processing organic matter that arrives in the passing water; organisms are adapted to remain attached to the solid medium as long as the habitat is suitable, but may detach and drift in the flowing water when conditions become adverse; the products of organic carbon processing by organisms at any point in the continuum comprise the carbon source (input) to the organisms occupying a position immediately downstream; availability of oxygen, nitrates, and other electron acceptors at any point in the continuum is determined by the amount passed along from upstream communities and by the rate of recharge from above; the biomass and its composition at any point in the continuum are determined by the rate and nature of the carbon supply, by the rate and nature of electron acceptor supply, and by the rate and composition of mineral nutrient supplies, all arriving from upstream (upgradient).

An aquifer microbial continuum is a succession of microbial community structure, aligned along the groundwater flow path in any aquifer. This succession constitutes a continuum of habitat conditions and species composition. The steady-state microbial community structure at any point in the continuum is a reflection of the inputs received from upgradient communities and through recharge from above (or below). When the inputs remain approximately constant, the microbial community structure will maintain an approximate steady state. The continuum can be disrupted by a change in the inputs at any point. The microbial community structure downgradient from the point of disruption will restructure in response to the changing habitat, and a succession of species may be observed at any point in the continuum as a new steady-state structure develops.

There are many examples of the microbial continuum in natural aquifers. A natural continuum of aquifer metabolism in a confined (closed) system was described about two decades ago.[29] Without the recharge of electron acceptors such as oxygen or nitrate, the groundwater became progressively more reducing. The microbial communities shifted through a succession of metabolic patterns in response to the available electron acceptors. Microbial carbon consumption drove the processes that were observed through the decreasing redox potential.

2.3.1.2.3 Engineering the Aquifer Microbial Continuum

Aquifer microbial communities adapt quickly and predictably to changes in the electron donor and acceptor loadings, and this gives us an opportunity to engineer the continuum to create an IRZ for the destruction of contaminants. When highly degradable organic carbon is injected into the groundwater flow, bacterial biomass increases and is sustained as long as the loading continues. In the simple example shown in Figure 2.12, glucose is added to the flowing groundwater, and the biomass of aerobic bacteria increases in response. Oxygen, the most competitive electron acceptor, diffuses into the aquifer from the overlying vadose zone, and the oxidized carbon leaves the aquifer in the form of carbon dioxide.

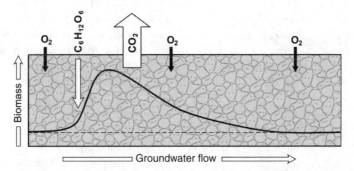

FIGURE 2.12 Response of the aquifer microbial continuum to the addition of a highly degradable electron donor, glucose ($C_6H_{12}O_6$), under aerobic conditions. Biomass rapidly increases above baseline levels and the oxidized carbon leaves the aquifer as carbon dioxide. When the added electron donor has been fully metabolized to carbon dioxide, the biomass returns to baseline levels, which are determined by the background loading of electron donors to the continuum.

Small additions of degradable carbon can be accommodated by the microbial continuum with minimal impact. The increased biomass is observed over a short segment of the continuum, and the associated increase in metabolic oxygen demand is accommodated easily by diffusion from the overlying vadose zone. It is important to note that dissolved oxygen levels in the aquifer decrease and the concentration gradient through the vadose zone increases in magnitude to support the additional demand. Aerobic metabolism of simple sugars, like glucose or fructose, and polysaccharides, like sucrose, can be accomplished within a single organism. The first metabolic phase is aerobic glycolysis, producing pyruvate. The pyruvate then enters the citric acid cycle, yielding carbon dioxide. The entire process yields 38 moles of ATP for every mole of glucose that enters the aerobic glycolysis metabolic pathway. ATP is the molecule in which organisms capture and internally transfer energy gained through metabolism.

The aerobic biomass that can develop in response to electron donor loading is limited by several factors, including space — bacteria occupy space on the aquifer soil matrix, and it is possible for populations to reach space-limited levels; dissolved oxygen — when the bacterial biomass reaches high levels, the diffusion of oxygen from above cannot sustain the electron acceptor demand of the biomass, and the population turns to alternative electron acceptors to support metabolism of the abundant electron donors; product buildup — by-products of metabolism, including carbon dioxide, increase to inhibitory levels.

When electron donor loadings are sufficient to induce consumption of alternative electron acceptors, the composition of the aquifer microbial community undergoes a fundamental shift. Microbial consumption of dissolved organic carbon drives a succession of habitat changes that lead to a continuum of decreasing redox potential and the associated progressive shift of terminal electron acceptors. The microbial communities that develop under high electron donor loads are described in Figure 2.13. These are anaerobic metabolic processes, and the several steps that occur in the processing of electron donors are carried out by distinct groups of organisms (unlike aerobic degradation, which can be completed with competitive energy yields by a single organism).

Many of the metabolic steps in anaerobic respiration are not the classic reduction–oxidation half-cell reactions in which one compound (the electron donor) is oxidized and a second compound (the electron acceptor) is reduced. In anaerobic systems, oxidation and reduction often occurs simultaneously through the action of enzymes on a single compound, a process known to chemists as disproportionation and to biologists as fermentation. The anaerobic metabolism that can be sustained by fermentation and by oxidation–reduction reactions that consume alternative electron acceptors yields much less energy than oxidative metabolism. Under anaerobic conditions, the oxidation of glucose yields only 2 moles of ATP per mole of glucose, clearly not a competitive strategy when oxygen is present.

The metabolic cascade depicted in Figure 2.13 provides numerous pathways for the anaerobic decomposition of electron donors along the aquifer microbial continuum. Reactions in the upper portion of the cascade consume electron acceptors, leaving reactions in the lower parts of the cascade to be supported by other, potentially less energy yielding, electron acceptors. The sequence of electron acceptors, ranked in approximate order of decreasing energy yield per reaction, is O_2, NO_3^-, Mn^{4+}, Fe^{3+}, SO_4^{2-}, and CO_2.

Microbial Reactive Zones

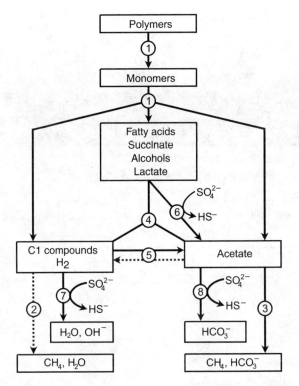

1. Primary fermenters
2. Hydrogen-oxidizing (hydrogenotrophic) methanogens
3. Acetate cleaving (acetoclastic) methanogens
4. Secondary fermenters (syntrophic)
5. Homoacetogenic bacteria
6. Group I sulfate reducers: non-acetate oxidizers
7. Facultative lithotrophic sulfate reducers
8. Group II sulfate reducers: acetate-oxidizers

FIGURE 2.13 Major metabolic pathways in the aquifer microbial continuum, when it is engineered to load electron donors at a rate that exceeds electron acceptor recharge. Dashed-line pathways (2 and 5) decrease in importance at temperatures below 20°C.[32]

The metabolic process occurs in several steps and, at any point in the cascade, the most likely next reaction is determined by the concentration of donors and acceptors, and the energy that competing organisms can gain from the reaction. The reaction that gains the most energy at any point in the cascade is thought to be the most competitive, therefore most likely to occur. For example, in Figure 2.13, hydrogenotrophic methanogens compete with lithotrophic sulfate reducers for hydrogen as an electron donor (these organisms will also require a source of carbon for growth). When hydrogen supplies are limited, the sulfate reducers are expected to out-compete the methanogens, because they gain more energy from the oxidation of hydrogen (they use a more energetic electron acceptor). It is important to note that when hydrogen is abundant, sulfate reducers and methanogens can coexist — they may compete for other resources, but the energy disadvantage of CO_2 as an electron acceptor is not a major factor.

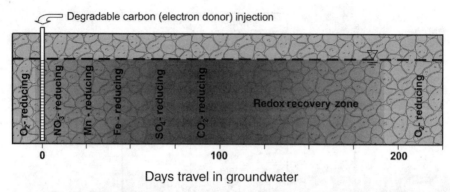

FIGURE 2.14 Zonation of dominant metabolic processes in the aquifer microbial continuum, when electron donor loading is sustained at levels that exceed electron acceptor recharge.

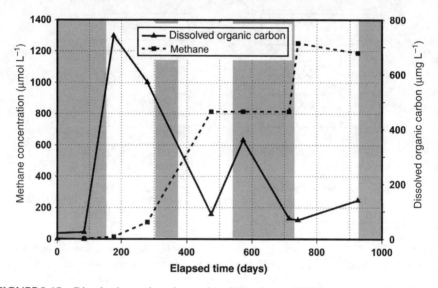

FIGURE 2.15 Dissolved organic carbon and methane observed 100 days downgradient from a biostimulation injection zone, where readily degradable organic carbon (molasses) was injected into the aquifer. During carbon loading periods (gray bands), injections were performed on a monthly basis at a rate sufficient to produce an average injection zone concentration of 6000 mg dissolved organic carbon (DOC) per liter.

When electron donor loading is sustained at high levels, a zonation of dominant metabolic processes forms in the continuum, each zone progressively more reducing, until the electron donor supply is exhausted. This is shown in Figure 2.14.

Based on field experience in aquifers flowing at velocities greater than 15 cm day^{-1}, the microbial continuum responds very quickly to the addition of highly degradable carbon. At a site where injection zone loading was maintained at approximately 6000 mg L^{-1} dissolved organic carbon (DOC), the DOC measured at a location 100 days downgradient from the injection zone spiked at the 6-month measurement, then steadily declined in time, as shown in Figure 2.15.

Microbial Reactive Zones

Many phylogenetic classifications of bacteria participate in the aerobic carbohydrate metabolism and the carbohydrate fermentation that occurs in the upgradient portions of an ERD zone. However, the portion of the reactive zone that carries out reductive dechlorination is populated by only a few metabolic classifications of bacteria. These groups behave very differently from one another, and include methanogens, sulfate-reducing bacteria, and dehalorespiring bacteria. They are discussed in greater detail below.

Ecological bacterial associations between dechlorinating microorganisms and other members of the microbial community play an important role in reductive dechlorination. This close association, called syntrophism, makes the isolation of the responsible organisms more difficult. Syntrophism is a common ecological interaction in anaerobic microbial communities, possibly an adaptation to the small enthalpy changes associated with anaerobic biotransformations.

The development of methanogenic conditions is also depicted in Figure 2.15. As will be discussed in a later section, the IRZs achieve their best reductive dechlorination performance when they reach methanogenesis. There may be other indicators of high-rate reductive dechlorination as well, linked to reactions that occur early in the metabolic cascade.

The consumption of carbon by the aquifer microbial continuum increases with time, as microbial biomass builds. This consumption increase can be observed in the decreasing half-life of dissolved organic carbon in the aquifer. Figure 2.16 shows the progression of dissolved organic carbon half-life values measured in a reductive dechlorination treatment. Prior to injection of highly degradable carbon into the aquifer (molasses, in this case), there was no net degradation of dissolved organic

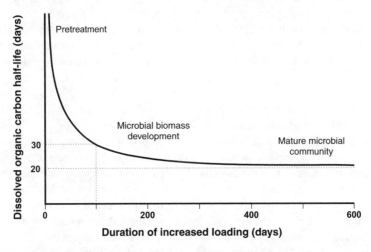

FIGURE 2.16 Response of the microbial community to increased dissolved organic carbon loading. In its pretreatment condition, dissolved organic carbon consumption is balanced by recharge, and the apparent half-life is extremely long. When dissolved organic carbon loading is increased dramatically, microbial biomass increases, and the half-life decreases rapidly. When the injection zone DOC is maintained at 6000 mg L^{-1}, the DOC half-life decreases to 20 days within the first year of operation.

carbon, yielding an apparently "infinite" half-life. In fact, the carbon consumption is balanced by recharge in this section of the aquifer, so there is no net degradation. When degradable carbon injections began, the apparent half-life rapidly decreased to 35 days. This was due to the buildup of aquifer bacteria that consumed the injected carbon through a wide range of metabolic pathways.

As injections continued, the dissolved organic carbon half-life declined slowly to approximately 20 days, a carbon consumption rate that has now been sustained for more than 400 days. The carbon consumption rate constant will vary from site to site.

There are many practical constraints on the engineering of the aquifer microbial continuum including:

- *Groundwater flow velocity.* The zonation depicted in Figure 2.14 depends on the movement of groundwater to spread the microbial succession over a long distance in the aquifer. At low groundwater flow velocities, the succession must occur in a fixed space, one species replacing another in time, at any particular location.
- *Aquifer alkalinity.* Many of the fermentation products are organic acids that can lower groundwater pH if the aquifer's buffering capacity is too low. There is much greater design latitude when the aquifer alkalinity is high (in excess of 100 mg L^{-1} as $CaCO_3$).
- *Aquifer microbial populations.* The aquifer must contain a sufficient bacterial population to perform the needed metabolic transformations, including groundwork reactions among the primary fermenters, and functional dechlorinating reactions among the secondary fermenters, sulfate reducers, methanogens, and acetogens. Microbial populations in near-surface aquifer formations have been sufficient to drive commercial scale–enhanced reductive dechlorination at each of more than 250 sites undertaken by the authors to date. The risk of inadequate bacterial populations is more likely at locations with low background microbial counts, such as might occur at great depths in aquifers with low organic recharge.

The most critical aspect of engineering the aquifer microbial continuum is achieving reductive dechlorination of alkenes, alkanes, and other organic contaminants in an aquifer. This is accomplished by encouraging microbial population shifts that favor organisms capable of reductive dechlorination. Fortunately, there are many organisms that can accomplish this task, located throughout the anaerobic metabolic cascade (refer to Figure 2.13).

The early part of the reductive continuum (corresponding to the upper portion of the anaerobic metabolic cascade) is populated by organisms that can function with oxygen or alternative electron acceptors. These species are termed facultative, and there are examples of reductive dechlorination occurring as a result of facultative bacteria using PCE as an alternative electron acceptor (*Enterobacter agglomerans*).[30]

Reductive dechlorination reactions have also been observed in *Clostridium bifermentans*.[31] *Clostridium* is a bacterial genus well known as a secondary fermenter in the anaerobic metabolic cascade (Figure 2.13), producing ketones as a product. Case study data show the accumulation of ketones (acetone and 2-butanone) in high-rate reductive dechlorination zones. This may be a signature of the beneficial involvement of *Clostridium*.

The most highly studied dechlorinating species occur late in the metabolic cascade. This is likely an artifact of the processes used to search for dechlorinating species, in which cultures are fed substrates far down the anaerobic metabolic cascade. This limits the search to species in the sulfate-reducing and methanogenic portion of the continuum. Methanogens, acetogens, and sulfate-reducing species all have been shown to participate in dechlorination of ethenes. Highly specialized bacterial species, which are totally reliant on chlorinated ethenes as well as unidentified consortium partners that supply key nutrients or cofactors, have also been discovered in the lower portion of the cascade.

At low to moderate carbon loading rates, fermentation reactions result primarily in the formation of C1 compounds and acetate.[32] When carbon loading is pushed to high rates, the primary fermenters deliver large quantities of fatty acids, feeding the central portion of the anaerobic metabolic cascade (Figure 2.13), enabling the secondary fermenters. This is the portion of the cascade inhabited by *Clostridium*, which can be a productive dechlorinator.[33]

Case studies in PCE- and TCE-contaminated aquifers show that the aquifer microbial continuum can be manipulated, through injection of highly degradable carbon, to carry out dechlorination reactions in a predictable and highly effective manner. Engineering of the aquifer microbial continuum to create an IRZ is now a frontline remedial technology, displacing the mass removal technology, aquifer sparging, as the presumptive remedy for many cleanup situations.

2.3.1.2.4 Methanogens

Methanogens are obligate anaerobic bacteria that are capable of reducing carbon dioxide to methane. Many methanogens use hydrogen as their electron donor while reducing carbon dioxide. These bacteria are called "autotrophic," since they use both an inorganic electron donor and acceptor for metabolism. Some species of methanogens can also reduce formate, carbon monoxide, methanol, various amines, and acetate to methane. Many of these methanogenic electron acceptors are produced within an enhanced reductive dechlorination system via fermentation.

Methanogens are also capable of degrading CAH compounds, although this process does not yield any metabolic benefit for the bacteria. This unproductive (for the bacteria) and fortuitous degradation is called "cometabolic." Each reductive dechlorination event leads to the release of a chloride ion resulting in the formation of a lesser chlorinated daughter product. Because of the fortuitous nature of each dechlorination event, the probability of a subsequent dechlorination event, and therefore the overall rate of reductive dechlorination, can vary with the number of chlorinated constituents present.

Examples of syntrophic association of dechlorinating bacteria within a methanogenic microbial consortium have been reported.[34] The dechlorinating bacteria *Desulfomonile tiedjei* was speculated to act as a scavenger relying on electron donors and vitamins from other organisms. Species also capable of cometabolic dechlorination include *Methanosarcina thermophila*, *M. mazei*, which can dechlorinate PCE to TCE, and *Methanobacterium thermoautotrophicum* and

Methanothrix soehngenii, both capable of reductively dechlorinating *cis*-DCE to chloroethane.[24,35]

Methanogens reductively dechlorinate compounds such as tetra- and trichloromethane, 1,2-DCA, and tetrachloroethene. Attempts to link the reductive dechlorination activity found in homoacetogens and methanogens with corrinoid-containing enzyme systems has not led to conclusive results. In some cases, strong indications of a link between reductive dechlorination and corrinoid-containing enzymes have been obtained, whereas other studies have indicated the absence of such a link. The purified corrinoid containing carbon monoxide dehydrogenase of *M. thermophila*, an acetoclastic methanogen, has been reported to dechlorinate TCE, mainly to *cis*-DCE and to traces of VC and ethane.[35]

2.3.1.2.5 Sulfate-Reducing Bacteria

Microorganisms that use sulfate as a terminal electron acceptor are called sulfate-reducing bacteria (SRBs). SRBs produce sulfide as a waste product while obtaining electrons from molecules such as alcohols or organic acids, both of which are end-products of carbohydrate fermentation.

$$SO_4^{2-} + 4H_2 + H^+ \rightarrow HS^+ + 4H_2O$$

Several anaerobic bacteria belonging to the genera *Desulfovibrio* and *Desulfotomaculum* can utilize SO_4^{2-} as the ultimate electron acceptor of energy metabolism. Sulfate-reducing bacteria have been shown to grow on H_2 and sulfate as the sole energy source.

Dehalorespirers (such as *Dehalococcoides ethenogenes* and *Dehalobacter restrictus*), strict anaerobes that gain energy from electron transfer to the chlorinated aliphatics, such as PCE, generally occupy a distinct niche in the middle of the redox range. The dehalorespiring microorganisms are flanked by iron- and sulfate-reducing bacteria at the positive and negative ends of their redox range, respectively.

Many designers of anaerobic IRZs have relied on SRBs and sulfide production during ERD implementation for the precipitation of certain metals. However, several of the SRB species can also be useful in the cometabolic reduction of CAHs. *Desulfitobacterium frappieri* and *D. tiedjei* are capable of degrading PCE to *cis*-DCE.[36,37] *Desulfitobacterium chlororespirans* has been shown to degrade other CAHs such as 3-chloro-4-hydroxybenzoate.[36] The idea of SRBs being capable of reductive dechlorination has been widely accepted (Table 2.1).

2.3.1.2.6 Fermentation

The production of H_2 by different microorganisms is intimately linked with their respective energy metabolisms. The production of H_2 is one of the specific mechanisms to dispose excess electrons through the activity of hydrogenase present in H_2-producing microorganisms.[38,39] Production of H_2 by obligate anaerobic microorganisms has optimum stoichiometry (1 : 4, with glucose as substrate) compared with facultative anaerobes (1 : 2), although the latter process is comparatively simpler than the former.[38]

Under natural conditions, fermentation is the process that generates the hydrogen used in reductive dechlorination. In the absence of externally available electron acceptors, many organisms perform internally balanced (different portions of the same substrate are oxidized and reduced) oxidation–reduction reactions of organic compounds with the release of energy, and this process is called *fermentation*. Since only *partial* oxidation of the carbon atoms of the organic compound occurs, fermentation yields substantially less energy per unit of substrate compared to oxidation reactions. (Oxidation reactions occur in those external electron acceptors that participate in the reaction.) For instance, the fermentation of glucose to ethanol and CO_2 has a theoretical energy yield of -57 kcal mol^{-1}, enough to produce about 6 ATPs. However, only 2 ATPs are produced, which implies that the organism operates at considerably less than maximum efficiency.[38]

In any fermentation reaction, there must be a *balance* between oxidation and reduction. In a number of fermentation reactions, electron balance is maintained by the production of molecular hydrogen, H_2. In H_2 production, protons (H^+) of the medium, derived from water, serve as electron acceptors. The energetics of hydrogen production are actually somewhat unfavorable, so that most fermentative organisms only produce a relatively small amount of hydrogen along with other fermentation products. Fermentation reactions that have pyruvate as an intermediate product have the potential of producing more H_2. Conversion of pyruvate to acetyl-CoA is an oxidation process and the excess electrons generated must either be used to make a more reduced end-product, or can be used in the production of H_2.

Fermentation by bacteria can also be important in controlling the biogeochemical environment of anaerobic aquifers. Bacterial fermentation can be divided into two categories:[6,40]

Primary fermentation: The fermentation of primary substrates such as sugars, amino acids, and lipids to yield acetate, formate, CO_2, and H_2, also yields ethanol, lactate, succinate, propionate, and butyrate. While primary fermentation often yields H_2, production of H_2 is not required for these reactions to occur.

Secondary fermentation or coupled fermentation: The fermentation of primary fermentation products such as ethanol, lactate, succinate, propionate, and butyrate, yielding acetate, formate, H_2, and CO_2. Bacteria that carry out these reactions are called *obligate proton reducers* because the reactions must produce hydrogen in order to balance the oxidation of the carbon substrates. These secondary fermentation reactions are energetically favorable only if hydrogen concentrations are very low (10^{-2} to 10^{-4} atm or 8000 to 80 nM dissolved hydrogen, depending on the fermentation substrate). Thus these fermentation reactions occur only when the produced hydrogen is utilized by other bacteria, such as methanogens that convert H_2 and CO_2 into CH_4 and H_2O. The process by which hydrogen is produced by one strain of bacteria and utilized by another is called *interspecies hydrogen transfer*. It should be noted that the terminal products of anaerobic decomposition, CH_4 and CO_2, respectively, are the most reduced and the most oxidized carbon compounds.[6,40]

There are a number of compounds that can be fermented to produce hydrogen (Figure 2.17). Within enhanced reductive dechlorination systems, two distinct processes are needed in order for dehalorespiration to occur: fermentation to produce an electron donor and uptake of the electron donor by the dehalorespiring bacteria. During primary or secondary fermentation, organic compounds are transformed to compounds such as

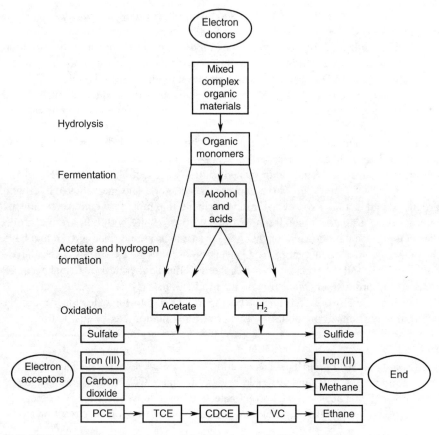

FIGURE 2.17 Steps in the process of biodegradation of PCE by reductive dechlorination. As shown, biodegradable organic matter is required as an electron donor to initiate the process. Different types of microbes are involved at each stage. The bottom step shows that PCE must compete for electrons with sulfate, iron, and carbon dioxide, meaning that a large amount of organic electron donors may be needed to supply enough electrons. *Note*: CDCE = *cis*-dichloroethene. (After McCarty, 1997.[188])

acetate, water, carbon dioxide, and dissolved hydrogen. Fermentable substrates can be biodegradable, nonchlorinated contaminants (i.e., BTEX — benzene, toluene, ethyl benzene, and xylenes), naturally occurring organic carbon, or added electron donors (i.e., sugar). Once fermentation has occurred, the dehalorespiring microbial consortia can utilize the hydrogen produced by fermentation for reductive dechlorination.

There is mounting evidence for dechlorination by hydrogenotrophic (hydrogen-utilizing) bacteria other than *Dehalococcoides ethenogenes* strain 195. At the time of this writing, several cultures contain *Dehalococcoides* or *Dehalococcoides*-like bacteria that are hydrogenotrophic dechlorinators, and among these are populations that dechlorinate *cis*-DCE and VC but not PCE. There are also indications of non-*Dehalococcoides* dechlorination of VC, although no new species have been isolated.

During the application of ERD technology, a highly reduced biogeochemical environment is created throughout the treatment zone. In addition, this zone contains an excess of organic carbon. Aquifer parameters, such as low permeability (10^{-5} cm s^{-1} or less), or low alkalinity (low buffering capacity), can result in the formation of excess organic acids in the groundwater via fermentation. Optimal sulfate reduction and methanogenesis will occur at pH values ranging from 6 to 8. Low pH conditions (<5) are detrimental to sulfate-reducing and methanogenic bacteria. However, fermentative organisms favor lower pH conditions, and therefore will outcompete both sulfate-reducing and methanogenic bacteria in this environment. This can result in the formation of low-molecular-weight organic by-products of fermentation, such as ketones, alcohols, and aldehydes. Fermentation reactions can be minimized by avoiding development of low pH zones, through balancing mineralogy, buffering capacity, groundwater flow velocity, and injection rate into the aquifer. Total organic carbon (TOC) and pH in groundwater should be sampled periodically, and if these or other indicators suggest the electron donor solution is overloading the system, the injection rate is reduced or a buffer introduced. Some fermentation is a necessary and natural part of ERD, however, control of the injection area is necessary because too much fermentation can be detrimental to ERD.

2.3.1.2.7 The Role of Hydrogen and Competition for Molecular Hydrogen

In environments where hydrogen is the most important electron donor for the dechlorination of chlorinated solvents, competition for the uptake of hydrogen between different types of microorganisms, such as methanogenic, homoacetogenic, sulfidogenic, and dechlorinating bacteria, becomes important. In several studies, it has been claimed that dechlorinating organisms have a higher affinity for molecular hydrogen than methanogens.[24,34,41] This indicates that the dechlorinating organisms are able to survive at lower hydrogen levels, but will possibly be outcompeted by other microorganisms when elevated hydrogen levels are present. These studies suggest that a more effective dechlorination may be achieved by using an electron donor such as propionate or butyrate, which generates low hydrogen concentrations during its fermentation. The speculation was that this would then create more favorable conditions for dechlorinating bacteria than for hydrogen-consuming methanogens.[24,34]

Recent studies have suggested that the type of substrate and the rate of fermentation may not have an impact on reductive dechlorination, and this has led to a new school of thought. A recent study showed the ability of four fermentable substrates to sustain PCE dechlorination long term (i.e., approximately 4 months).[34] The choice of organic substrates was based upon their rates of fermentation and the H_2 partial pressures that could be developed and maintained. Despite the difference in the resulting H_2 partial pressures (ranging approximately 1×10^{-5} to 3×10^{-3} atm), no long-term effect on dechlorination was observed. This result may indicate that either low H_2 partial pressures were not required to maintain a competitive dechlorinating community or several isolated PCE respiring bacteria do not utilize H_2 as an electron donor.[31,42] H_2 was not the source of PCE-reducing equivalents in all systems tested. Other laboratory and field studies have also suggested that the steady-state concentration of hydrogen is controlled by the type of bacteria utilizing the hydrogen and is almost completely independent of the rate of hydrogen production.

Although H_2 is a waste product of fermentation, it is a highly reduced molecule, which in turn makes it an excellent, high-energy electron donor. In this symbiotic relationship, the hydrogen-utilizing bacteria gain a high-energy electron donor, while for the fermenters the removal of hydrogen allows continuous fermentation to be energetically favorable.

In addition to methanogens, a wide variety of bacteria can utilize hydrogen as an electron donor: denitrifiers, Fe(III) reducers, sulfate reducers, and halorespirators. As discussed earlier, for dechlorination to take place, halorespirators must successfully compete against all these hydrogen utilizers.

Recently researchers have found that steady-state H_2 concentrations in the field are controlled by the type of bacteria utilizing the hydrogen.[16] For example, under nitrate-reducing concentrations, steady-state H_2 concentrations were less than 0.05; under Fe(III)-reducing conditions they were less than 0.2 to 0.8 nM; under sulfate-reducing conditions they were 1 to 4 nM; and under methanogenic conditions they were 5 to 14 nM. Thus it is clear that an increased rate of hydrogen production will result in increased halorespiration without affecting the competition between various bacteria for the available hydrogen. Attempting to stimulate halorespiration with poor fermentation substrates, as suggested in the past, may unnecessarily limit the amount of dechlorination taking place.

Microorganisms that can use H_2 as an electron donor are known as hydrogenotrophic bacteria. The importance of hydrogen utilization in ERD systems is becoming increasingly apparent. The take-home lesson from this discussion is clear: an increased rate of hydrogen production will result in increased halorespiration without affecting the competition between various bacteria for the available hydrogen. Attempting to stimulate halorespiration with poor fermentation substrates may unnecessarily limit the amount of dechlorination taking place. Attempts to limit hydrogen concentration in practical heterotrophic field systems may result in significant portions of the targeted zone not reaching sufficiently reducing conditions for optimum treatment, which can result in sites "stalling" at *cis*-DCE and VC. This "stalling" effect, caused by poor reducing conditions, has been attributed erroneously in the past to the apparent absence of *D. ethenogenes* within the reactive zones.

2.3.1.3 Biochemistry of Reductive Dechlorination

All organisms live by incorporating carbon into lasting physical structures such as leaves, roots, and seeds. Many organisms have an ability to move. The energy to build those structures and to power movements is captured through chemical reactions at the cellular level. These reactions are broadly described as the organism's metabolism, and the principal metabolic reactions are oxidation–reduction reactions.

In reduction–oxidation reactions, two compounds are transformed simultaneously — one is reduced (gains electrons) and the other is oxidized (loses electrons). In fermentation reactions, one compound undergoes both oxidation and reduction — a process called disproportionation. The compound that is reduced is called the electron acceptor (or oxidant) and the compound that is oxidized is termed the electron donor (or reductant). In aerobic systems, oxygen is typically the electron acceptor. In reductive dechlorination, chlorinated organic compounds such as perchloroethene accept electrons and are reduced, losing a chlorine atom and gaining a proton (hydrogen ion).

There are two important questions we can ask about any prospective chemical reaction. Does the reaction yield energy or consume energy? How fast does the reaction proceed?

The first question is one of chemical thermodynamics. In that analysis, we can determine whether any particular reaction can be profitably performed by an organism. In a profitable metabolic reaction, sufficient energy is released for the organism to capture it and harness it for structure, locomotion, or other energy demand of life.

The second question is one of chemical kinetics, and for organisms, the kinetics of interest are enzyme kinetics. Enzymes are large protein molecules that facilitate metabolic reactions by lowering energy thresholds for reactions that yield energy. Put simply, enzymes bind to reactants and force them into proximity so reactions can occur. The kinetics of the binding and reaction processes determine how fast a reaction proceeds.

Reductive dechlorination may also occur in unproductive reactions that are termed cometabolic. In some cases, cometabolic reactions can be maximized by controlling the aquifer microbial continuum to favor particular species or groups.

2.3.1.3.1 Chemical Thermodynamics

The maintenance and growth of any bacterial population is founded on its ability to extract energy from the chemical reaction of electron donors and acceptors. These reactions can be described in terms of the enthalpy or heat consumed or released and the change in entropy entailed in the conversion of reactants to products. This total energy change is termed the Gibbs free energy, and is defined more completely later. Reactions that lead to a net release of energy are termed exergonic, while reactions that require a net input of energy are termed endergonic. Calculations of the Gibbs free energy change for metabolic reactions can be used to determine whether reactions are feasible and whether organisms gain sufficient energy to sustain growth.

Among aerobic organisms, electron donors such as glucose are oxidized by oxygen, yielding a substantial energy gain. Anaerobic organisms also may utilize glucose as a food source, but the electron acceptors that are available do not provide the same high energy yield obtained in aerobic metabolism.

Gibbs free energy is defined by the change in enthalpy (H) and entropy (S) in a reaction, and can be calculated by the van't Hoff equation.

$$\Delta G = \Delta H - T\Delta S \quad (2.2)$$

The change in Gibbs energy for a reaction is calculated by the difference between the Gibbs energy of products and reactants:

$$\Delta G = \sum G^0_{products} - \sum G^0_{reactants} \quad (2.3)$$

where G^0 are values for the free energy of formation for individual compounds, referenced to a standard physical–chemical state. Values of ΔG for a reaction can be classified according to the following.

1. *Endergonic reactions.* Reactions which are energy consuming, as measured by an increase in Gibbs free energy, are termed endergonic. Endergonic reactions cannot occur spontaneously. Metabolic reaction sequences may include endergonic steps, but the overall reaction must be exergonic to proceed.
2. *Exergonic reactions.* Reactions which are energy releasing, as measured by a decrease in Gibbs free energy, are termed exergonic. Exergonic reactions may occur spontaneously.
3. *Energy-conserving reactions.* Energy-conserving reactions is a subset of the exergonic reactions. They comprise reactions with energy yield above the ATP formation threshold of approximately 20 kJ mol^{-1}.[32]

The metabolic energy yield of glucose consumption can be calculated under aerobic and anaerobic conditions using Equations (2.4) and (2.5) and values for Gibbs free energy of formation from Table 2.2.

TABLE 2.2
Values for Gibbs Free Energy of Formation for Selected Compounds

Compound	ΔG_f^0 (kJ mol^{-1})
O_2 (g)	0
CH_4 (g)	−50.53
CO_2 (g)	−394.36
H_2 (g)	0
N_2 (g)	0
H^+ (at pH 0)	0
H^+ (at pH 7)	−40.01
OH^- (aq)	−157.2
H_2O (aq)	−237.14
HCO_3^- (aq)	−586.8
NO_3^- (aq)	−111.3
NO_2^{2-} (aq)	−32.2
SO_4^{2-} (aq)	−744.5
H_2S (g)	−33.4
HS^- (aq)	12.1
Fe^{3+} (aq)	−4.7
Fe^{2+} (aq)	−78.9
Cl^- (aq)	−131.2
CH_2CH_2 (aq)	81.43
$CHClCH_2$ (aq)	59.65
$CHClCHCl$ (aq)	27.8
CCl_2CHCl (aq)	25.41
CCl_2CCl_2	27.59
$C_6H_{12}O_6$ (s)	−910.6

Source: Data from References 31, 48, 93, and 182.

Aerobic: $C_6H_{12}O_6(s) + O_2(g) \rightarrow 6CO_2(g) + 6H_2O(g) \quad \Delta G = -2879 \text{ kJ mol}^{-1}$ (2.4)

Anaerobic: $C_6H_{12}O_6(s) \rightarrow 3CO_2(g) + 3CH_4(g) \quad \Delta G = -425 \text{ kJ mol}^{-1}$ (2.5)

This comparison shows the limited energy available to anaerobes and explains the typically lower biomass observed in anaerobic systems when electron donors are abundant.

For an organism to survive on the metabolism of any electron-donor–electron-acceptor pair, the organism must be capable of conducting its reduction and oxidation reactions in an energy-conserving manner. Anaerobic metabolism has limited energy gain potential, as shown in the comparison above, and the use of Gibbs free energy in determining whether any reaction system supports growth in chemotrophic anaerobes can be outlined.[43] Researchers have recognized the energy required for the formation of ATP as a threshold value, below which energy released from metabolic reactions cannot be captured by an organism:

$$ADP + P_i \rightarrow ATP \quad \Delta G = +60 \text{ kJ mol}^{-1} \quad (2.6)$$

where ADP is adenosine diphosphate and P_i is inorganic phosphate.

All organisms utilize ATP to store and transfer energy captured in metabolic reactions. Bacteria consume 20 mmol of ATP in the production of 1 g of biomass.[43] They also estimated that the formation of ATP from ADP and inorganic phosphate consumes between 42 and 50 kJ mol^{-1} when the ATP-generating process is fully reversible. The ATP formation energy sets a threshold energy yield for a metabolic reaction to be energy conserving.

Adding a 20 kJ mol^{-1} heat loss estimate to the energy demand of ATP formation, it was estimated that 70 kJ mol^{-1} is required for ATP formation, 60 kJ mol^{-1} under the best circumstances.[32] The author also updated the notion of the minimal energy gain to support growth: ATP synthesis is driven by active (energy-demanding) transport of protons across cells or mitochondrial membranes. Because three protons must pass the membrane for every ATP formed, the minimum energy gain from catabolic metabolism is 20 kJ mol^{-1}. Smaller free energy changes cannot be "captured" in support of bacterial growth.

2.3.1.3.2 Corrections for Nonstandard Conditions

Calculations of Gibbs free energy by Equation (2.2) describe the thermodynamics of a reaction at standard physical–chemical conditions (pH = 0, each constituent at 1 M or 1 atm, 298 K). However, standard conditions are substantially different from typical aquifer conditions, and laboratory studies have generally been conducted at temperatures and substrate concentrations far above those observed in field conditions. For example:

> The dechlorination metabolism of *Dehalococcoides ethenogenes* has been studied at 35°C.[39,42] This temperature is dramatically higher than aquifer temperatures at most

industrial sites. In the U.S. midwest, for example, aquifer temperatures range from 10 to 15°C. Can *D. ethenogenes* accomplish the same metabolic reactions at 15°C?

Many studies of dechlorination in methanogenic cultures are conducted at 30 to 35°C. Can these cultures perform comparable dechlorination at 10°C?

There is no absolute frame of reference for free energy so all measurements are made relative to a reference state. For physical chemistry studies, the reference state is 298 K, at 1 atm pressure, with each of the chemical constituents at 1 M or 1 atm. For biochemical reactions, the defined standard state conditions for Gibbs free energy includes pH = 7 and temperature = 25°C (298 K). Values of Gibbs free energy calculated relative to the biochemical reference state are denoted $\Delta G^{0\prime}$. The Gibbs free energy of formation for elements in their most stable form is zero.

The Gibbs free energy for a reaction at nonstandard conditions is described by Equation (2.7), where A and B are reactants and C and D are products. The free energy change is determined by the standard free energy of the reaction (calculated by Equation (2.5)), the reaction temperature, reaction stoichiometries, and the concentrations of products and reactants, as shown in Equation (2.7).

$$\Delta G = \Delta G^0 + RT \ln \left[\frac{[C]^c [D]^d \ldots}{[A]^a [B]^b \ldots} \right] \qquad (2.7)$$

Table 2.3 shows free energies of reaction for several dechlorination reactions, as well as key metabolic reactions of the aquifer microbial community. The standard free energy of each reaction, $\Delta G^{0\prime}$, was calculated using Equation (2.7), using values for free energy of formation of reactants and products given in Table 2.2.

Data for dechlorination reactions under nonstandard conditions show that energy yields exceed the ATP formation minimum of 20 kJ mol^{-1} under most temperature and pH conditions examined. The smallest energy gains are associated with dechlorination of *cis*-DCE and the largest with dechlorination of PCE, for any specified physical–chemical conditions.

Thermodynamic analysis shows that each of the chlorinated alkenes in the PCE-to-ethene degradation sequence can be used as alternative electron acceptors with energy yields far above the ATP formation threshold. Under nonstandard conditions, including low H_2 levels, low temperatures, and low pH, the reactions remain strongly exergonic. This shows that the differences in utilization of these compounds as electron acceptors are more likely related to enzyme kinetics than to chemical thermodynamics.

The ΔG^0 values for the reductive dehalogenation of halogenated aliphatic compounds fall into the same range as those for aromatics. For both types of compounds, microorganisms are known to grow dependent on the hydrogenolysis of the carbon–chlorine bond. The amount of energy released upon defluorination of chlorofluorocarbons is less than the amount available from dechlorination. Because C–Cl bonds generally provide a greater net energy gain than C–F bonds, the preferred transformation route of chlorofluorocarbons under anaerobic conditions is dechlorination rather than defluorination. The amount of energy available per dechlorination step is fairly constant. It is therefore unclear as to why reductive dechlorination of highly chlorinated aliphatics, such as PCE, appears easier and proceeds faster than VC (Figure 2.18).

TABLE 2.3
Gibbs Free Energy Yield for Reactions in Standard Biochemical Conditions (298 K, 1 atm, pH 7) and at Four Nonstandard Conditions (a through d) of Interest in Aquifer Formations

Reaction	$\Delta G^{0'}$ kJ mol^{-1}	ΔG^a kJ mol^{-1}	ΔG^b kJ mol^{-1}	ΔG^c kJ mol^{-1}	ΔG^d kJ mol^{-1}
Reductive Dechlorination Reactions					
PCE to TCE					
$C_2Cl_4 + H_2 \rightarrow C_2HCl_3 + H^+ + Cl^-$	−173	−162	−156	−171	−150
TCE to cis-DCE					
$C_2HCl_3 + H_2 \rightarrow C_2H_2Cl_2 + H^+ + Cl^-$	−169	−157	−152	−167	−145
cis-DCE to VC					
$C_2H_2Cl_2 + H_2 \rightarrow C_2H_3Cl + H^+ + Cl^-$	−139	−128	−122	−137	−115
VC to ethene					
$C_2H_3Cl + H_2 \rightarrow C_2H_4 + H^+ + Cl^-$	−149	−138	−132	−148	−126
Native Microbial Metabolism					
Hydrogenotrophic methanogenesis					
$CO_2 + 4H_2 \rightarrow CH_4 + 2H_2O$	−130	−130	−62	−130	−130
Acetoclastic methanogenesis					
$CH_3COO^- + H_2O \rightarrow CH_4 + HCO_3^-$	−31	−31	−31	−31	−31
Lithotrophic sulfate reduction					
$SO_4^{2-} + 4H_2 + H^+ \rightarrow HS^- + 4H_2O$	−152	−163	−83	−154	−94
Acetate-oxidizing sulfate reduction					
$CH_3COO^- + SO_4^{2-} \rightarrow HS^- + 2HCO_3^-$	−48	−48	−48	−48	−48
pH	7	5	7	7	6
H_2 (atm)	1.0	1.0	0.001	1.0	0.001
Temperature (°C)	25	25	25	10	10
PCE (M)	1	1	1	1	1
TCE (M)	1	1	1	1	1
cis-DCE (M)	1	1	1	1	1
VC (M)	1	1	1	1	1

2.3.1.3.3 Enzyme Kinetics

Reductive dechlorination reactions are controlled by enzymes and enzyme cofactors that are produced by aquifer bacteria, and the rate at which reactions occur is determined by the availability of enzymes and cofactors and by the kinetics of their reactions with the halogenated organics.

Enzymes are proteins that bind reactants in metabolic processes and bring the reactants into close proximity in a spatial orientation that maximizes the probability that a reaction will occur. Enzymes cannot overcome thermodynamic barriers to reactions — in other words, enzymes cannot make an energy-consuming (endergonic) reaction into an energy-yielding (exergonic) reaction. They can, however, dramatically increase the rate at which a reaction occurs by lowering the activation energy barrier that may prevent an exergonic reaction from occurring at a meaningful

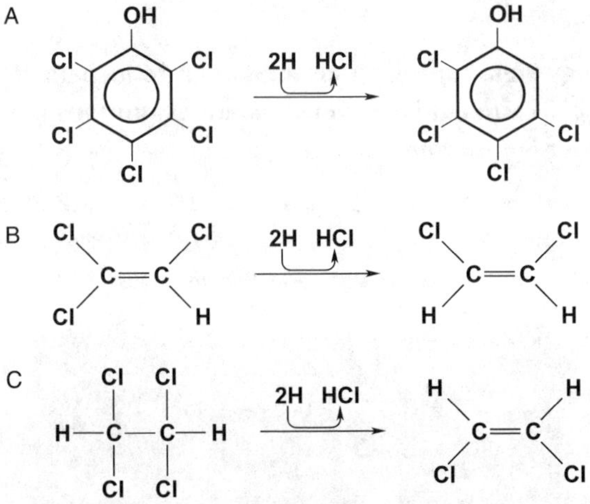

FIGURE 2.18 Examples of three types of reductive dechlorination: (A) hydrogenolysis of pentachlorophenol to tetrachlorophenol; (B) hydrogenolysis of TCE to *cis*-DCE; (C) dehydrodehalogenation of 1,1,2,2-TeCA to *cis*-DCE.

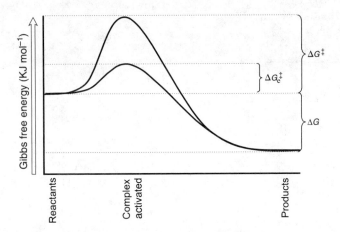

FIGURE 2.19 Effect of enzyme participation on the activation energy of a reaction. The activation energy of a reaction, ΔG, is dramatically reduced by the intervention of an enzyme in the reaction, as depicted by the lower activation energy, $\Delta G_e^{\ddagger}$. The overall reaction energy yield, ΔG, is unaffected.

rate. Figure 2.19 shows the role of an enzyme in the reduction of activation energy in a reaction profile.

Enzymes that catalyze oxidation–reduction reactions typically are joined by cofactors. The cofactors carry the oxidation or reduction capacity, and the role of the enzyme is to bring the cofactor and the target molecule (known as the substrate) into proximity. Enzymes that catalyze the reduction of substrates are known as reductases, and they join with a coenzyme that bears the reducing agent.

For reductases, a common coenzyme is nicotinamide adenine dinucleotide, or NAD^+. Its reduced form is NADH. The redox-active portion of the NADH molecule is nicotinic acid (niacin), which transfers two hydrogen atoms per reaction. The reducing capacity of the oxidized form, NAD^+, is reinstated through energy-capturing reactions such as glycolysis. Another class of enzyme cofactor that supports reductase reactions is the flavins, such as riboflavin (vitamin B_2). The redox-active portion of the riboflavin molecule is a hydroquinone, $FADH_2$.

Enzymes are extreme-molecular-weight compounds of amino acids and functional groups, synthesized by the RNA of cells. The molecular weight of PCE reductase has been estimated at 61,000 g mol^{-1}. Clearly, a single bacterium can only contain a few copies of this molecule. Their three-dimensional structure includes substrate-specific binding sites for the reactants that fit like a lock and key. When the reaction is complete, the product(s) no longer fits the binding site, and is expelled.

Enzymes vary in their substrate specificity. Some, such as soluble methane monooxygenase (sMMO) are not highly specific, and can participate in many reactions that support groundwater cleanup. Although its primary function is the oxidation of methane, sMMO can also oxidize a wide range of low-molecular-weight compounds, including methyl-*tert*-butyl ether (MTBE) and 1,4-dioxane. Other enzymes, such as a perchloroethene reductase (PCErdase) isolated from the dechlorinating bacterium *Dehalococcoides ethenogenes*, can only reduce perchloroethene. The PCErdase of other species, such as *Clostridium bifermentans*, reduces a much broader range of substrates.

Researchers have provided an analysis of two reductase enzymes isolated from *D. ethenogenes*.[44] PCErdase and TCErdase both appeared to rely on a corrinoid cofactor, such as vitamin B_{12}, as the reducing agent. Corrinoids are large cobalt-bearing molecules that are structurally similar to the heme group of hemoglobin. It was also observed that the dechlorination reactions of *D. ethenogenes* were accelerated by the addition of vitamin B_{12}, which is consistent with the earlier findings.[42,44]

Accordingly, the TCErdase isolated from cultures of *D. ethenogenes* dechlorinated a wide range of ethenes, including *cis*-DCE, *trans*-DCE, 1,1-DCE, and VC.[44] The relative enzymatic dechlorination rates were

$$cis\text{-DCE} > TCE \gg VC \qquad (2.8)$$

The earlier study showed that *cis*-DCE was dechlorinated at up to 12 μmol min^{-1} mg^{-1} protein and TCE was dechlorinated at approximately 5 μmol min^{-1} mg^{-1} protein.[44] VC dechlorination was dramatically slower, at 0.04 μmol min^{-1} mg^{-1} protein. This contrasts with the available energy as calculated in Table 2.3, in which TCE provides the most and *cis*-DCE provides the least energy yield of the three electron acceptors, indicating that the reaction rates are controlled by enzyme kinetics rather than free energy yields.

Another result was a very rapid sequencing of the reduction steps. *cis*-DCE created by the dechlorination of TCE was immediately available to the enzyme system for further dechlorination. This observation lays the groundwork for later discussion on degradation signatures — the isolated peaks of both field-scale and microcosm dechlorinations are inconsistent with *D. ethenogenes* Strain 195 enzyme kinetics, providing a strong basis for arguing that more than one bacterial species participates in the dechlorination process.

Field-scale and enrichment culture degradation behavior represents the superposition of several variables, including

- Bacterial biomass and nutritional status
- Available enzyme mass
- Available coenzyme mass
- Rate of coenzyme replenishment
- Enzyme kinetics

It is important to keep in mind that fast kinetics is not usable if the bacteria cannot generate significant biomass in the field. Conversely, cometabolic processes with low per-unit-biomass degradation rates may provide substantial dechlorination in the field, if the species involved reach high biomass levels.

2.3.1.3.4 Cometabolism

A number of enzyme and coenzyme reactions may occur which lead to the destruction of contaminants, but which do not support growth or maintenance of all aquifer bacterial populations. These reactions, collectively, are termed cometabolic.

Cometabolic reactions can play a significant role in the dechlorination of ethenes in aquifers. These reactions can be engineered *in situ* by stimulating the production of bacteria that are known to generate cometabolizing enzymes and cofactors.

There are many mechanisms and patterns of cometabolism that lead to reductive dechlorination, and they can be assembled into two broad groups, based on their utility in a remedial technology.

Many enzyme systems can metabolize multiple related substrates, some of which can support the growth of the bacteria, while others fall below the energy conservation threshold. The energy conservation threshold is twofold. First, the organism must gain at least 20 kJ mol^{-1} to capture the energy released in the form of ATP. Second, the reaction kinetics must be fast enough to allow meaningful energy harvest. Thermodynamically favorable reactions may occur too slowly to support microbial growth. This is the case for VC reduction by *D. ethenogenes* strain 195.

For example, TCE reductive dehalogenase dechlorinates all three substrates in the perchloroethene degradation chain: TCE, *cis*-DCE, and VC.[44] Yet, as described by other researchers, the dechlorination of TCE and *cis*-DCE are growth supporting, while the dechlorination of VC cannot support growth and is, therefore, cometabolic.[42]

The cometabolic dechlorination of VC by *D. ethenogenes* strain 195 is not helpful in remediation, because it cannot be utilized to support remedial efforts in a real aquifer. At the point in the aquifer microbial continuum where VC degradation must occur, no perchloroethene remains to fuel the growth of *Dehalococcoides*, and we certainly cannot inject any.

Some enzymes and coenzymes perform a wide range of reactions, including reactions which appear entirely unrelated to the organism's metabolism. It is from these reactions that we can develop a functional cometabolic attack on aquifer contamination.

Functional cometabolism comprises enzymatic and coenzymatic reactions that can be induced by the promotion of bacterial populations through management of habitat variables *other than the injection of co-contaminants such as perchloroethene*. There are three potential strategies for implementation of functional cometabolism.

Cyclical enzyme induction. Bacteria synthesize enzymes in response to a number of factors, including the availability of substrate. If a particular substrate is unavailable, there will be no synthesis for its dependent enzymes. When the substrate is present, binding specificity may be so high that cometabolic reactions are extremely improbable. In these cases, cometabolic contaminant destruction might be maximized by cyclical feeding of the primary substrate. During the feeding phase, bacterial biomass builds and enzyme production increases. The primary substrate is then cut off. Bacterial enzyme production continues for a period, during which cometabolic reactions occur at a productive rate. With time, bacterial production of the enzyme ceases and a new feeding phase must be started.

One example of this strategy occurs in cometabolism by the aerobic reactions of monooxygenases. Methane-oxidizing bacteria (methanotrophs) synthesize a soluble methane monooxygenase (sMMO) in response to dissolved methane. sMMO can oxidize a broad range of substrates, but it has a high substrate specificity. Cyclical induction is needed to maximize the oxidation of contaminants in this cometabolic system. Trichloroethene, 1,4-dioxane, MTBE, tertiary butyl alcohol, are but a few examples of substrates that can be oxidized by sMMO. Perchloroethene cannot be oxidized by sMMO or related enzymes.

Steady-state enzyme induction. In some cases, enzyme specificity is low, and the presence of primary substrate does not exclude meaningful rates of reaction with the target contaminants. In these cases, continuous feeding of primary substrate will maximize contaminant destruction.

Coenzyme induction. The presence of reduced coenzymes has been shown to dramatically stimulate the action of metabolically dechlorinating bacteria. Promotion of bacterial production of cofactors by species that do not participate directly in the reductive dechlorination process constitutes a cometabolic system. Moreover, coenzymes can directly dechlorinate, as has been shown for cobalamin, cofactor F_{430}, and hematin.[45,46] Table 2.4 shows the relative second-order rate constants for dechlorination of PCE, TCE, *cis*-DCE, and VC.[45]

The rate constants observed follow a pattern that differs greatly from what might be expected from dechlorinating bacteria. Several authors have shown the diminishing energy that can be gained by dechlorinators in the stepwise reduction of PCE to ethene.[45] The decrease is so great that others have described VC dechlorination as cometabolic, dependent on PCE at moderate levels.[42] The cob(I)alamin kinetic rate minimum at *cis*-DCE leads to an expectation that *cis*-DCE will build up in the system dominated by cob(I)alamin, while VC will build up in the system driven by a metabolic dechlorination.

The authors have observed a similar pattern in full-scale field applications where high electron donor loading was applied and methanogenic conditions were

TABLE 2.4
Relative Second-Order Kinetic Rate Constants for Dechlorinating Reactions of Cob(I)alamin[45]

Dechlorination Reaction	pH		
	7.0	8.0	9.0
PCE → TCE	5000	5000	5000
TCE → cis-DCE	100	100	100
cis-DCE → VC	1	0.1	0.01
VC → ethene	~100	~10	~1
Ethene → ethane	~100	~10	~1

TABLE 2.5
Literature Values for Rates of PCE Dechlorination

Bacterial Species or Strain	PCE Dechlorination Rate $\mu mol\ h^{-1}\ mg^{-1}$ Protein
Clostridium bifermentans DPH-1	0.43
Dehalobacter restrictus PER-K23	1.0
Dehalospirillum multivorans	4.5
Desulfomonile tiedjei	0.02
KB-1 enrichment cultures	1.0
Methanosarcina sp.	9.8×10^{-5}

Note: Dechlorination rates were normalized to mg protein, a reflection of biomass and nutritional status of the cultures.[31]

sustained. Case Study 1 shows data from such a site in the U.S. midwest. Alkalinity at that site is high, and aquifer pH was not decreased by organic acid development outside the immediate vicinity of the injection wells.

The importance of cobalt in the formation of corrinoid coenzymes and the evidence that corrinoid metabolism may be a key factor in dechlorination reactions provide strong arguments that reductive dechlorination injections should be augmented with cobalt minerals to provide maximal support of this process.

Cometabolic strategies can be a major contributor to enhanced reductive dechlorination remedies, but laboratory research that can be used to guide development of cometabolic strategies has been limited.

2.3.1.4 Dechlorinating Bacterial Species and Consortia

Early efforts to clarify the mechanism of microbial reductive dechlorination concluded that this process was essentially accidental and of no benefit to the responsible organisms. Because the molar concentration(s) of chloroethenes in groundwater systems is low, except in the most contaminated source areas, it was assumed that these compounds are not sufficiently present to serve as primary substrates under *in situ* conditions. Consistent with this assumption, the first pure cultures shown to be capable of reductive dechlorination were methanogens, sulfate reducers and homo-acetogens. Based on these observations, reductive dechlorination was originally viewed as an anaerobic cometabolism brought about by the accidental interaction of chloroethenes with enzymes and reduced cofactors produced by the microorganisms for other metabolic purposes.[6]

Many dechlorinating populations have been isolated and identified in the past 20 years. Early work identified broad groups of organisms that dechlorinate through the use of enrichment cultures. The methanogens, for example, were recognized as potential dechlorinators through methanogenic enrichment cultures.[47] Finding organisms that can dechlorinate man-made molecules, as well as similar, naturally occurring molecules, is no surprise. It is somewhat surprising, however, that organisms have been identified that appear to be entirely reliant on man-made compounds such as the chlorinated alkenes, which have been dispersed in the environment for only 50 years.

2.3.1.4.1 Evolutionary Relationships

Life on Earth can be classified into three major domains: the Bacteria, the Archaea, and the Eucarya. As shown in Figure 2.20, these domains are believed to have evolved from a common ancestral group of "proto-bacteria."

The characteristic that distinguishes Archaea from the other domains is their cytoplasmic membrane composition. The Archaea membranes are comprised of ether-linked biopolymers, while Bacteria and Eucarya membranes consist of ester-linked biopolymers. The ether-linked biopolymers of the Archaea appear to provide a survival advantage at high temperatures.

2.3.1.4.2 Dechlorinating Species

On a global scale the distribution of well-understood bacteria such as methanogens and SRBs is assumed to be universal by microbiologists.[1,48] Since both of these bacteria are obligately anaerobic, larger concentrations are expected in anaerobic

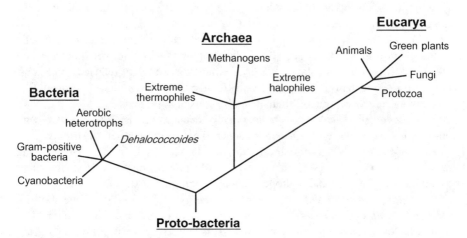

FIGURE 2.20 Evolutionary relationships between the three major domains of life on Earth: the Bacteria, the Archaea, and the Eucarya. The number of groups within each domain has been simplified, and spatial relationships on the graphs are not meant to imply evolutionary relationships within the domains. All three domains are represented in aquifer biota.

environments. Even in aerobic soil environments, anaerobic "microsites" provide for the survival of obligately anaerobic bacteria. Additionally, some types of SRBs can form spores under adverse conditions which germinate upon the reestablishment of suitable growth conditions.

Microbiologists have only recently begun to research the distribution of dehalorespiring bacteria. Perhaps the first assemblage of information on dehalorespiration distribution was published in February 2002.[49] These researchers examined soil samples from eight widely scattered CAH contaminated sites in both temperate and subtropical zones within the U.S. The samples were analyzed for 16S rDNA sequences specific to the genus *Dehalococcoides*. In addition, microcosms were established with soil samples from each site to determine whether each bacterial community could achieve complete CAH dechlorination. 16S rRNA consistent with *Dehalococcoides* was found at all eight sites. The presence of *Dehalococcoides* 16S rRNA was also highly correlated with complete dechlorination of CAHs. Groundwater samples collected from eight CAH-contaminated sites across the U.S. and Europe where CAH dechlorination proceeded to completion contained *Dehalococcoides* 16S rRNA. The results of this study initially support widespread distribution of one genus of bacteria that exhibits the ability to rapidly and completely dechlorinate CAHs using the process previously described as dehalorespiration.

Integration of advanced bacterial culture methods with the ability to "RNA fingerprint" the isolated colonies has led to a flurry of studies reporting the isolation of dechlorinating bacteria species. The following sections report a few of the findings, but the list is not exhaustive. These reports encompass both cometabolic and metabolic (energy-conserving) dechlorination capabilities.

Acetobacterium woodii is involved in cometabolic degradation of PCE to TCE at very low rates, and degradation of CT to methylene chloride and carbon dioxide at high rates.[50,51] This is a homoacetogenic bacterium that produces acetate as part of its metabolic pathway. Homoacetogenic bacteria thrive in very reduced conditions and grow both organotrophically or lithotrophically by either fermentation of sugars or reduction of carbon dioxide to acetate with hydrogen as an electron donor.

Clostridium bifermentans DPH-1 is another bacterium capable of cometabolic degradation of PCE, TCE, *cis*-DCE, VC, and several other chlorinated alkenes and alkanes.[31] This is a fermentative bacterium that thrives in very reduced conditions. During the fermentation of sugars, *Clostridium* species can produce hydrogen, carbon dioxide, acetone, butanol, ethanol, and various organic acids.

Dehalobacter restrictus can metabolically dechlorinate PCE to *cis*-DCE with only hydrogen as the electron donor (hence the specific epithet "*restrictus*").[41,52] *Dehalococcoides ethenogenes* strain 195 was shown to exhibit energy-conserving dechlorination of PCE to VC, and cometabolic dechlorination of VC to ethene under a limited range of PCE and hydrogen concentrations.[42,53–56] Recently, other cultures of *Dehalococcoides* have demonstrated energy-conserving VC metabolism. *Dehalococcoides ethenogenes* could not be grown in culture on electron donor, electron acceptor, and mineral nutrients alone. This suggests it grows only as part of a microbial consortium.

Dehalospirillum multivorans was shown to dechlorinate PCE to *cis*-DCE in culture with hydrogen as the electron donor, and acetate as the carbon source. This bacterium could also use lactate, pyruvate, ethanol, formate, and glycerol as electron donors. When grown on pyruvate and PCE, the main fermentation products were lactate, acetate, DCE, and hydrogen.[57]

Desulfitobacterium chlororespirans Co23 was isolated in 1996 and this bacterium could utilize 2,3-dichlorophol as an electron acceptor with lactate, hydrogen, pyruvate, formate, butyrate, and crotonoate as electron donors. Pyruvate supported homoacetogenic growth in the absence of an electron acceptor. Researchers further characterized the reductive dehalogenase from this bacterium, showing it dechlorinated chlorophenols but did not study chlorinated ethenes.[58]

Desulfitobacterium sp. strain PCE-1 bacterium could utilize PCE as an electron acceptor in the presence of L-lactate, pyruvate, butyrate, formate, succinate, or ethanol as the electron donor. PCE was transformed to TCE and small amounts of *cis*-DCE and *trans*-DCE. This bacterium is also capable of fermentative growth with pyruvate as the substrate.[36]

Desulfitobacterium sp. strain Y51 was documented to be capable of rapid dechlorination of PCE to *cis*-DCE with pyruvate and fumarate as electron donors. Lactate and formate also supported growth, but at a low rate. Further analysis of the PCE reductive dehalogenase showed capability to dehalogenate various other chlorinated ethanes.[59]

Desulfomonile tiedjei was described in 1990, and additional characterization of its dechlorinating behavior was reported later.[37,60,61] It is a sulfate-reducing species that dechlorinates PCE cometabolically in the presence of 3-chlorobenzoate.

Desulfuromonas chloroethenica is capable of PCE reduction to *cis*-DCE, associated with acetate oxidation for this species. This organism could also grow with pyruvate as an electron donor.[62,63]

Desulfuromonas sp. strain BB1 requires acetate, not hydrogen as an electron donor in the reductive dechlorination of PCE and TCE. *Desulfuromonas* species are sulfur-reducing bacteria, unable to reduce sulfate to sulfide, but they can reduce elemental sulfur to sulfide.[64]

Enterobacter sp. was documented to have the dechlorination capability of this facultative species. It was able to dechlorinate high concentrations of PCE to *cis*-DCE following growth on a variety of carbon sources. PCE dehalogenation required the absence of oxygen, nitrate, and high concentrations of fermentable compounds such as glucose.[30]

Methanosarcina sp. strain DCM showed low-rate dechlorination of PCE in cultures of an unidentified *Methanosarcina* sp., later named strain DCM.[47,65,66] This species could dechlorinate PCE while growing on methanol, acetate, methylamine, and trimethylamine. Reductive dechlorination of PCE occurred only during methanogenesis. *Methanosarcina mazei* showed a low-rate dechlorination of PCE for this species while growing on methanol. TCE was the sole end-product.[65]

Dechlorinating by cell exudates of the species *M. thermophila* showed that the degradation was likely driven by porphorinogens.[67] The purified carbon monoxide dehydrogenase enzyme from this bacterium mainly transformed TCE to *cis*-DCE; however, trace amounts of *trans*-DCE, VC, and ethane were also formed.[67] Of the homoacetogenic bacteria surveyed, *Sporomusa ovata* was the only species among eight tested that could dechlorinate PCE. TCE was the only product observed. Methanol was used as the electron donor.[51]

2.3.1.4.3 Dechlorinating Enrichments and Consortia

Early studies on the reductive dechlorination of chlorinated ethenes were performed on enrichment cultures, many of them methanogenic.[68] Since then, a significant portion of the scientific effort has shifted to work with isolated species cultures. However, important information is still being gained from broader enrichment culture studies, as is shown below.

2.3.1.4.3.1 Methanogenic Enrichment Studies

Numerous methanogenic enrichment studies have been conducted since the early work of the mid-1980s. Methanogenic cultures are generally shown to completely dechlorinate any of the chlorinated ethene compounds.

The destruction of chlorinated ethenes in methanogenic cultures seeded with fresh water sediments was studied recently.[69,70] They injected radioactively labeled TCE and VC into the methanogenic cultures and found that radioactive carbon dioxide and methane were produced, in addition to the more commonly observed ethene and ethane.

Complete dechlorination of PCE, TCE, *cis*-DCE, and VC at high rates has been demonstrated. It was also shown that complete dechlorination of PCE in methanogenic cultures is possible by using radiolabeled solvent precursors.[68] Reductive dechlorination of PCE to *cis*-DCE and 1,2-dichloropropene to propene in methanogenic enrichment cultures has been described. The cultures were fed acetate and hydrogen as potential electron donors.[71] Dechlorination of PCE in sulfate-reducing methanogenic cultures was studied and it was concluded that the methanogens are critical to the dechlorination process in that setting.[72–74]

Researchers have studied the *Dehalococcoides*-containing KB-1 culture, establishing subcultures that exhibited widely varying capacities for dechlorination of each compound in the PCE-to-ethene sequence.[75] One of the subcultures was capable of energy-conserving metabolism of VC, in contrast to *D. ethenogenes* (strain 195).[56] The KB-1 consortium contains a number of bacterial genera that may contribute to the dechlorination processes.

Complete dechlorination of PCE was demonstrated, including what appeared to be energy-conserving reduction of VC, in a consortium that did not contain *Dehalococcoides*. The authors suggested that bacteria in the genus *Spirochaetes* may have participated in the dechlorination process.[92]

2.3.1.4.3.2 Nonmethanogenic Enrichments

In repeated subcultures, enrichments can be obtained that have lost their methanogen populations, but retain dechlorinating capabilities.

A series of nonmethanogenic enrichments from Michigan river sediments was developed recently.[85] Initially, these cultures fully dechlorinated PCE to ethene, but subsequent subcultures were obtained that lost their ability to dechlorinate PCE. 16S rRNA analyses showed that the dominant ribotypes were associated with *Azoarcus* sp. (nitrogen fixers), *Acidaminobacter hydrogenoformans*, *Clostridium limosum*, and *C. butyricum*. VC dechlorination was closely correlated with the presence of *Azoarcus*.

2.3.1.4.3.3 Other Dechlorinating Consortia

In most of the metabolic and cometabolic pathways that lead to destruction of chlorinated alkenes, the alkenes are electron acceptor. Recently, metabolic systems have been observed that incorporate VC as electron donor in anaerobic respiration.

Oxidation of VC to CO_2 with bio-available Fe(III) in the role of electron acceptor has been documented.[69]

Complete dechlorination is possible in cultures that had been initiated as methanogenic cultures and which lost methanogenesis when PCE was added.[76,77] Indirect evidence suggests dechlorination was associated with growth of the homoacetogens. Later work showed dechlorinating abilities in cultures of homoacetogens.[51]

A study of mixed anaerobic enrichment cultures that contained approximately 90% Eubacteria and 10% Archaea (note that methanogens are among the Archaea) showed very high rates of VC dechlorination, and the data suggested a membrane-bound, energy-conserving metabolic process was at work.[78] The authors observed that dechlorination reactions may have been driven by transition metal enzyme cofactors.

Of the many isolated species shown to dechlorinate the ethene-family compounds, none have been shown to complete the entire reduction sequence to ethene in energy-conserving metabolism. Conversely, methanogenic and other bacterial mixed cultures were capable of completing the entire reduction sequence, producing ethene at high rates.

The extrapolation of all these laboratory studies to field conditions is constrained by the fact that aquifer temperatures are dramatically lower than any of the studies cited above, and that aquifer chemistries are much more complex than the simple media in which these studies were conducted. Reestimation of energy conservation to obtain "field" metabolic capabilities can be done. While it is unlikely that we can sufficiently define the biogeochemical habitat to get a precise result, we can clearly see trends such as the reduction of energy yields at lower temperatures. Reactions that occur at lab temperatures and reagent strengths may not occur in aquifers.

Due to the prominence given to *D. ethenogenes* and, consequently, the most convenient target for molecular characterization techniques, a more comprehensive evaluation to accurately discriminate between known dechlorinators or targeting of dehalogenase specific gene sequences have not been pursued.[6] A special emphasis on identifying the target sequences of dechlorinators capable of reducing DCE and VC to ethene or other nonchlorinated end-products is required. The fact that *D. ethenogenes* was not present in two of three environmental enrichments capable of complete dechlorination of PCE to ethene[85] suggests that molecular characterizations based solely on the presence/absence of *D. ethenogenes* may underestimate the indigenous capacity for complete dechlorination of PCE or TCE. The evidence of the existence of chlororespiring microbial communities in pristine environments[85] raises the possibility that their apparent absence at some sites is more appropriately attributed to unfavorable environmental conditions than to environmental scarcity.

2.3.1.5 Review of Field-Scale Bioaugmentation Studies

Most of the literature surrounding the bioaugmentation approach to reductive dechlorination has been drawn from laboratory microcosm and column studies. At the time of the writing of this chapter, only three field-scale studies of bioaugmented reductive dechlorination had been published in refereed journals.[79–81] These studies all relied on groundwater withdrawal, amendment, and reinjection to support precise microbial habitat control in the groundwater. These were also all pilot-scale operations, focused on dissolved-phase contaminants, not "source area" contaminant loads.

A quantitative understanding of microorganism migration in groundwater systems is critical to predict the dissemination of injected microorganisms during bioaugmentation. The key event that retards the movement of microorganisms in the saturated zone with respect to convective water flow is the interaction between microorganisms and the matrix surfaces. This interaction may result in adhesion and concomitant retardation. Interactions are determined by the surface thermodynamics of the microorganism and the matrix. Whereas the nature of the matrix substratum surface may be considered temporarily invariant, the nature of the bacterial cell surfaces is a function of its physiological state. It should be noted that the culturing-induced physiological state of the cultured isolates will be very different from that of the free-living native microbial population. This aspect will certainly have an impact on the survivability and transport of these injected microorganisms.

2.3.1.5.1 Dover Air Force Base

The first study reported the results of bioaugmentation field trials for the dechlorination of TCE.[80] The work was preceded by microcosm studies.[82] The Dover team utilized an anaerobic dechlorinating culture obtained from the U.S. Department of Energy Pinellas site in Largo, Florida. The culture contained a strain of *D. ethenogenes*, but according to the authors, "the abundance of this organism was low relative to other phylogenetic types found by 16S rRNA analysis of the culture...."[80]

Groundwater was extracted, amended, and reinjected to maintain control of chemistry in the bioaugmentation zone. The aquifer was highly aerobic prior to the pilot study and carried TCE at concentrations near 45 μmol L^{-1} on a groundwater flow of approximately 0.05 m day^{-1}. Concentrations of *cis*-DCE were approximately 10 μmol L^{-1} prior to the pilot study, and VC and ethene were below detection.

The experimental system consisted of three injection and three extraction wells, spaced in parallel flow pathways. The three flow paths were approximately 17 m long, and they were spaced approximately 6 m apart. Travel time along each flow line was approximately 60 days. The central flow line was intended to be the experimental treatment (receiving the bioaugmentation culture) and the two outer flow lines were intended to serve as control pathways (biostimulation only). According to the authors again, the controls failed due to inadequate isolation, so there was no comparison to biostimulation in this experiment.[80]

Lactate was added to the amended groundwater as the primary electron donor. The lactate concentration in the injection stream was initially 40 mg L^{-1}, then was increased to 130 mg L^{-1} due to insufficient transport. Pre-bioaugmentation injections were maintained for 269 days. Throughout the first 424 days, groundwater from three extraction wells was merged, amended with lactate, and divided for reinjection among the experimental and control flow lines. According to the authors, the merger of the extracted groundwater was carried on too long, and they claim that bacteria transported along the experimental path contaminated the control pathways. No 16S rRNA study was reported to confirm this, and the claim that dechlorination in the controls was due to cross-contamination remains controversial.

Complete dechlorination to ethene was observed in both the experimental and control zone monitoring wells after inoculation with the Pinellas culture.

The control flow line behaved as would have been predicted for a biostimulation-only approach — without the inoculum, the control plot experienced a longer lag phase prior to the onset of complete dechlorination. Because of the potential cross-contamination due to inadequate experimental design, the Dover data cannot be used to determine the relative contribution of bioaugmentation and biostimulation at the study site.

2.3.1.5.2 Schoolcraft, Michigan Site

The second study was on field-scale work using *Pseudomonas stutzeri* strain KC bioaugmentation for the dechlorination of CT.[79] At the Schoolcraft, Michigan site, groundwater is contaminated by chloroform at concentrations up to 0.13 μmol L^{-1} (20 μg L^{-1}). The aquifer is very productive in this area, with a saturated thickness of 18 m and a hydraulic conductivity ranging from 1.2×10^{-2} cm s^{-1} in the upper zone to 4.6×10^{-2} in the lower zone. Groundwater velocities were estimated to be at 15 cm day^{-1}.[79]

In addition to CT contamination, the Schoolcraft aquifer is highly contaminated by nitrates, with median values at all depths of the aquifer between 40 and 42 mg L^{-1}. This provides a significant reservoir of alternative electron acceptors that must be consumed before effective remedial dechlorination can proceed.

The authors' strategy was designed to achieve dechlorination without formation of chloroform, which can be toxic to aquifer bacteria at high concentrations. It is unclear why there was a concern regarding chloroform toxicity, when the total mass of CT was not sufficient to generate chloroform toxicity (300 μg L^{-1}; Mark Harkness personal communication), even if it was all dechlorinated at once. As is typical for bioaugmentation, the Schoolcraft team pumped groundwater, amended it, and reinjected it into the target formation. In this case, 15 wells drilled to a 27-m depth were spaced at 1-m intervals, to enable the precise control of aquifer chemistry needed to support the bioaugmentation process. Approximately 65% of the groundwater flux was used in the amendment/reinjection process.

The principal amendments were pH adjustment with NaOH to 8.2 (from 7.6 background pH), addition of phosphate as a nutrient at 10 mg L^{-1}, and addition of acetate as an electron donor. The acetate concentration was limited, to prevent sulfate reduction and chloroform formation. Acetate at a concentration of 100 mg L^{-1} was determined to be too high, and was reduced to 50 mg L^{-1} after 314 days system operation.

The extraction, amendment, and reinjection system was operated for baseline monitoring and tracer testing for 72 days, then preinoculation amendments were started. The injection wells were inoculated twice with *P. stutzeri* strain KC on two occasions — day 117, and days 200 through 201.

Pre-inoculation groundwater pH adjustment and recirculation caused a dramatic increase in aqueous-phase CT levels, to 0.31 μmol L^{-1} (42 μg L^{-1}). CT concentrations declined following the inoculations, and ranged from 0.007 to 0.05 μmol L^{-1} (1 to 8 μg L^{-1}) after 300 days of treatment system operation, based on data from monitoring wells 1 m downgradient from the line of injection wells.

Bacterial transport through the aquifer was severely limited by adhesion to aquifer solids in both studies. As anticipated by other researchers, bacterial movement was constrained to less than 3 m from the injection points in both studies.[83]

Although the Schoolcraft study authors suggest the bioaugmentation system was efficient, it would not be judged as cost-effective in a competitive commercial setting. There are three substantive limitations on this type of bioaugmentation system that prohibit its commercialization:

- Pump and treat requirement — withdrawal, amendment, and reinjection of 65% of the aquifer flux combines the most objectionable elements of groundwater containment and biostimulation systems.
- The very short effective radius of these injection wells and the resulting requirement to space injection and withdrawal wells at 1-m intervals to a depth of 27 m is not commercially viable.
- High-frequency monitoring and injection make the system operation very labor-intensive, again increasing operating costs to noncompetitive levels.

To a great extent, the bioaugmentation challenges faced by the Schoolcraft team were greater than those in the Dover and Kelly Air Force Base studies, due to the selection of a bacterium that requires very finely controlled chemical conditions to achieve reductive dechlorination of CT. Although the *Dehalococcoides* consortia deployed in the Dover and Kelly Air Force Base studies require extensive habitat control relative to biostimulation, they are apparently robust when compared to *P. stutzeri* strain KC.

2.3.1.5.3 *Kelly Air Force Base*

The third available study described microcosm and field-scale use of the *Dehalococcoides*-like KB-1 culture in a pilot-scale treatment at Kelly Air Force Base.[81] Their experimental design was similar to the other bioaugmentations, with a pre-inoculation period of biostimulation, followed by inoculation with dechlorinating culture and continued biostimulation.

The Kelly Air Force Base study area was an aerobic aquifer, contaminated with PCE at approximately 5 μmol L^{-1} (800 μg L^{-1}) a small amount of TCE (0.2 μmol) and 2 μmol *cis*-DCE. The aquifer flow velocity was approximately 0.9 m day^{-1} and a groundwater capture and reinjection system was established that recirculated groundwater in the test plot at a rate of one pore volume (64,000 L) every 8.5 days.

Like the Schoolcraft study, the Kelly AFB test was run in recirculation mode for 89 days to establish flow volumes and run tracer testing. Biostimulation was initiated on Day 89 by the injection of methanol and acetate. According to the authors, methanol was chosen because it was the electron donor for the KB-1 culture and acetate was added as a more general electron donor to drive the aquifer into anaerobic conditions.[81] Two aspects of the biostimulation strategy stand out:

- First, the selection of electron donors severely restricts the participation of known dechlorinating species in the study. Many of the known dechlorinators (e.g., *Desulfitobacterium*) rely on fatty acids as electron donors. The limitation on the range of available electron donors biases the study significantly toward the hydrogenotrophic dechlorinators.
- Second, the biostimulation phase was run for only 87 days. Typically, biostimulation sites do not begin to dechlorinate fully until after 180 days. Two control plots were run, but only for 216 days and 117 days, respectively. The authors do not indicate whether electron donor injections began at Day 0 for each of the control plots, but neither control plot was run long enough to observe typical onset of complete dechlorination in biostimulation.

The short duration of the biostimulation phase and the severe limitations on the breadth of electron donor availability caused this study to highlight performance of the KB-1 inoculum, relative to other known declorinating bacteria.

KB-1 inoculum consists of a consortium of the *Dehalococcoides*-containing culture KB-1 and other bacteria that are essential to survival of the culture, but remain only partially identified.[84] Thirteen liters of the culture was injected under anaerobic gas blanket to protect against the toxic effects of oxygen exposure. The density of *Dehalococcoides* in the culture was 10^9 cells L^{-1}, yielding a total inoculum of 1.3×10^{10} cells. The total aquifer volume was 64,000 L, or 6.4×10^7 ml, which yielded an average inoculum density of 200 cells ml^{-1} throughout the study volume. 16S rRNA testing indicated that the initial distribution was limited to very near the injection wells, although the culture was observed throughout the test volume by the end of the study. This indicates up to 9 m of travel over 142 days, under a pumping-induced flow of 17 pore volumes.

Significant dechlorination was observed after Day 250. Although the chlorinated alkene data was presented only in graphical format, it can be seen that drinking water standards were not achieved at the completion of the study at Day 320.

Authors of the Kelly Air Force base study observed that *Dehalococcoides* was responsible for dechlorination they observed, and they asserted:

> "To date, all cultures that dechlorinate chlorinated ethenes completely to ethene have been found to contain a phylogenetically close relative of *Dehalococcoides*…"

It is important to note that other researchers subsequently observed complete dechlorination in bacterial cultures that did not contain *Dehalococcoides*. For example, Flynn et al.[85] showed complete dechlorination by cultures isolated from river sediments in Michigan. 16S rRNA analysis showed no *Dehalococcoides*-related bacteria in isolations that dechlorinated VC. Instead, the cultures contained *Azoarcus* sp., *Acidaminobacter hydrogenoformans*, *Clostridium limosum*, and *C. butyricum*. This is strong evidence for non-*Dehalococcoides* dechlorination of VC, although the responsible species has not been identified.

Although the Kelly study was interpreted to indicate biostimulation was not workable at the site, the biostimulation period was very short, especially considering the aerobic nature of the aquifer prior to treatment. There was a 176-day pretreatment period: 89 days of the recirculation dedicated to flow and tracer testing, with no electron donor additions, followed by 87 days of electron donor addition. It is possible that dechlorination would have been observed through biostimulation alone, if electron donor injections had been applied over a longer period.

2.3.1.5.4 Bachman Road Site Studies

The most experimentally complete comparison to date of biostimulation and bioaugmentation was reported recently. The study site was Bachman Road Residential Well Site near Oscoda, Michigan. Extensive aquifer contamination studies have been conducted at the site. The experimental design was similar to other bioaugmentation studies, with the study area encompassed in a groundwater extraction and reinjection system, with electron donors and other amendments injected into the recirculation stream.[86]

Prior to the experimental work, the aquifer was contaminated with PCE, TCE, cis-DCE, and VC. The authors did not report pretreatment ethene concentrations, which would have been valuable in understanding preexisting dechlorination behavior. cis-DCE comprised more than 50% of the chlorinated alkene mole fraction in the pretreatment samples. Initial aqueous-phase VOCs in the control/biostimulation plot were double the starting concentrations in the bioaugmentation plot (10 µM vs. 5 µM).

The study area was divided into a control/biostimulation plot and a bioaugmentation plot. Injection and withdrawal wells were spaced approximately 2 m apart, and groundwater was withdrawn and reinjected at a rate sufficient to isolate the entrained groundwater from the flow of the aquifer. Tracer studies were used to confirm separation between the control and treatment sectors, and to estimate the groundwater flow velocity under the pumping regime (12 cm day^{-1}). In the control/biostimulation plot, pumping was conducted for 140 days of "control" observation. During this period, VOCs were observed in monitoring wells and aquifer soil samples were collected for bacterial analysis. After completion of control-phase observations, lactate was injected to serve as the electron donor for a 121-day biostimulation test.

The bioaugmentation treatment cell was pretreated with lactate injections for 29 days, to establish reducing conditions suitable for the bioaugmentation inoculum. The bioaugmentation process was then operated for 182 days after inoculation, giving a total 211 days of lactate injection.

Both the biostimulation and bioaugmentation treatments yielded substantial dechlorination, with most of the chlorinated ethenes in aqueous phase converted to ethene at the conclusion of the reported studies. Neither of the treatments was shown to achieve MCL values for VC, which is surprising. The rate of dechlorination at the time of the last reported observation suggested that, eventually, dechlorination would reach completion.

As expected, the bioaugmented plots dechlorinated at a higher rate than the biostimulation plots. Since the experiments were not run to the point of achieving drinking water standards for VC (2 µg L^{-1} or 0.03 µM), it cannot be determined how long either process may have to run to achieve the standards. At the conclusion of observations, lactate injections at the bioaugmentation plot had been running for 72 days, and the biostimulation injections had been running for 121 days. Considering the fact that the bioaugmentation plot started with half the aqueous-phase chlorinated alkene concentration, it is difficult to justify the cost of bioaugmentation inoculum and the groundwater recirculation system it requires.

Prior to beginning the experiments, groundwater and aquifer soil samples were collected to characterize the pretreatment microbial populations. Polymerase chain reaction (PCR) analyses failed to confirm the presence of *Dehalococcoides*-related bacteria in the aquifer prior to the start of experimental activities. Intermittent positive results were observed during the pretreatment recirculation phase of the study, and PCR analyses for *Dehalococcoides* became steadily positive during the biostimulation phase of the project. These results gave a clear indication that biostimulation can encourage the propagation of *Dehalococcoides* from a pretreatment population too small to register in PCR analyses of aquifer soils. It is important to note that aquifer soil samples are much more likely to yield evidence of soil bacteria than are groundwater samples.

2.3.1.6 Summary of Bioaugmentation

The bioaugmentation studies published to date provide insight into the performance of microcosm-derived strategies for *in situ* bioremediation. In sites that do not bear large anaerobic microbial biomass, preparation of the site by electron donor injection, followed by inoculation with a dechlorinating culture, has the potential to shorten the lag time for biomass buildup observed in biostimulation projects. However, these studies show clearly that bioaugmentation will be an expensive approach to cleanup.

There are two very significant limitations on the cost-effectiveness of bioaugmentation. The first is the limitation of bacterial transport in aquifers by their adhesion to aquifer soils, and the second is the requirement that key aquifer habitat variables be controlled within precise limits to protect the injected bacterial species from competition for limited resources. These limitations drove all four studies to construct groundwater withdrawal, amendment, and reinjection systems that superimposed major mechanical systems on the IRZs.

2.3.1.7 Review of Field-Scale Biostimulation Studies

Because biostimulation has been a commercial process for nearly 10 years, it attracts much less attention from the academic researchers. Data collected from operating systems typically are not extensive enough, and the practitioners do not have the funding to support publication in peer-reviewed journals. Researchers are likely to prefer working with known species rather than uncharacterized microbial consortia in order to more easily develop the repeatability that is essential to hypothesis testing and theory building. As a result, the volume of literature that has developed around bioaugmentation far exceeds that for biostimulation. Drawing conclusions regarding the state of the two technologies from comparisons of their respective literature would leave the incorrect perception that bioaugmentation is the leading approach to enhanced reductive dechlorination. The following case study data illustrate the capacity of biostimulation in field-scale applications. These case studies have been drawn from more than 200 datasets to illustrate the range of site conditions in which the technology has been applied and the range of dechlorination patterns we have observed.

CASE STUDY 1

The Case Study 1 site lies in a glaciated region of the U.S. in a high-carbonate aquifer. Groundwater flow velocity is approximately 1 ft day^{-1}, and the contaminated formation is approximately 8 ft in thickness. The site was contaminated at various times by oils and solvents that were deposited in the formation during a long period of high regional groundwater pumping, which significantly depressed the aquifer surface elevation. Aquifer pumping patterns changed after the oils and solvents were deposited, and groundwater elevations rebounded, inundating a large mass of residual contamination.

Now, chlorinated alkenes are leaching from the residual nonaqueous-phase liquid mass at concentrations exceeding 1000 μg L^{-1}. PCE and TCE are both present in the groundwater as source material, and *cis*-DCE and VC have been observed as degradation products. Soil organic matter is high throughout the source area, with TOC levels ranging from 2000 to 6000 mg kg^{-1}.

An IRZ was constructed in September 1999 to employ enhanced reductive dechlorination through biostimulation. Molasses was selected as the most cost-effective electron donor for the site. Initial injections were biweekly, then shifted to monthly after eight biweekly events. The injection zone dissolved organic carbon level has been sustained near 6000 mg L^{-1} during active injection periods. There have been three periods of 90 days or more during which injections were suspended.

The reactive zone consisted of 13 injection wells, aligned slightly off-perpendicular to groundwater flow. Monitor wells were location approximately 100 days upgradient and 100 days downgradient from the reactive zone injection wells, as shown in Figure 2.21(A).

Figure 2.21(B) describes concentrations of each compound in the PCE-to-ethene sequence for the Case Study 1 site. This constitutes an ideal biostimulation for the destruction of PCE, TCE, and their dechlorination products. The aquifer has a high organic carbon content (f_{oc} = 0.005), so the PCE and TCE were largely contained in the sorbed phase prior to enhancement of reductive dechlorination. The dechlorination process began on Day 0, with molasses injections that raised the average injection zone dissolved organic carbon concentration to 6000 mg L^{-1}. The monitoring location shown in Figure 2.21(A) is approximately 100 days downgradient from the injection zone, and DOC concentrations ranged between 100 and 200 mg L^{-1} at the monitoring location.

Within 6 months, PCE and TCE were below detection, while *cis*-DCE had reached its peak value. Ethene had peaked and essentially disappeared by the end of the first year of operation. VC degradation rates exceeded those of *cis*-DCE by such a margin that VC concentrations never rose above trace levels (0.05 to 0.2 μmol L^{-1}).

This pattern — rapid, complete dechlorination without VC buildup — has been consistently observed at biostimulation sites when injection zone carbon loading can be sustained at levels that yield 100 mg L^{-1}, 100 days downgradient, without substantial pH declines. High-carbonate aquifers accept these carbon loads easily, while low-carbonate aquifers require the addition of carbonate buffers to keep pH above 5.5.

The rapid dechlorination of VC has only been observed at sites with very high organic carbon loadings. Sites that receive lesser carbon dosage rates may also fully dechlorinate the ethenes, but significant VC peaks occur during the degradation. This

Microbial Reactive Zones

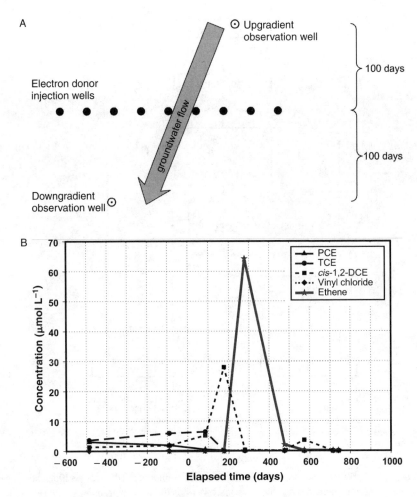

FIGURE 2.21 (A) Reaction zone layout for Case Study 1. (B) Archetypical biostimulation enhancement of reductive dechlorination in a barrier-style *in situ* reactive zone. Data are from a site in a glaciated region, where dissolved organic carbon in the injection zone is sustained at an average 6000 mg L^{-1}, and concentrations 100 days downgradient average 150 mg L^{-1}. The monitor well is located approximately 100 days downgradient from the injection zone. Biostimulation was initiated at elapsed time = 0 days.

distinction leads us to the formation of a working hypothesis on the reductive dechlorination process:

> At low carbon loading rates, primary fermenters produce C1 compounds and acetate as the main products. This can support dehalorespiring species that rely on hydrogen as an electron donor, with the formation of a strong VC peak as dechlorination nears completion.
>
> At high carbon loading rates, the C1 and acetate pathways are saturated, forcing a significant portion of the carbon through the fatty acid synthesis pathways. Under these

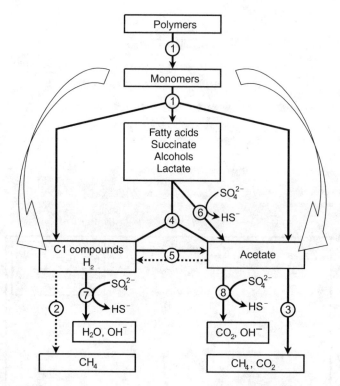

1. Primary fermenters
2. Hydrogen-oxidizing (hydrogenotrophic) methanogens
3. Acetate cleaving (acetoclastic) methanogens
4. Secondary fermenters (syntrophic)
5. Homoacetogenic bacteria
6. Group I sulfate reducers: non-acetate oxidizers
7. Facultative lithotrophic sulfate reducers
8. Group II sulfate reducers: acetate-oxidizers

FIGURE 2.22 Illustration of a working hypothesis of enhanced reductive dechlorination. At low electron-donor loading rates, the outer pathways of the anaerobic metabolic cascade are dominant. This favors the dehalorespiring bacteria that rely on hydrogen as electron donor. Under these conditions, a strong vinyl chloride peak is observed as dechlorination nears completion. When electron donor loading is high, the outer pathways are saturated, and a significant portion of the carbon flows through the central pathways.[32]

conditions, additional dechlorinators are encouraged, and VC dechlorination proceeds at a rate that suppresses peak development. Ketone formation (acetone and 2-butanone) often accompanies the high-rate dechlorination processes that suppress the VC peak.

Figure 2.22 illustrates the working hypothesis.

Figure 2.23 shows the dissolved organic carbon concentrations reaching a monitoring well located 100 days downgradient from the Case Study 1 injection zone. When carbon loading in the injection zones is sufficient to achieve methanogenesis, with at least 100 mg DOC/L remaining, the VC degradation rate exceeds the cis-DCE degradation rate by at least 20-fold, and the VC peak is fully suppressed. This degradation signature is discussed more completely in Chapter 5.

Microbial Reactive Zones

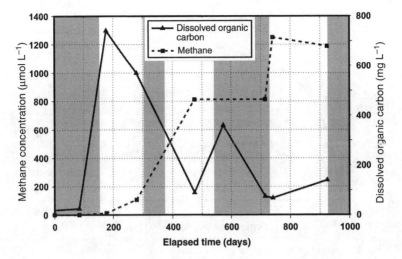

FIGURE 2.23 Dissolved organic carbon and methane observed 100 days downgradient from a biostimulation injection zone, where readily degradable organic carbon (molasses) was injected into the aquifer. During carbon loading periods (gray bands), injections were performed on a monthly basis at a rate sufficient to produce an average injection zone concentration of 6000 mg dissolved organic carbon (DOC) per liter.

Another pattern in this data, the increasing concentration of total alkenes, is commonly observed as reductive dechlorination proceeds. In Case Study 1, total alkenes increased more than six-fold as the alkenes were dechlorinated from PCE and TCE to ethene. There is not an increase in total alkene mass in the system. The increasing dissolved-phase concentration was a result of stepwise decreases in the organic carbon partition coefficient as chlorine atoms were successively removed from the alkene backbone. Equation (2.9) was derived from organic carbon partition calculations, to describe the portion of each chlorinated ethene that will reside in sorbed phase, as a function of the soil organic carbon fraction.

$$\frac{\text{Mass}_{\text{sorbed}}}{\text{Mass}_{\text{total}}} = \frac{(C_w \times K_{oc} \times f_{oc} \times \rho_w) \quad \text{mg m}^{-3}}{(C_w \times K_{oc} \times f_{oc} \times \rho_w) + (C_w \times \theta) \quad \text{mg m}^{-3}} \quad (2.9)$$

Figure 2.24 shows the percentage sorption for PCE, TCE, *cis*-DCE, and VC as a function of soil organic carbon fraction, from 0 to 0.01. At the Case Study 1 site, organic carbon fractions averaged approximately 0.005 in the study area. Under these conditions, approximately 86% of the PCE and 70% of the TCE were in sorbed phase at the beginning of the enhanced reductive dechlorination process. Dechlorination yielded compounds of lower organic carbon partition coefficients, increasing the total alkenes in aqueous phase. Total alkene levels did not decrease until onset of ethene decomposition, which appears to have been inhibited by the presence of *cis*-DCE.

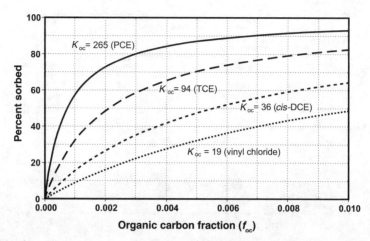

FIGURE 2.24 Relationship between organic carbon fraction in aquifer soil and the sorption of chlorinated ethenes.

Case Study 2

The second case study was selected to show the effects of lower carbon loadings on the dechlorination of VC. Although the site is achieving complete dechlorination, as demonstrated by the formation of ethene, there is a noticeable VC peak. Consistent with the working hypothesis for reductive dechlorination, the rate of VC degradation appears to be sensitive to carbon loading rates. At the Case Study 2 site, the carbon dose has not migrated equally across the treatment zone and the resulting inequality of dissolved organic carbon distribution is matched by an inverse inequality of VC peak formation.

The Case Study 2 site lies in a glaciated region of the U.S. in moderate- to high-permeability glacial soils. Enhanced reductive dechlorination has been applied to a plume consisting of TCE and its degradation products, with a horizontal groundwater velocity of approximately 1.5 ft day^{-1}. The carbon injection system consists of 20 wells, spaced at 20-ft intervals. Figure 2.25(A) shows the injection and monitoring well layout for Case Study 2.

Electron donor (molasses) injections were initially undertaken at 2-week intervals, then were decreased in frequency to monthly. Dissolved organic carbon concentrations quickly increased in MW-A and MW-B, as shown in Figure 2.25(B). By Day 200, dissolved organic carbon levels in both monitoring wells had fallen to less than the pretreatment concentrations. Microbial biomass increases as a result of the increased carbon loading to the aquifer microbial continuum. If the carbon loading is stopped, the large biomass can suppress dissolved organic carbon to levels below pretreatment steady-state levels. This was an indication that the carbon injection rate was insufficient to maintain a 100-day reducing zone.

After Day 200, carbon mass loading was doubled from the earlier levels in an attempt to reestablish the 100-day reducing zone. These efforts were effective for the area represented by MW-A, but not for the zone of MW-B, as shown in Figure 2.25(B). The area under each curve was integrated using the trapezoidal rule. The carbon

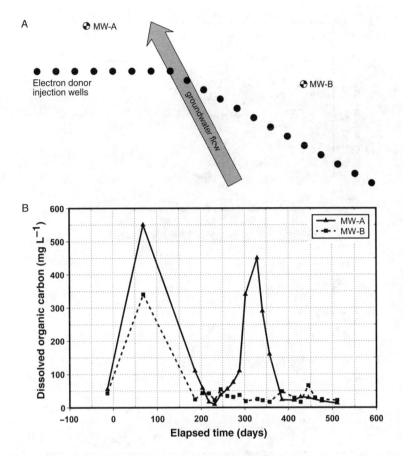

FIGURE 2.25 (A) Reactive zone layout for Case Study 2. The injection system consists of 22 injection wells, spaced at 20 ft intervals. Groundwater velocity is approximately 1.5 ft day^{-1}. (B) Comparison of dissolved organic carbon loading in monitor wells downgradient from the injection zone at the Case Study 2 site. Initial dissolved organic carbon concentrations at MW-B were sufficient but concentrations decreased to below 50 mg L^{-1}, and increased carbon injection has not raised the levels in MW-B. MW-A shows much higher loads after 300 days.

exposure at MW-A was 95,600 mg day L^{-1}, and the area under the MW-B curve was 47,400 mg day L^{-1}, less than half of the MW-A exposure.

The difference in carbon loading rate appears to have resulted in a notably lower dechlorination rate for VC in MW-B than observed for MW-A. Figure 2.26(A) shows chlorinated ethene and ethene concentrations for MW-A. At this location, *cis*-DCE levels dramatically exceeded the pretreatment TCE level, indicating a significant sorbed mass as in Case Study 1. Unlike Case Study 1, VC degradation was slow enough to accumulate more than half the observed *cis*-DCE level. Ethene formation was observed to occur simultaneously with VC formation, indicating that the VC was immediately available for dechlorination. VC levels fell to detection limits after Day 300, and ethene levels are approaching similar lows.

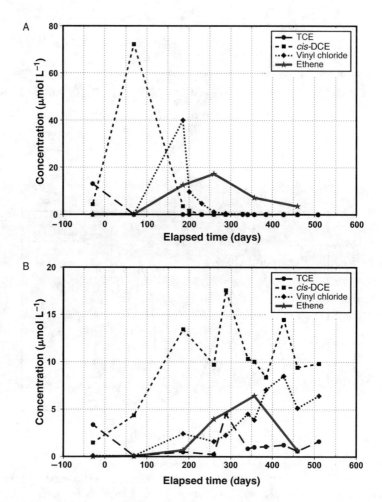

FIGURE 2.26 (A) Decreases in chlorinated alkene concentrations in groundwater at the Case Study 2 site, MW-A. This location received greater dissolved organic carbon than at the MW-B location. (B) Chlorinated alkene concentrations in groundwater at the Case Study 2 site, MW-B. Dissolved organic carbon loading has been minimal at this location.

Chlorinated ethene and ethene data for MW-B are shown in Figure 2.26(B). The initial carbon loading was sufficient to rapidly dechlorinate TCE, but TCE levels were not consistently suppressed. This is an indication of inadequate development of the reducing zone (refer to Figure 2.14), consistent with the low dissolved organic carbon levels observed in MW-B after Day 200. MW-B has not responded to increased carbon injection rates, and TCE levels have returned to more than one third of the pretreatment levels. Since Day 350, *cis*-DCE and VC concentrations have been maintained at levels fivefold and threefold higher than TCE levels, respectively. During that period, ethene levels dropped to less than 1 μmol L^{-1}.

Microbial Reactive Zones

The relationship between dissolved organic carbon and chlorinated ethene concentrations reaching the downgradient observation points in Case Study 2 were consistent with the working hypothesis of enhanced reductive dechlorination, described in the analysis of Case Study 1, above. In addition, the dechlorination performance in Case Study 1 was not matched in Case Study 2, even in MW-A. The fact that in Case Study 1 the dissolved organic carbon levels were maintained at more than 100 mg L^{-1} in the reactive zone for more than 100 days entrainment time may indicate a critical threshold in the efficiency of these systems.

This leads to the postulation of an additional element of the working hypothesis on the reductive dechlorination process:

> When carbon injection rates are sufficient to sustain dissolved organic carbon concentrations significantly above background levels after 100 days entrainment in the reactive zone, the dechlorination of vinyl chloride proceeds at rates more than 20-fold greater than those for cis-DCE, circumventing the formation of a vinyl chloride concentration peak.

Efforts are currently under way to re-evaluate groundwater flow characteristics in the Case Study 2 reactive zone. Design assumptions on both the direction and velocity of groundwater flow may have been incorrect, leading to an inability to establish sufficient carbon entrainment in the MW-B sector. An additional monitor well (not shown) has received less carbon exposure than MW-B. Groundwater flow maps for the area are based on widely spaced sample locations, and groundwater surface elevation maps show anomalous patterns in the treatment area.

CASE STUDY 3

The third case study was selected to show the effects of one-time injection of a large mass of slow-degrading carbon in a poorly buffered aquifer. After 400 days of observation, the Case Study 3 site had not begun to dechlorinate past cis-DCE, which appears to be related to exhaustion of the carbon source. Decline of cis-DCE concentrations and transformation to VC took place only after a second injection of organic carbon. Figure 2.27(A) shows the Case Study 3 reactive zone layout.

Case Study 3 was planned to be a one-time injection of cheese whey and high-fructose corn syrup, due to site-construction-related activities, to provide a longer residence time for electron donors in the injection zone than can be obtained with molasses or other more mobile materials. The groundwater velocity is approximately 1.5 ft day^{-1} at the site. However, due to the exhaustion of TOC, a second injection of organic carbon took place at around 400 days after the first injection.

Figure 2.27(B) shows the concentrations of dissolved organic carbon in the monitoring well located immediately downgradient of the injection zone (based on groundwater velocities, the monitoring well is only about 6.5 days downgradient from the nearest injection points). Concentrations in the monitoring well were initially 5800 mg L^{-1} and declined to 37 mg L^{-1} at Day 420. Methane levels increased to more than 500 μmol L^{-1}, indicating strongly reducing conditions were achieved.

Alkalinity at the Case Study 3 site is low, and aquifer pH dropped rapidly after electron donor was injected. The pretreatment pH was 7.1, but decreased to 3.7 by Day 36. After 6 months, the pH had risen to just less than 5.0, and had reached 6.8, 249 days

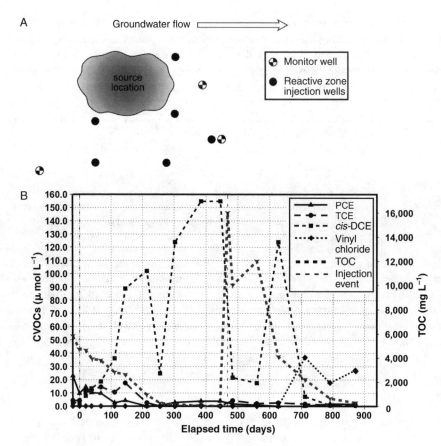

FIGURE 2.27 (A) Case Study 3 reactive zone layout. (B) Dissolved alkenes in groundwater at a monitoring location within the injection zone at the site of Case Study 3 in relation to the concentration of TOC.

after the injection. pH decreases to less than 5.0 are undesirable, and may be inhibitory to reductive dechlorination processes. The chemical thermodynamics analysis provided earlier showed that reductive dechlorination energy yields are only slightly inhibited at low pH. However, enzyme kinetics may be significantly altered by pH declines.

Completion of the reductive dechlorination sequence (e.g., appearance of ethene VC and past *cis*-DCE) is taking place for the Case Study 3 only after the second injection of TOC. Figure 2.27(B) shows a steady increase in *cis*-DCE through Day 420, along with significant declines in PCE and TCE. The eightfold increase in total alkenes is consistent with the other case studies, indicating a large portion of the pretreatment chlorinated ethenes resided in sorbed phase. A steady decline of *cis*-DCE was apparent when the TOC concentrations increased again.

What is not consistent with previous case studies is the failure to achieve dechlorination of *cis*-DCE when methanogenic conditions were achieved even with one injection of TOC. There are several possibilities that may explain this:

- The initial pH excursion may have inhibited growth of critical microbial populations.
- Methanogenic conditions were not sustained long enough to provide the optimum growth conditions required to develop the necessary microbial consortium for complete reductive dechlorination.
- Complete exhaustion of TOC within the IRZ might have affected the delicate balance of the microbial ecology.

Reestablishing the methanogenic conditions by supplementing adequate levels of TOC has led to the transformation of *cis*-DCE to VC and ethene. Hence, the level of TOC itself may play a contributing factor in addition to helping to establish the methanogenic conditions.

Another aspect of Case Study 3 is the observation of methyl ethyl ketone (MEK) formation. MEK has been observed at other sites where highly productive dechlorination reactions were occurring, and it may be associated with the presence of *Clostridium* sp. or other ketone-producing fermenters that also dechlorinate ethenes. MEK was observed at a maximum of 3200 $\mu g\ L^{-1}$ on Day 249. At other times, MEK concentrations were below the reporting limit of 100 $\mu g\ L^{-1}$.

The injected carbon source consisted of corn syrup and cheese whey that undergoes fermentation to produce by-products such as alcohols, acids and ketones, and carbon dioxide. Ketones are also formed as intermediates in the bacterial metabolism of short-chain hydrocarbons such as alkanes and alkenes, and can be utilized as primary carbon and energy source by a number of diverse bacteria. Studies of anaerobic bacteria such as facultative nitrate reducers, sulfate reducers, and fermentative bacteria have demonstrated that acetone metabolism proceeds by a carboxylation reaction.[87–89] Acetone is the simplest of the ketone family; however, degradation of higher ketones follows the same carboxylation reaction requiring carbon dioxide for the reaction to proceed.[87–89] Collectively, these lab studies show that we can expect degradation of acetone and higher ketones in an anaerobic setting.[88,91]

Additional information was obtained from the Case Study 1 site in southwestern Ohio, in a high-carbonate aquifer. Groundwater flow velocity is approximately 1 ft day^{-1}, and the contaminated formation is approximately 8 ft in thickness. The site was contaminated at various times by oils and solvents that were deposited in the formation during a long period of high regional groundwater pumping, which significantly depressed the aquifer surface elevation. Aquifer pumping patterns changed after the oils and solvents were deposited, and groundwater elevations rebounded, inundating a large mass of residual contamination.

Injection zone wells were sampled during the first year of operation at approximately 180 and 265 days post injection initiation. Ketones including acetone, methyl ethyl ketone (MEK), and methyl isobutyl ketone (MIBK) were detected during these sampling events. Two separate injection lines were sampled, and they contained comparable ketone concentrations. Four downgradient wells were sampled during the same time period located at 30, 100, 150, and 300 days downgradient of the injection lines.

MEK was selected as the representative compound for the ketone analysis as it was detected at elevated concentrations at several locations. An analysis of MEK concentrations vs. distance from the injection line was performed with the injection line being distance = 0. The concentrations of MEK in injection line wells were averaged to obtain one value for analysis. An additional sampling event is presented in this analysis that excluded injection well concentrations. This sampling event took place on day = 300 post injection.

MEK concentrations declined with distance from the injection line in all sampling events, as shown in Figure 2.28. MEK was detected at part per million concentrations in the injection wells and concentrations 150 days or further downgradient were either nondetect or in the part per billion range.

Ketone compounds can naturally be produced through the fermentation of sugar. Published laboratory studies demonstrate the anaerobic degradation of ketone compounds. The case study presented indicates that ketones produced during the application of ERD technology will be degraded as the groundwater travels away from the injection area.

CASE STUDY 4

Case Study 4 was selected to show the results of microcosm enrichments using molasses, with a comparison of biostimulation-only and biostimulation plus bioaugmentation using the KB-1 consortium that includes *D. ethenogenes*-like bacteria.

Case Study 4 is the site of TCE contamination in groundwater without evidence of significant dechlorination. Aquifer samples were collected for analysis of 16S rRNA by PCR and comparison with RNA primers that are characteristic of *D. ethenogenes* and related dechlorinating bacteria. These tests were negative for the presence of *D. ethenogenes* or related bacteria. This is a scenario for which some recent claims suggest biostimulation will not succeed due to a lack of known dechlorinating species.[49] The viewpoint among biostimulation practitioners is that dechlorination will be a little slow at first if no degradation products were found prior to

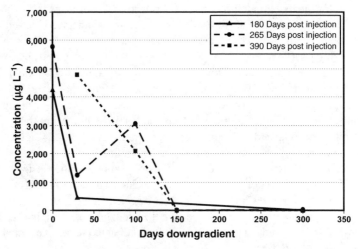

FIGURE 2.28 Methyl ethyl ketone concentrations vs. distance. Three sampling events are included in the analysis.

carbon loading. Dechlorination rates will be enhanced within a short time frame as appropriate bacterial populations build in response to carbon loading into the aquifer microbial continuum and resulting shift in the biogeochemical conditions.

Microcosms were established with aquifer soils and groundwater amended with TCE at a 21 μmol L^{-1} concentration. Triplicate microcosms were prepared for sterile controls, TCE-only controls, and experimental cultures for biostimulation-only and biostimulation plus bioaugmentation. However, there were no provisions made to control the pH within the microcosms. Molasses was used as the electron donor for all experimental treatments. Two concentrations were targeted: 300 mg TOC per liter and 1500 mg TOC per liter. Measured values for dissolved organic carbon were lower than targeted.

No changes in TCE were observed in sterile or unamended controls, but two of the biostimulation trials showed significant dechlorination. Of these, one showed near-complete dechlorination of *cis*-DCE, which contradicts the expectations of bioaugmentation proponents. Figure 2.29(A) shows the pattern of TCE degradation observed in one of the microcosms that was treated with molasses only. The results depicted in Figure 2.29(A) are consistent with the viewpoint that many species and consortia are capable of completely dechlorinating TCE, and that the bioaugmentation process shortens the lag time for dechlorination. However, dechlorination is expected to proceed, even when one particular species is absent from the formation.

Analysis of the dissolved organic carbon and methane concentrations show that the onset of *cis*-DCE dechlorination coincides with the onset of methanogenesis, as seen in Figure 2.29(B). After the Day 82 sampling, methane began a dramatic rise while *cis*-DCE began a rapid dechlorination to VC. Dissolved organic carbon levels fell to extreme low levels by Day 144, and the culture has not achieved the rapid dechlorination of VC observed at other sites. Additional molasses was injected into the microcosm, and the sampling program has been extended to determine whether dechlorination can be completed.

As expected, bioaugmentation shortened the lag time to achieve dechlorination, as shown in Figure 2.29(C) (note the shorter time span for the bioaugmentation graphs, relative to the biostimulation graphs). Unlike the bioaugmentation-only microcosm, the KB-1-augmented culture showed simultaneous dechlorination of *cis*-DCE and VC. Rapid dechlorination of VC is unlike *D. ethenogenes* strain 195, which can only dechlorinate VC cometabolically.[56]

As in the biostimulation microcosm, onset of *cis*-DCE dechlorination coincided with the beginning of methanogenesis. However, the rate of methane formation in the KB-1-amended culture never achieved the rate observed in the biostimulation trial, as can be seen through comparison of Figure 2.29(B) and (D). This is suggestive of a greater role for methanogens in dechlorination observed during the biostimulation trial. It is also suggestive of competitive interaction between KB-1 bacteria and the methanogens for limited hydrogen.

The bioaugmented culture utilized a smaller portion of the dissolved organic carbon in the process of achieving complete dechlorination. Efficient use of electron donors is an argument raised in favor of bioaugmentation. However, the electron donor is one of the smaller cost elements of reductive dechlorination, and the cost of inoculum, plus the capital and operating cost for the withdrawal, amendment, and

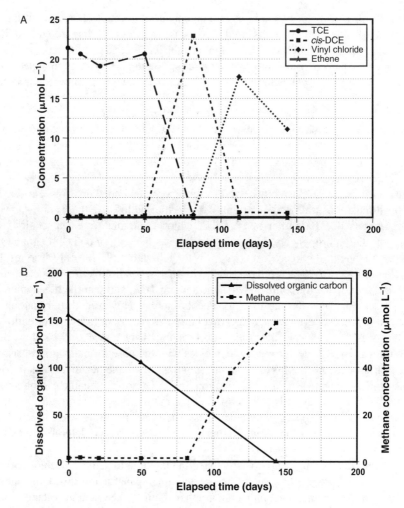

FIGURE 2.29 (A) Results of molasses-only treatment of a microcosm constructed with aquifer soil that tested negative for *Dehalococcoides* or related 16S rRNA. (B) Dissolved organic carbon and methane concentrations in a microcosm treated with molasses only.

reinjection of groundwater needed to support the precise habitat control of bioaugmentation far outweigh the cost of inefficient use of electron donors.

The authors share a concern with others that the genomics approach is very sensitive to DNA extraction efficiencies and the limits of detection achievable with currently available gene probes.

At one of the sites in Colorado (where the authors and their colleagues are implementing an IRZ system for PCE dechlorination), microbial communities from seven sediment-boring samples and three subsequent composite samples were screened for the presence of *D. ethenogenes* by the targeted gene detection approach. The analysis failed to produce a positive result. This result "indicated" a site-specific absence of these organisms and suggested that complete anaerobic reductive dechlorination of

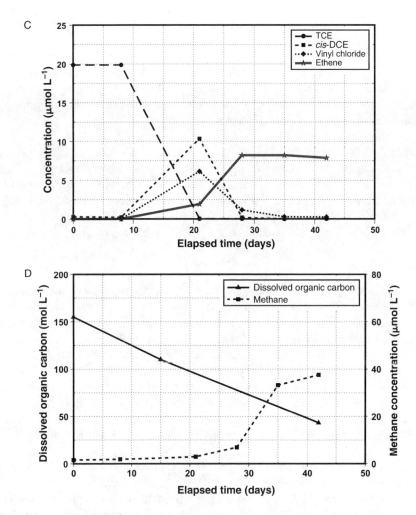

FIGURE 2.29 (continued) (C) Results of combined biostimulation and bioaugmentation for the dechlorination of TCE. (D) Dissolved organic carbon consumption and methane concentration patterns in microcosms treated with molasses plus KB-1 inoculum.

PCE to ethene cannot take place at this site. However, the full-scale IRZ system being implemented at this site in Colorado has been able to achieve complete reductive dechlorination from PCE all the way to ethene.

The absence of *D. ethenogenes* in the initial run of host samples raised concerns about the ability to extract DNA from the site samples, a critical factor in these molecular probe analyses. Therefore, three additional composite samples were submitted for analysis using various modifications to the DNA extraction protocols to increase extraction efficiency. One of the composite samples was spiked with DNA extract from *D. ethenogenes* that served as a positive control with a site-specific sample. These results still yielded negative results on the site composite samples with the exception of one positive, from which the spiked *D. ethenogenes* DNA could be detected with a fairly strong signal. Literature reports indicate that the sensitivity of these primers is in the range of 10^3 cells ml^{-1} or gram of sample.

Two conclusions the authors arrived upon from these analyses are that *D. ethenogenes* may be absent from the site and complete reductive dechlorination can still be achieved without the presence of this highly touted organism. On the other hand, *D. ethenogenes* may be present within the indigenous microbial community and its levels are so low as to be nondetectable.

Case Studies 5 through 13, which follow, present data and a brief discussion from many IRZ systems across the U.S. We have selected the locations of these systems across the country to counter the argument that "dechlorinating bacteria are not ubiquitous ... and hence bioaugmentation is necessary at most of the sites." The data presented below and the authors' experience accumulated from 250 sites strongly indicate that biostimulation — proper biogeochemical management of native microbial communities — alone can normally achieve complete reductive dechlorination.

CASE STUDY 5

This case study site lies in Indiana and was selected to demonstrate the reductive dechlorination of TCA. The injection area of this site was concentrated in the middle of a former underground storage tank basin. Enhanced reductive dechlorination was applied to a plume consisting of TCE and 1,1,1-TCA and their respective degradation products, with a horizontal groundwater velocity of approximately 0.4 ft day^{-1}. The carbon injection system consisted of three wells, spaced at 15-ft intervals.

Molasses was selected as the most cost-effective electron donor for the site, and injections were performed at 2-week intervals for 5 months starting in May 2002. The dechlorination process began immediately, with molasses injections that raised the average injection zone TOC concentration to 4000 mg L^{-1}. The monitoring location was approximately 120 days downgradient from the injection zone, and TOC concentrations ranged between 85 and 300 mg L^{-1} at the monitoring location. Within the 5-month injection period, 1,1,1-TCA concentrations decreased over 99% at a monitoring well located 120 days downgradient from the injection zone, as shown in Figure 2.30(A) and Figure 2.30(B), and concentrations had peaked and declined by the end of the 5-month study. It is important to note that complete dechlorination was achieved during the onset of methanogenic conditions.

CASE STUDY 6

Figure 2.31(A) and Figure 2.31(B) describe the enhanced solubilization of adsorbed and NAPL phase as a result of the implementation of an ERD system within an IRZ. This effect significantly increases the dissolved phase concentrations of the chlorinated ethenes within the IRZ. The authors coined the phrase "DNAPL mining" to describe this effect.

The site from which the performance data are presented in Figure 2.31(A) and Figure 2.31(B) is located in South Carolina. The groundwater velocity is in the range of 60 to 100 ft per year. Presentation of the data in this format is another method available to describe the system perfomance. The DNAPL mining phenomenon is described in detail in the discussion of Case Study 9.

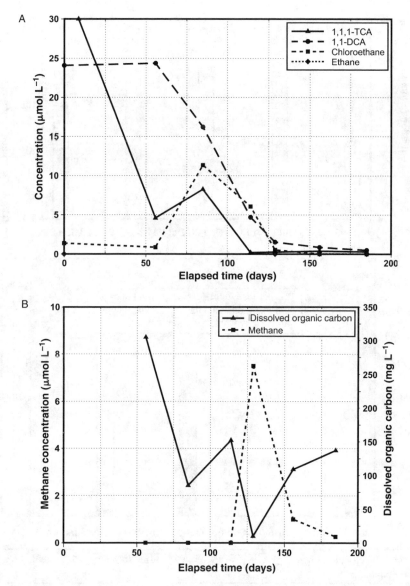

FIGURE 2.30 (A) Performance data for a site in Indiana from a monitoring well, 120 days downgradient. (B) TOC and methane data from the same well.

Case Study 7

This site was a former metal plating facility and was contaminated with TCE and Cr(VI). Very few daughter products were present prior to injection of molasses. A grid-like IRZ injection system was installed throughout the entire plume (2 acres in size) and injection of molasses began during the first quarter of 1996. Figure 2.32 and Figure 2.33 present the degradation and remediation of TCE and the formed daughter products in

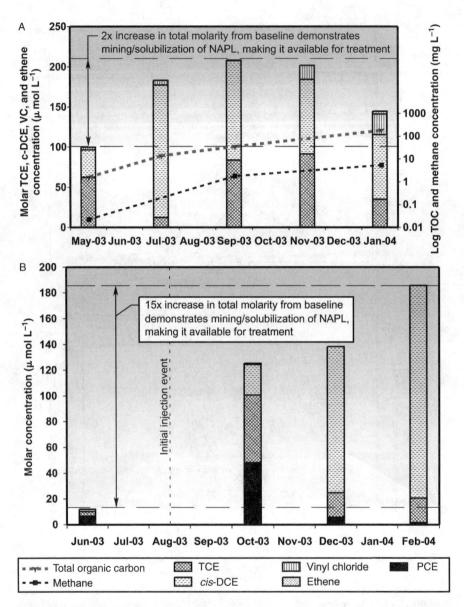

FIGURE 2.31 (A) Degradation trends demonstrating solubilization and treatment of NAPL at a site in South Carolina. (B) Degradation trends demonstrating solubilization site in Alabama.

Microbial Reactive Zones

a single well and in terms of average concentrations throughout the entire plume. Figure 2.32 describes the highest concentration of TCE found at the site and its decline during the implementation of the IRZ. The increased concentration of TCE at this well, at 130 μm (17 ppm), is a result of the microbial surfactant effect. The REDOX

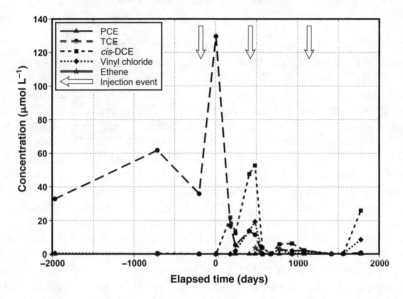

FIGURE 2.32 Performance data with concentrated batch injections at a site in California.

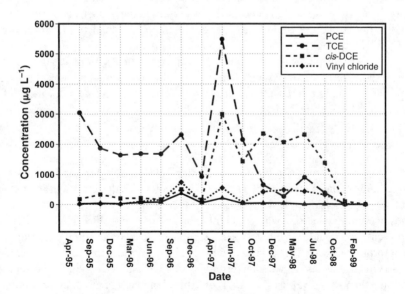

FIGURE 2.33 Average treatment results from the site in California.

potential within the plume was maintained at lower than -250 mV via batch injections of molasses. The TOC concentrations were always maintained above 200 ppm. This was an IRZ system implemented in 1996 during the infancy of this technology from a field implementation perspective and the authors have learned a lot since that time frame. Today, we would inject the organic carbon at more frequent intervals and at lower concentrations to maintain more uniform TOC levels within the IRZ.

CASE STUDY 8

Figure 2.34 presents data collected from an IRZ established in a fractured, low-permeability bedrock formation in New Jersey. Dissolved PCE concentrations in groundwater were observed at concentrations up to 90 mg L^{-1} (we are using mg L^{-1} here to make the readers easily appreciate the performance of this system under DNAPL conditions in a fractured bedrock environment). Shortly after molasses injections started, PCE concentrations dropped faster and TCE and *cis*-DCE concentrations increased. Within a 2-month period PCE and TCE continued to decrease rapidly while *cis*-DCE concentrations increased to 160 mg L^{-1}, which on a molar basis is much higher than the initial chlorinated alkene concentrations. This is due to the enhanced desorption of the sorbed mass by biosurfactants and cosolvency by fermentation products formed within the reactive zone. Reduced K_{oc} values for *cis*-DCE also keep more mass of this compound in the dissolved phase. A very important observation to note again is that, even at these high concentrations of chlorinated ethenes, complete transformation of PCE took place in the presence of methane up to 12 mg L^{-1}. This evidence indicates that the notion of competitive inhibition to reductive dechlorination by methanogenic bacteria is only important in microcosms, not in full-scale ERD applications. We saw clear evidence of complete reductive dechlorination at very high concentrations of PCE with biostimulation, alone, at this site. We are seeing some VC and we attribute that to the very high concentrations present in the aqueous and sorbed phase and also due to the fluctuations in TOC concentrations. Complete transformation of PCE is evidenced by ethene concentrations up to 12 mg L^{-1}.

At this site, the plume was very long and a pump and treat system was in place prior to the installation of the IRZ system. The primary objective was to implement a containment IRZ curtain to bifurcate the plume. Figure 2.34(B), describe the locations of the containment IRZ curtain and the bifurcation of the plume in a short time frame.

CASE STUDY 9

Performance data from another highly contaminated site in Alabama are shown in Figure 2.35(A) and Figure 2.35(B). These are shown mostly to present the authors' level of confidence in implementing ERD systems within highly contaminated source zones, including locations where a highly adsorbed or emulsified NAPL phase may be present. You could immediately see a significant level of accumulation caused by the desorption effects. This accumulation always takes place in the form of *cis*-DCE for two main reasons: rapid transformation of PCE and TCE within the reducing zone, and the low K_{oc} value of *cis*-DCE. A 10-fold increase in molar concentration of the chlorinated VOCs was seen as a result of the "mining" of the DNAPL within the IRZ. This effect is described in detail in Chapter 5.

Microbial Reactive Zones

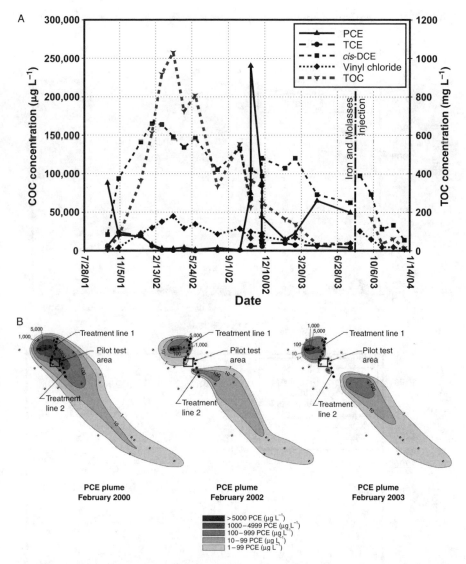

FIGURE 2.34 Performance data from a site in New Jersey where flowing DNAPL was recovered from a bedrock environment.

This field pilot study is ongoing and we see positive results with increasing concentrations of VC and the onset of ethene formation along with the decline in *cis*-DCE concentrations.

The injection of an abundant source of easily degradable organic carbon during the application of ERD typically results in a rapid and large increase in the population of microorganisms in the treatment zone. As in any microbiological system, this large population increase will also result in an increase in production of natural *biosurfactants* and *bioemulsifiers* by the microorganisms. Natural biosurfactants result in desorption of the chlorinated contaminants adsorbed to the aquifer media.

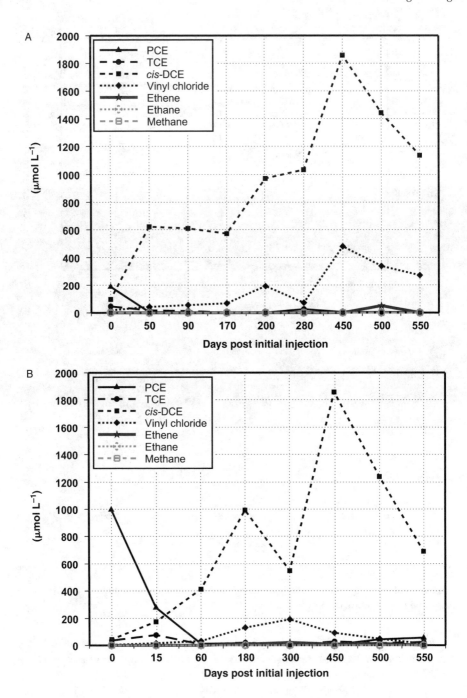

FIGURE 2.35 (A) Performance data from a highly contaminated DNAPL site in Alabama. (B) Performance data from the same highly contaminated site in Alabama.

Microbial Reactive Zones

To assimilate less soluble substrates, such as chlorinated solvents, microorganisms require a large contact area between themselves and the contaminant. They achieve this by emulsifying the adsorbed contaminants into an aqueous phase. Most microbes synthesize and excrete chemicals that promote such emulsification. These excreted chemicals fall into two main groups: biosurfactants and bioemulsifiers. In addition, the fermentation by-products such as alcohols and ketones also increase the solubilization of the adsorbed contaminant mass by cosolvency effects.

Biosurfactants reduce the interfacial tension between water and the chlorinated contaminant so that the chlorinated contaminant is easily microemulsified into the water phase. These microemulsion droplets are known to be smaller than microbial cells. Some bacterial glycolipids are extremely effective surfactants. In addition to enhancing the "mobilization" of the contaminants by microemulsions, biosurfactants can also increase apparent solubilities by partitioning the contaminants into surfactant micelles.

This desorption, or natural surfactant effect, is observed in many biological treatment processes as an increase in the constituent levels in the treatment zone and, in some cases, downgradient of the treatment zone. In some cases, the constituent concentrations in the treatment zone may remain unchanged, due to increased solubilization of the contaminants, for a short period even when biodegradation end-product data support the conclusion that sufficient mass is being degraded by the ERD processes.

The production of surfactants that facilitate the partitioning of contaminants from the DNAPL to the dissolved phase (thus resulting in enhanced biodegradation) has received considerable attention recently. The success of this approach depends on enhancing and maintaining biodegradation rates faster than the rate of mass transfer from NAPL to the dissolved phase. The authors' experience in implementing ERD systems within an IRZ at highly contaminated sites via the injection of very cheap organic substrates (such as molasses and cheese whey) precludes the need to inject very expensive "designer" substrates just for cosmetic reasons.

CASE STUDY 10

Performance data collected from an IRZ site in Indiana are presented in Figure 2.36(A) and Figure 2.36(B). The monitoring well is located 60 days of travel time downgradient. It is clear to see the complete transformation of TCE within 6 months. The initial increase of cis-DCE is due to biosurfactant effects and reduced K_{oc} values. We could see a direct correlation between the need of the onset of sustained methanogenic conditions and complete reductive chlorination. When TOC and methane levels dropped, there was an accumulation of VC. Hence, it is very important to maintain sustained methanogenic conditions within the reactive zone to achieve complete reductive dechlorination.

CASE STUDY 11

Figure 2.37(A) and Figure 2.37(B) present the performance data from another ERD system implemented in Pennsylvania. The monitoring well is located 30 days of travel time downgradient of the injection zone. We again saw the "classic" accumulation of cis-DCE due to the enhanced levels of biosurfactants and cosolvents. This case study illustrates an example of maintaining a minimum level of TOC and sustained methanogenic conditions.

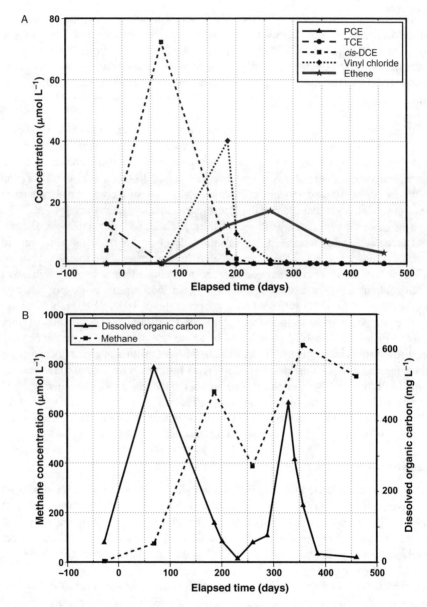

FIGURE 2.36 (A) Performance data from another site in Indiana. (B) TOC and methane concentrations from the same well.

CASE STUDY 12

Figure 2.38 presents the data from an IRZ system implemented in South Carolina for dechlorinating chlorinated methanes, including CT and chloroform. We saw complete reductive dechlorination of CT via chloroform and methylene chloride. At present the authors have initiated an ERD system in West Virginia for CT transformation at concentrations of 60 to 100 mg L^{-1}.

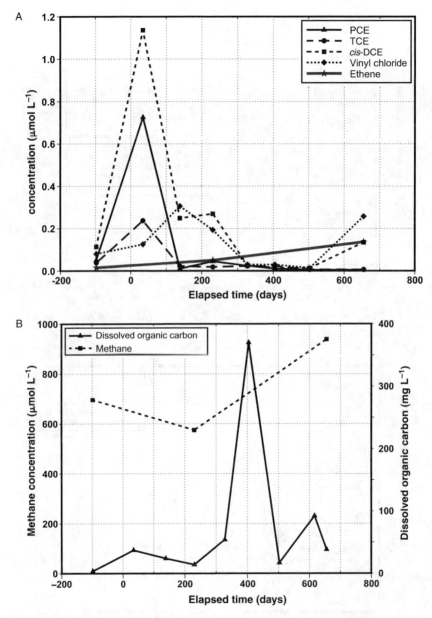

FIGURE 2.37 (A) Performance data from a site in Pennsylvania. (B) TOC and methane data from the same well.

Case Study 13

Figure 2.39 presents the performance data collected from an IRZ system in New Jersey. As expected, we saw an initial increase in TCE and *cis*-DCE due to desorption by surfactant and cosolvency effects. When concentrations of TOC

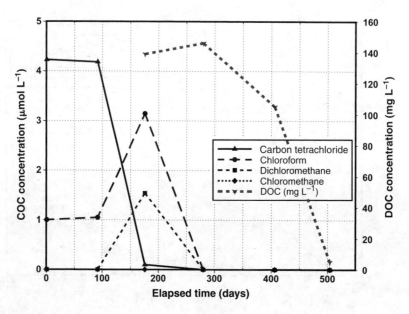

FIGURE 2.38 Performance data for reduction of chlorinated methane(s) from a site in South Carolina.

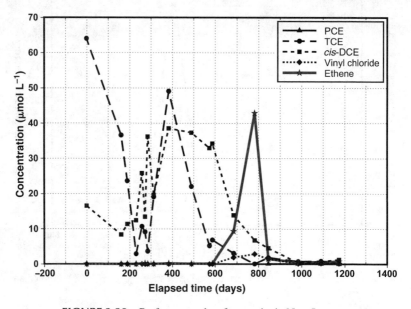

FIGURE 2.39 Performance data from a site in New Jersey.

dropped significantly below a threshold level at this well 175 days of travel time downgradient, we did not see substantial reductions in both TCE and *cis*-DCE levels (Figure 2.40).

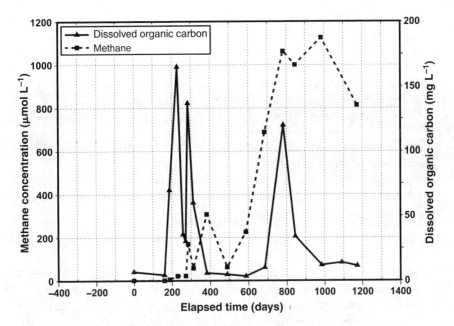

FIGURE 2.40 TOC and methane concentrations from a site in New Jersey.

The onset of methanogenic conditions was the beginning of the decline of TCE and *cis*-DCE. However, complete reductive dechlorination was observed without any accumulation of VC or ethene. When the TOC levels were increased in the injection wells, methanogenic conditions in the downgradient zone were enhanced. At this point we could see an accumulation of VC and significant accumulation of ethene.

The initial transformation of *cis*-DCE (without VC and ethene) was probably due to abiotic reactions, given the high Fe concentrations in the formation. The Fe minerals responsible for this transformation may be the oxidized species as opposed to FeS minerals, evidenced by the redox recovery of this zone during that period (very low TOC and methane). It should be noted that this monitoring well is 175 days (of travel time) downgradient of the injection zone.

The onset of highly reducing and sustained methanogenic conditions triggered the biotic transformation of *cis*-DCE to VC and ethene. High concentrations of ethene (based on molar mass balance) at this downgradient location indicates that, at these highly reducing conditions, complete reductive dechlorination is achieved by biotic transformations. As discussed earlier, the "trigger" here was the onset of sustained methanogenic conditions. In the absence of the abiotic reactions seen during the initial period, this site would have been "branded" as the classic site stalling due to the absence of *D. ethenogenes*.

2.3.1.8 Evaluating Degradation Signatures

Numerous enzymes and cofactors have been shown to dechlorinate the chlorinated alkenes PCE, TCE, *cis*-DCE, and VC. Each of these reactions occurs in laboratory culture at a rate determined by the enzyme or cofactor kinetics, the microbial biomass and the metabolic status of the microbial populations. Dechlorination rates vary dramatically among the microbial chemistries documented in the scientific literature, and it seems likely that patterns observed in the laboratory will scale up into similar patterns that will be observable when large microbial populations dechlorinate solvents in contaminated aquifers. Table 2.6 provides rate data for the dechlorination of PCE in laboratory cultures. The data are normalized to $\mu mol\ h^{-1}\ mg^{-1}$ protein.

To test the hypothesis that degradation kinetic signatures will emerge from aquifers under treatment with reductive dechlorination, it was first necessary to develop a calculation system that captures the sequential dechlorination of perchloroethene and trichloroethene, one chlorine at a time.

For enzyme or cofactor reactions that dechlorinate one chlorine atom at a time, the reaction sequence is described by Equation 2.10. The kinetic rate constant for each reaction is independent of the others.

$$PCE \xrightarrow{k_1} TCE \xrightarrow{k_2} DCE \xrightarrow{k_3} VC \xrightarrow{k_4} ETH \xrightarrow{k_5} Ethane \qquad (2.10)$$

For this analysis, we will assume that the chemical reaction rates are controlled by the concentration of each chlorinated alkene, and the respective kinetic rate constant. Reactions that meet these conditions are termed first-order reactions, and Equation (2.11) describes the rate of PCE degradation as a function of the kinetic rate constant and the initial PCE concentration.

$$\frac{dPCE}{dt} = -k_1 \cdot PCE \qquad (2.11)$$

Equation (2.11) is a differential equation, and the solution to Equation (2.11) is the familiar first-order degradation calculation shown in Equation (2.12).

$$PCE = PCE_0 \cdot e^{-k_1 \cdot t} \qquad (2.12)$$

The rate of change of TCE is the net of its production rate (equal to the PCE dechlorination rate) and its dechlorination rate, as shown in Equation 2.13. The solution to the differential given in Equation (2.13) is shown in Equation (2.14).

TABLE 2.6

Degradation Rate Constants Used in Conjunction with the Solutions to Equations (2.13) through (2.17), to Calculate Sequential Degradation of Perchloroethene to Ethene for Figure 2.29(A) and (C)

	Degradation Rate Constant (day^{-1})				
Figure	k_1	k_2	k_3	k_4	k_5
2.41	0.20	0.05	0.12	0.0004	0.0
2.42(B)	0.014	0.035	0.042	6.9	0.025

The solutions to differential Equations (2.15) through (2.17) are more complicated than for Equation (2.13), and are not provided here.

$$\frac{dTCE}{dt} = k_1 \cdot PCE - k_2 \cdot TCE \tag{2.13}$$

$$TCE = \frac{k_1 \cdot PCE_0}{(k_2 - k_1)} (e^{-k_1 t} - e^{-k_2 t}) + TCE_0 \cdot e^{-k_2 t} \tag{2.14}$$

$$\frac{dDCE}{dt} = k_2 \cdot TCE - k_3 \cdot DCE \tag{2.15}$$

$$\frac{dVC}{dt} = k_3 \cdot DCE - k_4 \cdot VC \tag{2.16}$$

$$\frac{dETH}{dt} = k_4 \cdot VC - k_5 \cdot ETH \tag{2.17}$$

The solution to this series of differential equations provides a valuable tool to study dechlorination patterns in laboratory and field data. Figure 2.41 shows the results of sequential degradation when perchloroethene has a starting concentration of 100 μmol L^{-1}, and the remaining compounds in the sequence are initially at zero. Table 2.6 shows the key equation parameters for Figure 2.42(A) and Figure 2.42(B) and each of the subsequent figures.

The degradation rate constants used to calculate sequential degradation in Figure 2.41 were based on the relative enzyme kinetic constants observed for PCErdase and TCErdase extracted from *D. ethenogenes*.[44] The resulting concentration peaks of TCE and *cis*-DCE are superimposed as in the previous studies, suggesting the sequential degradation calculation is in general agreement with degradation signatures of the enzyme kinetic studies.

Case Study 1 provides a field dataset from which degradation rate constants can be extracted. The extracted rate constants can then be used in the sequential degradation calculation to simulate actual field degradation patterns. First, the dataset must be normalized to eliminate the effects of unequal sorption among compounds in the degradation sequence.

Aqueous-phase concentrations of the chlorinated alkenes showed an increasing trend in the Case 1 data as the degradation proceeded. It was recognized that the dechlorination process reduced the strength of solvent adsorption to soil organic matter, causing the apparent increase in total alkene concentrations. To counteract this effect, the Case 1 data were normalized to obtain estimates of the total aquifer concentration for each constituent.

For each compound, the total aquifer concentration was estimated from the aqueous-phase concentration, C_w, the organic carbon fraction, f_{oc}, the organic carbon partition coefficient, K_{oc}, the soil bulk density, ρ_b, and the volumetric water content, θ.

$$C_{tot} = (C_w \cdot K_{oc} \cdot f_{oc} \cdot \rho_b) + (C_w \cdot \theta) \tag{2.18}$$

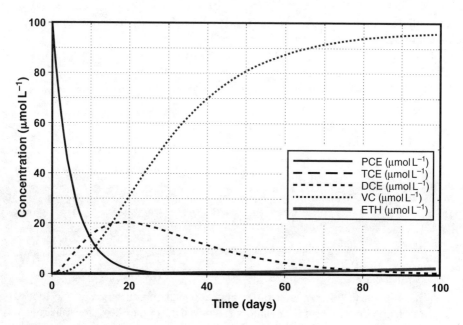

FIGURE 2.41 Sequential degradation of chlorinated ethenes, with an initial concentration of 100 μmol L^{-1} perchloroethene.

Calculations were performed using Equation (2.18) to normalize the Case 1 data. Conservation of mass was assumed, which required that the total alkene estimate remain stable until degradation of the final compound (ethene) was under way. Values for f_{oc} and K_{oc} for ethene were adjusted by trial and error until a relatively stable estimate for total alkenes was obtained. With the large number of variables that affect this calculation, a more rigorous approach is not available. Figure 2.42(A) shows the resulting pattern of degradation, which does not match the first-order sequential degradation shown in Figure 2.41.

In the degradation pattern that emerged from the Case 1 remedial system, the degradation rates for TCE, *cis*-DCE, and ethene were not constant over time. The TCE degradation rate constant, for example, was lower during the first 100 days than during the second 100 days. Degradation of *cis*-DCE appeared to be completely inhibited until TCE levels dropped to very low levels, and degradation of ethene appears to have been inhibited in the presence of *cis*-DCE. Combined, these data suggest the passing of threshold events during the degradation process.

Solutions to the differential equations for sequential degradation require that degradation rates are constant over time. This means that products of degradation are themselves immediately available for degradation. The degradation pattern observed in the Case 1 remedial system exhibited variable degradation rates for TCE, *cis*-DCE, and ethene. As a result, it was necessary to invoke two rate constants for each of those compounds: one representing prethreshold degradation and a second representing postthreshold degradation. The results of the variable-rate degradation are shown in Figure 2.42(B).

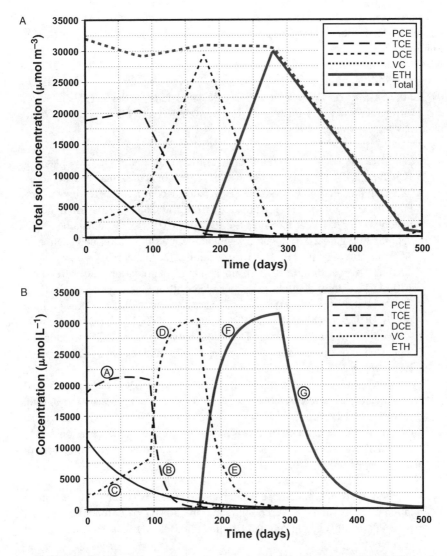

FIGURE 2.42 (A) Estimated total soil concentrations in the sequential dechlorination process at the Case 1 site. Total soil concentrations were estimated from aqueous-phase concentrations, K values for each compound, and estimates of soil organic content. (B) Simulation of the sequential degradation pattern, PCE through ethene, using initial concentrations and dual kinetic rate constants extracted from the Case 1 dataset. Curve segments A through G are explained in the accompanying text.

The results of calculations shown in Figure 2.42(B) and the Case Study 1 data on which they were based are a dramatic departure from the continuous, sequential degradation signature described in Figure 2.41. To obtain a simulation that matched the observed degradation signature, it was necessary to abandon the direct solutions to the differential equations representing the PCE-to-ethene sequence, because the solutions were based on constant first-order rates of degradation for each compound in the sequence. To obtain agreement between simulated concentration patterns seen in Figure 2.42(B) with observations from Case Study 1 (shown in Figure 2.42(A)), it was necessary to use two rate constants for three of the compounds: TCE, *cis*-DCE, and ethene. The differential equations were solved iteratively, with values for dPCE/dt, dTCE/dt, dDCE/dt, dVC/dt, and dETH/dt calculated daily for the 500-day interval.

The degradation rate adjustments that were required to obtain a match with the observed degradation signature for Case Study 1 were as follows:

> *PCE.* PCE followed conventional first-order degradation as described in Equation (2.12), with a degradation rate constant of approximately 0.14 day^{-1}.
>
> *TCE.* TCE did not follow first-order degradation at constant rate. During the first 90 days, the value for k_2 (Equations (2.10), (2.13), and (2.15)) was approximately 0.0032 day^{-1}. k_2 then increased 25-fold to 0.08 for the remaining observations. In Figure 2.42(B), curve segment A shows the slower degradation of TCE and curve segment B shows degradation at the higher rate.
>
> *DCE.* *cis*-DCE accumulated with no apparent degradation until after Day 170. At that time, degradation proceeded with a kinetic rate constant, k_3, approximately equal to 0.042 day^{-1}. Curve segment C shows a slow buildup of *cis*-DCE, matching the slow degradation of TCE. Curve segment D shows an increased rate of *cis*-DCE accumulation when TCE degradation increased to its higher rate. *cis*-DCE degradation did not begin until approximately Day 170, as shown in curve segment E.
>
> *VC.* VC degradation was significantly higher than *cis*-DCE degradation, and no threshold events were observed. As a result of its high degradation rate, relative to *cis*-DCE degradation, no accumulations of VC were observed. To explain the observed pattern, it was necessary for the VC degradation rate to exceed the *cis*-DCE by at least 20-fold. Calculations supporting Figure 2.42(B) were made with k_4 at 0.9 day^{-1}.
>
> *Ethene.* Like *cis*-DCE, there was an accumulation of ethene with no apparent degradation which, for ethene, lasted approximately 280 days. The accumulation phase is shown as curve segment F in 2.42(B). After that, degradation proceeded with a k_5 approximately equal to 0.025 day^{-1}, as shown on curve segment G.

From these data, we can conclude that there are threshold events that precede TCE, *cis*-DCE, and ethene degradation. We can also see that the degradation rate constant for VC is very high — much higher than for *cis*-DCE.

Although the fact that there were threshold events during Case 1 degradation is clear from the data, the nature of the thresholds cannot be determined. One set of conjectures that is consistent with the data and that has been invoked by other researchers in studies of *D. ethenogenes* is the competitive inhibition of degradation by more chlorinated alkenes: PCE partially inhibits TCE degradation at levels above 3000 μmol L^{-1}; TCE fully inhibits *cis*-DCE degradation at levels above 300 μmol L^{-1}, and *cis*-DCE fully inhibits ethene degradation at levels above 300 μmol L^{-1}. These are conjectures only, although they are consistent with observations at the Case Study 1 site and possibly others.

However, the authors also feel that these threshold events could be triggered merely by the shifting of the microbial ecology within the reactive zone. Microbial population shift or growth leading to complete dechlorination can be attributed to the following:

- Microbial populations have the benefit of a long acclimatization time to the parent compound(s), such as PCE or TCE. Hence, the populations required for dechlorinating the parent compounds are already developed. When optimum biogeochemical conditions are created, the degradation product *cis*-DCE accumulates because the microbial community was never exposed to this compound as an energy source or cometabolic substrate.
- Once the microbial population is exposed to the daughter compounds, species specifically responsible for dechlorinating, *cis*-DCE and VC, have the ability to grow and multiply. Optimum biogeochemical conditions and availability of substrate could be the specific triggers that lead toward complete reductive dechlorination.
- It is a fact that any level of reductive dechlorination beyond *cis*-DCE prior to carbon loading significantly reduces the lag time required to achieve complete dechlorination. This may be due to the reason explained above, where creation of optimum biogeochemical conditions is the only trigger required to enhance the reductive dechlorination rates.
- Based on the evaluation of data from more than 200 ERD sites, the authors conclude that formation of methanogenic conditions is a required trigger to achieve complete reductive dechlorination, specifically for the steps *cis*-DCE → VC → ethene.

The Case 1 signature is distinct from enzyme systems documented for *D. ethenogenes* and for intact cultures.[39,91]

- No VC degradation inhibition was observed. Inhibition of VC degradation was observed by precursor compounds, which was not observed in the Case 1 study.[39] Others have shown that VC degradation was very slow compared to *cis*-DCE.[44] The opposite was true during Case 1 studies.
- Limited formation of *trans*-DCE was observed. The pretreatment *cis*-DCE concentration was 1.8 μmol L^{-1}, and *trans*-DCE was 0.1 μmol L^{-1}. During TCE degradation, the *trans*-DCE concentration remained relatively low, peaking at 0.38 μmol, while *cis*-DCE increased to a peak value of 28 μmol L^{-1}. The *cis*-DCE to *trans*-DCE production ratio was greater than 90 to 1. It was shown that *trans*-DCE formation occurred at the same rate as *cis*-DCE formation in enzyme-level studies, and indicated that *trans*-DCE formation occurred, but they did not provide specific data.[39,44]

2.3.1.9 Summary and Conclusions

University-level funded research on enhancement of reductive dechlorination has focused almost exclusively on single-species isolates and energy-conserving metabolism. Some authors of these studies recommend adoption of these isolates at tools in a bioaugmentation approach to aquifer cleanup. In many cases, these bacteria have been deployed at field scale in systems that demanded groundwater pumping and reinjection, a long-discredited remedial strategy that now may be excused as a "high-tech" approach.

The fact that single-species studies dominate the literature threatens to obscure the long-standing field-scale accomplishments of broad microbial consortia through the biostimulation of native aquifer populations. Although inelegant, relative to a single-species attack with exquisite control of habitat chemistry, biostimulation of native populations through injection of electron donors at rates that exceed the recharge of electron acceptors is a predictable, reliable, and most importantly, cost-effective method of groundwater cleanup.

There are a number of concerns related to bioaugmentation, but the greatest is the fact that all the bioaugmentation studies published in peer-reviewed journals have been coupled with groundwater pumping and reinjection to date. This is required to maintain the aquifer chemistry within narrow limits required to support the injected bacterial culture. In addition, underground injection control permits may be difficult to obtain for the injection of unidentified bacteria. The microbial cultures currently available for bioaugmentation of chlorinated alkenes contain bacteria that are essential to the complete dechlorination process, but which have not been isolated or identified. This reduces the likelihood of injection permitting in many states.

There are circumstances in which the cost and risk associated with bioaugmentation are justified. However, any site that has ongoing reductive dechlorination (as shown by the presence of degradation products) can be treated very cost-effectively by biostimulation.

Commercial interests in the U.S. and Canada are now promoting bioaugmentation using a bacterial species that was initially isolated from an anaerobic enrichment culture developed from wastewater treatment plant sludge. Now named *D. ethenogenes* strain 195, the bacterium dechlorinates PCE and TCE to vinyl chloride in energy conserving metabolism but can only cometabolize VC in the presence of PCE. Other strains of *Dehalococcoides*-like bacteria have also been identified, but no single species has been identified that can fully dechlorinate PCE without the participation of consortium partners. Commercial bioaugmentation inocula that fully dechlorinate PCE or TCE to ethene contain numerous bacterial species, and their dechlorination abilities are not due to *D. ethenogenes*, alone.

The focus on *D. ethenogenes* and *Dehalococcides*-like bacteria has diverted attention from the benefits of alternative dechlorination processes such as cometabolic and abiotic reduction pathways and anaerobic oxidation mechanisms (an alternative to *D. ethenogenes* for *cis*-DCE and VC dechlorination). Other critical in situ remedial technology issues such as the development of reliable reagent distribution methods and the development of strategies that attack sorbed and non-aqueous-phase solvent mass are receiving too little attention.

Dehalococcoides, when present, is a welcome participant in the reductive dechlorination process. In fact, there is evidence that this species is present as a "naturally occurring" member of the microbial community in many solvent-contaminated aquifers. However, it is not cost-effective to design an enhanced reductive dechlorination project strictly to meet the narrow preferred habitat of this bacterium. In the field, much higher dechlorination rates can be achieved by promoting broader microbial participation, as has been shown in Case Study 1.

"High-performance" reductive dechlorination can only be engineered when the rate of electron-donor consumption exceeds the rate of electron-acceptor recharge. The carbon source must be highly mobile and highly degradable, and injected at rates commensurate with the overall flux of groundwater, electron acceptors, and CAHs that are moving through the treatment zone. The organic acids that form as breakdown compounds must be buffered by aquifer carbonates or by the addition of carbonates and bicarbonates to the injection mix. Certain site hydrogeologic characteristics require modification of the high-performance approach, but most can still be treated by ERD systems that are more cost-effective than alternative approaches.

Microbially catalyzed processes that take place during ERD are the result of the interactions between the biochemical conditions created in the engineered IRZ, and its complex and dynamic microbial community. Ecological selection pressures define the site-specific ERD microbial community. These pressures include changing availability of electron acceptors, pH, redox potential, electron-donor availability, and the chemical nature and concentration of the contaminants themselves. The ERD microbial community is capable of surviving and thriving in the conditions that develop within the IRZ. In this chemically and microbially complex environment, there is not any one degradation mechanism, nor is there one bacterium that is completely "responsible" for the biodegradation process. It is a broad community of microorganisms and a variety of mechanisms that bring about the desired degradation of chlorinated alkenes and alkanes.

The message that reductive dechlorination system designers should gain from most of the industry experience and peer-reviewed literature is that biostimulation approaches, alone, can achieve complete dechlorination of PCE in a predictable and efficient manner. Most importantly, the biostimulation approach can be readily scaled-up to field application. Conversely, the limitation of the remediation system to a particular species cannot guarantee complete dechlorination, and field application will require uneconomical control of the microbial habitat to sustain the selected species in the face of naturally occurring competition.

The authors have discouraged microbial analyses, especially the genetic assay for *D. ethenogenes*, due to the lack of correlation between pre-treatment test results and the ability to later achieve complete reductive dechlorination through biostimulation for ERD. We have encountered multiple instances in which a negative test result for *D. ethenogenes* was received, yet biostimulation still resulted in complete dechlorination of PCE and TCE. The authors would like the industry, especially the regulatory community, to demystify *D. ethenogenes*, particularly the notion that this species is absolutely essential for complete dechlorination.

The authors have repeatedly observed that through the addition of very cheap organic substrates such as molasses, cheese whey, and corn syrup, complete dechlorination of millimolar concentrations of PCE and TCE can be achieved, even in large-scale field applications. A very important observation to note is that complete reductive dechlorination is achieved most successfully when the system reaches methanogenic conditions. Hence, we strongly believe that the "apparent" need for slow-release "designer" organic substrate(s) to avoid the competitive inhibition of dechlorination by methogenic bacteria is false. Field-scale data from many sites are presented below to prove this point.

As an extension of the above point, the authors have also observed that the buildup of VC can be completely avoided if the redox conditions are sufficiently reduced in the sulfate-reducing to methanogenic conditions. We have also observed that there is a minimum TOC value associated with each site to maintain the biochemistry to avoid the buildup of VC. We have also observed that VC shows up only rarely when there is an abundance of ferric ions in the soil formations. Data from a few sites are presented to substantiate these conclusions.

2.3.1.10 Abiotic Reductive Dechlorination

ERD involves the introduction of an easily degradable substrate that enhances and alters the *in situ* microbial community as the system is driven anaerobically. For reductive dechlorination to be thermodynamically favorable, the oxidation–reduction potential (ORP) must be sufficiently low, with neither dissolved oxygen (DO) nor nitrate available as terminal electron acceptors. Remediation engineers have usually interpreted the removal of contaminants as the product of biological processes. Biologically, the chlorinated solvent may act as a primary electron acceptor via dehalorespiration, or be cometabolized under reducing conditions, including sulfate reduction and methanogenic environments.[88]

However, a series of recent articles have demonstrated that abiotic degradation processes, involving newly formed solid surfaces, especially iron sulfides and "green rust" minerals, may account for a significant portion of the mode of action of these already installed IRZ systems. Since these abiotic processes result in different transformation products, understanding their prevalence and efficiency is critical to the correct assessment of the performance of full-scale ERD systems. Specifically, if abiotic processes are not taken into account, sites may be judged to have "stalled" at DCE when they are actually proceeding to complete mineralization through abiotic processes. At some sites where the conditions are right to prevent stalling, we observe enhanced treatment from PCE to DCE without the classically expected breakdown products of further biodegradation — VC and ethene. Some interpret these data as stalling. However, a recent series of papers have indicated that these systems are probably not stalled, but rather are degrading DCE by an alternate, abiotic mechanism that proceeds to different products.[89,90,94] In particular, a variety of iron- and sulfur-bearing mineral species take part in degradation reactions with CAHs that occur at the mineral–water interface. Abiotic reaction conditions favor transformation of CAHs by dichloroelimination rather than by sequential hydrogenolysis; consequently, this pathway is desirable in that the production of comparatively more toxic daughter products is circumvented.

Several iron-bearing soil minerals have been studied as reactive materials for abiotic reductive dechlorination.[95–98] Iron sulfides, green rust, and iron oxides (magnetite) are among the most important. These minerals occur in aqueous iron-reducing and/or sulfate-reducing environments. Green rust (Fe_3O_4) compounds are compositionally variable, mixed valence Fe(II)/Fe(III)-layered hydroxides. The relevant iron-sulfide phases include disordered mackinawite, $Fe_{1+x}S$; mackinawite, $Fe_{1+x}S$; cubic iron sulfide, FeS; hexagonal pyrrhotite, $Fe_{1-x}S$; greigite, Fe_3S_4; smythite, Fe_9S_{11}; marcasite, orthorhombic FeS_2; and pyrite, cubic FeS_2. Pyrrhotite and pyrite represent the thermodynamically stable phases at the temperatures and pressures of early diagenesis. Disordered mackinawite, mackinawite, and greigite are metastable with respect to pyrite and/or stoichiometric pyrrhotite but are considered to be the principal precursor phases to pyrite.[90]

When ERD is implemented in an IRZ system, the necessary predecessors for formation of reactive iron sulfices are abundant:

- As the system becomes more anaerobic reduced iron species are formed, both from iron containing soil minerals, and from aqueous-phase iron (III).
- Sulfides are formed under sulfate-reducing conditions from aqueous sulfate. At least one commonly used ERD substrate, blackstrap molasses, has the additional benefit of containing roughly a percent by weight as sulfur — initially in both sulfate and sulfide forms but ultimately transformed into sulfide. Naturally occurring iron would be expected to react with H_2S or HS^- and initially produce reactive iron monosulfides, especially mackinawite and greigite.

On the downgradient edge of the anaerobic zones formed in ERD systems, conditions return gradually to aerobic as the organic substrate is exhausted. In these zones, the iron that was reduced and mobilized is reprecipitated as fresh iron oxyhydroxides and oxides. Thus, formation of abundant iron sulfide and GR mineral surfaces is, almost invariably, the result of implementation of ERD, especially when using molasses as the substrate.

Known abiotic degradation products of CAHs include chloroacetylenes and acetylene. These abiotic iron minerals can be formed *in situ* without the need for costly trenching technologies. Initial calculations suggest that these reaction rates are sufficient to achieve substantial treatment over thousands of feet to 1 mi of reactive zone when percentage levels are formed.[90] These conditions are practically achievable in full scale, plume-wide enhanced bioremediation systems.

Performance data from an ERD system implemented within an IRZ are presented in Figure 2.43(A) and (B). In Figure 2.43(A), MW-7 is located at a distance of about 30 ft upgradient with a travel time of approximately 52 days. Monitoring wells MW-1 to GGM-02-11X are located at distances of 15 to 200 ft. Even though the designers of this system picked the TOC injection dosage to create a reactive zone of 100 ft downgradient of the injection zone, reducing conditions were created even at a distance of 200 ft.

Total CAH concentration (all in the form of PCE) of 25.4 µM was flowing into the reactive zone (Figure 2.43(B)). There was an increase in CAH concentrations in the monitoring wells immediately downgradient of the reactive zone, resulting from desorption by the formed microbial surfactants. It is obvious that progressive mass destruction of CAHs was taking place within the IRZ. It is equally important to note that the sulfate concentrations were declining from the injection well (100s of ppm — not shown in the figure) and then reaching the background concentrations in the recovery zone. This indicates strongly sulfate-reducing conditions immediately downgradient of the injection zone.

The 90% mass destruction that took place within the reactive zone (Figure 2.43(B)) can be attributed only partially to the biotic transformations — especially from PCE → TCE → *cis*-DCE. However, the disappearance of *cis*-DCE (without the formation of VC and ethene) can be explained only by the abiotic reactions. Strong

A

9/22/2003	MW-7	IW	MW-1	MW-2	MW-3	MW-5	G6M-02-10X	G6M-02-11X
Distance*	−30	0	15	30	45	75	100	200
Travel time**	−52	0	26	52	78	129	172	345
PCE ($\mu g L^{-1}$)	4,200		4,000	1,800	670	960	**1**	590
TCE ($\mu g L^{-1}$)	3		230	540	1,100	7	**1**	31
DCE ($\mu g L^{-1}$)	**1**		1,400	1,900	1,900	530	720	37
Ethene ($ng L^{-1}$)	27		7	**2.5**	12	**2.5**	400	**2.5**
Methane ($\mu g L^{-1}$)	58		7,100	34,000	22,000	33,000	28,000	1,200
Sulfate ($mg L^{-1}$)	16		10	**2.9**	2	4.4	**2.4**	8
Fe^{2+} ($mg L^{-1}$)	**1**		68	86	89	150	180	**1**
As ($mg L^{-1}$)	**2.5**		320	340	460	360	490	**2.5**
TOC ($mg L^{-1}$)	**2.5**		69	14	6	26	200	19
pH (su)	3.4		4.9	6.0	6.1	6.3	6.6	6.6
ORP (mV)	171		−36	−115	−138	−101	−156	146
PCE (μM)	25		24	11	4	6	0.006	4
TCE (μM)	0.025		1.750	4.110	8.371	0.056	0.008	0.236
DCE (μM)	0.010		14.440	19.598	19.598	5.467	7.427	0.382
Ethene (μM)	0.001		0.000	0.000	0.000	0.000	0.014	0.000
Total micromolar Equivalent	25.4		40.3	34.6	32.0	11.3	7.5	4.2

Notes:
* Distance relative to injection well transect, parallel to groundwater flow.
** Based on advective groundwater flow rate.
IW Injection well transect.
2.5 Bold values represent half of the detection limit.

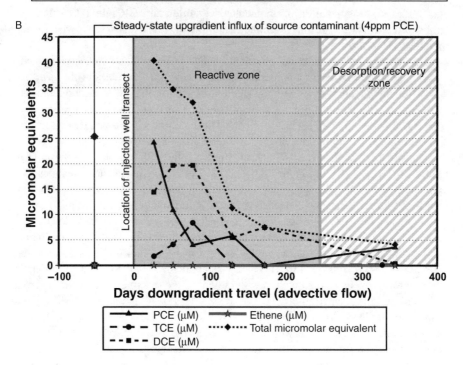

FIGURE 2.43 (A) Performance obtained from an IRZ system in Massachusetts, where the authors predict that abiotic reactions are responsible for complete reductive dechlorination. (B) Contaminant trends downgradient an IRZ barrier at the system described above.

Microbial Reactive Zones

sulfate-reducing conditions (observed by the SO_4^{2-} disappearance) immediately downgradient of the reactive zone and the relatively lower concentrations of Fe^{2+} would have resulted in the formation of FeS in that zone. The biggest contributor to abiotic mass destruction within this zone is probably the precipitated FeS minerals. Even though there is a net concentration of 4.2 μM present (as PCE) in the furthest monitoring well in the recovery zone, the DCE leaving the reactive zone (GGM-02-10X) must have been transformed abiotically by the reprecipitated iron oxyhydroxides and oxides. It is also important to note that the As solubilized within the reducing zone has reprecipitated within the recovery zone.

More than 90% (closer to 100% if we exclude the background PCE levels in the farthest downgradient monitoring well) mass destruction of the CAHs without any accumulation of VC and ethene was rather surprising. This can be explained only if the abiotic transformations were instantaneous. The FeS mineral generation must have very little lag time and dechlorination reactions must be very rapid to see no VC or ethene within the IRZ.

It has been hypothesized that water table fluctuations are a significant contributor to increased oxygen transfer from air to water for the recovery zone downgradient of the reactive zone to reestablish aerobic conditions (Figure 2.43(B)). It has been reported that zones with entrapped air saturations of up to 16% are formed during the ground-water table fluctuations.[99] Other natural processes that aid in restoring aerobic conditions in downgradient recovery zones include replenishment of DO as a result of diffusion from air in the unsaturated zone, recharge with DO rich rainwater, and trapped air bubbles.

2.3.2 IN SITU PRECIPITATION OF HEAVY METALS

Despite its widespread use, the term "heavy metal" does not have a rigorous scientific basis or a chemical definition. Although many of the elements listed as "heavy metals" have high specific gravities, major exceptions to this rule remain. The list below shows the heavy metals that are of significant environmental concern along with commonly occurring light alkali and alkali earth metals for comparison.[100] The numbers in parentheses represent the specific gravity of each metal.

Regulated heavy metals:	Cr (7.19), Co (8.90), Ni (8.90), Cu (8.96), Zn (7.13), Ag (10.5), Cd (8.65), Hg (13.6), Ti (11.9), Pb (11.4)
Regulated metalloids:	As (5.78), Se (4.79), Sb (6.69)
Commonly occurring light metals:	Na (0.47), Mg (1.74), K (0.86), Ca (1.55), Al (2.70)

Strictly from a chemistry viewpoint, heavy metals constitute transition and posttransition elements along with metalloids, namely, arsenic and selenium.[100] They are indeed significantly heavier than sodium, calcium, and other light metals.

These heavy metal elements often exist in different oxidation states in the subsurface. The reactivities, ionic charges, and solubilities of these metals in water vary widely. For their short- and long-term impacts, the maximum permissible concentrations of these heavy metals in drinking water as well as in groundwater are closely regulated through legislation. Nevertheless, barring the exceptions of Cd, Hg, and Pb, heavy metals are also required micronutrients. Toxicity and inhibitory effects of these elements are, thus, largely a function of concentration. However, nonessential heavy metal elements are considered to be inhibitory at all concentrations.[100]

The presence of metals in the subsurface environment can be in many forms. Metals can be present in elemental form, ionic form and/or organometallic form. Distribution of the commonly encountered metals in the subsurface can be categorized as follows:

- Elemental form
  - Mercury
  - Lead
  - Gold, silver, and the other noble metals
  - Metal alloys: brass (copper and zinc); bronze (copper, tin, and zinc); nickel–cadmium
- Ionic form
  - Arsenic: As(III) arsenite, As(V) arsenate
  - Chromium: trivalent Cr(III); hexavalent Cr(VI)
  - Iron: ferrous Fe(II); ferric Fe(III)
  - Copper, lead, zinc, cadmium: Cu^{1+}, Cu^{2+}, Pb^{2+}, Zn^{2+}, Cd^{2+}
  - Mercury: Hg^{1+}, Hg^{2+}
- Organometallic form
  - Dimethyl mercury: $Hg(CH_3)_2$
  - Dimethyl arsenic: $As_2(CH_3)_4$
  - Tetraethyl lead: $Pb(C_2H_5)_4$
  - Metal cyanide complexes: $Hg(CN)_2$; $Zn(CN)_4^{-2}$; $Cu(CN)_2^{-1}$; $Fe(CN)_6^{-4}$

Common range of concentrations of naturally encountered metals in the subsurface environment is shown in Table 2.7.

In order to understand the fate of heavy metals in the soil–water system, it is important to understand the general characteristics of soil and the chemistry of heavy metals in an aqueous solution. The speciation and fate of metals in the natural environment as well as their separation and control within engineered systems are ultimately governed by the specific chemical characteristics of each of these heavy metals. Such characteristics also dictate the biochemical reactions of metals as nutrients or toxicants. In order to develop an insight, let us consider the electronic configurations of a light metal cation (Ca^{2+}) and a heavy metal cation (Cu^{2+}).

$$Ca^{2+}: 1s^2 2s^2 p^6 3s^2 3p^6$$

$$Cu^{2+}: 1s^2 2s^2 2p^6 3s^2 3p^6 3d^9$$

Note that Ca^{2+} has the noble gas configuration of krypton, that is, its outermost electron shell is completely filled, and the octet formation is satisfied. Thus, Ca^{2+} is not a good electron acceptor and, hence, a poor Lewis acid. Ions like Ca^{2+} are not readily deformed by electric fields and have low polarizabilities. They are referred to as "hard" cations, and they form only outer sphere complexes with aqueous-phase ligands containing primarily oxygen-donor atoms.[100]

TABLE 2.7
Common Concentration Range of Metals in Soils (mg kg^{-1})[181]

Element	Range	Average
Antimony (Sb)	2–10	—
Arsenic (As)	1–50	5
Barium (Ba)	100–3000	430
Beryllium (Be)	0.1–40	6
Cadmium (Cd)	0.01–0.7	0.06
Chromium (Cr)	1–1000	100
Cobalt (Co)	1–40	8
Copper (Cu)	2–100	3
Lead (Pb)	2200	10
Mercury (Hg)	0.02–0.3	0.03
Nickel (Ni)	5–500	40
Selenium (Se)	0.1–2	0.3
Silver (Ag)	0.01–5	0.05
Tin (Sn)	2–200	1
Vanadium (V)	20–500	100
Zinc (Zn)	10–300	50

Source: Data from Reference 180.

In contrast, the transition metal cation Cu^{2+} or Cu(II) has an incomplete d-orbital and contains electron clouds more readily deformable by electric fields of other species. In general, these ions are fairly strong Lewis acids and tend to form inner sphere complexes with ligands in the aqueous phase. Electrostatically, Ca^{2+} and Cu^{2+} are identical, that is, both Ca^{2+} and Cu^{2+} have two charges. However, Cu(III) is a stronger Lewis acid or electron acceptor and a relatively "soft" cation. Metal cations can be classified as hard, borderline, and soft cations. Note that most of the heavy metals of interest fall under "borderline" and "soft." In general, the toxicity of metals increases as one moves from hard cations to borderline, and then to soft. Relative affinities of these metal ions to form complexes with O-, N-, and S-containing ligands vary widely. While hard cations prefer oxygen-donating ligands (Lewis bases), borderline and soft cations exhibit higher affinities toward nitrogenous and sulfurous species.

Classification of selected metal cations and the specific characteristics of each type are described as follows:[99]

Hard cations. Na^{1+}, K^{1+}, Mg^{2+}, Ca^{2+}, Al^{3+}, Be^{2+}, etc. Spherically symmetric and electronic configurations conform to inert gases; form complexes only with ligands containing oxygen donor atoms; weak affinity toward ligands with nitrogen and sulfur donor atoms; of these, beryllium is the only one considered to be toxic at low concentrations.

Borderline cations. Fe^{3+}, Cu^{2+}, Pb^{2+}, Fe^{2+}, Ni^{2+}, Zn^{2+}, Co^{2+}, Mn^{2+}. Spherically asymmetric, and electronic configurations do not conform to inert gases; form inner sphere complexes with O- and N-atom containing ligands; moderate affinity toward S-atom containing ligands; all are toxic except Fe and Mn.

Soft cations. Hg^+, Cu^+, Hg^{2+}, Ag^+, Cd^{2+}. Spherically asymmetric, and electronic configurations do not conform to inert gases; high affinity toward S-atom containing ligands; most toxic among the heavy metals.

In the aquatic environment, heavy metals may be classified into at least two different categories:

- In true solution as free or complexed ions (e.g., Ca^{2+}, Fe^{2+}, K^+ or $Zn(OH)_4^{2+}$, $Ca(P_2O_7)^{2-}$).
- In particulates from adsorption onto other particles (e.g., mineral sediments or cations sorbed to sediments), or incorporation into biomass of living organisms and inorganic precipitates such as hydroxides, carbonates, sulfides, and sulfates.

In a soil–water system, the fate of heavy metals is directly related to their states of identity and the existing biogeochemical conditions. The free and complexed metal ions may be removed from solution by adsorption and precipitation mechanisms, while the particulate heavy metals may be transformed by their own dissolution and filtration mechanism of soils. In principle, the concentration of heavy metals in an aqueous system is controlled by the congruent and incongruent solubility of various oxides, carbonates, sulfates, and sulfides.

Metal precipitates in soil systems represent a selective accumulation of at least two or more constituent ions into an organized solid matrix that is often crystalline in nature. The process by which this selective accumulation occurs to form a distinct solid phase is termed precipitation. A *precipitate* can be considered a particulate phase that separates from a continuous medium. The fact that solid phases form in soil–water systems means that the overall free energy of formation is negative for the combined physical–chemical processes operating during the period of formation. The actual steps leading to the formation of a separate solid phase, however, must occur at the microscale level: the joining together of the constituent ions or molecules that will eventually be recognized as a distinct separate phase.[101,102] Under classical nucleation theory, three steps are generally considered necessary for those microscale processes to result in the formation of crystals that will persist and survive over relatively long periods of time: nucleus formation, crystalline formation, and crystal (precipitate) formation.[101,102]

Precipitation reactions often do not lead to immediate formation of the most stable form, the precipitated compound. The first solids that form are generally more chemically active than the stable precipitates that form over very long periods, and they have higher solubilities. These are called amorphous crystalline masses and have very high surface area to mass ratios. These very fine crystal masses mature into larger and mature crystals over long time spans. It may be necessary to sustain the biogeochemical conditions that generated the precipitation reactions for an extended period, to allow the formation of a stable crystal structure.

Complexation reactions are also important in determining the saturation state of groundwater. A complex is an ion that forms by combining simpler cations, anions, and sometimes molecules. The cation or central atom is typically one of the metals,

Microbial Reactive Zones

and the anions, often called ligands, include many of the common inorganic species found in groundwater, such as S^{2-}, CO_3^{2-}, SO_4^{2-}, PO_4^{3-}, NO_3^-, Cl^-. The ligand might also be an organic molecule such as an amino acid.

Presence of metals in an aqueous system is impacted by the presence and availability of ligands. Ligands are anions or molecules that form coordination compounds or complexes with metals. Inorganic ligands of primary importance in natural waters are F^-, Cl^-, SO_4^{2-}, OH^-, HCO_3^-, CO_3^{2-}, HPO_4^{2-}, NH_4^+, and in anoxic waters HS^- and S^{2-}. Nitrate (NO_3^-) does not form strong complexes with metals.

2.3.2.1 Principles of Heavy-Metal Precipitation

The mechanisms that can be used to reduce the concentrations of heavy metals dissolved in groundwater are transformation and immobilization. These mechanisms can be induced by both abiotic and biotic pathways. Abiotic pathways include oxidation, reduction, hydration, sorption, and precipitation. Examples of biotically mediated processes include reduction, oxidation, precipitation, biosorption, bioaccumulation, organo-metal complexation, and phytoremediation. In this chapter we discuss only immobilization mechanisms induced by precipitation within an IRZ.

Dissolved heavy metals can be precipitated out of solution through various precipitation reactions to form insoluble sulfide, hydroxide, or carbonate precipitates. A divalent metallic cation is used as an example in these reactions:

$$\text{Hydroxide precipitation:} \quad Me^{++} + 2OH^- \rightarrow Me(OH)_2 \downarrow \quad (2.19)$$

$$\text{Sulfide precipitation:} \quad Me^{++} + S^{2-} \rightarrow MeS \downarrow \quad (2.20)$$

$$\text{Carbonate precipitation:} \quad Me^{++} + CO_3^{2-} \rightarrow MeCO_3 \downarrow \quad (2.21)$$

Theoretical behavior of solubility of these precipitation mechanisms is shown in Figure 2.44(A) and Figure 2.44(B).

Hydroxide precipitation occurs when the pH of an aqueous system containing dissolved metal ions is raised to some optimum level for a specific metal. The optimum pH is different for each metal and, often, for different valence states of the same metal. It may also vary for a specific metal ion with the presence of other species in solution, with the redox potential, and with the aging of the hydroxide. Aging effects can be explained as the slow conversion of an active, crystalline matrix to a more inactive amorphous form.

Most metal sulfide precipitates are less soluble than hydroxides at the pH ranges encountered in groundwater systems (arsenic is an exception). It is evident from Table 2.8 that sulfide precipitation is effective. The low solubilities required for highly toxic metals are often achievable only by precipitation as sulfides because the metal sulfides have solubilities several orders of magnitude lower than hydroxides throughout the pH range. Their solubilities also are not as sensitive to changes in pH.

In certain cases, metal carbonates are less soluble than their corresponding hydroxides. This is true in the case of barium, cadmium, and lead. Carbonate ion concentration in a system depends on both CO_2 partial pressure and pH. The carbonate species, CO_3^{2-}, dominates at pH values larger than 10.3. The pH at which carbonate

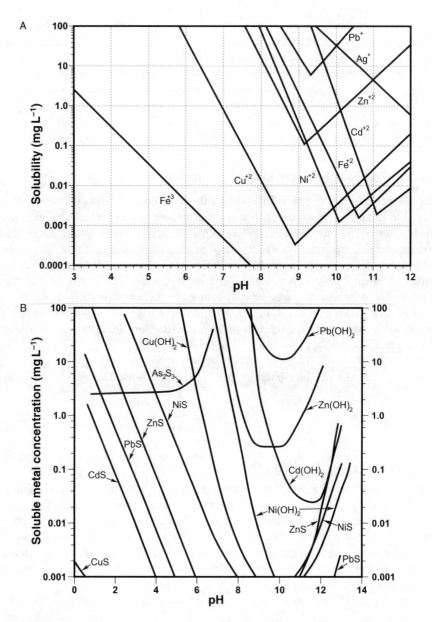

FIGURE 2.44 (A) Solubilities of metal hydroxides with pH. (B) Solubilities of metal hydroxides and metal sulfides as a function of solution pH.[105]

precipitation occurs depends on the solubility products of the carbonate and hydroxide species, and the CO_2 concentration. It should be noted that precipitation by addition of carbonates is not practiced widely. One problem is that the metal carbonates are decomposed at low pH.

TABLE 2.8
Solubility Products of Some Metal Sulfides[104,110] and the Difference between the Metal Sulfide and Hydroxide

Metal	K_{sp} of MeS	Difference Factor
Cadmium	3.6×10^{-29}	3×10^8
Chromium	None	—
Iron	1.1×10^{-10}	5×10^3
Copper	8.5×10^{-45}	1×10^{11}
Lead	3.4×10^{-28}	3×10^8
Mercury	2.0×10^{-49}	6×10^{17}
Nickel	1.4×10^{-24}	1×10^7
Zinc	1.2×10^{-23}	3×10^8
Silver	1.0×10^{-51}	5×10^{12}

$$\text{Difference factor} = \frac{\text{Me(OH)}_2 \text{ solubility (mg L}^{-1})}{\text{MeS solubility (mg L}^{-1})}$$

Hydroxide and sulfide precipitation of heavy metals have been used successfully in conventional industrial waste-water systems. Lime ($Ca(OH)_2$) or other alkaline solutions such as potash (KOH) are used as reagents for hydroxide precipitation. Sodium sulfide (Na_2S) is normally used as the reagent to form extremely insoluble metallic sulfide precipitates. Injection of these chemical reagents into the contaminated aquifer to create a reactive zone will precipitate the heavy metals out of solution. However, injection of a reactive, pH-altering chemical reagent into the groundwater may be objectionable from a regulatory point of view. Obtaining the required permits to implement chemical precipitation may be difficult. Furthermore, the metallic cations precipitated out as hydroxide could be resolubilized slightly as a result of any significant shift in groundwater pH.

Under reducing conditions, heavy metal cations can be removed from solution as sulfide precipitates if sufficient sulfur is available. In systems containing a sufficient supply of sulfur, neutral to mildly alkaline pH and low redox conditions are most favorable for the precipitation of many heavy metals. While excess sulfide ion is necessary for the precipitation reaction, the excess must be kept to a minimum so that free sulfide removal is not required. In addition, maintaining the pH on the mildly alkaline side will also prevent the evolution of H_2S.

Precipitation as sulfides is considered the dominant mechanism limiting the solubility of many heavy metals. Sulfide precipitation is particularly strong for metals exhibiting so-called "B-character," such as Cu(I), Ag, Hg, Cd, Pb, and Zn; it is also an important mechanism for transition elements such as Cu(II), Ni(I), Co(II), Fe(II), and Mn(II).[103] Two situations can be distinguished in natural systems during sulfide precipitation conditions: the existence of a certain sulfide precipitation capacity (SPC) or (when exceeding the SPC) the accumulation of free sulfide (as H_2S or HS^-) in the aqueous phase. At excess sulfide concentrations, solubility of some metals can be increased by the formation of thio complexes. However, the stability of these complexes is still questionable. Possible pathways of metal precipitate interactions are shown in Figure 2.45.

One interesting exception in sulfide precipitation is that of chromium, which does not precipitate as the sulfide, but as the hydroxide. Therefore, chromium removal will be controlled by hydroxide precipitation, and hence by pH. Precipitation as chromic hydroxide $Cr(OH)_{3(s)}$ has long been the traditional approach for treating Cr(VI)-laden industrial wastewater streams.

The sulfide ions necessary to mediate sulfide precipitation can be directly injected into a reactive zone in the form of sodium sulfide (Na_2S). However, the sulfide ion (S^{2-}) is one of the most reduced ions and its stability within the reactive zone is short-lived. It will be converted to sulfate (SO_4^{2-}) very quickly in the presence of oxidizing conditions within the contaminated plume. Addition of a very easily biodegradable organic substrate, such as carbohydrates, will enhance the formation of reduced, anaerobic conditions by depleting the available oxidation potential. The presence of carbohydrates serves two purposes: microorganisms use it as their growth substrate by depleting the available oxygen, and they use it as an energy source for the reduction of sulfate to sulfide. Theoretically, 3 mg L^{-1} of sulfate produces 1 mg L^{-1} of sulfide. In reality, more sulfate must be reduced because two of the chemical forms of sulfide are easily removed from solution.

Indirect microbial transformation of metals can occur as a result of sulfate reduction when SRBs oxidize simple carbon substrates with sulfate serving as the electron acceptor. The net result of the process is the production of hydrogen sulfide (H_2S) and alkalinity (HCO_3^-). Sulfate reduction is strictly an anaerobic process and proceeds only in the absence of oxygen. The process requires a source of carbon to support microbial growth, a source of sulfate, and a population of SRBs. Dilute black strap molasses solution is an ideal feed substrate for this purpose since typical black strap molasses contains approximately 20% sucrose, 20% reducing sugars, 10% sulfated ash, 20% organic nonsugars, and 30% water.

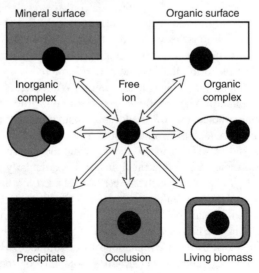

FIGURE 2.45 Heavy metal interactions in an aquifer matrix.

Microbial Reactive Zones

Whether formed biotically or abiotically, the metal sulfides result from an interaction between the metal ion and sulfide ion:

$$Me^{2+} + S^{2-} \rightarrow MS\downarrow \tag{2.22}$$

It is the source of the sulfide that determines whether a biological agent is implicated in metal sulfide formation. If the sulfide results from bacterial sulfate reduction or from bacterial mineralization of organic compounds, it is obviously of biotic origin. If it is derived from volcanic activity, it is generally of abiotic origin. Because of their relative insolubility, the metal sulfides form readily at ordinary temperatures and pressures by interaction of metal ions and sulfide ions. Table 2.8 lists solubility products for some common simple sulfides[103,104] and the difference between metal sulfide and hydroxide solubilities.

The following calculation shows that relatively low concentrations are needed to form metal sulfides by reacting with H_2S at typical concentrations that can be formed in an anaerobic IRZ. Let us examine, for instance, the case of iron. The dissociation constant for iron sulfide (FeS) is

$$[Fe^{2+}][S^{2-}] = 1 \times 10^{-19} \tag{2.23}$$

The dissociation constant for H_2S is

$$[S^{2-}] = 1.1 \times 10^{-22} \frac{[H_2S]}{[H^+]^2} \tag{2.24}$$

since

$$\frac{[HS^-][H^+]}{[H_2S]} = 1.1 \times 10^{-7} \tag{2.25}$$

and

$$\frac{[S^{2-}][H^+]}{[HS^-]} = 1 \times 10^{-15} \tag{2.26}$$

Therefore,

$$[Fe^{2+}] = \frac{[H^+]^2}{[H_2S]} \times \frac{1 \times 10^{-19}}{1.1 \times 10^{-22}} = \frac{[H^+]^2}{[H_2S]}(9.1 \times 10^2) \tag{2.27}$$

About 5.08×10^{-3} mg of Fe^{2+} per liter (9.1×10^{-8} M) will be precipitated by 3.4 mg of hydrogen sulfide per liter (10^{-4} M) at pH 7. The unused H_2S will ensure reducing conditions, which will keep the iron in the ferrous state. Since ferrous sulfide is one of the more soluble sulfides, it can be seen that metals whose sulfides have even smaller solubility products will form even more readily at lower H_2S concentrations.

The distribution of sulfide between various phases is similar to that of the carbon dioxide/ bicarbonate/carbonate interactions. Equation (2.25) indicates that at neutral pH, the distribution between H_2S and HS^- will be even. Hydrogen sulfide, H_2S, is a gas of intermediate solubility, and may be stripped from solution. Higher H_2S production per unit volume of the anaerobic IRZ system and a lower pH favor stripping of H_2S from the groundwater system.

Metal sulfides have been generated in laboratory experiments utilizing H_2S from bacterial sulfate reduction. It has been reported that sulfides of Sb, Bi, Co, Cd, Fe, Pb, Ni, and Zn were formed in a lactate broth culture of *Desulfovibrio desulfuricans* to which sulfate and salts of selected metals had been added. Metal toxicity to *D. desulfuricans* depends in part on the concentration of the metallic ion in question. Obviously, for the corresponding metal sulfide to be formed, the metal sulfide must be even more insoluble than the starting compound of the metal. In addition to Fe and Mn, metals such as Cu, Ag, Cd, Pb, Zn, Ni, and Co can also be precipitated as metallic sulfides. Precipitated metallic sulfides will remain in an insoluble, stable form, unless the subsurface redox conditions change dramatically.

The production of alkalinity from sulfate reduction, denitrification, and other reactions causes an increase in pH, which can result in metal precipitation through the formation of insoluble metal hydroxides or oxides. This process follows the reactions

$$Me^{2+} + SO_4^{2-} + 8H^+ + 8e \rightarrow MeS_{(s)} + 4H_2O \qquad (2.28)$$

Consumption of H^+ ions in the system results in an increase in pH.

$$Me^{2+} + 2H_2O \rightarrow Me(OH)_2 \downarrow + 2H^+ \qquad (2.29)$$

Another aspect of the behavior of dissolved metals in groundwater systems is that metal cations always attract a hydration shell of water molecules by electrostatic attraction to the positive charge of the cation. This happens because the water molecules are polar.

$$Me^{+n} \xrightarrow{H_2O} Me(H_2O)_x^{+n} \qquad (2.30)$$

where $x = 6$ for most cations.

Hydrated metal ions behave as acids by donating protons (H^+) to H_2O molecules, forming the acidic H_3O^+ hydronium ion. The process can continue stepwise up to n times to make a neutral metal hydroxide. For example, with Fe^{3+} it takes three proton transfer steps to form ferric hydroxide.[105]

$$Fe(H_2O)_3^{3+} + 3H_2O \rightarrow Fe(OH)_3 \downarrow + 3H_3O^+ \qquad (2.31)$$

With each step, the hydrated metal is progressively deprotonated, forming polyhydroxides and becoming increasingly insoluble. Eventually, the metal precipitates as a low solubility hydroxide.

2.3.2.2 Chromium Precipitation

Chromium belongs to group VIB of the periodic table. It has three valence states, $+2$, $+3$, and $+6$, but Cr(III) and Cr(VI) are the most common. Cr^{+6} is acidic, forming chromates $(CrO_4)^{-2}$ and dichromates $(Cr_2O_7)^{-2}$, while the other valence states are basic. The efficient and cost-effective reduction of Cr^{+6} is a two-step process: reduce the chromium to the trivalent state and then precipitate it as the chromic hydroxide $Cr(OH)_3$. Microbial transformation of Cr(VI) to Cr(III) has only recently been realized. In such a process, Cr(VI) may serve as an electron acceptor for the oxidation of an electron donor.

Many facultative anaerobes are capable of reducing Cr(VI) to Cr(III) under appropriate conditions. Chromium-reducing bacteria belong to a variety of genera such as *Achromobacter, Aeromonas, Agrobacterium, Bacillus, Desulfovibrio, Enterobacter, Escherichia, Micrococcus,* and *Pseudomonas*.[106] Although chromium-reducing bacteria are widespread, high cell densities are required for significant Cr(VI) reduction to occur. The oxidizing power of Cr(VI) generates its toxic effects toward bacterial cells. At high concentrations of Cr(VI), in addition to the enzymatic mechanism of Cr(VI) reduction, Cr(VI) may directly interact with various intracellular reducing agents inside cells once it has penetrated the cells.

In situ microbial reduction of dissolved hexavent chromium Cr(VI) to trivalent chromium Cr(III) yields significant remedial benefits because trivalent chromium Cr(III) is less toxic, water insoluble and thus nonmobile, and precipitates out of solution. In fact, it has been stated that the natural attenuation of Cr(VI) to the reduced Cr(III) form within an aquifer is a viable groundwater remediation technique.

In situ microbial reduction of Cr(VI) to Cr(III) can be promoted by injecting a carbohydrate solution, such as dilute molasses solution. The carbohydrates, which consist mostly of sucrose, are readily degraded by the heterotrophic microorganisms present in the aquifer, thus depleting all the available dissolved oxygen present in the groundwater. Depletion of the available oxygen present causes reducing conditions to develop as aquifer bacteria consume alternate electron acceptors. The mechanisms of Cr(VI) reduction to Cr(III), under the induced reducing conditions can be (1) likely a microbial reduction process involving Cr(VI) as a terminal electron acceptor for the metabolism of carbohydrates, (2) an extracellular reaction with by-products of sulfate reduction such as H_2S, and (3) abiotic oxidation of the organic compounds including the soil organic matter such as humic and fulvic acids.[107]

Cr(VI) is known to be reduced both aerobically and anaerobically in different bacterial systems. In anaerobic systems, membrane preparations reduce Cr(VI), and Cr(VI) has been shown to serve as a terminal electron acceptor. Aerobic reduction of Cr(VI) has been found to be associated with soluble proteins. The enzymatic basis for aerobic chromate reduction is not known, but it has been proposed that chromate may be reduced by a soluble reductase enzyme with a completely unrelated primary physiological role. Based on the diversity of Cr(VI) reducing microorganisms in soil, provision of a suitable electron donor such as molasses may be sufficient and the ORP within the IRZ need not be reduced to -250 to -300 mV as is the case during ERD applications.[108,109] The experience of the authors at many chromium precipitation sites indicates that the ORP does not have to be reduced to the levels that are required for reductive dechlorination, but more around the denitrifying or iron-reducing conditions (Figure 2.46).

The primary end-product of Cr(VI) to Cr(III) reduction process is chromic hydroxide $(Cr(OH)_3)$, which readily precipitates out of solution under alkaline to

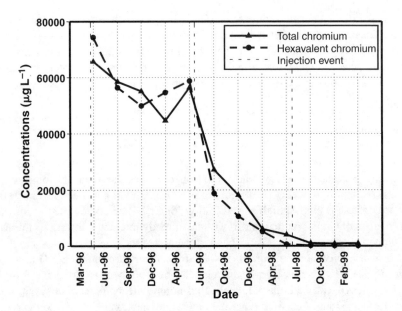

FIGURE 2.46 Site 5: average treatment results for total and hexavalent chromium.

moderately acidic and alkaline conditions. To ensure that this process will provide both short- and long-term effectiveness in meeting groundwater cleanup objectives, the chromium precipitates must remain immobilized within the soil matrix of the aquifer, and shall not be subject to $Cr(OH)_3$ precipitate dissolution or oxidation of Cr(III) back to Cr(VI) once groundwater conditions revert back to its natural conditions. Based on the results of significant research being conducted on the *in situ* chromium reduction process, it is readily apparent that the $Cr(OH)_3$ precipitate is essentially an insoluble, stable precipitate, immobilized in the soil matrix of the aquifer. Figure 2.47(A) and (B) describe the solubility of Cr^{3+} and the solubility of $Cr(OH)_3^{(s)}$ is affected on the acid side more than the alkaline side.

Contrary to the numerous natural mechanisms that cause the reduction of Cr(VI) to Cr(III), there appear to be only a few natural mechanisms for the oxidation of Cr(III). Indeed, only two constituents in the subsurface environment (dissolved oxygen and manganese dioxide) are known to oxidize Cr(III) to Cr(VI).[110] The results of studies conducted on the potential reaction between dissolved oxygen and Cr(III) indicate that dissolved oxygen will not cause the oxidation of Cr(III) under normal groundwater conditions. However, studies have shown that Cr(III) can be oxidized by manganese dioxides, which may be present in the soil matrix. However, only one phase of manganese dioxides is known to oxidize appreciable amounts of Cr(III) and this process is inversely related to groundwater pH. Hence, the oxidation of Cr(III) back to Cr(VI) in a natural aquifer system is highly unlikely.

The $Cr(OH)_3$ precipitate has an extremely low solubility (solubility product, $K_{sp} = 6.7 \times 10^{-31}$), and thus very little of the chromium hydroxide is expected to remain in solution. It has been reported that aqueous concentration of Cr(III), in equilibrium with $Cr(OH)_3$ precipitates, is around 0.05 mg L^{-1} within the pH range of

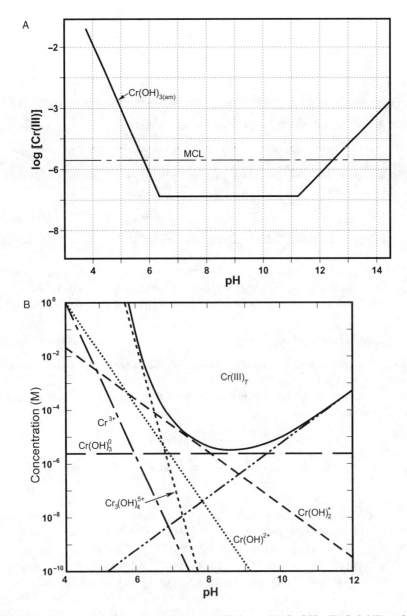

FIGURE 2.47 (A) Cr(III) concentration in equilibrium with Cr(OH). (B) Solubility of precipitated $Cr(OH)_{3(s)}$ (25°C). (Adapted from Stumm and Morgan.[189])

5 to 12 (Figure 2.47(A) and Figure 2.47(B)). The pH range of natural aquifer systems will be within 5 to 12, and, hence, the potential for the chromic hydroxide to resolubilize is unlikely. Furthermore, FeS has the ability to reduce Cr(VI).

Dissolved Cr(VI) can be also precipitated as $Cr(OH)_3$ in a reactive zone by the injection of ferrous sulfate solution at appropriate concentrations. It is safe to use, inexpensive (as chemical reducing agents go), and often produces additional benefits by coprecipitating other heavy metals. Cr(VI) exists as chromate, CrO_4^{2-}, under neutral or alkaline conditions and dichromate, $Cr_2O_7^{2-}$, under acidic conditions. Both species react with ferrous ion:

Acidic conditions: $Cr_2O_7^{2-} + 6Fe^{2+} + 14H^+ \rightarrow 2Cr^{3+} + 6Fe^{3+} + 7H_2O$ (2.32)

Neutral or alkaline condition:
$$CrO_4^{2-} + 3Fe^{2+} + 4H_2O \rightarrow Cr^{3+} + 3Fe^{3+} + 8OH^- \quad (2.33)$$

Both Cr(III) and Fe(III) ions are highly insoluble under natural conditions of groundwater (neutral pH or slightly acidic or alkaline conditions).

$$Fe^{3+} + 3OH^- \rightarrow Fe(OH)_3\downarrow \quad (2.34)$$
$$Cr^{3+} + 3OH^- \rightarrow Cr(OH)_3\downarrow \quad (2.35)$$

The addition of ferrous sulfate into the reactive zone may create acidic conditions and hence the zone downgradient of the ferrous sulfate injection zone may have to be injected with soda ash or caustic soda to bring the pH back to neutral conditions.

2.3.2.3 Nickel Precipitation

When considered in isolation, nickel sulfide (NiS) and other metal sulfides such as cadmium or lead sulfide are relatively insoluble in aqueous phase. The solubility product for amorphous nickel sulfide dissociation (α-NiS) is 3×10^{-21} and the dissociation equation is written as

$$NiS_{(s)} \rightarrow Ni^{2+} + S^{2-}; \quad K_{sp} = 3 \times 10^{-21} \quad (2.36)$$

The dissolved nickel, Ni^{2+}, can be isolated to show its predicted concentration as a function of sulfide ion (S^{2-}) and the K_{sp}. The E_H–pH relationship for Ni is presented in Figure 2.48.

A nickel-containing plating solution was released to shallow groundwater underneath a commercial plating facility in the U.S. midwest, creating a dissolved nickel source zone that fed a plume of more than 1000 mg L^{-1} Ni^{2+} outside the building footprint. Migration of nickel-contaminated groundwater reached off-site, with concentrations at the site perimeter in excess of the regulatory limit of 0.1 mg L^{-1} for drinking water protection. The migration of high-concentration groundwater indicated that the aquifer matrix cation exchange capacity was exhausted, and that the persistent nickel source comprised the plating solution saturating spill point aquifer soils, as well as adsorbed nickel mass adsorbed to the aquifer matrix and dissolved nickel residing in static groundwater along the flow path. Removal of plating solution from the spill point alone would not eliminate nickel migration (additional discussion of contaminant source distribution and migration patterns can be found in Chapter 5).

A biological reactive zone was established to remove dissolved nickel from the groundwater, forming durable precipitates. Molasses injections were used to provide a carbohydrate source to drive the aquifer microbial community into anaerobic

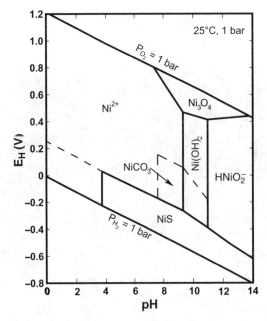

FIGURE 2.48 E_H–pH diagram for part of the system Ni-O-H-S-C. Assumed activities for dissolved species are: Ni = 10^{-4}, S = 10^{-3}, C = 10^{-3} (all in mol L^{-1}) at 298 K and 1 bar.

metabolism and to supply sulfate that supports SRBs. In combination, these injected reagents provide a low-redox, high-sulfide chemical environment, in which nickel sulfide can precipitate. Nickel sulfide is an insoluble salt with equilibrium concentrations far below 1 µg L^{-1} at neutral aquifer pH. Because the sulfide anion is a strong base, sulfide scavenging occurs at low pH, drawing nickel sulfide into solution. Additional information on the chemistry of nickel sulfide precipitation and reactive zone strategies for dissolved metal plumes is provided in Chapters 3 and 5.

The biological reactive zone injections succeeded in providing the low redox and available sulfide required to precipitate nickel, although the production of organic acids by the anaerobic aquifer microbial community drove the pH to low levels in the early phases of the reactive zone development. This generated a partial rebound of dissolved nickel concentrations each time the pH declined, as seen in Figure 2.49, which shows nickel concentrations in groundwater immediately outside the plating building footprint. Sodium carbonate buffer was added along with molasses during reactive zone injections after Day 1200. This stabilized the pH and allowed the durable nickel sulfide precipitate to form.

2.3.2.4 Arsenic Precipitation

Arsenic is actually classified as a nonmetal or metalloid, although it is grouped with the metals for most environmental purposes. Arsenic, although not abundant in the Earth's crust, is widely distributed around the Earth in more than 150 arsenic-bearing minerals. It is ubiquitous in small amounts, with average concentrations ranging from

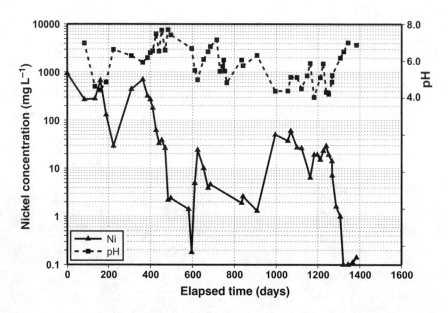

FIGURE 2.49 Strong dependence of Ni precipitation on pH.

7 to 11 mg kg^{-1} in American soils.[111] Arsenic can occur in several valence states, but it is primarily found as an oxyanion of trivalent arsenite (As^{3+}) or pentavalent (As^{5+}) (Figure 2.50). Because arsenic is labile, redox state (E_H) and pH are the most important factors controlling its speciation and chemical form. This makes it highly susceptible to biogeochemical reactions in the subsurface. Under oxidizing conditions, arsenate predominates in the form of $H_2AsO_4^-$ at low pH (<7) and $HAsO_4^{2-}$ at high pH (>7). Under reducing conditions, arsenite predominates in the form of non-ionic H_3AsO_3 at a pH below 8.

In the subsurface, the solubility/mobility of geogenic arsenic is typically controlled by sorption to other minerals. The most common companions of arsenic in natural sediments are iron, manganese, and sulfur. Of these, iron plays the most significant role and since iron concentrations in most sediments typically range from 0.5 to 5% by weight, or 5000 to 50,000 mg kg^{-1}, the role of iron in controlling geogenic arsenic mobility is significant. Naturally occurring hydrous ferric oxide (HFO) particles have an adsorption affinity for both arsenate and non-ionic arsenite. Non-ionic arsenite can sorb to HFO particles through Lewis acid–base interactions, with anionic arsenate sorption occurring through coulombic (electrostatic) and Lewis acid–base (electron pair) interactions (Figure 2.51).

The biogeochemistry of arsenic in sediments, as mentioned earlier, is complicated by the strong binding of As(V) to minerals like FeOOH. Hence, bacterial reduction of Fe(III) to Fe(II) should release any bound As(V) and make it available for further chemical or biological reduction. The question then arises whether these changes in speciation were caused by chemical reactions or direct biological reductions. The investigation into the number of species of microorganisms capable of dissimilatory arsenate

Microbial Reactive Zones

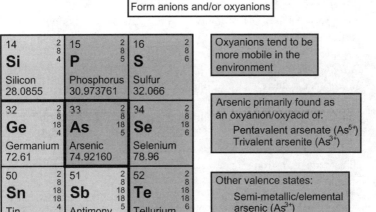

FIGURE 2.50 Properties of arsenic.

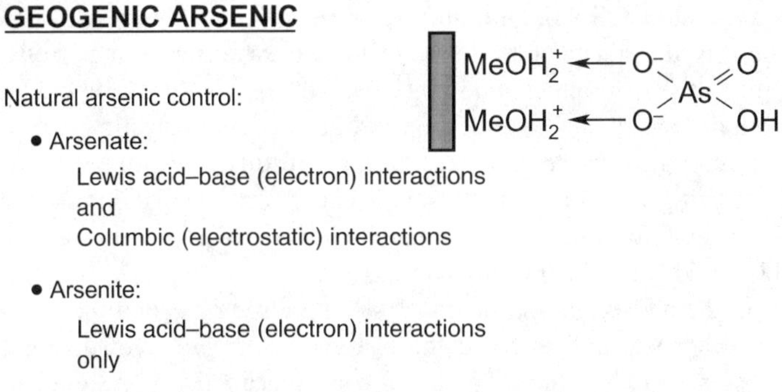

GEOGENIC ARSENIC

Natural arsenic control:

- Arsenate:
  Lewis acid–base (electron) interactions
  and
  Columbic (electrostatic) interactions

- Arsenite:
  Lewis acid–base (electron) interactions
  only

FIGURE 2.51 Geogenic characteristics of arsenic.

reduction reveals a great diversity. The realization that arsenate is indeed a suitable electron acceptor and is readily available in both natural and contaminated environments suggests that even more unrelated species will be discovered.[112]

Because iron is labile, geochemical changes can significantly affect its arsenic sorption capacity. The three primary triggers that can result in a release of geogenic arsenic are as follows:

- *Development of high pH in an aerobic/oxidizing groundwater environment.* HFO particles are amphoteric ion exchangers dependent on pH.[113]

At low pH, HFO particles possess a positive charge. This charge decreases as pH increases. Under normal conditions, the point of zero charge for HFO particles is a pH of approximately 8.5. Above this point, the HFO particles become negatively charged and they reject both arsenate and arsenite.

- *Introduction of high concentrations of competing anions.* Both arsenate and arsenite are relatively strong ligands (electron donor/Lewis base), forming inner-sphere complexes with HFO particles (electron acceptor/Lewis acid). Anions such as chloride and sulfate are poor ligands that form weak complexes with HFO particles.[114] Consequently, both arsenate and arsenite are preferentially sorbed. However, anions such as phosphate, bicarbonate, and silicate are also strong ligands and in sufficient concentrations can decrease arsenic sorption.
- *Development of reducing conditions.* Reducing environments can solubilize arsenic in two ways. The first involves the reduction of arsenate to arsenite, which is less strongly sorbed. The second involves the reduction and dissolution of ferric iron, releasing sorbed arsenic.

Of the above, the third reason is most relevant to natural or engineered reducing environments. There is evidence that some control can be realized through interaction of the arsenic with dissolved (ferrous) iron and reactive sulfide. However, it is expected that the greatest measure of control will be provided through adsorption, incorporation, and coprecipitation with HFO floc once the ambient aerobic and oxidizing poise is restored. Solid-solution partition coefficients (K_d) describing the retardation of dissolved arsenate and arsenite to advective transport range from 2 L mg^{-1} for clean quartz to greater than 500,000 L mg^{-1} for HFO.[111] Despite the fact that the actual sorption capacity of natural soils is likely to be many times less than that reported for laboratory-grade HFO particles, the results confirm the extent to which naturally occurring metal oxides and clays can control geogenic arsenic mobility under aerobic conditions. In fact, the research indicates that the total sorption capacity of the soil is unlikely to be a limiting factor.

Based on the above, the migration of soluble arsenic beyond the boundary of a reducing zone appears unlikely, and the restoration of aerobic/oxidizing conditions beyond an engineered or naturally reducing zone appears likely to sufficiently restore background arsenic concentrations in groundwater (Figure 2.52(A) and Figure 2.52(B)).

Based on available empirical data and published research, it is reasonable to expect the solubilization of arsenic within an engineered or naturally reducing zone is a transient phenomenon that will be limited to the boundaries and that will reverse once the original aerobic and oxidizing poise of the aquifer is restored. However, additional data and research confirming this effect need to be collected. Other areas of focus include:

- *The size of engineered or naturally reducing zones.* The size of these zones is typically established by equilibrium between the amount of organic carbon present, the availability/recharge of electron acceptors, microbial consumption of the carbon, dispersion, and dilution. As the carbon moves through an aquifer, its concentrations diminish due to microbial consumption, dispersion, and dilution, eventually allowing microbial activity to return to pretreatment levels. It is at or just beyond this point that the geochemical environment will recover its aerobic/oxidizing poise because the utilization of electron acceptors falls short of the natural recharge.

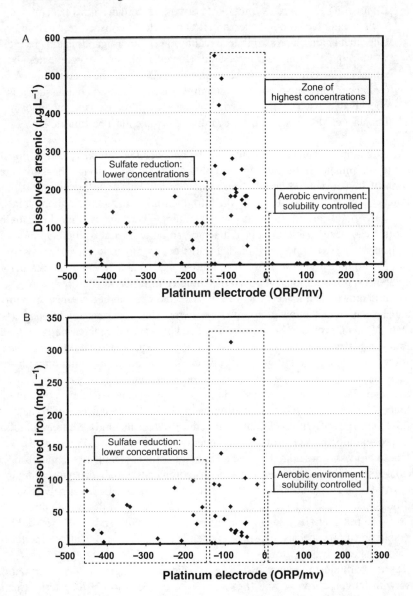

FIGURE 2.52 (A) Arsenic concentrations as a function of biogeochemical environment showing removal at even mildly oxidized conditions. (B) Iron concentrations as a function of biogeochemical environment.

- *The time required for an engineered reducing zone to dissipate.* Upon completion of remedial efforts, the reducing conditions within an IRZ will dissipate. Without a continuing source of degradable organic carbon, the influx of dissolved oxygen and other electron acceptors will eventually allow the natural aerobic/oxidizing poise to be reestablished, and should result in the immobilization of any dissolved arsenic solubilized within the IRZ.
- *Competitive sorption.* As previously discussed, certain anions can compete with arsenic for sorption sites on HFO particles. Published research indicates that anions such as sulfate and chloride are poor ligands and therefore would not be preferred over arsenic. However, phosphate, carbonate, and silicate possess qualities similar to arsenic giving them the potential to have a major impact on sorption. Although this is not likely to be an issue, it is conceivable that a source of ferric iron could be introduced to counteract an imbalance. More specific data regarding this topic must be collected.

Arsenic mobility can also be controlled by *enhanced anaerobic reductive precipitation* that stimulates indigenous microbiological organisms through the engineered addition of low-cost, food-grade electron donors and sulfate to precipitate metals. Arsenic mobility can be controlled by the introduction of degradable carbohydrates, a sulfur source, and ferrous iron salts into aquifers. There are two mechanisms by which this can occur. First, dissolved arsenic can be removed by coprecipitation in FeS phases by SRBs operating on naturally occurring or injected ferrous iron. Addition of soluble iron *in situ* for remediation is well established and thus regulatory acceptance for this additional application should be routine.

Natural biogeochemical cycles of Fe and As are linked and Fe chemistry controls As contamination in shallow groundwater systems. Under aerobic conditions Fe(III) and Mn(IV) oxides remove As from the dissolved phase. Under increasingly reducing conditions, SRBs can remove dissolved As by sequestering it in Fe sulfides. Dissolved arsenic can be removed by coprecipitation and/or precipitation with iron sulfides. However, the arsenic sulfide solids (orpiment and realgar) precipitation will have to take place at low pH values and kinetic stability can be achieved only at low pH values. *In situ* microbial precipitation methods for As are considered to be cost-effective, under the right conditions, compared to other methods. Sulfate reduction and coprecipitation of As in iron sulfide minerals can be achieved (Figure 2.53).

These processes are complex, and an example of this complexity is the observation that arsenic also can form stable aqueous complexes with hydrogen sulfide; iron chemistry is the key to controlling this undesirable process. Under highly reducing conditions arsenic can exist as $As_2S_{3(s)}$, $AsS_{(s)}$, and $FeAsS_{(s)}$. If these solid phases are exposed to a relatively oxidizing environment, sulfides will tend to be oxidized into more soluble sulfate species, causing geochemical leaching or oxidative dissolution of arsenic in groundwater.

There is sufficient evidence available to believe that engineered coprecipitation of As with biogenic iron sulfide (FeS) or pyrite can lead to the long-term kinetic stability required within the reactive zones. Arsenic and other metals present in the pyrite at concentrations of several hundred parts per million had coprecipitated with pyrite and has been shown to have long-term stability.

Thermodynamics provides useful insight into the equilibrium chemistry of inorganic arsenic species. In oxygenated waters, As(V) is dominant, existing in anionic forms of either $H_2AsO_4^-$, $HAsO_4^{2-}$, or AsO_4^{3-} over the pH range of 5 to 12, which covers the range encountered in natural groundwater. Under anoxic conditions, As(III) is stable, with nonionic (H_3AsO_3) and anionic ($H_2AsO_3^-$) species dominant below and above pH 9.22, respectively. In the presence of sulfides, precipitation of AsS (realgar) or As_2S_3 (orpiment) may remove soluble As(III) and exert considerable control over trace arsenic concentrations. The thermodynamic reduction of As(V) to As(III) in the absence of oxygen could be chemically slow and may require bacterial mediation. As noted in the previous section, injection of dilute solution of blackstrap molasses will create the reducing conditions for As(V) to be reduced to As(III) and also provide the sulfide ions for As(III) to precipitate as As_2S_3. These reactions are described by the following equations.

Reduction of As(V) to As(III) under anaerobic conditions:

$$HAsO_4^{2-} \rightarrow HAsO_2 \qquad (2.37)$$

In the presence of S^- under anaerobic conditions:

$$HAsO_2 + S^- \rightarrow As_2S_3\downarrow \qquad (2.38)$$

Within oxygenated zones in the aquifer, oxidation of ferrous ion (Fe(II)) and Mn(II) leads to formation of hydroxides that will remove soluble As(V) by coprecipitation or adsorption reactions. The production of oxidized Fe–Mn species and subsequent precipitation of hydroxides are analogous to an *in situ* coagulation process for removing As(V).

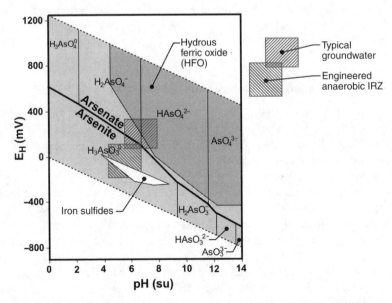

FIGURE 2.53 E_H–pH diagram of aqueous As species (As–O_2–H_2O, 25°C, 1 bar).

Contaminated groundwaters normally contain arsenites and arsenates, the ratio being dependent on the biogeochemical conditions of the groundwater system. Between them arsenates are more amenable to removal by a majority of reaction mechanisms. Preoxidation of arsenites to arsenates followed by selective sorption onto the surfaces of microparticles of hydrated Fe oxide (HFO) is the underlying mechanism for excellent As(V) removal. However, the effectiveness of HFO particles varies markedly for As(III) removal. Combined reactions of As(III) oxidation to As(V) and the introduction of Fe(III) can be accomplished by implementing the Fenton's reaction within an IRZ. Pre-oxidation followed by injection of ferric chloride ($FeCl_3$) can also achieve the same objectives.

2.3.2.5 Cadmium

The dominant cadmium (Cd) solution species in groundwater at neutral pH values is Cd^{2+}. Both precipitation and adsorption reactions control dissolved Cd concentrations. Several researchers have indicated that $CdCO_3$ limits Cd solution concentrations in alkaline soils. $Cd_3(PO_4)_2$ has also been reported to be a solubility controlling solid. Cadmium can also drop out of solution as CdS.

$$Cd^{2+} + CO_3^{2-} \rightarrow CdCO_3^{(s)} \qquad (2.39)$$
$$Cd^{2+} + S^{2-} \rightarrow CdS_{(s)} \qquad (2.40)$$

Performance data from a cadmium precipitation project the authors worked on are presented in Figure 2.54. This full-scale application consisted of 20 injection wells, and molasses solution, varying in strength from 1 : 20 to 1 : 200, was injected periodically. Figure 2.54(A) and Figure 2.54(B) show the baseline Cd distribution and post-treatment Cd concentrations. It should be noted that remediation goals were achieved in less than 18 months.

2.3.2.6 Aquifer Parameters and Transport Mechanisms

Redox processes can induce strong acidification or alkalinization of soils and aquifer systems. Oxidized components are more acidic (SO_4^{2-}, NO_3^{-}) or less basic (Fe_2O_3) than their reduced counterparts (H_2S, NH_3). As a result alkalinity and pH tend to increase with reduction and decrease with oxidation. Carbonates are efficient buffers in natural aquifer systems in the neutral pH range.

Many events can cause changes in redox conditions in an aquifer. Infiltration of water with high dissolved oxygen concentration, fluctuating water table, excess organic matter, introduction of contaminants that are easily degradable, increased microbial activity, deterioration of soil structure can impact the redox conditions in the subsurface. However, there is an inherent capacity to resist redox changes in natural aquifer systems. This inherent capacity depends on the availability of oxidized or reduced species. Redox buffering is provided by the presence of various electron donors and electron acceptors in the aquifer.

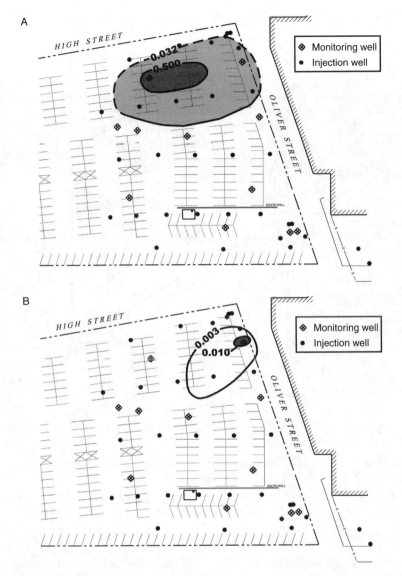

FIGURE 2.54 (A) Distribution of cadmium, July 1998. (B) Distribution of cadmium, January 2000.

An engineered IRZ has to take into consideration how the target reactions will impact the redox conditions within and downgradient of the reactive zone, in addition to degrading the contaminants with the available residence time. Furthermore, careful evaluation should be performed regarding the selectivity of the injected reagents toward the target contaminants and the potential to react with other compounds or aquifer materials. Careful monitoring, short term and long term, should be performed

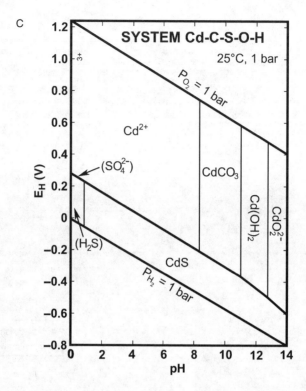

FIGURE 2.54 (continued) (C) E_H–pH diagram of cadmium.

to determine whether the natural equilibrium conditions can be restored at the end of the remediation process. In some cases modified biogeochemical equilibrium conditions may have to be maintained over a long period of time to prevent the rebound of contaminant concentrations.

2.3.2.7 Contaminant Removal Mechanisms

As noted earlier, the mechanisms used to reduce the toxicity of dissolved contaminants can be grouped into two major categories: transformation and immobilization. Examples of some of these mechanisms have been discussed earlier. Conversion of chlorinated organic compounds to innocuous end-products such as CO_2, H_2O, and Cl^- either by biotic or abiotic reaction pathways is an example of transformation mechanisms. Precipitation of Cr(VI) as $Cr(OH)_3$ by either abiotic or biotic reaction pathways and subsequent filtration by the soil matrix is an example of immobilization mechanisms.

It can be assumed, in most cases, that the end-products of transformation mechanisms will result in dissolved and gaseous species. It can also be assumed that the impact of these end-products on the natural redox equilibrium will be short term. If the impact is expected to be significant, it can be controlled by limiting the reaction kinetics and the transport of the end-products away from the reaction zone. Dilution and escape of dissolved gases will also help restore the natural equilibrium conditions in the reaction zone.

Immobilization mechanisms, which include heavy metals precipitation reactions, in reality transform the contaminant into a form (precipitate) that is much less soluble. In addition, transport of dissolved heavy metals in groundwater should also be considered as a two-phase system in which the dissolved metals partition between the soil matrix and the mobile aqueous phase.

Metal precipitates resulting from an IRZ may move in association with colloidal particles or as particles themselves of colloidal dimensions. The term colloid is generally applied to particles with a size range of 0.001 to 1 μm. The transport of contaminants as colloids may result in unexpected mobility of low solubility precipitates. It is important to remember that the transport behavior of colloids is determined by the physical/chemical properties of the colloids as well as the soil matrix.

Generally, when fine particles of colloid dimensions are formed flocculation naturally occurs unless steps are taken to prevent it. Even when the primary precipitates are of colloid dimensions, if they form larger lumps a stable dispersed transport cannot take place. These larger flocs will settle on the soil matrix.

Metal precipitates may be pure solids (e.g., PbS, ZnS, $Cr(OH)_3$) or mixed solids (e.g., $(Fe_x, Cr_{1-x})(OH)_3$, $Ba(CrO_4, SO_4)$). Mixed solids are formed when various elements coprecipitate or interact with aquifer materials.

Colloidal precipitates larger than 2 μm will be removed by sedimentation in the low flow conditions common in aquifer systems. Colloidal precipitates are more often removed mechanically in the soil matrix. Mechanical removal of particles occurs most often by straining, a process in which particles can enter the matrix, but are caught by the smaller pore spaces as they traverse the matrix.

Colloidal particles below 0.1 μm will be subjected more to adsorptive mechanisms than mechanical processes. Adsorptive interactions of colloids may be affected by the ionic strength of the groundwater; ionic composition, quantity, nature, and size of the suspended colloids; geologic composition of the soil matrix; and flow velocity of the groundwater. Higher levels of total dissolved solids (TDSs) in the groundwater encourages colloid deposition.

In aquifer systems with high Fe concentrations, the amorphous hydrous ferric oxide can be described as an amphoteric ion exchange media. As pH conditions change, it has the capacity to offer hydrogen ions (H^+) or hydroxyl ions (OH^-) for a cation or an ion exchange, respectively. Adsorption behavior is primarily related to pH (within the typical range of 5.0 to 8.5), and at typical average concentrations in soil, the iron in a cubic yard of soil is capable of adsorbing from 0.5 to 2 lb of metals as cations or metallic complexes. This phenomenon is extremely useful for the removal of As and Cr.

2.3.3 IN SITU DENITRIFICATION

Nitrogen can form a variety of compounds due to its different oxidation states. In the natural ecosystem, most changes from one oxidation state to another are biologically induced. The following nitrogen forms are of interest in relation to the subsurface environment (Table 2.9).

Nitrogen in the environment can be introduced by many anthropogenic activities in addition to the natural sources of nitrogen. Waste produced by humans and animals are important sources of nitrogen in any area characterized by significant human or animal populations. Nitrates from such waste can exhibit the characteristics of either point or nonpoint source pollution.

Other common sources of nitrogen in the subsurface environment are derived from fertilizer use, testing and usage of explosives, and industrial usage. Nitrogen is the most common element used as a fertilizer supplement. Nitrate's high solubility and low sorptivity make it an excellent fertilizer component. Nitrate is a major element in the manufacture of explosives, which primarily uses ammonium nitrate as a raw material. Without proper management, waste streams that contain high concentrations of ammonium nitrate can cause groundwater quality degradation. Nitrogen compounds are also used extensively in industrial settings in the form of anhydrous ammonia, aqua ammonia, nitric acid, ammonium nitrate, and urea. Plastic manufacturing, metal processing, textile industry, acid production, refrigeration, pulp, and paper are some of the industries that use significant amount nitrogen compounds.

The neutral, molecular ammonia exists in equilibrium with the ammonium ion, the distribution of which depends upon the pH and temperature of the biogeochemical system; in fact, very little ammonia exists at pH levels less than neutral. Transformation of nitrogen compounds can occur through several mechanisms, including fixation, ammonification, synthesis, nitrification, and denitrification.

Ammonification refers to the change from organic nitrogen to the ammonium form. In general, ammonification occurs during decomposition of animal and plant tissue and animal fecal matter and can be expressed as follows:

$$\text{Organic Nitrogen} + \text{Ammonifying Microorganisms} \rightarrow NH_3/NH_4^+ \quad (2.41)$$

Nitrification refers to the biological oxidation of ammonium ions under aerobic conditions by the chemoautotrophic organisms called nitrifiers. Two specific chemoautotrophic bacterial genera are involved, using inorganic carbon as their source of cellular carbon:

$$NH_4^+ + O_2 \xrightarrow[\text{bacteria}]{\textit{Nitrosomonas}} NO_2^- + O_2 \xrightarrow[\text{bacteria}]{\textit{Nitrobacter}} NO_3^- \quad (2.42)$$

TABLE 2.9
Nitrogen Forms Present in the Subsurface Environment

Nitrogen Compound	Formula	Oxidation State
Ammonia	NH_3	-3
Ammonium ion	NH_4^+	-3
Nitrogen gas	N_2	0
Nitrite ion	NO_2^-	$+3$
Nitrate ion	NO_3^-	$+5$

The transformation reactions are generally coupled and proceed rapidly to the nitrate form.

In situ denitrification can be accomplished by organisms belonging to the genera *Micrococcus, Pseudomonas, Denitrobacillus, Spirillum, Bacillus, Achromobacter, Acinetobacter, Gluconobacter, Alcaligens,* and *Thiobacillus*, which are present in the groundwater environment. Denitrifying organisms will utilize nitrate or nitrite in the absence of oxygen as the terminal electron acceptor for their metabolic activity. If any oxygen is present in the environment, it will probably be used preferentially. The energy for the denitrifying reactions is released by organic carbon sources that act as electron donors. The microbial pathways of denitrification include the reduction of nitrate to nitrite and the subsequent reduction of nitrite to nitrogen gas.

$$NO_3^- \rightarrow NO_2^- \rightarrow N_2\uparrow \qquad (2.43)$$

In biological wastewater treatment processes employing denitrification, a cheap, external carbon source such as methanol is added as the electron donor. It has long been known that NO_3^- can be converted to N_2 gas in anaerobic groundwater zones in the presence of a labile carbon source.

In situ microbial denitrification is based on the same principle as conventional biological wastewater treatment systems, except that it is carried out in the subsurface by injecting the appropriate organic carbon source. Since methanol could be an objectionable substrate from a regulatory point of view, sucrose or sugar solution is an optimum substrate to be injected. By amendment injection, the rate of denitrification is greatly accelerated and nitrate is converted to nitrogen gas within the reactive zone. This approach has the potential of remediating sizeable nitrate plumes in groundwater systems.

It should be noted that in the hierarchy of redox reactions, NO_3^- is the most favored electron acceptor after dissolved oxygen. Hence, considerable attention should be focused in maintaining the redox potential in the optimum range, so that Mn(IV), Fe(III), sulfate reduction conditions or methanogenic conditions are not formed in the subsurface. Furthermore, since denitrification is a reduction reaction, alkalinity and pH tend to increase in the aquifer. Since the end-product N_2 gas will escape into the vadose zone and, hence, the aquifer system is not a closed system, increased alkalinity will be observed in the groundwater. If the NO_3^- concentration is not very high, this concern will be short-lived.

2.4 EMERGING CONTAMINANTS

Emerging environmental contaminants include a wide variety of chemicals that have, thus far, been largely outside of the scope of environmental regulation both at the federal and local jurisdictions. Some of these contaminants are derived product(s) consumed by humans or animals for health or cosmetic reasons. There are other emerging contaminants introduced into the environment as a result of the efforts to enhance the efficiencies of gasoline consumption, automobile traffic, military exercises, etc.

Some of the emerging contaminants, such as MTBE, are removed from the perspective of an "emerging" contaminant as soon as clear understanding develops regarding the fate and transport of that contaminant in the environment. In addition, when certain technologies are proven to be effective in treating these contaminants, the label "emerging contaminant" is removed from the public's perception.

The authors want to discuss the developing remediation strategies for a few emerging contaminants that pose significant challenges to the remediation industry today: 1,4-dioxane, perchlorate, explosives, nitrosodimethylamine (NDMA), and radionuclides.

2.4.1 Perchlorate Reduction

Perchloric acid and its salts have been used extensively in a number of commercial applications such as wet digestions, organic syntheses, electropolishing of metals, animal feed additives, explosives, pyrotechnics, missile propellants, and herbicides. Perchlorate (ClO_4^-) is the soluble anion associated with the solid salts of ammonium, potassium, and sodium perchlorate.

Ammonium perchlorate is used as an energetics booster or oxidant in solid propellant for rockets and missiles. Ammonium perchlorate is also used in certain fireworks, the manufacture of matches, as a component of air bag inflators, and in analytical chemistry to preserve ionic strength. Large-scale production of ammonium perchlorate began in the U.S. in the mid-1940s. Ammonium perchlorate has a limited shelf life, and must be periodically replaced in munitions and rockets. This had led to the disposal of large volumes of this compound since the 1940s in Nevada, California, Utah, and other states.

The anion perchlorate poses potential environmental concerns because its ionic radius and charge are similar to that of iodine, which allows perchlorate to competitively block thyroid iodine uptake.[116,117] The first concerns about perchlorate in the environment surfaced in 1896 when sodium and potassium salts of perchlorate found in Chilean potassium nitrate deposits were noted to be harmful to certain agricultural crops.[117] These deposits (Chilean saltpeter), have been used in the manufacture of some chemical fertilizers.

Since the discovery of the naturally occurring perchlorate in Chilean saltpeter, additional natural occurrences of perchlorate have been identified in natural potash-bearing evaporite, hanksite, and playa crust samples.[118] All of these perchlorate sources are in arid climates and associated with evaporite deposits. Since naturally occurring concentrations of perchlorate are too low to be of any economic benefit, it is not surprising that until the recent emergence of concern about perchlorate as a contamiant, few efforts have been made to map its distribution in geological materials.

While the mechanism for the natural formation of perchlorate is not well understood, it has been hypothesized that perchlorate can be formed in the atmosphere by the reaction of tropospheric ozone with chlorine.[119,120] It is unlikely that this mechanism would produce substantial quantities of perchlorate that would reach the surface via precipitation, as the concentration of reactive chlorine existing in the troposphere is low. However, it has been postulated that the reaction between the ozone and halogenated acid gases and chlorinated volatile organics produced in the combustion zone of forest fires (favored by the high heat of the fire near ground surface) could lead to perchlorate formation at the soil particle surfaces or on the surfaces of atmospheric particulates created by the forest fire itself. Perchlorate produced by this mechanism could reach the ground surface through wet deposition and could be transported by the oxic rainwater.

The postulated chemical mechanisms that would produce perchlorate are oxidative in nature: reaction of ozone with chlorine is an example of an oxidation reaction. However, oxidized materials that are subsequently buried become isolated from the oxygen in the atmosphere, and they may become anaerobic where reduced minerals or organic carbon are available. As an example, a natural process could be envisioned where perchlorate was naturally produced by ozone produced from forest fires buried by rapid sedimentation from runoff after the fire, precipitated and concentrated as a salt under evaporative conditions, and then would become subject to biological reduction and dechlorination by bacteria utilizing naturally occurring organic carbon and perchlorate as reductant and oxidants, respectively. Such an example shows a system where naturally formed perchlorate did not survive because of a natural biogeochemical process.[121]

Natural geologic settings that would preserve or have preserved perchlorate over long time frames can be called "permissive geological units." Two examples of such settings include:

- *Oxidation poise setting.* This would involve deposition of perchlorate within oxidized materials with other natural materials that contribute oxidation equivalents over a long time frame (nitrate deposits are examples of this, where nitrate would tend to oxidize any reductants within the system, including reduced sulfide, iron, manganese, and organic carbon). In such systems, it is not required that the naturally occurring oxidant be at a higher formal oxidation potential than perchlorate, but rather that the oxidant be either in great molar excess (percentage-level nitrate vs. ppm-level perchlorate), or that it be physically or geochemically in a more reactive position (such as layered around the perchlorate) within the formation. Geologic units in contact with the atmosphere where rainfall is also limited could also comprise an oxidation poise setting, where oxygen from the atmosphere supplies the oxidation equivalents.[121]
- *Biogeochemical inhibitor setting.* This would involve deposition of perchlorate within a geologic formation that contains other materials that generally prevent biological oxidation–reduction reactions from occurring. Examples of this mechanism could include direct inhibition mechanisms, such as the presence of specific compounds in the formation that would act as poisons for biological electron transport, preventing perchlorate reduction by biochemical inhibition, or indirect inhibition mechanisms, such as low water availability in a brine or evaporite materials, which generally inhibit all biological processes.[121]

These two speculative mechanisms may explain the distribution of naturally occurring perchlorate in some regions of the U.S.

Most of the anthropogenic perchlorate contamination in the groundwater appears to have come from the legal discharge decades ago of then unregulated waste effluents containing high levels of ammonium perchlorate. Although ammonium perchlorate was released initially, the salt is highly soluble and dissociates completely to ammonium and perchlorate ions upon dissolving in water:

$$NH_4ClO_4 \rightarrow NH_4^+ + ClO_4^- \qquad (2.44)$$

It is likely that most of the ammonium has been biodegraded and the cation is now best viewed as mostly Na^+ or possibly H^+, especially where perchlorate (ClO_4^-) levels are below 100 ppb. At those sites where contamination dates back decades, very little (if any) ammonium has been found.[122]

The persistence of perchlorate in aquifers results primarily from a combination of aerobic conditions and lack of an electron donor. A number of bacteria that synthesize nitrate reductases are capable of dissimilatory reduction of perchlorate.[122,123] Many mixed cultures have reduced perchlorate, chlorate, chlorite, nitrate, nitrite, and sulfate under the right conditions. Inhibition of perchlorate reduction also has been observed in the presence of other substrates, particularly chlorate, chlorite, and sulfate.[122] Chlorate reductase has been isolated from microorganisms that also possess nitrate reductase. Although most perchlorate strains may be denitrifying facultative anaerobes, not all denitrifiers are (per)chlorate reducers. Simultaneous reduction of NO_3^- and ClO_4^- has been demonstrated in many laboratory and field studies.[122,123]

Microbially catalyzed perchlorate reduction is metabolically promising because the chlorine atom within the perchlorate molecule is in its highest oxidation state (+7). Hence, the reduction of perchlorate is highly thermodynamically favorable. Thermodynamic data indicate that perchlorate is a strong oxidant and could provide a large amount of energy to microorganisms as an electron acceptor. Some researchers have claimed that chlorite (ClO_2^-) is the microbial perchlorate (ClO_4^-) reduction end-product because it has been demonstrated that the microbially mediated transformation of ClO_2^- to Cl^- is a dismutation reaction that yields no energy for perchlorate-reducing microorganisms.[124,125]

Denitrifying microorganisms that have the enzyme chlorate dismutase can use the denitrification pathway to reduce (per)chlorate, and there is evidence in some microorganisms that an enzyme system may exist that is not involved in NO_3^- reduction and is also able to reduce ClO_3^- and ClO_4^-.[126,127] So far, perchlorate-degrading isolates reported have been mostly heterotrophic, and require organic substrates for synthesizing cellular materials. However, the isolation of a novel chemolithoantotrophic, perchlorate-reducing, hydrogen-oxidizing bacterium (*Dechloromonas* sp. strain HZ) was reported recently.[128] This strain also grew using acetate as the electron donor and chlorate, nitrate, or oxygen as an electron acceptor. The study of this species being able to degrade ClO_4^- without the presence of organic carbon may lead to easily acceptable approaches for the treatment of drinking water.

The conversion of chlorine in perchlorate to chloride requires the overall transfer of eight electrons. The sequence of intermediates involved in perchlorate reduction is as follows:

$$ClO_4^- \rightarrow ClO_3^- \rightarrow ClO_2^- \rightarrow O_2^+ \quad Cl^- \qquad (2.45)$$
$$\text{(perchlorate)} \quad \text{(chlorate)} \quad \text{(chlorite)} \quad \quad \text{(chloride)}$$

Although oxygen is produced, it is used by the perchlorate-respiring bacteria and does not accumulate in solution.[126] The breakdown intermediates ClO_3^- and ClO_2^- do not accumulate to measurable levels (Figure 2.55(A)).[129]

In situ bioremediation, via an IRZ, appears to be the most economically feasible, fastest, and easiest means of dealing with perchlorate-laden groundwater at all concentrations. Microbial transformation of perchlorate to chlorite occurs in the absence of oxygen as a result of anaerobic respiration. Anaerobic respiration is an energy-yielding process in which the oxidation of an electron donor, such as an easily degradable organic substrate, is coupled with the reduction of an electron acceptor, such as perchlorate and chlorate.

Implementation of an IRZ with the introduction of easily biodegradable electron donors such as hexoses, acetate, or lactate (without the presence of other electron acceptors such as SO_4^{2-}) should be able to reduce the concentrations of ClO_4^- present in the groundwater. A tenfold reduction of perchorate was achieved in column experiments at residence times of less than 48 h. Laboratory column experiments have demonstrated that perchlorate degradation can be achieved at influent levels ranging from 0.1 to 1000 mg L^{-1}.[126] The effluent levels achieved were in the range of 0.005 mg L^{-1}, which is lower than the state of California drinking-water action level of 0.018 mg L^{-1}. The authors and their colleagues are currently involved in many field-scale projects to implement *in situ* biodegradation of perchlorate.

Since most of the perchlorate plumes are decades old and also because of perchlorate's high solubility we are dealing with extremely large groundwater plumes of this contaminant. Hence it is very important to select the cheapest electron donor to create an IRZ to achieve perchlorate degradation *in situ*. The persistence of perchlorate itself is enhanced by the *oxidative poise* available within these plumes. Hence it is equally important to select the cheapest electron donor to overcome the oxidative poise within these large plumes.

There are many reports in the literature using methanol, acetate, acetic acid, whey, corn syrup, lactate, lactate polymer, and vegetable oil as the source of the degradable organic carbon to create the mildly reducing IRZs for perchlorate degradation. Many practitioners seem to claim that acetate/acetic acid is the best source of organic carbon for the supply of electron donor within the reactive zone. However, in the opinion of the authors and their colleagues, this seems to be a flawed approach.[130]

The flaw here is that just because acetate is the substrate that is left over from natural anaerobic environments does not mean it is a very good carbon source. It is left over from the initial degradation reactions that are occurring because it is not a good carbon source, either for cell yield or as an electron donor. Fundamentally the perchlorate process is a reduction process, with electron transfer occurring from the acetate, through the bacteria, to the perchlorate. As an electron source, acetate is a very poor electron donor. On an equimolar basis, it supplies about one third of the electrons of ethanol (another two-carbon molecule).

The authors have presented the data from one of many field-scale projects being implemented by them for perchlorate removal within an IRZ (Figure 2.55(A) and Figure 2.55(B)).

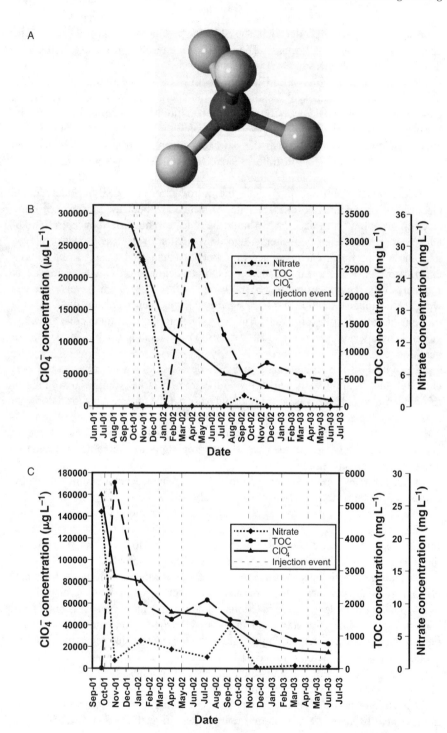

FIGURE 2.55 (A) Perchlorate. (B) Performance data from an IRZ system for perchlorate degradation. (C) Performance data from an IRZ system for perchlorate degradation.

Microbial Reactive Zones

The most important lesson to be learned from these two figures is that complete biodegradation of perchlorate within anaerobic IRZs can be achieved. The cleanup goals required within the current regulatory framework can be also achieved within a reasonably short time frame. In addition, it becomes very clear that the redox conditions have to be maintained at a "mildly reducing range." When the TOC levels are increased, thus leading to very reducing environments, perchlorate reduction rates are significantly decreased or the transformation is almost stalled.

Figure 2.56(A) describes the operational basis for implementing an IRZ for perchlorate degradation. The most important aspect is to maintain the redox conditions around the denitrifying to Fe(III) reducing range. Hence, the injection regime of organic carbon has to be maintained in such a way as to maintain the biogeochemical conditions in a mildly reducing range. The redox conditions and the organic carbon dosage required to achieve reductive dechlorination is drastically different than that required for perchlorate degradation. Figure 2.56 shows that the concentration of organic carbon required is significantly lower compared to other microbially reducing IRZs. However, since the perchlorate plumes tend to be significantly bigger than the plumes of other contaminants, the total amount of organic carbon required will still be high. It is also important to note that the optimum redox conditions required for ClO_4^- removal also lie within the optimum range for Cr^{6+} and NO_3^- removal (Figure 2.56).

Contamination and disposal of perchlorate at many of the sites being evaluated today took place years or decades ago. Yet, a significant level of soil contamination is encountered at many of these sites. In spite of its high solubility, perchlorate has been

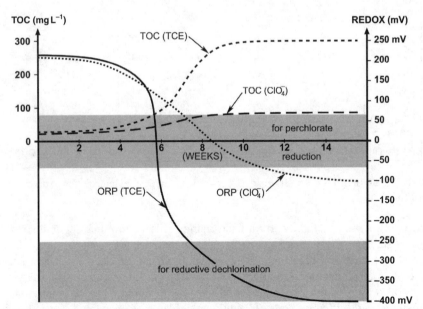

FIGURE 2.56 (A) Biogeochemical operating principles of an IRZ system for perchlorate degradation.

Area 1

Depth (ft bgs)	Perchlorate concentrations in soil ($\mu g\,kg^{-1}$)					
	Historic 9/12/2003	Baseline 12/3/2003	1 week 1/7/2004	6 weeks 2/10/2004	19 weeks 5/10/2004	
5	510	1,800	4,900	2,200	770	
10	1,700	250	3,600	3,400	<10	
20	310	4,600	4,300	21	<10	
30	13,000	170	290	3,700	920	
40	100	39	<10	<10	<10	
50	--	--	10	<10	--	
60	--	--	11	<10	--	
70	--	--	16	--	--	
Totals:	15,620	6,859	13,090	9,321	1,690	89% reduction
Remaining TOC in pore water (mg L^{-1})					2,100	

Area 2

Depth (ft bgs)	Perchlorate concentrations in soil ($\mu g\,kg^{-1}$)				
	Baseline 12/3/2003	1 week 1/7/2004	6 weeks 2/10/2004	19 weeks 5/10/2004	
5	85	54	170	<10	
10	440	66	23	<10	
20	3,200	3,300	<10	<10	
30	9,700	330	990	<10	
40	<10	<10	<10	<10	
50	--	<10	<10	--	
60	--	<10	<10	--	
Totals:	13,425	3,750	1,183	0	100% reduction
Remaining TOC in pore water (mg L^{-1})				110	

FIGURE 2.56 (continued) (B) Field demonstration of biological remediation of perchlorate present in soil.

observed to adsorb slightly to variable-charge soils in low pH environments.[131] Recent studies indicate that soils with a high anion-exchange capacity may adsorb perchlorate under low pH conditions.[132] However, there is no clear evidence whether adsorption of perchlorate can affect its transport under these conditions. The same study also showed that biodegradation can significantly affect the transport of perchlorate in different soils. Conditions needed for biodegradation in soils include anaerobic conditions and the presence of sufficient carbon. The presence of perchlorate-degrading bacteria also is an important factor, since in some materials, such as the dredge tailings used in this study, indigenous perchlorate-degrading microbes take a long time to develop a population large enough to reduce perchlorate, even when a carbon source is available.[132]

A field pilot study for the degradation of perchlorate present in the soil was implemented by the authors. Reasonably sized soil plots were selected for the injection of corn syrup into the zones where perchlorate was present within the soil matrix (Figure 2.56(B)). Significant to complete degradation was achieved within these pilot plots within a period of 6 weeks. This pilot study demonstrated that perchlorate contamination present in soil can be biodegraded *in situ* by injecting an appropriate organic carbon and maintaining optimum biogeochemical conditions. Based on the success of this pilot study the authors have initiated other *in situ* biogradation systems for the removal of perchlorate contamination present in soil.

Redox conditions required for perchlorate degradation are similar to those for NO_3^- reduction. The lag time for microbial perchlorate reduction was much longer than the lag time for NO_3^- reduction in soils not previously exposed to perchlorate. Growth of microbial populations able to degrade perchlorate and/or the induction of enzymes able to degrade perchlorate may strongly affect whether perchlorate or NO_3^- is used first as a terminal electron acceptor. The results of this research have implications for *in situ* remediation schemes for perchlorate-contaminated groundwater and for agricultural fields contaminated by perchlorate-tainted irrigation water.[132]

2.4.2 1,4-DIOXANE

1,4-Dioxane (dioxane), a cyclic ether also known as diethylene dioxide, has been used in a wide range of industrial applications (Figure 2.57). It is highly water soluble and not readily degradable under ambient aquifer conditions, either chemically or biologically. As a consequence, it moves freely through aquifer formations, traveling at nearly the speed of groundwater.[133] 1,4-Dioxane has not received the same level of attention as petroleum hydrocarbons or chlorinated alkenes. However, it has received a significant level of regulatory attention recently. This compound's chemical characteristics and its low compliance concentration combine to mount a very difficult remediation design challenge.

The authors have considered several potential remedial strategies for 1,4-dioxane, and each is discussed in detail below. Among these, *in situ* chemical oxidation has been proven at bench scale and is deployable at some sites, while a mass-transfer mode of phytoremediation has been demonstrated and may be applied for both soil and shallow groundwater cleanup. Aerobic biodegradation has been demonstrated at bench scale, but no successful field deployments have been reported. Reductive biotransformation may be thermodynamically feasible, and there is field data from aquifers with very low redox potentials that is suggestive of dioxane degradation.[134] Additional development work is required at both field and laboratory scales before 1,4-dioxane remedies can be managed consistently.

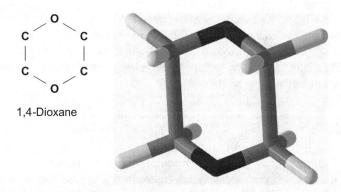

FIGURE 2.57 Diagrammatic and structural views of 1, 4-dioxane.

2.4.2.1 Prospects for Mass Transfer

Aquifer sparging is a mass transfer–based remedy that *cannot* be effectively applied to 1,4-dioxane. The success of sparging (*in situ* stripping) for any compound is determined by its Henry's law coefficient and its molecular weight (representing its aqueous-phase diffusion rate). Figure 2.58 shows the relative spargeability of 1,4-dioxane and many other compounds. Based on that analysis, the compound is not spargeable.

The inability to sparge 1,4-dioxane was confirmed during bench-scale ozone treatability experiments that are discussed in detail later. A sparge control was established to ensure that dioxane reductions measured in ozone-sparging trials were entirely related to oxidation and not to mass transfer. A 300 ml min^{-1} oxygen stream (without any ozone) was sparged into a 1-L test chamber containing dioxane at 9 mg L^{-1}. After 40 min of sparging, there was no detectable reduction of 1,4-dioxane. This was comparable to 42 days of conventional sparging that would be expected to yield greater than 90% reduction for xylenes. These results are consistent with the relative positions of these compounds on the spargeability diagram (Figure 2.58).

Soil vapor extraction is a mass transfer–based remedy that has limited application to 1,4-dioxane. The high water solubility of dioxane (miscible) and its low organic carbon partition coefficient (3.5 L kg^{-1}) lead to the expectation that most vadose zone dioxane contamination will reside in soil pore water (up to 15% of the vadose zone bulk density). To recover dioxane from the vadose zone, a soil vapor extraction system must also recover soil moisture. This is accomplished by installation of a site surface seal and injection of air containing lower relative humidity than the soil (at soil temperature).

Phytoremediation has now been shown to be effective for 1,4-dioxane in shallow groundwater and vadose zone applications has been reported. Substantial mass transfer of dioxane in hydroponic studies using cuttings of hybrid poplars has been reported.[135,136] These studies indicated that dioxane removal occurred from both soil and groundwater, and in one study the dioxane was transferred at high rates to the atmosphere through transpiration. Others have summarized the current status of

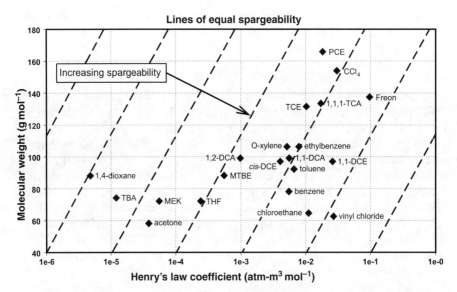

FIGURE 2.58 Relative spargeability of organic compounds. The lines of equal spargeability were developed from comparisons of contaminant reduction rates at sites where sparging was applied to several compounds simultaneously. Those observations showed that Henry's law alone could not predict the relative rates of contaminant decrease. For example, perchloroethene is sparged more slowly than *cis*-DCE, despite the fact that the Henry's law constant for PCE is more than fivefold higher.

phytoremediation, indicating that hydrophilic compounds such as 1,4-dioxane are likely to be mobilized in the plant's transpiration stream and displaced from belowground to the atmosphere.[137] 1,4-Dioxane is rapidly photo-oxidized in the atmosphere.

2.4.2.2 Biological Remedies

1,4-Dioxane, like most ether compounds, is not highly biodegradable. Evidence for biodegradation is mainly from aerobic systems, although there is limited evidence for reductive degradation of ethers with high molecular weight. Metabolic degradation is the consumption of a compound by the species as an electron donor or an electron acceptor, with enough energy captured to support growth. In the case of aerobic metabolism, 1,4-dioxane serves as an electron donor and as a source of carbon for growth. Cometabolic degradation occurs when bacterial growth cannot be supported by 1,4-dioxane, but enzymes are present that can destroy it. In these cases, the growth of bacterial communities must be supported by other electron donors and carbon sources, which may be naturally occurring or injected in support of remedial efforts.

2.4.2.2.1 Metabolic Aerobic Biodegradation Potential

Four bacterial species are known to degrade 1,4-dioxane. *Amycolata* sp. strain CB1190, *Mycobacterium vaccae*, and two strains of *Rhodococcus* sp. have been identified that can degrade dioxane alone, or in combination with tetrahydrofuran.

Scalable commercial processes have not been developed from these genera to date, although many industrial wastewater treatment plants receiving 1,4-dioxane are populated by microbial consortia that are very effective in its degradation.[138–140] If scaling-up is to proceed, it would most likely be in bioreactor settings where operating conditions can be closely controlled.

It was shown that *Amycolata* strain CB1190 is capable of aerobic growth with 1,4-dioxane as the sole carbon and energy source.[141] They also showed that CB1190 is capable of growth on tetrahydrofuran as well as other linear ethers. Others have used *Amycolata* strain CB1190 in bioaugmentation experiments and showed that this species' dioxane degradation rate is increased by the presence of cosubstrates such as tetrahydrofuran and poplar root extract. Their experiments suggested that mass-transfer phytoremediation by hybrid poplars can be enhanced by the bacterial degradation of dioxane in the root zone. It was also reported that *Mycobacterium vaccae* is capable of degrading a large number of groundwater contaminants, including 1,4-dioxane and a number of substituted benzene rings.[142] It has been shown that six strains of *Rhodococcus* were capable of tetrahydrofuran degradation and two of those strains could also degrade 1,4-dioxane.[143]

2.4.2.2.2 Cometabolic Aerobic Biodegradation Potential

Studies have shown that 1,4-dioxane was cometabolically degraded in the presence of tetrahydrofuran by a consortium of aerobic bacteria. No dioxane degradation occurred without the tetrahydrofuran.[144] Several unpublished studies have examined monooxygenase enzymes and their capacity to degrade dioxane, MTBE, and other recalcitrant compounds. These compounds are generally susceptible to degradation by soluble monooxygenases that catalyze oxidation of low molecular weight n-alkanes. sMMO and soluble propane monooxygenase (sPMO) are two such examples that have been shown to degrade MTBE, dioxane, and chlorinated alkenes. These enzyme systems are found among methane- and propane-degrading bacteria and can be induced in laboratory studies by aerobic addition of the alkane gas and in some cases by its corresponding alcohol (e.g., propanol can stimulate sPMO production among propane-oxidizing bacteria). The low solubility of these gases and their explosive nature have limited field trials. However, the fieldwork that has been completed to date failed to achieve evidence of propane degradation or stimulation of sPMO production. This suggests that indigenous populations of the methane or propane degraders are not widespread in aquifers and bacterial seeding would be required in many instances.

An alternative to the soluble monooxygenase approach is to stimulate degradation of structural analogs of dioxane.[144] Enzymes often exhibit degradation activity (although reduced) for analogs of their primary substrate. Acetaldehyde is a candidate stimulant, because dioxane is essentially two acetaldehyde molecules linked end to end, forming a ring. In fact, recent laboratory studies led to a suspicion that either acetaldehyde or acetic acid may stimulate degradation of dioxane. Tetrahydrofuran (THF) is another analog that may stimulate fortuitous degradation of dioxane, as in the *Rhodococcus*. The problem with these cometabolic degradation stimulants is that they represent health risks of their own and may not be injectable at all locations.

Bioreactor systems for *ex situ* treatment of dioxane offer an opportunity to deploy cometabolic degradation without the risks associated with *in situ* distribution of THF or acetaldehyde. In a "negative growth" reactor system, a microbial consortium is acclimated to degradation of THF, then switched to a dioxane-bearing wastewater stream. Dioxane is degraded, although the microbial population declines due to lack of an effective food source. During the negative growth phase of the first reactor, a second reactor is in the preparation phase, being fed a stream containing THF.

2.4.2.2.3 Anaerobic Biodegradation Potential

Anaerobic degradation of low-molecular-weight ethers has not been directly observed, and a recent report casts doubt on the search for anaerobic degradation.[145] However, three factors give encouragement that some reduction may occur in low-redox environments.

- First, field observations of highly anaerobic contaminant plumes in groundwater suggest that dioxane attenuation rates exceed what can be expected from dilution and dispersion alone.
- Second, molecular hydrogen that is generated in highly reducing conditions may chemically reduce compounds such as dioxane, enabled by the catalytic activity of soluble metals or soil minerals. The apparent hydrogenation of chlorinated alkenes in low-redox groundwater suggests the presence of molecular hydrogen as well as the presence of an adequate catalyst.
- Third, reductive biodegradation of high-molecular-weight ethers has been observed from *Pseudomonas* sp. experiments.[146] Comparable reductive degradation of dioxane has not been reported to date although attenuation patterns at some sites with highly reducing environments are suggestive of rates that exceed what can be expected through dilution and dispersion alone.

Bioaugmentation for other contaminants has proven to be very costly, and biostimulation generally achieves equal contaminant reductions after an acclimation period. The success of biostimulation depends on the presence of suitable bacterial populations as well as our ability to manage the aquifer microbial habitat to promote their growth and contaminant destruction behaviors. The studies on *Amycolata*, *Mycobacterium*, and *Rhodococcus* will be most beneficial if the strains that function well in the degradation of dioxane are found to be widely distributed in aquifers and can be stimulated through aquifer habitat management.

2.4.2.2.4 Reductive Destruction

Reductive destruction may be operable as a source area or barrier zone treatment. The critical question for this approach is to determine its kinetic feasibility. Low-redox conditions that develop during enhanced reduction applications are believed to develop sufficient molecular hydrogen to support reduction of dioxane. However, it is not known whether appropriate hydrogenation catalysts are present in aquifer systems. These questions can be resolved in a field trial that develops a strongly reducing zone in a dioxane-contaminated aquifer through injection of rapidly degradable carbon. The potential cost savings associated with a reductive barrier (relative to an oxidant barrier system) merit the investment to make an initial feasibility determination.

There are indications from the physical chemistry literature that 1,4-dioxane will be resistant to reducing agents such as the dithionate ion: experiments on chemical reduction often are performed in a water–dioxane medium, and results indicated substantial resistance to reduction in aqueous-phase. There are examples of the use of dioxane as a reduction test medium for reductant-susceptible compounds.

A reductive alternative that might be considered for bench trials is zero-valent iron, which can be applied as a barrier system in the field.

2.4.3 EXPLOSIVES

Explosives are materials which, when suitably initiated, result in the rapid release of energy. Detonation of a solid explosive generates expanding hot gases, and this expansion creates a shock wave that exerts high pressures on the surroundings, causing an explosion. Explosives generally have high nitrogen and oxygen contents, which aid the formation of the gaseous products, typically including carbon dioxide, carbon monoxide, nitrogen, and water vapor. The history of explosives can be defined from the first mixture of gunpowder to the industrialization of explosives manufacturing to provide both greater power and control.[147]

The manufacture, use, and disposal of explosives from military operations have resulted in extensive contamination of soils and groundwater.[148,149] Of the nearly 20 different energetic compounds used in conventional munitions by the military today, hexahydro-1,3,5-trinitro-1,3,5-triazine (RDX) and octahydro-1,3,5,7-tetranitro-1,3,5,7-tetrazocine (HMX) are the most powerful and the most commonly used. RDX and HMX are acronyms that stand for *royal demolition explosive* and *high melting explosive*, respectively. Another commonly used military explosive is 2,4,6-trinitrotoluene. Figure 2.59 illustrates the molecular structure of these three compounds.

The release of these contaminants, resulting in significant soil and groundwater contamination, has been associated with various commercial and military activities including manufacturing, waste discharge, testing and training, open burning, and open detonation. It has been estimated that during RDX manufacturing up to 12 mg L^{-1} can be discharged into the environment in process wastewaters. In addition to dilution and dispersion, processes important to groundwater transport of explosives include biotic and abiotic transformations, covalent bonding to soil organic matter, and adsorption by soils. There are some locations in our country where the soil concentrations of TNT and RDX exceed the cleanup levels of 17.2 and 5.2 mg kg^{-1}, respectively, by orders of

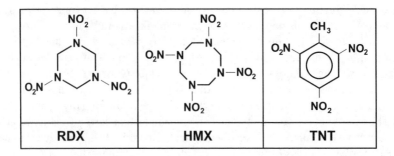

FIGURE 2.59 Molecular structure of three explosives.

magnitude. The two major targets for remediation are RDX and TNT, both of which were manufactured in vast quantities during the 20th century.

In addition to their destructive capacity, explosives commonly have toxic effects on ecological systems. All explosives exhibit harmful effects to varying degrees; however, toxicological exposure limits have not been determined for explosive compounds. In addition, explosive compounds are also considered to be highly recalcitrant compounds.

A characteristic of explosives is the presence of nitro ($-NO_2$) groups on the molecule (Figure 2.59). Major factors affecting fate and transport of TNT in the subsurface are transformation, sorption, and irreversible soil binding. Although TNT reductive transformation has been known for some time, only recently have TNT reductive transformation products been routinely measured in laboratory and field studies.[150] Possible TNT transformations include reduction of one, two, or all three of the nitro-moieties to amines, and coupling of amino transformation products to form dimers. Formation of the two monoamino transformation products, 2-amino-4,6-dinitrotoluene (2ADNT) and 4-amino-2,6-dinitrotoluene (4ADNT), is favored, and they are typically observed in TNT-contaminated soils and groundwater. Since the diamino products are energetically more difficult to form, they are observed less frequently and typically at lower concentration than the monoamino products. The triamino product is rarely observed not only because it is more energetically difficult to form, but also because once formed, it does not persist, but is likely to be immobilized by chemical reactions with soil components or by microbial degradation. The reductions can be enzymatically catalyzed or, under proper reducing conditions, abiotic.

The amino transformation products are amenable to several attenuation mechanisms in soils. These include covalent bonding to functional groups on soil organic matter such as described for similar amines, reactions at mineral surfaces, sequestration, and reversible adsorption. Although these mechanisms for attenuation of TNT have received attention in laboratory and field studies in the last several years, little data have been reported on their occurrence in the field beyond detection of the amino transformation products in soil and groundwater and declines in TNT concentrations over time.[150,151] Hence, there is a school of thought that TNT tends to undergo only transformation reactions rather than mineralization. The transformation products bind to the soil and become unextractable.

While TNT biodegradation is problematic, toluene, benzene, and phenols with one or two nitro groups on the molecule are more subject to biodegradation and mineralization. Initial monooxygenation or dioxygenation reactions can lead to the release of $-NO_2$ groups and substitution of $-OH$. This can lead to mineralization. In some cases, the nitro group is reduced to an amino group, and the resulting amino aromatics accumulate as dead-end products.[151]

In principle, TNT should be biodegradable by a two-stage reductive/oxidative process (Figure 2.60). Due to the stability of the aromatic ring and the electron-withdrawing properties of the nitro group, microbial action on compounds such as TNT generally proceeds via reduction of the nitro groups, reducing them successively to nitroso, hydroxylamino, and amino groups.[147] Candidates for key metabolites in such a pathway are the products of reduction of the nitro groups or the

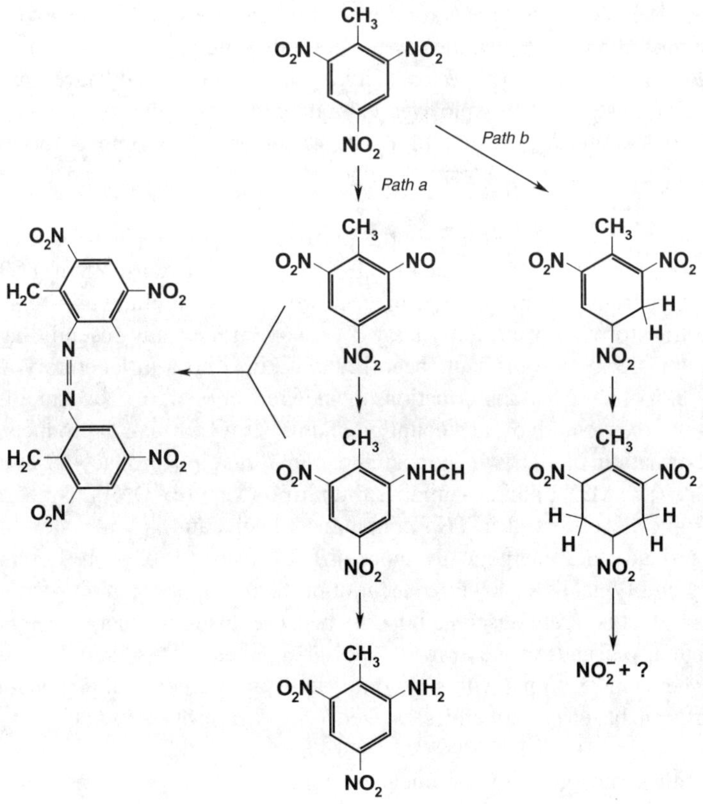

FIGURE 2.60 Microbial transformation of TNT. *Path a*. TNT can undergo reduction of the nitro groups through nitroso, hydroxylamino, and amino derivatives. Each of the nitro groups may undergo reduction, although triaminotoluene only forms under anaerobic conditions. Azo and azoxy dimers can form from nitroso and hydroxylamino dimerization. *Path b*. TNT can also undergo hydride addition to form hydride– and dihydride–Meisenheimer complexes. Nitrite is produced from this reaction, but other products are unidentified.[183,184]

aromatic ring system. For example, the hydroxylamino toluenes could be substrates for mutases or hydroxyaminolyases, which generate aminophenols or catechols. The amino derivatives, in particular, are also very stable and adsorb very strongly to soil, factors that strongly hinder further breakdown in the subsurface environment. In addition, the hydroylamino and amino derivatives can dimerize to form azo and azoxy dimers, which are resistant to further metabolism.[152]

Such compounds and the triaminotoluene from complete reduction of TNT or its hydride complexes harbor structural features and sufficient chemical reactivities to be potential parent compounds of complete, mainly oxidation pathways. The potential function of TAT as a key metabolite in a two-stage reductive/oxidative process is limited, however, by its high reactivity toward oxygen, which gives rise to extensive misrouting and finally the generation of dark high-molecular-weight products.[151,153,154]

The research conducted thus far shows tentatively that biodegradation of RDX and HMX under both aerobic and anaerobic conditions is feasible. Unlike TNT, which is subject to biotransformation rather than mineralization, the two cyclic nitramines RDX and HMX undergo ring cleavage following initial transformation of the molecules by anaerobic sludge. Under reducing conditions, small organic molecules (HCHO, HCOOH, CO, CO_2) and small nitrogen-containing products (NO_2^-, NO_3^-, CN^-, and NH_3) are produced. Mineralization of either RDX or HMX under anaerobic conditions is extensive (60% with anaerobic sludge). Under aerobic conditions the biodegradation of the two explosives is clearly far from being understood and certainly further research is needed. In general, the current understanding of the biodegradation of cyclic nitramines, particularly with respect to the microbial populations and enzymes involved, is incomplete.[150,151,154] It should be noted that RDX can also be broken down by alkaline hydrolysis under the right conditions. The mechanism by which this reaction is proposed to occur involves the abstraction of a proton by the alkali, with the simultaneous loss of nitrite (NO_2), to form an unstable intermediate that further degrades.[155]

RDX contains chemical groups, such as the nitramine group, which were not previously found in nature. It could be proposed that the novelty of these compounds would mean that the environmental microflora would not possess enzymes that could degrade or transform them. However, laboratory, field, and composting studies indicate microorganisms are able to remove the nitramine explosives.

RDX is much more amenable to biodegradation than TNT (Figure 2.61(A), Figure 2.61(B), and Figure 2.61(C)). With no aromaticity it appears to be able to undergo several types of reaction. Under anaerobic conditions, reduction of the nitro groups forms nitroso intermediates that subsequently break down further.[156] Under aerobic conditions, nitrite has been seen to accumulate. Unlike the nitrate ester and nitro-aromatic explosives, there has been no identification of enzymes responsible for any of the reactions that RDX undergoes. This interesting and fundamental area remains to be investigated.[147]

There are only a few reports of microbial activity on HMX.[147] Products of the anaerobic degradation of HMX include nitroso derivatives, indicating a mechanism similar to the anaerobic degradation of RDX, and also at a slower rate than RDX. Aerobic removal of HMX has also been reported.[157] The reduced microbial activities against these compounds may be due to the lower solubility than RDX, the greater steric effects within transition states, and the higher bond-dissociation energy of the $N-NO_2$ bond.

One clear conclusion that can be reached from the literature reports is that once the ring in cyclic nitramine cleaves, the resulting degradation products are expected to be thermally unstable and to hydrolyze readily in water. Such abiotic reactions would compete with other biological reactions during attempted biodegradation. If the biodegradation reactions are considerably slower, it would be impossible to evaluate the degradation process at the microbial level unless some innovative research is undertaken to either slow down the abiotic processes or enhance the enzymatic ones.[154]

Recent studies indicate that hydrogen is a key factor in stimulating the anaerobic biotransformation of TNT, RDX, and HMX. It was shown that electron donors that produce H_2 support the biodegradation of these compounds.[158] The authors' experience

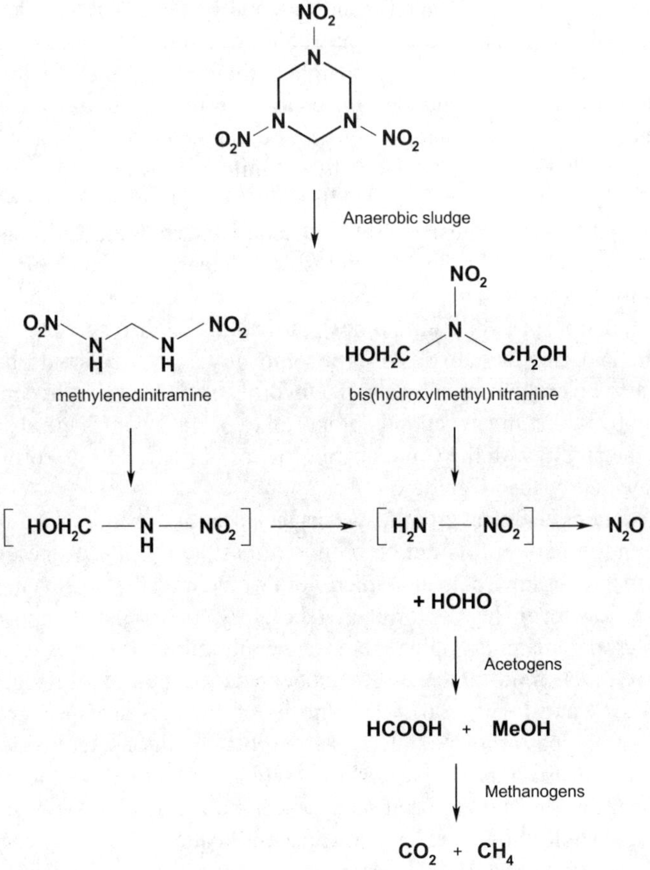

FIGURE 2.61 (A) Second putative pathway for RDX degradation by anaerobic sludge. Compounds identified include methylenedinitramine, *bis*(hydroxylmethyl)nitramine, formaldehyde, formate, methanol, nitrous oxide, methane, and carbon dioxide. Two hypothetical intermediates are also shown in brackets. Traces of nitrogen gas and nitrite were also detected, along with some soluble, nonextractable degradation products.[184,185]

in field tests indicates that TNT, RDX, and HMX degradation can be achieved within anaerobic reducing IRZs created with molasses as the electron donor.

The field experience of the authors and their colleagues at ARCADIS and theoretical explanations for some of their observations are summarized later.[147]

The capacity constant for sorption of RDX into soil is relatively low indicating RDX is not extensively sorbed by soil and sorption was reported to be nearly irreversible.[159] It was also reported that because the desorption isotherm lies above the sorption isotherm and the capacity constants for desorption are significantly higher than for sorption so there is considerable sorption–desporption hysteresis. Therefore, it appears that a proportion of the RDX may be irreversibly bound to the soil and

FIGURE 2.61 (continued) (B) Proposed mechanism of RDX biodegradation by *Rhodococcus* sp. strain DN22. Denitration as an enzymatic first step creates unstable intermediates that undergo ring cleavage. NO_2^-, N_2O, NH_3, HCHO, and CO_2 were identified as products of RDX degradation, as well as the dead-end product $C_2H_3N_3O_3$. Hypothetical components of the pathway are shown in brackets.[184,186] (C) Putative pathway for the anaerobic biodegradation of RDX via nitroso intermediates. Compounds identified include the three nitroso derivatives, formaldehyde, methanol, and hydrazine, and the two dimethylhydrazines. The hydroxylamino derivatives and ring cleavage products are hypothetical intermediates and are shown in brackets.[184,187]

desorption should not be significant after addition of molasses, as seen in other microbial IRZs due to the microbial surfactant and cosolvency effects.

Similarly, TNT can be reversibly sorbed by soils. However, reactions that remove TNT from solution and bind its transformation products to soil in an unextractable manner can be mistaken for adsorption. The reaction products of TNT, particularly hydroxylaminodinitrotoluene and triaminotoluene, have been demonstrated to bind irreversibly to soil components. Substantially fewer data are available on the sortion of HMX; however, it is apparently sorbed less than TNT by soils. In summary, soil–water partitioning coefficients for RDX and HNX ($K_d < 1$) are less than those reported for TNT ($K_d > 4$), indicating that the nitramines are more mobile than TNT and capable of migration through subsurface soil to groundwater supplies. TNT transformation products bind irreversibly to soil and are immobilized.

The threshold concentrations at which these compounds become toxic in aqueous solution cannot be achieved as they are relatively insoluble in water and cannot reach toxic levels. At 20 to 25°C the reported solubilities are TNT 140 mg L^{-1}; RDX 42 mg L^{-1}; and HMX 5 mg L^{-1}. However, the toxicity of these explosives when adsorbed by soil is another matter. In general, TNT is the most toxic of these explosives to soil microorganisms, while RDX and HMX are also toxic at certain concentrations.

Under anaerobic conditions, at redox potentials below -200 mV the transformation of TNT to triaminotoluene (TAT) has been reported. However, further degradation of this compound has not been conclusively proven. There are some reports that hydroxylated toluene derivatives form but these reports have not yet been substantiated. In general, the anaerobic mode of attack on TNT looks more promising as a process leading to mineralization because aerobic attack ends up forming dead-end products.[147,160]

It should be noted that RDX and HMX are compounds that do not possess any carbon to carbon bonds and are therefore metabolized by a specific group of bacteria, methylotrophs, which have the metabolic capacity to assimilate and metabolize these compounds.[147,160] This may account for the suggestions that RDX degradation is primarily a cometabolic process as only certain members of any soil-borne microbial population will be able to utilize RDX and HMX as a source of carbon and an electron donor thus growing at their expense. These explosives may be more readily biodegraded as a nitrogen source in the presence of a readily degradable carbon source. Hence, it may be important to choose an organic substrate with the lowest C : N ratio to create the reducing IRZ for explosives degradation. Based on the field experience of the authors and known metabolic breakdown pathways of these compounds, it may be important to maintain the reducing IRZ under redox potentials associated with sulfate-reducing or methanogenic conditions.

2.4.4 RADIONUCLIDES

The release of radionuclides into the environment is a subject of intense public concern. Although significant quantities of radionuclides were released as a consequence of nuclear weapons testing in the 1950s and 1960s and via accidental releases (such as in Chernobyl), the major burden of anthropogenic environmental radioactivity is from the controlled discharge of process effluents produced by industrial activities related to the generation of nuclear power. Radionuclide-containing wastes are also produced at all steps in the nuclear fuel cycle, from low-level, high-volume radioactive effluents produced during mining operations to the intensely radioactive plant, fuel, and liquid wastes produced from reactor operation and fuel reprocessing.[161,162]

2.4.4.1 Uranium

Uranium ore bodies form where precipitation of uranium from groundwaters is geochemically favored. These conditions include a mildly basic pH and reducing conditions. Uranium is transported through groundwater in the oxidized U^{6+} redox state as a stable uranium carbonate complex ($UO_2(CO_3)_3^{4-}$ or $UO_2(CO_3)_2^{2-}$), and when waters become sufficiently reducing it is converted to uraninite (UO_2) in the reduced insoluble U^{4+} redox state. The conditions necessary for uranium deposition are also favorable to deposition of other elements whose solubility is redox sensitive. These elements include the heavy metals arsenic, selenium, and vanadium.

The most commonly used uranium mining technique is referred to as *in situ* leach. This method involves drilling a well field into an ore body, injecting an oxidizing agent (such as hydrogen peroxide or compressed air) to oxidize uranium to the soluble U^{6+} state in the presence of carbonate (frequently introduced as ammonium carbonate) to form the stable soluble uranium carbonate complexes. The well field is then pumped and the uranium is recovered aboveground by ion exchange methods. In addition to uranium, this method also mobilizes other elements whose solubilities are redox sensitive.

As mentioned earlier, uranium is a naturally occurring radionuclide that is naturally available as ^{238}U, the most abundant isotope, plus ^{235}U, and minor quantities of ^{234}U are present as a decay product of ^{238}U. Still, natural uranium contains only 0.7% of fissile ^{235}U, and enrichment to 3% ^{235}U is required for use in power generation.

Under ordinary aerobic conditions, uranium exists in the +6 oxidation state. In the absence of carbon, the uranyl (UO_2^{+2}) ion is found in acid to slightly basic conditions. Under strongly basic conditions, the mixed oxidation state complex U_3O_8 exists. In the presence of carbon, the uranyl ion exists only under clearly acidic conditions. Above a pH of about 5, carbonates prevail. The solid uranyl carbonate exists over a fairly narrow pH range of 5 to 6 with uranyl carbonate anions existing from near neutral to strongly basic conditions. Throughout this range, uranium remains in the +6 oxidation state. An E_H–pH diagram for uranium is provided as Figure 2.62.

Remediation of uranium in groundwater using an *in situ* bioremediation process has been proposed in the past. These studies report that injection of an organic carbon source into a typical uranium plume will result in uranium precipitation as insoluble uranium oxides, and that sulfides can be coprecipitated with the uranium to provide long-term uranium stability.[163–165] Others have shown that uranium reduction can be rapidly achieved in an aquifer containing excess organic carbon and soluble uranium where acetate, lactate, and formate (organic carbon sources) is supplied. The importance of excluding nitrate and denitrification intermediates from the aquifer following uranium reduction to prevent remobilization of the uranium was also demonstrated.[166]

It has been shown that SRBs are abundant in groundwater in a zone containing high concentrations of uranium at the Shiprock, New Mexico, site, and that these bacteria are capable of both uranium reduction and sulfate reduction and iron sulfide

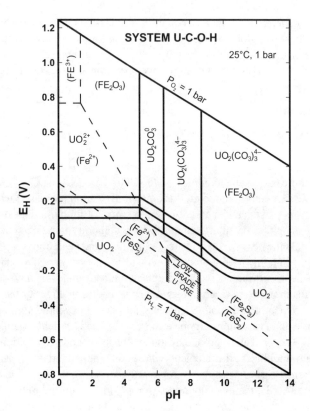

FIGURE 2.62 E_H–pH diagram of uranium.

precipitation. Microbial reduction of uranium was shown to result in soluble uranium concentrations as low as 1 pCi L^{-1}.[167] Other column studies have shown that excess iron sulfide is an optimal material to provide a redox buffer to prevent oxidative dissolution of uranium. It was also stated that the more iron sulfide present, the higher the stability of uraninite, the mineral form that has to be created to maintain uranium stability. A maximum concentration of 29 pCi L^{-1} dissolved uranium was formed during re-oxidation of freshly precipitated uranium where a 10^4 molar excess iron sulfide was precipitated along with the uranium.[168]

Other studies show that uranium reduction proceeds rapidly in the presence of excess organic carbon, and that sulfate reduction will also co-occur along with uranium reduction if sufficient organic carbon is added and sulfate is available.[169] Accelerated oxidation experiments performed in column studies documented uranium stabilization for simulated "hundreds of years" where iron sulfide had been codeposited with the uranium. Numerous other recent papers have documented that uranium removal and sulfate reduction follows the injection of sulfate and organic carbon into groundwater containing uranium, as well as the ability of these systems to prevent remobilization of uranium at concentrations of concern.

2.4.4.1.1 Sulfate Reduction and the Role of Iron Sulfides

Sulfate-reducing bacteria have been utilized to perform *in situ* bioremediation for a wide variety of contaminants, including hydrocarbons, chlorinated solvents, and heavy metals. To perform sulfate reduction, an electron donor, typically organic carbon, and sulfate (which acts as the electron acceptor) must be present. Many SRBs have been shown to be capable of uranium reduction as well.[167,169] The process developed by the authors and their colleagues primarily relies on SRBs to take soluble sulfate and iron and reduce the sulfate to sulfide (Figure 2.63(A) and Figure 2.63(B)). Sulfide then chemically reacts with iron to make iron sulfides. The sulfate reduction/iron sulfide formation process is a process that occurs in soils and sediments of lakes, rivers, swamps, and estuaries; it is a nearly universal process wherever oxygen and other electron acceptors can be excluded or minimized. SRBs are nearly ubiquitous bacteria. SRBs are often active in clay lenses in otherwise aerobic aquifers, and they are also abundant in root zones where photosynthetic exudates are produced or plant biomass is degraded.

In order to initially activate the sulfate reduction process, sulfate will have to be added with organic carbon if the existing soluble sources of sulfate are limited to a few hundred milligrams per liter. In order to create iron sulfides in the percent range (more than 10,000 g iron sulfide per kilogram aquifer material) necessary to obtain the molar ratios with uranium needed for uranium stability, it is not unusual to inject sulfate to create concentrations in the plume of a few grams per liter. Sulfate concentrations, both in the soluble and the solid phase, will be critical for determining the required sulfate addition rate.

In the absence of iron (and other oxidized minerals), sulfate reduction in aquifers will lead to accumulation of aqueous sulfide (HS^-). Where oxidized iron minerals (such as hematite) are abundant (range 1 to 3%), sulfide will react with the iron, reducing the ferric iron to ferrous iron. Ferrous iron formed in this way is relatively soluble; however, additional sulfide formed will quickly react with the ferrous iron, and at pH greater than 6.5, iron sulfide minerals will rapidly precipitate. The temporary increase in iron concentrations will last for a period of a few months. The pH will have to be monitored to ensure that the iron sulfide precipitates from solution. The presence of the IRZ at the forward edge of the plume will prevent iron migration beyond the treatment area.

Iron sulfide has been recognized as being critical to maintaining uranium stability in groundwater during bioremediation[168,170] as well as in natural uranium ore deposits. A very strong correlation between uranium and sulfur was found, indicating both a role for sulfur in depositing the uranium as well as in maintaining its deposition.[171] It is critical to note that the sulfide continues to perform a stabilizing function in many uranium deposits, which has been measured at millions of years old. These deposits are called "kinetically stable" where the sulfur acts to control uranium stability.[172] In uranium ore geology terms, it is important to create a "regionally reduced" host aquifer within the contaminated area. In these geologic conditions, a very small fraction, typically less than 10^{-8} of the uranium in the ore deposit is made soluble per year.[173]

Iron sulfide has also been recognized as an important redox buffer for several situations that are instructive for many uranium plumes. The data from a study where reactivity of naturally occurring pyrite where nitrate was injected was studied showed that this iron sulfide source, even though aged over geologic time scales, was still

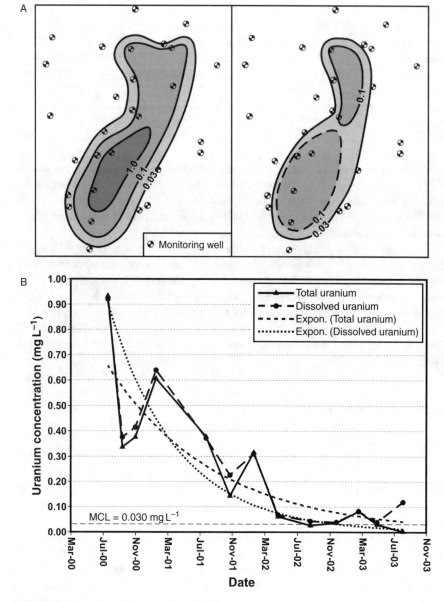

FIGURE 2.63 (A) Significant reductions achieved in dissolved uranium concentrations by implementing an anaerobic IRZ system. (B) Performance data from the above system.

reactive toward maintaining *in situ* reducing conditions. Nitrate reduction was rapid (half-life of reduction of 2 days in a sandy aquifer matrix), leading to the oxidation of pyrite to ferric iron and sulfate minerals, which deposited as jarosite and natroalunite. Others have shown that in aquifers receiving agricultural runoff oxygen and nitrate in the runoff were reduced by iron sulfide when infiltrated runoff reached the deeper aquifer.[174] Iron sulfide, reduced iron compounds in addition to iron sulfides (including siderite), and bulk organic matter can all provide redox buffering in aquifers receiving agricultural runoff.[152] These examples are very relevant for the "resident farmer" scenario, indicating that even under agricultural runoff scenarios the uranium can be maintained insoluble by iron sulfide.

An additional factor to maintain uranium stability even in conditions where iron sulfide has been exhausted in the aquifer is the residual iron oxides that form after iron sulfides oxidize. Ferric iron oxides sorb uranium with strong binding energy (bidentate and tridentate inner-sphere complexes).[175] It has been shown that these iron oxides could maintain very low dissolved uranium concentrations (less than 30 pCi L^{-1}).[176] Continued reactivity of hydrous ferric oxides formed in this way for more than 30 pore volumes has also been demonstrated.[177] This means that in many remediated uranium plumes the iron oxides that will form as oxygen enters the plume area (transported by diffusion, in rainwater and in groundwater) will minimize uranium from remobilizing.

As discussed above, fresh amorphous Fe(III) oxide is a very strong sorbent for a variety of metals including uranium. This amorphous Fe(III) oxide is a *precursor of many natural forms of crystalline Fe(III) oxides* that adsorb or incorporate into structures many trace metals.[175] Hence, selective anaerobic bio-oxidation of Fe(II) may be an additional means of "ensuring" the capping off and completing the attenuation of heavy metals and radionuclides in a reducing environment, allowing the system to naturally revert to an oxic state while preventing or minimizing remobilization of the previously reduced precipitates. As described previously, this mechanism takes place as a result of the precipitation of Fe(III) (hydr.) oxides over already immobilized precipitates of uranium and forming an insoluble barrier that *crystallizes* with time.

The discussion above provides the information needed to identify the necessary characteristics of a stable, fully reduced zone. To remain stable over long periods of time a reduced zone must contain a variety of reduced compounds after treatment, including some combination of the following:

- Iron sulfides (ranging from amorphous FeS to pyrite); to ensure very low soluble uranium concentrations over long periods of time, the concentration of iron sulfides must be several orders of magnitude greater than the concentration of uranium in the reducing zone
- Elemental sulfur
- Residual reduced organic carbon either incorporated in cellular biomass or stored by microorganisms
- Reduced uranium compounds (UO_2 and potentially US_2)
- Potentially a variety of other reduced sulfur, manganese, iron, and trace mineral compounds

In this zone, the reoxidation and remobilization of uranium will be limited by the oxygen available to react with the precipitated uranium. The available oxygen will be controlled by the presence of stored, reduced compounds emplaced in the aquifer by the treatment process.

In relative terms, expressed in molar ratios of uranium to all of the other reduced compounds stored in the aquifer, the potential oxidation of uranium will be very low compared to the potential oxidation of iron, sulfur, and other reduced species. Utilizing FeS compounds alone, more than 10^4 moles of FeS will be present for every mole of UO_2. Any proposed remediation plan should anticipate the introduction of oxygen via natural pathways and provide for sufficient reduced compounds to exhaust these sources of oxygen. As the aquifer materials are exposed to oxygen, FeS would oxidize at least as rapidly as the precipitated UO_2 and consume the oxygen. Because the ratio of iron sulfide to uranium is so large, a very limited amount of oxygen will be available to react with uranium. Because UO_2 will be precipitated first during treatment, the FeS precipitate would be laid down over the top of the UO_2 as an FeS coating. FeS will therefore be exposed to the oxygen in the groundwater before uranium-containing precipitates would be exposed. A small amount of the uranium in the aquifer will mobilize very slowly, as the FeS is depleted, and because there is so much more FeS in the aquifer material, the uranium will only mobilize at concentrations substantially less than the allowable regulatory standards.

For long-lived radionuclides such as uranium (which has a half-life greater than 245,000 years for all common isotopes), the nuclear regulatory commission (NRC) requires assurance that uranium will not remobilize exceeding a certain concentration over a period of 1000 years. Hence, the success criteria for permanent immobilization by precipitation of uranium within an IRZ should be defined in a way to incorporate the technological advances achieved by the authors and his colleagues. This criteria should be developed in such a way that the maximum groundwater concentration of a precipitated radionuclide will not exceed the required concentration, established by NRC, at selected downgradient compliance wells. The authors' experience, from their laboratory and field work, leads us to believe that at the rate of any dissolution reactions under any foreseeable environmental condition, (i.e., pH 4 to 9, and ORP, -400 to $+400$ mV) will be such that this requirement can be achieved. With respect to the 1000-year requirement, it can only be shown by properly implemented biogeochemical modeling using models such as PHREEQ.

2.4.4.2 Technetium-99 (^{99}Tc)

The isotope ^{99}Tc is a product of nuclear fission reactions and is formed in kilogram quantities during nuclear reactions and has been released into the environment during weapons testing and the disposal of low- and intermediate-level radioactive wastes. ^{99}Tc has been found in groundwaters at sites where nuclear wastes have been reprocessed or stored. The stable pertechnate (TcO_4^-) anion has high environmental mobility and is taken up into the food chain as an analog of sulfate.

Due to the low solubility of reduced Tc, the redox chemistry of Tc is crucial in governing its mobility (Figure 2.64). As described above, several recent studies have shown that ^{99}Tc can be removed from aqueous solution via the reduction of pertechnate to insoluble, low valence forms. U(VI) and Tc(VII) can act as respiratory

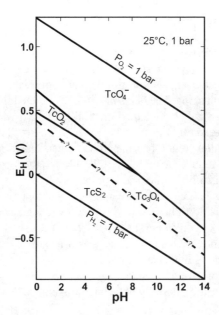

FIGURE 2.64 E_H–pH diagram of technetium.

electron acceptors, and direct enzymatic reduction of U(VI) and Tc(VII) has been reported. On the basis of broad-metal reductase activity against high valence metals (Cr(VI), Fe(III), Mn(IV), and U(VI)), SRBs may be able to reduce Tc(VII) enzymatically. This hypothesis is also supported by the fact that the pertechnate anion is bioavailable as a sulfate analog and may therefore be a surrogate electron acceptor for anoxic growth. Hence, microbial reduction of the pertechnate ion, preferably as a sulfide precipitate, should be a preferred method for remediating ^{99}Tc contaminated groundwater systems.[178,179]

^{99}Tc has a half-life of 2.13×10^5 years. Reference to the E_H–pH diagram for the system Tc-S-O-H shows that the anion TcO_4^- is the stable species under ordinary aerobic conditions. As the reductive precipitation technology is implemented and the system moves toward anaerobic, reductive conditions, this ion is reduced first to TcO_2 and then to the mixed oxidation state oxide, Tc_3O_4. Finally, as reductive conditions stabilize and sulfides are generated, the stable compound TcS_2 becomes the dominant species. In the absence of sulfide, Tc_3O_4 is reduced to the hydroxide, $Tc(OH)_2$. TcS_2 is known to be stable in the presence of sulfide. The stability of the TcS_2 species is well documented under sulfur-present conditions in laboratory experiments, and may further retard and retain any escaping Tc from breached radioactive waste containers.

An even more emphatic conclusion regarding a slightly different technetium sulfide stochiometry Tc_2S_7 has been presented. Pertechnetate oxyanion, $^{99}TcO_4^-$, a potentially mobile species in leachate from a breached radioactive waste repository, was removed from a brine solution by precipitation with sulfide, iron, and ferrous sulfide at normal environmental pHs. Hematite (Fe_2O_3) and goethite (FeOOH) were the dominant minerals in the precipitate obtained from the TcO_4^- ferrous iron reaction. The observation of small particle size and poor crystallinity of the minerals formed in the presence of Tc suggested that the Tc was incorporated into the mineral structure after reduction to a lower valence state. Amorphous ferrous sulfide, an initial phase precipitating in the TcO_4^- ferrous iron sulfide reaction, was transformed to goethite and hematite (Fe_2O_3) on aging. The black precipitate obtained from the TcO_4^- sulfide reaction was poorly crystallized technetium sulfide (Tc_2S_7) which was insoluble in both acid and alkaline solution in the absence of strong oxidants. The results suggested that ferrous- and/or sulfide-bearing groundwaters and minerals in host rocks or backfill barriers could reduce the mobility of Tc through the formation of less soluble Tc-bearing iron and/or sulfide minerals.

It is likely that the Tc(IV) oxides are more stable than previously predicted and, hence, they are less likely to be oxidized to TcO_4^-(aq) under moderately reducing conditions. The authors of this report have revised earlier calculations done to predict the solution concentrations of technetium species in a vault as a function of the oxidation conditions in model groundwaters; however, we will need to verify the performance of this technology in the laboratory for technetium since mobile humic substances can enhance the solubility of technetium oxides.

2.4.4.3 Plutonium

Figure 2.65 shows a detailed E_H–pH diagram for plutonium. Here, the solid compound PuO_2 occupies the largest area of stability. It is estimated that the solubility of undissociated $Pu(OH)_4$ is at 10^{-17} M in this largest stability region. For comparison, a maximum permissible concentration in drinking water is 0.08 ppb ($10^{-9.5}$ M). Thus,

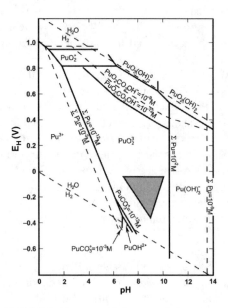

FIGURE 2.65 E_H–pH diagram of plutonium.

the conventional view until recently was that only under extreme oxidizing conditions or reducing conditions, in the presence of low pH, does the soluble plutonium species become possible. Recently, however, two routes of mobilization, colloidal transport and oxidation of PuO_2 by water, have been discovered. The colloidal transport of plutonium for a distance of 1.3 km in Nevada was explained with convincing evidence of transport of colloidal plutonium in a deep groundwater aquifer with a maximum flow velocity of 80 m year^{-1}. They are not able to determine if the colloids are pure PuO_2 or Pu sorbed onto clays. In carefully performed and lengthy experiments, it was found that plutonium dioxide is slowly oxidized by water to form a mixed oxide. The oxidized fraction of the plutonium is in the more mobile +6 state.

As a result of activities ranging from nuclear weapons testing to plutonium production to nuclear accidents, significant quantities of soils and groundwater systems have been contaminated with plutonium (Pu) and other actinides. Migration of Pu from waste disposal basins had been thought to be insignificant, given that Pu was believed to be particle reactive, and hence would be transported in groundwater at a much slower rate than in cases of clearly mobile radionuclides. Recently, however, concerns have increased due to speculation that Pu migration might be enhanced through colloid-facilitated transport.

Reductive metal precipitation technology may be able to counteract the routes of mobilization. The reducing environment provided by reagent injection may be sufficient to reverse the slow oxidation described. While this reduction is unlikely to be fully irreversible, the slow rate of the oxidative reaction may allow this reduction to be a feasible treatment technology if applied on a periodic "maintenance" basis. It is also theoretically possible that the reductive process could remove colloidal metals as well — by mechanical entrapment by the increased quantity of biomass, by direct electrostatic attraction to the new biomass surfaces or by altering the overall charge balance of the system. If reductive technology was applied to Pu we would have to be cautious about pH since the Pu^{3+} ion is predominant at low pH especially under reducing conditions.

Microbially enhanced chemisorption of heavy metals process postulates that predeposition of the phosphate crystal of a "benign" metal promotes the subsequent deposition of a metal that does not precipitate easily. Immobilized cells of *Citrobacter* sp. removed plutonium from solution using this technique that used biologically produced phosphate ligand for metal phosphate bioprecipitation. Removal of this transuranic radionuclide was enhanced by prior exposure of the biomass to lanthanum in the presence of organophosphate substrate to form cell-bound $LaPO_4$.

2.4.5 N-NITROSODIMETHYLAMINE (NDMA)

N-nitrosodimethylamine is commonly known as NDMA (Figure 2.66). It is a yellow liquid that has no distinct odor. NDMA is formed by the nitrosation of dimethylamine

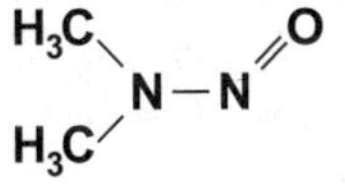

FIGURE 2.66 *N*-nitrosodimethylamine — NDMA.

(DMA) with a nitrite ion (NO_2^-). Short- or long-term exposure of animals to water or food containing NDMA is associated with serious health effects, such as liver disease and death.

NDMA is used primarily in research, but it has been used in the production of 1,1-dimethylhydrazine for liquid rocket fuel and a variety of other industrial applications: a nematocide, a plasticizer for rubber, in polymers and copolymers, a component of batteries, a solvent, an antioxidant, and a lubricant additive. NDMA was reported to be present in a variety of foods, beverages, and drugs, and in tobacco smoke; it has been detected as an air pollutant and in treated industrial wastewater, treated sewage in proximity to a 1,1-dimethylhydrazine manufacturing facility, deionized water, high nitrate well water, and chlorinated drinking water. NDMA is unintentionally formed during various manufacturing processes at many industrial sites, and in air, water, and soil from reactions involving other chemicals call alkylamines. Alkylamines are both natural and man-made compounds that are found widely distributed throughout the environment.

At present, the only effective treatment method for removal of NDMA from water is ultraviolet photolysis and chemical oxidation. Carbon adsorption and air stripping have not been effective. Photolysis of NDMA also occurs in sunlight. Although zero-valent iron catalyzes NDMA transformation by hydrogenation, the relatively slow kinetics of the reaction suggested that it would not be a cost-effective treatment option. A review article, stated that bioremediation could hold significant potential for the *in situ* remediation of NDMA-contaminated water.

There is no clear evidence for biodegradation of NDMA in groundwater. However, the potential for bacteria to degrade NDMA exists, based upon the similarity of bacterial monooxygenase systems to the cytochrome P-450 enzymes in plants and animals that catalyze the oxidation of NDMA. Two studies observed the complete mineralization of NDMA by undefined consortia, and biodegradation of NDMA has been reported in anaerobic and aerobic incubations of native microbial soil consortia with half-lives ranging from 12 to 55 days. Although some of these studies documented degradation intermediates, they were unable to identify the responsible microorganisms or elucidate degradation pathways.

REFERENCES

1. Chappelle, F.H., *Groundwater Microbiology and Geochemistry*, John Wiley & Sons, New York, 2001.
2. Wilson, J.T. and B.H. Wilson, Biotransformation of trichloroethylene in soil, *Appl. Environ. Microbiol.*, 49, 242–243, 1985.
3. Wackett, L.P. and C.P. Hershberger, *Biocatalysis and Biodegradation*, ASM Press, Washington, D.C., 2001.
4. Schwarzenbach, R.P., P.M. Gschwend, and D.M. Imboden, *Environmental Organic Chemistry*, 2nd ed., Wiley Interscience, New York, 2003.
5. Michelsen, D.L. and M. Lofti, Oxygen microbubble injection for *in situ* bioremediation: Possible field scenario, in *Biological Processes: Innovative Waste Treatment Technology Series*, Vol. 3, H.M. Freeman et al., Eds., Technomic Publishing, Lancaster, PA, 1993.
6. Wiedemeier, T.H. et al., *Natural Attenuation of Fuels and Chlorinated Solvents in the Subsurface*, John Wiley & Sons, New York, 1999.

7. McCarty, P.E. and L. Semprini, Groundwater treatment for chlorinated solvents, in *Handbook of Bioremediation*, R.D. Norris, et al., Eds., Lewis Publishers, Boca Raton, FL, 1994.
8. Finneran, K.T. and D.R. Lovely, Anaerobic *in situ* bioremediation, in *MTBE: Remediation Handbook*, E.E. Mouer and P.T. Kostecki, Eds., Amherst Scientific Publishers, Amherst, MA, 2003.
9. Cunninghan, J.A. et al., Enhanced *in situ* bioremediation of BTEX-contaminated groundwater by combined injection of nitrate and sulfate, *Environ. Sci. Technol.*, 35, 1663–1670, 2001.
10. Lovely, D.R. et al., Humic substances as a mediator for microbially catalyzed metal reduction, *Acta Hydrochim. Hydrobiol.*, 26, 152–157, 1998.
11. Scott, D.T. et al., Quinone moieties act as electron acceptors in the reduction of humic substances by humics-reducing microorganisms, *Environ. Sci. Technol.*, 32, 2984–2989, 1998.
12. Stefan, R., Envirogen Corporation, personal communication, 2000.
13. Hartsman, S., and J.A.M. de Bont, Aerobic vinyl chloride metabolism in *Mycobacterium aunum*, *Appl. Environ. Microbiol.*, 58, 1220–1226, 1985.
14. Wilson, J.T., Aerobic *in situ* bioremediation, in *MTBE: Remediation Handbook*, E.E. Moyer, and P.T. Kostecki, Eds., Amherst Scientific Publishers, Amherst, MA, 2003.
15. Salanitro, J.P. et al., Perspectives on MTBE biodegradation and the potential for *in situ* aquifer bioremediation, in *Southwestern Regional Conference of NGWA*, Anaheim, CA, June 3–4, 1998.
16. Kolhatkar, R. et al., Evaluating natural biodegradation of MTBE at multiple UST sites, *API Petroleum Conference*, Anaheim, CA, November 2000.
17. Loeffler, F.E. et al., Diversity of dechlorinating bacteria, in *Dechlorination: Microbial Processes and Environmental Applications*, M.M. Haggblom and I.D. Bossert, Eds., Kluwer Academic Publishers, Boston, 2003.
18. Isidorov, V.A., *Organic Chemistry of the Earth's Atmosphere*, Springer-Verlag, Berlin, 1990.
19. Pereira, W.E., C.E. Rostad, and H.E. Taylor, Mount St. Helens, Washington, 1980 volanic eruption: characterization of organic compounds in ash samples, *Geophys. Res. Lett.*, II, 111–130, 1980.
20. Castro, C.E. and N.O. Belser, Biodehalogenation. Reductive dehalogenation of the biocides ethylene dibromide, 1,2-dibromo-3-chloropropane, and 2,3-dibromobutane in soil, *Environ. Sci. Technol.*, 2(10), 779–783, 1968.
21. French, A.L. and Hoopingarner, R.A., Dechlorination of DDT by membranes isolated from *Escherichia coli*, *J. Econ. Entomol.*, 63, 756–759, 1970.
22. Heritage, A.D. and I.C. MacRae, Identification of intermediates formed during the degradation of hexachlorocyclohexanes by *Clostridium sphenoides*, *Appl. Environ. Microbiol.*, 33, 1295–1297, 1977.
23. Ohisa, N., M. Yamaguchi, and N. Kurihara, Lindane degradation by cell-free extracts of *Clostridium rectum*, *Arch. Microbiol.*, 125, 221–225, 1980.
24. Fantroussi, S.E., H. Naveau, and S.N. Agathos, Anaerobic dechlorinating bacteria. *Biotechnol. Prog.*, 14(2), 167–188, 1998.
25. Graham, D.W. and T.P. Curtis, Ecological Theory and Bioremediation, in *Bioremediation: A Critical Review*, I.M. Head et al., Eds., Horizon Scientific Press, Norfolk, U.K., 2003.
26. Mortland, M.M. and W.D. Kemper, Specific surface, in *Methods of Soil Analysis, Part 1*, American Society of Agronomy, Madison, WI, 1965.
27. Torsvik, V., J. Goksoyr, and F.L. Daae, High diversity in DNA of soil bacteria. *Appl. Environ. Microbiol.*, 56, 782–787, 1990.

28. Torsvik, V., K. Salte, R. Sorheim, and J. Goksoyr, Comparison of phenotypic diversity and DNA heterogeneity in a population of soil bacteria, *Appl. Environ. Microbiol.*, 56, 776–781, 1990.
29. Champ, D.R., J. Gulens, and R.E. Jackson, Oxidation–reduction sequences in ground water flow systems, *Can. J. Earth Sci.*, 16, 12–23, 1979.
30. Sharma, P.K., and P.L. McCarty, Isolation and characterization of a facultatively aerobic bacterium that reductively dehalogenates tetrachloroethene to *cis*-1,2 dichloroethene, *Appl. Environ. Microbiol.*, 62(3), 761–765, 1996.
31. Chang, Y.C., M. Hatsu, K. Jung, Y.S. Yoo, and K. Takmizawa, Isolation and characterization of a tetrachloroethylene degrading bacterium, *Clostridium bifermentans* DPH-1, *Biosci. Bioeng.*, 89(5), 489–491, 2000.
32. Schink, B., Energetics of syntrophic cooperation in methanogenic degradation, *Microbiol. Mol. Biol. Rev.*, 61, 262–280, 1997.
33. Ohisa, N., M. Yamaguchi, and N. Kurihara, Lindane degradation by cell-free extracts of *Clostridium rectum*, *Arch. Microbiol.*, 125, 221–225, 1980.
34. Mohn, W.W. and T.M. Tiedje, Microbial reductive dehalogenation, *Microbiol. Rev.*, 56, 482–507, 1991.
35. Holliger, C., C. Regeard, and G. Dietert, Dehalogenation by anaerobic bacteria, in *Dehalogenation: Microbial Processes and Environmental Applications,* M.M. Haggblom and I.D. Bossert, Eds., Kluwer Academic Publishers, Boston, 2003.
36. Gerritse, J., V. Renard, T.M. Pedro-Gomes, P.A. Lawson, M.D. Collins, and J.C. Thomson, *Desulfitobacterium* sp. strain PCE1, an anaerobic bacterium that can grow by reductive dechlorination of tetrachloroethene or ortho-chlorinated phenols, *Arch. Microbiol.*, 165, 132–140, 1996.
37. Townsend, G.T. and J.M. Sulfita, Characterization of chloroethylene dehalogenation by cell extracts of *Desulfomonile tiedjei* and its relationship to chlorobenzoate dehalogenation, *Appl. Environ. Microbiol.*, 62(8), 2850–2853, 1996.
38. Nandi, R. and S. Sengupta, Microbial production of hydrogen: an overview, *CRC Rev. Microbiol.*, 24, 61–84, 1998.
39. Tandoi, V., T.D. Stefano, P.A. Bowser, J.M. Gossett, and S.H. Zinder, Reductive dehalogenation of chlorinated ethenes and halogenated ethanes by a high rate anaerobic enrichment culture, *Environ. Sci. Technol.*, 28, 973–979, 1994.
40. Brock, T.D. and M.T. Madigan, *Biology of Microorganisms*, 5th ed., Prentice Hall, Englewood Cliffs, NJ, 1985.
41. Wild, A., R. Hermann, and T. Leisinger, Isolation of an anaerobic bacterium which reductively dechlorinates tetrachloroethene and trichloroethene, *Biodegradation*, 7, 506–511, 1996.
42. Maymo-Gatell, X., I. Nijenhuis, and S.H. Zinder, Reductive dechlorination of *cis*-1,2-dichloroethene and vinyl chloride by *Dehalococcoides ethenogenes*, *Environ. Sci. Technol.*, 35, 516–521, 2001.
43. Thauer, R.K., K. Jungermann, and K. Decker, Energy conservation in chemotrophic anaerobic bacteria, *Bacteriol. Rev.*, 41, 100–180, 1977.
44. Magnuson, J.K., R.V. Stern, J.M. Gossett, S.H. Zinder, and D.R. Burris, Reductive dechlorination of tetrachloroethene to ethene by a two-component enzyme pathway, *Appl. Environ. Microbiol.*, 64(4), 1270–1275, 1998.
45. Glod, G., W. Angst, C. Hollinger, and R.P. Schwarzenbach, Corrinoid-mediated reduction of tetrachloroethene, trichloroethene, and trichlorofluoroethene in homogeneous aqueous solution: reaction kinetics and reaction mechanisms, *Environ. Sci. Technol.*, 31(1), 253–260, 1997.
46. Gantzer, C.J. and L.P. Wackett, Reductive dechlorination catalyzed by bacterial transition metal coenzymes, *Environ. Sci. Technol.*, 25(4), 715–722, 1991.

47. Fathepure, B.Z. and S.A. Boyd, Reductive dechlorination of perchloroethylene and the role of methanogens, *FEMS Microbiol. Lett.*, 49, 149–156, 1988.
48. Brock, T.D. and Madigan, M.T. *Biology of Microorganisms*, 6th ed., Prentice Hall, New York, 1991.
49. Hendrickson, E.R., J.A. Payne, R.M. Young, M.G. Starr, M.P. Perry, S. Fahnestock, D.E. Ellis, and R.C. Ebersole, Molecular analysis of *Dehalococcoides* 16S ribosomal DNA from chloroethene-contaminated sites throughout North America and Europe, *Appl. Environ. Microbiol.*, 68(2), 485–495, 2002.
50. Egli, C., T. Tschan, R. Scholtz, A.M. Cook, and T. Leisinger, Transformation of tetrachloromethane to dichloromethane and carbon dioxide by *Acetobacterium woodii*, *Appl. Environ. Microbiol.*, 54, 2819–2824, 1988.
51. Tersenbach, D.P., and M. Blaunt, Transformation of tetrachloroethylene to trichloroethylene by homoacetogenic bacteria, *FEMS Microbiol. Lett.*, 123(1–2), 213–218, 1994.
52. Holliger, C., D. Hahn, H. Harmsen, W. Ludwig, W. Schumacker, and B. Tindall, F, *Dehalobacter restrictus* gen. nov. and sp. nov., a strictly anaerobic bacterium that reductively dechlorinates tetra- and trichloroethene in an anaerobic respiration, *Arch. Microbiol.*, 169, 313–321, 1998.
53. Maymo-Gatell, X., V. Tandoi, J.M. Gossett, and S.H. Zinder, Characterization of an H_2-utilizing enrichment culture that reductively dechlorinates tetrachloroethene to ethene in the absence of methanogenesis and acetogenesis, *Appl. Environ. Microbiol.*, 61, 3928–3933, 1995.
54. Maymo-Gatell, X., Y.-T.Chien, J. Gossett, and S. Zinder, Isolation of a bacterium that reductively dechlorinates tetrachloroethene to ethene, *Science*, 276, 1568–1571, 1997.
55. Maymo-Gatell, X., T. Anguish, and S.H. Zinder, Reductive dechlorination of chlorinated ethenes and 1,2-dichloroethane by "*Dehalococcoides ethenogenes*" 195, *Appl. Environ. Microbiol.*, 65(7), 3108–3113, 1999.
56. Maymo-Gatell, X., I. Nijenhuis, and S.H. Zinder, Reductive dechlorination of *cis*-1,2-dichloroethene and vinyl chloride by *Dehalococcoides ethenogenes*, *Environ. Sci. Technol.*, 35, 516–521, 2001.
57. Scholtz-Muramatsu, H., A. Neuman, M. Messmer, E. Moore, and G. Diekert, Isolation and characterization of *Dehalospirillum multivorans* gen. nov., sp. Nov., a tetrachloroethene-utilizing, strictly anaerobic bacterium, *Arch. Microbiol.*, 163, 48–56, 1995.
58. Loeffler, F.E., R.A. Sanford, and J.M. Tiedje, Initial characterization of a reductive dehalogenase from *Desulfitobacterium chlororespirans* Co23, *Appl. Environ. Microbiol.*, 62(10), 3809–3813, 1996.
59. Suyama, A., R. Iwakiri, K. Kai, T. Tokunaga, N. Sera, and K. Furukawa, Isolation and characterization of *Desulfitobacterium* sp. strain Y51 capable of efficient dehalogenation of tetrachloroethene and polychloroethanes, *Biosci. Biotechnol. Biochem.*, 65(7), 1474–1481, 2001.
60. DeWeerd, K.A., F. Concannon, and J.M. Suflita, Relationship between hydrogen consumption, dehalogenation, and the reduction of sulfur oxyanions by *Desulfomonile tiedjei*, *Appl. Environ. Microbiol.*, 57, 1929–1934, 1991.
61. Fathepure, B.Z. and J.M. Tiedje, Reductive dechlorination of tetrachloroethylene by a chlorobenzoate-enriched biofilm reactor, *Environ. Sci. Technol.*, 28(4), 746–752, 1994.
62. Krumholz, L.R., R. Sharp, and S.B. Fishbain, A freshwater anaerobe coupling acetate oxidation to tetrachloroethylene dehalogenation, *Appl. Environ. Microbiol.*, 62, 4108–4113, 1996.
63. Krumholz, L.R., *Desulfuromonas chloroethenica* sp. nov. uses tetrachloroethene and trichloroethene as electron acceptors, *Int. J. Syst. Bacteriol.*, 47, 1262–1263, 1997.

64. Loeffler, F.E., Q. Sun, J. Li., and J.M. Tiedje, 16S rRNA gene-based detection of tetrachloroethene-dechlorinating *Desulfuromonas* and *Dehalococcoides* species, *Appl. Environ. Microbiol.*, 66, 1369–1374, 2000.
65. Fathepure, B.Z., J.P. Nengu, and S.A. Boyd, Anaerobic bacteria that dechlorinate perchloroethene, *Appl. Environ. Microbiol.*, 53(11), 2671–2674, 1987.
66. Fathepure, B.Z. and S.A. Boyd, Dependence of tetrachloroethylene dechlorination on methanogenic substrate consumption by *Methanosarcina* sp. strain DCM, *Appl. Environ. Microbiol.*, 54(12), 2976–2980, 1988.
67. Baeseman, J.L. and P.J. Novak, Effects of various environmental conditions on the transformation of chlorinated solvents by *Methanosarcina thermophila* cell exudates, *Biotechnol. Bioeng.*, 75(6), 634–641, 2001.
68. Freedman, D.L. and J.M. Gossett, Biological reductive dechlorination of tetrachloroethylene to ethylene under methanogenic conditions, *Appl. Environ. Microbiol.*, 55(9), 2144–2151, 1989.
69. Bradley, P.M. and F.H. Chapelle, Methane as a product of chloroethene biodegradation under methanogenic conditions, *Environ. Sci. Technol.*, 33, 653–656, 1999.
70. Bradley, P.M., F.H. Chapelle, and J.T. Wilson, Anaerobic mineralization of vinyl chloride in Fe(III)-reducing, aquifer sediments, *J. Contam. Hydrol.*, 31(1–2), 111–127, 1998.
71. Loeffler, F.E., J.E. Champine, K.M. Ritalahti, S.J. Sprague, and J.M. Tiedje, Complete reductive dechlorination of 1,2-dichloropropane by anaerobic bacteria, *Appl. Environ. Microbiol.*, 63(7), 2870–2875, 1997.
72. Cabirol, N., F. Jacob, J. Perrier, B. Fouillet, and P. Chambon, Interaction between methanogenic and sulfate-reducing microorganisms during dechlorination of a high-concentration of tetrachloroethylene, *J. Gen. Appl. Microbiol.*, 44(4), 297–301, 1998.
73. Cabirol, N., F. Jacob, J. Perrier, B. Foullet, P. Chambron, Complete degradation of high concentrations of tetrachloroethylene by a methanogenic consortium in a fixed-bed reactor, *J. Biotechnol.*, 62(2), 133–141, 1998.
74. Cabirol, N., R. Villemur, J. Perrier, F. Jacob, B. Fouillet, and P. Chambon, Isolation of a methanogenic bacterium, *Methanosarcina* sp. strain FR, for its ability to degrade high concentrations of perchloroethylene, *Can. J. Microbiol.*, 44(12), 1142–1147, 1998.
75. Duhamel, M., S.D. Wehr, L. Yu, H. Rizvi, D. Seepersad, S. Dworatzek, E.E. Cox, and E.A. Edwards, Comparison of anaerobic dechlorinating enrichment cultures maintained on tetrachloroethene, trichloroethene, *cis*-dichloroethene and vinyl chloride, *Water Res.*, 36, 4193–4202, 2002.
76. DiStefano, T.D., J.M. Gossett, and S.H. Zinder, Reductive dechlorination of high concentrations of tetrachloroethene to ethene by an anaerobic enrichment culture in the absence of methanogenesis, *Appl. Environ. Microbiol.*, 57(8), 2287–2292, 1991.
77. DiStefano, T.D., The effect of tetrachloroethene on biological dechlorination of vinyl chloride: potential implication for natural bioattentuation, *Water Res.*, 33(7), 1688–1694, 1999.
78. Rosner, B.M., P.L. McCarty, and A.M. Spormann, *In vitro* studies on reductive vinyl chloride dehaolgenation by an anaerobic mixed culture, *Appl. Environ. Microbiol.*, 63(11), 4139–4144, 1997.
79. Dybas, M.J., D.W. Hyndman, R. Heine, J. Tiedje, K. Linning, D. Wiggert, and T. Voice, Development, operation and long-term performance of a full-scale biocurtain utilizing bioaugmentation, *Environ. Sci. Technol.*, 36, 3635–3644, 2002.
80. Ellis, D.E., E.J. Lutz, J.M. Odom, R.J. Buchanan, C.L. Bartlett, M.D. Lee, M.R. Harkness, and K.A. DeWeerd, Bioaugmentation for accelerated *in situ* anaerobic bioremediation, *Environ. Sci. Technol.*, 34, 2254–2260, 2000.
81. Major, D.W., M.L. McMaster, E.E. Cox, E.A. Edwards, S.M. Dworatzek, E.R. Hendrickson, M.G. Starr, J.A. Payne, and L.W. Buonamici, Field demonstration of

successful bioaugmentation to achieve dechlorination of tetrachloroethene to ethene, *Environ. Sci. Technol.*, 36(23), 5106–5116, 2002.
82. Harkness, M.R., A.A. Bracco, M.J. Brennan, K.A. Deweerd, and J.L. Spivak, Use of bioaugmentation to stimulate complete reductive dechlorination of trichloroethene in Dover soil columns, *Environ. Sci. Technol.*, 33, 1100–1109, 1999.
83. DeFlaun, M.F., S.R. Oppenheimer, S. Streger, C.W. Condee, and M. Fletcher, Alterations in adhesion, transport, and membrane characteristics in an adhesion deficient Pseudomonad, *Appl. Environ. Microbiol.*, 65(2), 759–765, 1999.
84. Duhamel, M., S.D. Wehr, L. Yu, H. Rizvi, D. Seepersad, S. Dworatzek, E.E. Cox, and E.A. Edwards, Comparison of anaerobic dechlorinating enrichment cultures maintained on tetrachloroethene, trichloroethene, *cis*-dichloroethene and vinyl chloride, *Water Res.*, 36, 4193–4202, 2002.
85. Flynn, S.J., F.E. Loeffler, and J.M. Tiedje, Microbial community changes associated with a shift from reductive dechlorination of PCE to reductive dechlorination of *cis*-DCE and vinyl chloride, *Environ. Sci. Technol.*, 34, 1056–1061, 2000.
86. Lendvay, J.M., F.E. Loeffler, F.M. Dollhopf, M.R. Aiello, G. Daniels, B.J. Fathepure, M. Gebhard, R. Heine, R. Helton, J. Shi, R. Krajmalnik-Brown, D.L. Major, Jr., M.J. Barcelona, E. Petrovskis, R. Hickey, J.M. Tiedje, and P. Adriens, Bioreactive barriers: A comparison of bioaugmentation and biostimulation for rchlorinated solvent remediation, *Environ. Sci. Technol.*, 37, 1422–1432, 2003.
87. Platen, H. and B. Schink, Anaerobic degradation of acetone and higher ketones via carboxlylation by newly isolated denitrifying bacteria, *J. Gen. Microbiol.*, 135, 883–891, 1989.
88. Bradley, P.M. and Chappelle, F.H., Anaerobic mineralization of vinyl chloride in Fe(III) reducing aquifer sediments, *Environ. Sci. Technol.*, 30, 2084–2086, 1996.
89. Wilkin, R.T., "Reactive Minerals in Aquifers, Formation Processes and Quantitative Analyses," presented at the 2003 AFCEE Technology Transfer Conference, San Antonio, TX, extended abstract published in proceedings.
90. Wilkin, R.T. and H.L. Barnes, Impact of soil chemistry on reductive dechlorination, *Geochim. Cosmochim. Acta*, 60, 4167–4179, 1996.
91. Magnuson, J.K., R.V. Stern, J.M. Gossett, S.H. Zinder, and D.R. Burris, Reductive dechlorination of tetrachloroethene to ethene by atwo-component enzyme pathway, Appl. Environ. Microbiol., 64(4), 1270–1275, 1998.
92. Rosetti, S., L.L. Blackall, M. Majone, P. Hugenholtz, J.J. Plumb, and V. Tandoi, Kinetic and phylogenetic characterization of an anaerobic dechlorinating microbial community, *Microbiology*, 149, 459–469, 2003.
93. Dolfing, J. and D.B. Janssen, Estimates of Gibbs free energies of formation of chlorinated aliphatic compounds, *Biodegradation*, 5, 21–28, 1994.
94. Wilson, J.T. and M. Ferrey, "Abiotic Reactions May Be the Most Important Mechanism in Natural Attenuation of Chlorinated Solvents," presented at the 2003 AFCEE Technology Transfer Conference, San Antonio, TX, extended abstract published in proceedings.
95. Kriegman-King, M.R. and M. Reinhard, Transformation of carbon tetrachloride in the presence of sulfide, biotite, and vermiculite, *Environ. Sci. Technol.*, 26, 2198–2206, 1992.
96. Butler, E. and K. Hayes, Kinetics of the transformation of trichloroethylene and tetrachloroethylene by iron sulfide. *Environ. Sci. Technol.*, 33, 2021–2027, 1999.
97. Haschke, L., Reduction of perchloroethylene, *Environ. Sci. Technol.*, 34, 1851–1859, 1999.
98. Lee, W. and B. Batchelor, Abiotic reductive dechlorination of chlorinated ethylenes by iron-bearing soil minerals. 2. Green rust, *Environ. Sci. Technol.*, 36(24), 5348–5354, 2002.

99. Williams, M.D. and M. Oostrom, Oxygenation of anoxic water in a fluctuating water table system: an experimental and numerical study, *J. Hydrol.*, 230, 70–85, 2000.
100. Sen Gupta, A.K., *Environmental Separation of Heavy Metals: Engineering Processes*, Lewis Publishers, Boca Raton, FL, 2002.
101. Sparks, D.L., Ed., *Soil Physical Chemistry*, 2nd ed., Lewis Publishers, Boca Raton, FL, 1998.
102. Hong, J. et al., Effects of REDOX processes on acid producing potential and metal mobility in sediments, in *Bioavailability: Physical, Chemical and Biological Interactions*, Hamlink, J.E. et al., Eds., Lewis Publishers, Boca Raton, FL, 1994.
103. Ehrlich, H.L., *Geomicrobiology*, Marcel Dekker, New York, 1981.
104. Conner, J.R., *Chemical Fixation and Solidification of Hazardous Wastes*, Van Nostrand Reinhold, New York, 1990.
105. Weiner, E.R., *Applications of Environmental Chemistry*, Lewis Publishers, Boca Raton, FL, 2000.
106. Wang, Y.T., Microbial reduction of chromate, in *Environmental Microbe-Metal Interactions*, D.R. Lovely, Ed., ASM Press, Washington, D.C., 2000.
107. Bader, J.L. et al., Aerobic reduction of hexavalent chromium in soil by indigenous microorganisms, *Bioremediation J.*, 3, 201–212, 1999.
108. Palmer, C. and R. Puls, *Natural Attenuation of Hexavalent Chromium in Groundwater and Soils*, U.S. EPA, Office of Research and Development, OSWER, EPA/540/S-94/505,1994.
109. Gary, L., and D. Rai, Kinetics of chromium (III) oxidation to chromium (VI) by reaction with manganese dioxides, *Environ. Sci. Technol.*, 21, 1187–1193, 1987.
110. Suthersan, S., *Natural and Enhanced Remediation Systems*, Lewis Publishers, Boca Raton, FL, 2001.
111. Smedley, P.L. and D.G. Kinniburgh, A review of the source, behaviour, and distribution of arsenic in natural waters, *Appl. Geochem.*, 17, 2001.
112. Oremland, R.S. and J. Stolz, Dissimilatory reduction of selenate and arsenate in nature, in *Environmental Microbe-Metal Interactions*, D.R. Lovely, Ed., ASM Press, Washington, D.C., 2000.
113. Vance, D.B., Arsenic: chemical behavior and treatment, *Natl. Environ. J.*, May/June, 1995.
114. Sen Gupta, A.K. and J.E. Greenleaf, Arsenic in subsurface water: its chemistry and removal by engineered processes, in *Environmental Separation of Heavy Metals: Engineering Processes*, Lewis Publishers, Boca Raton, FL, 2002.
115. Steffan, R.J. et al., *In situ* and *ex situ* biodegradation of MTBE and TBA in contaminated groundwater, in *Petroleum Hydrocarbons and Organic Chemicals in Groundwater: Prevention, Detection, and Remediation Conference and Exposition*, National Ground Water Association, Houston, TX, 2001.
116. Renner, R., Reduction of perchlorate in natural systems, *Environ. Sci. Technol.*, 32, 210A, 1998.
117. Nzengung, V.A. et al., Plant-mediated transformation of perchlorate into chloride, *Environ. Sci. Technol.*, 33, 1470–1478, 1999.
118. Orris, G.J. et al., Preliminary analyses for perchlorate in selected natural materials and their derivation products, USGS Open File Report 03-314, 2003.
119. Simonaitis, R. and J. Heicklen, Perchloric acid — Possible sink for stratospheric chlorine, *Planetary and Space Science*, 23(II), 1567–1569, 1975.
120. Erickson, G.E., The Chilean nitrate deposits, *Am. Sci.*, 71, 366–374, 1983.
121. Harrington, J., ARCADIS, personal communication, 2003.
122. Giblin, T. et al., Removal of perchlorate in groundwater with a flow-through bioreactor, *J. Environ. Qual.*, 29, 578–583, 2000.

123. Logan, B.E., A review of chlorate and perchlorate respiring microorganisms, *Bioremed. J.*, 69–80, 1998.
124. Coates, J.D. et al., Ubiquity and diversity of dissimilatory (per)chlorate reducing bacteria, *Appl. Environ. Microbiol.*, 12, 5234, 1999.
125. Rikken, G.B.M. et al., Transformation of (per)chlorate into chloride by a newly isolated bacterium: reduction and dismutation, *Appl. Microbiol. Biotechnol.*, 45, 420, 1996.
126. Herman, D.C. and W.T. Frankenberger, Jr., Microbially-mediated reduction of perchlorate in groundwater, *J. Environ. Qual.*, 27, 750–754, 1998.
127. Wallace, W. et al., Identification of an anaerobic bacterium which reduces perchlorate and chlorate as *Wolinella succinogens*, *J. Ind. Microbiol.*, 16, 68–72, 1996.
128. Zhang, H. et al., Perchlorate reduction by a novel chemolithoautotrophic hydrogen-oxidizing bacterium, *Environ. Microbiol.*, 4(10), 570–576, 2002.
129. Attaway, H. and M. Smith, Reduction of perchlorate by an anaerobic enrichment culture, *J. Ind. Microbiol.*, 12, 408–412, 1993.
130. Harrington, J., ARCADIS, personal communication, 2004.
131. Ji, G. and X. Kong, Adsorption of chloride, nitrate, and perchlorate by variable charge soils, *Pedosphere*, 2, 317–326, 1992.
132. Tipton, D.K. et al., Transport and biodegradation of perchlorate in soils, *J. Environ. Qual.*, 32, 40–46, 2003.
133. Priddle, M.W. and R.E. Jackson, Laboratory column measurement of voc retardation factors and comparison with field values, *Ground Water* 29(2), 260–266, 1991.
134. Potter, S., ARCADIS, personal communication, 2003.
135. Aitchison, E.W., S.L. Kelley, P.J.J. Alvarez, and J.L. Schnoor, Phytoremediation of 1,4-dioxane by hybrid poplar trees, *Water Environ. Res.*, 72(3), 313–321, 2000.
136. Ying, O.Y., Phytoremediation: modeling plant uptake and contaminant transport in the soil-plant-atmosphere continuum, *J. Hydrol.*, 266(1–2), 66–82, 2002.
137. Dietz, A.C. and J.L. Schnoor, Advances in phytoremediation, *Environ. Health Perspect.*, 109(2), 163–168, 2001.
138. Raj, C.B.C., N. Ramkumar, A.H.J. Siraj, and S. Chidambaram, Biodegradation of acetic, benzoic, isophthalic, toluic and terephthalic acids using a mixed culture: effluents of PTA production, *Process Saf. Environ. Protect.*, 75(B4), 245–256, 1997.
139. Roy, D., G. Anagnostu, and P. Chaphalkar, Biodegradation of dioxane and diglyme in industrial-waste environmental science and engineering & toxic and hazardous substance control, *J. Environ. Sci. Health Part A-Environ. Sci. Eng. Toxic Hazard.*, 29(1), 129–147, 1994.
140. Roy, D., G. Anagnostu, and P. Chaphalkar, Analysis of respirometric data to obtain kinetic coefficients for biodegradation of 1,4-dioxane environmental science and engineering & toxic and hazardous substance control, *J. Environ. Sci. Health Part A-Environ. Sci. Eng. Toxic Hazard.*, 30(8), 1775–1790, 1995.
141. Parales, R.E., J.E. Adamus, N. White, and H.E. May, Degradation of 1,4-dioxane by an actinomycete in pure culture, *Appl. Environ. Microbiol.*, 60(12), 4527–4530, 1994.
142. Burback, B.L. and J.J. Perry, Biodegradation and biotransformation of groundwater pollutant mixtures by *Mycobacterium vaccae*, *Appl. Environ. Microbiol.*, 59(4), 1025–1029, 1993.
143. Bernhardt, D. and H. Diekmann, Degradation of dioxane, tetrahydrofuran and other cyclic ethers by an environmental *Rhodococcus* strain, *Appl. Microbiol. Biotechnol.*, 36(1), 120–123, 1991.
144. Zenker, M.J., R.C. Borden, and M.A. Barlaz, Mineralization of 1,4-dioxane in the presence of a structural analog, *Biodegradation*, 11(4), 239–246, 2000.

145. U.S. EPA, Anaerobic Biodegradation Rates of Organic Chemicals in Groundwater: A Summary of Field and Laboratory Studies, U.S. Environmental Protection Agency, Office of Solid Waste, Washington, D.C., June 1999.
146. White, G.F., N.J. Russell, and E.C. Tidswell, Bacterial scission of ether bonds, *Microbiol. Rev.*, 60(1), 216–232, 1996.
147. Ross, I., ARCADIS (U.K.), personal communication, 2004.
148. Pennington, J.C., Explosives, in *Environmental Availability of Chlorinated Organics, Explosives, and Metals in Soils*, W.C. Anderson et al., Eds, American Academy of Environmental Engineers, Annapolis, MD, 1999, 85–109.
149. Spain, J.C., *Biodegradation of Nitroaromatic Compounds*, Plenum Press, New York, 1995.
150. Pennington, J.C. and J.M. Brannon, Environmental fate of explosives, *Thermochim. Acta*, 384, 163–172, 2002.
151. Rittmann, B.E. and P.L. McCarty, *Environmental Biotechnology*, McGraw-Hill, New York, 2001.
152. Duque, E. et al., Construction of a *Pseudomonas* hybrid strain that mineralizes 2,4,6-trinitrotoluene, *J. Bacteriol.*, 175, 2278–2283, 1993.
153. Spain, J.C. et al., Eds., *Biodegradation of Nitroaromatic Compounds and Explosives*, Lewis Publishers, Boca Raton, FL, 2000.
154. Hawari, J., in *Biodegradation of Nitroaromatic Compounds and Explosives*, J.C. Spain et al., Eds., Lewis Publishers, Boca Raton, FL, 2000.
155. Hoffsommer, J.C., Kinetic isotope effects and intermediate formation for the aqueous alkaline homogeneous hydrolysis of 1,3,5-triaza-1,3,5-trinitrocyclohexane (RDX), *J. Phys. Chem.*, 81, 380–385, 1977.
156. McCormick, N.G. et al., Biodegradation of hexahydro-1,3,5-trinitro-1,3,5-triazine, *Appl. Environ. Microbiol.*, 42, 817–823, 1981.
157. Harkins, V.R., et al., Aerobic biodegradation of high explosives; phase I — HMX, *Bioremed. J.*, 3, 285–290, 1999.
158. Adrian, N.R. et al., Stimulating the anaerobic biodegradation of explosives by the addition of hydrogen or electron donors that produce hydrogen, *Water Res.*, 37, 3499–3507, 2003.
159. Sheremata, T.W. et al., The fate of cyclic nitramine explosive RDX in natural soil, *Environ. Sci. Technol.*, 35, 1037–1040, 2001.
160. Lutes, C., ARCADIS, personal communication, 2003.
161. Lovely, D.R., Ed., *Environmental Microbe-Metal Interactions*, ASM Press, Washington, D.C., 2000.
162. Lloyd, J.R. and L.E. Macaskie, Bioremediation of Radionuclide Containing Wastewaters in *Environmental Microbe–Metal Interactions*, D.R. Lovely, Ed., ASM Press, Washington, D.C., 2000.
163. Lovely, D.R. et al., Microbial reduction of uranium, *Nature*, 350, 413–416, 1991.
164. Lovely, D.R. and Y. Philips, Reduction of uranium by *Desulfovibrio desulfuricans*, *Applied and Environ. Microbiol.*, 58, 850–856, 1992.
165. Lloyd, J.R. and L.E. Macaskie, Bioremediation of radio-nuclide containing wastewaters, in *Environmental Microbe-Metal Interactions*, D.R. Lovely, Ed., ASM Press, Washington, D.C., 2000.
166. Senko, E.L. et al., Geochemistry of radionuclides, *Appl. Environ. Microbiol.*, 290–303, 2002.
167. Chang, G. et al., Diversity and characterization of sulfate-reducing bacteria in groundwater at a uranium mill tailings site, *Appl. Environ. Microbiol.*, 67, 3149–3160, 2001.

168. Abdelouas, A.W. et al., Biological reduction of uranium in groundwater and subsurface soil, *Sci. Total Environ.*, 25, 21–35, 2000.
169. Spear, J.L. et al., Modeling reduction of U (VI) under variable sulfate concentrations by sulfate-reducing bacteria, *Appl. Environ. Microbiol.*, 66, 3711–3721, 2000.
170. Abdelouas, A.W. et al., Oxidative dissolution of uraninite precipitated on Navajo sandstone. *J. Contam. Hydrol.*, 36, 353–375, 1999.
171. Leventhal, J. and E. Santos, Relative importance of organic carbon and sulfide sulfur in a Wyoming roll-type uranium deposit. Open File Report (U.S. Geological Survey) 81–580, 1981.
172. Park, C.F., Jr. and J. Guilbert, *The Geology of Ore Deposits*, W.H. Freeman, New York, 1986.
173. Waste Isolation Systems Panel, *Innovative Approaches for Redioactive Waste Management*, 1983.
174. Tesoriero, L. et al., Precipitation of heavy metals in biotic systems, *Appl. Environ. Microbiol.*, 576–586, (2000).
175. Lack, J. et al., Immobilization of radionuclides and heavy metals through anaerobic bio-oxidation of Fe(II), *Appl. Environ. Microbiol.*, 68, 2704–2710, 2002.
176. Ferris, F. et al., Retention of strontium, cesium, lead and uranium by bacterial iron oxides from a subterranean environment, *Appl. Geochem.*, 15, 1035–1042, 2000.
177. Martin, T. and H. Kempton, *In situ* stabilization of metal-contaminated groundwater by hydrous ferric oxide: an experimental and modeling investigation, *Environ. Sci. Tech.*, 34, 3229–3234, 2000.
178. Lloyd, J.R. et al., Reduction of technetium by *Desulfovibrio desulfuricans*: biocatalyst characterization and use in a flowthrough bioreactor, *Appl. Environ. Microbiol.*, 65, 2691–2696, 1999.
179. Istok, J.D. et al., *In situ* bioreduction of technetium and uranium in a nitrate-contaminated aquifer, *Environ. Sci. Technol.*, 38, 468–475, 2004.
180. Pyrih, R.Z., Recognizing the natural attenuation of metals, in *Proceedings of IBS's 4th Ann. Conf. Nat. Attenuation*, Pasadena, CA, December, 1998.
181. Vannote, R.L., G.W. Minshall, K.W. Cummins, J.R. Sedel, and C.E. Cushing, The river continuum concept, *Can. J. Fish. Aquatic Sci.*, 37, 130–137, 1980.
182. Lide, D.R., *CRC Handbook of Chemistry and Physics*, 77th ed., CRC Press, Boca Raton, FL, 1997.
183. Tan, E.L. et al., Mutagenicity of TNT and its metabolites formed during composting, *J. Toxicol. Environ. Health*, 36, 165–175, 1992.
184. Ross, I., ARCADIS, personal communication, 2003.
185. Hawari, J. et al., Characterization of metabolites during biodegradation of RDX with municipal anaerobic sludge, *Appl. Environ. Microbiol.*, 66, 2652–2657, 2000.
186. Duran, R., New shuffle vectors for *Rhodococcus* sp. R 312, a nitrile hydrase producing strain, *J. Basic. Microbiol.*, 38, 101–106, 1998.
187. McCormick, N.G. et al., Biodegradation of hexahydro-1,3,5-trinitro-1,3,5-triazine. *Appl. Environ. Microbiol.*, 42, 817–823, 1981.
188. McCarty, P.L., Breathing with chlorinated solvents, Science, 276(5318), 1521–1522, 1997.
189. Stumm, W. and J.J. Morgan, *Aquatic Chemistry*, John Wiley & Sons, New York, 1996.

3 Chemical Reactive Zones

3.1 INTRODUCTION TO CHEMICAL REACTIVE ZONES

Nothing about *in situ* chemical reactive zones is subtle. The most reliable reaction mechanisms are very simple and, generally, very energetic. The reactions engage many nontarget compounds, and nontarget reaction losses typically dominate the reagent budget. The fluid volume that must be injected to achieve complete contact between reagents and contaminants is often quite large, and the injection points must be spaced closely to overcome physical limitations on the infusion of injected fluids. Despite these challenging attributes, chemical *in situ* reactive zones have been very successful and play an important role in soil and groundwater cleanup.

Groundwater containment strategies of the 1970s and 1980s and the physical removal methods developed in the 1980s and 1990s (e.g., aquifer sparging) have not always succeeded in achieving satisfactory contaminant removal from source zones or from sorbed-phase mass in downgradient segments of aquifers. The limitations on early remedial strategies drove a transition from containment and physical removal methods, to *in situ* treatment strategies, which could achieve greater effectiveness through destruction, rather than removal, of contaminants. The desire for contaminant destruction naturally kindled an interest in chemical reaction mechanisms. There is a wide-ranging repertoire of reaction mechanisms that were developed for water supply and wastewater treatment systems, and many of the reactions now used in chemical reactive zones are "borrowed" from those aboveground systems. The major differences for their use in aquifers are the very large volumes that are engaged in reactions and the loss of reaction control that occurs in their migration from aboveground reactor systems to reactions on the loose in the subsurface.

The popularity of chemical reactive zones for treatment of aquifer contamination increased dramatically, beginning in the mid-1990s. Although there is no central base of statistics on reactive zone applications, from publication of field applications in conference proceedings, it appears that oxidation has become the dominant chemical reactive zone strategy, with permanganate and Fenton's reagent methods leading the way. Of the chemical reducing zone strategies, zero-valent iron appears to be most popular.

Chemical reactive zones have been deployed in source zone treatments, reactive barrier, and comprehensive site treatments, for metals and organics. Because chemical reactive zone treatments are relatively fast acting but costly, relative to biological reactive zone methods or monitored natural attenuation, they can be used to greatest advantage in projects that have tight time constraints. Chemical reactive zones can also be used for rapid source zone mass reduction, which can be coordinated with other reactive zone or natural attenuation methods to produce a comprehensive treatment program for a large-scale contaminant plume. Table 3.1 summarizes the chemical reactive zone treatment mechanisms and deployment modes that have been used in commercial-scale applications.

This chapter describes briefly the underlying chemistry that is deployed when we use chemical reactive zone technologies. We introduce the topic of chemical thermodynamics, which is used to determine whether a reaction is energetically feasible. We also describe chemical kinetics, the study of reaction rates and the variables that affect them — factors such as temperature and concentration of reactants. Thermodynamic feasibility has been determined for most reactions of interest, so we place a much greater emphasis on the effects of chemical kinetics on chemical reactive zone designs. We conclude the section on chemical reactivity with a description of the reaction mechanisms most commonly deployed in chemical reactive zone applications. The chapter continues with detailed descriptions of the major chemical reactive zone technologies, including *in situ* chemical oxidation and reduction methods, nucleophilic substitution reactions, and precipitation reactions. We conclude this chapter with discussions of limitations on the deployment of chemical reactive zones and the management of by-product formation.

3.2 CHEMICAL REACTIVITY

The microbially mediated reactions examined in Chapter 2 were fundamentally similar to the mechanisms we can deploy in chemical reactive zones — reductions and oxidations, in particular. The important difference is that microbial reactions are catalyzed by enzymes that significantly decrease the activation energy for reactions,

TABLE 3.1
Table of Processes That Have Been Deployed in Chemical Reactive Zone Treatments

Chemical Treatment mechanism	Applications		Reactive Zone Treatment Mode		
	Metals	Organics	Source zone	Reactive barrier	Comprehensive treatment
Precipitation	✓		✓	✓	✓
Chemical oxidation	✓	✓	✓	✓	✓
Chemical reduction	✓	✓	✓	✓	
Nucleophilic substitution		✓			✓

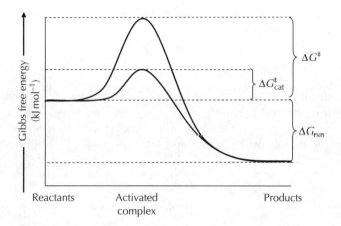

FIGURE 3.1 Effect of enzyme participation on the activation energy of an exergonic reaction. The activation energy of a reaction, $\Delta G^\ddagger$, is dramatically reduced by the intervention of an enzyme in the reaction, as depicted by the lower activation energy, $\Delta G^\ddagger_{cat}$. The overall reaction energy yield, ΔG_{rxn}, is unaffected by the presence of a catalyst.

increasing the rate at which many reactions occur, especially at low reactant concentrations (Figure 3.1 shows the effect of catalyst participation on the activation energy of an energy-producing reaction). The catalytic systems built by microbial communities are also very persistent, relative to the chemical reagents used in reactive zone methods. Chemical systems, lacking the catalytic capacities of biological systems, must operate at much higher energy levels to achieve the same level of treatment. This is accomplished through the use of more energetic reagents (e.g., hydroxyl radical), along with the use of very high reagent concentrations.

This section provides a brief review of chemical reactivity, including chemical thermodynamics and reaction kinetics. The objective is to describe some of the key variables that control chemical reactivity in aquifers, allowing the reader to critically examine reaction systems that might be considered for deployment and to troubleshoot systems that have not met remedial objectives. The greatest emphasis is placed on reaction kinetics, because the performance of chemical reactive zones is most often controlled by the concentrations of target compounds and treatment reagents that can be established in field applications. Finally, the treatment of chemistry fundamentals is not intended to be exhaustive, and we recommend Stumm and Morgan[5] and Schwarzenbach et al.[12] as additional reading for the advanced readers.

3.2.1 Oxidation States

The most common *in situ* chemical reactive zone mechanisms involve oxidation or reduction of target compounds. It is useful to recognize the oxidation states of molecules involved in these reactions, particularly among the organic molecules that might be targets of either reductions or oxidations. The oxidation state of a molecule is calculated from the oxidation numbers of its component atoms, according to the following guidelines (after Kotz and Treichel).[1]

1. Atoms of elements in their pure states (e.g., H_2 and O_2) are assigned oxidation state 0.
2. Single-element ions (e.g., Fe^{2+} and Cl^-) are assigned an oxidation number equaling the charge on the ion.
3. The oxidation number for F is always 1 in compounds with other elements.
4. Hydrogen is assigned +1 and oxygen is assigned −2 with the following exceptions:
   a. Metal hydrides — hydrogen in metal hydrides is assigned −1 (e.g., calcium hydride, CaH_2).
   b. Peroxides — oxygen in peroxides is assigned −1 (e.g., hydrogen peroxide, H_2O_2).
5. Chlorine, bromine, and iodine are assigned −1, except when they are compounded with oxygen or fluorine (in perchlorate (ClO_4^-), the chlorine carries the oxidation number +7).
6. The sum of oxidation numbers for atoms in a compound equals the net charge on the compound. We can use this guideline, combined with those above, to determine oxidation numbers for atoms that have not been specified. For example, in the permanganate ion (MnO_4^-), the oxygens total −8, so the manganese is +7 (calculated as $-1 = x + 4(-2)$), which is represented as Mn(VII).

The oxidation state of each carbon on an organic molecule can be calculated by these guidelines, and the overall oxidation state for an organic molecule can be characterized by a mean oxidation number for the carbons.[2] The value of this analysis for organic molecules is in the initial screening of potential chemical mechanisms for treatment.

Calculation of the mean oxidation number for carbon compounds follows the aforementioned rules, with the one addition:

7. Carbon atoms are neutral with respect to oxidation number in carbon–carbon bonds, regardless of the number of bonds (single, double, or triple). For example, each carbon atom in acetylene (HC≡CH) has the oxidation state (−1), balancing the hydrogens (+1) that each carries.

Carbon can assume oxidation numbers ranging from −4 to +4. To calculate the mean oxidation number for a molecule, the oxidation number is calculated for each carbon atom, then the oxidation numbers are averaged for all carbon atoms in the molecule. Here are some examples:

Carbon tetrachloride (CCl_4). Four chlorines at −1 each indicate that the carbon in this molecule has an oxidation state +4. That is the maximum for carbon, so additional oxidation is not likely to be productive (it is possible to oxidize carbon tetrachloride to carbon dioxide, but the energy input would be quite large, and by-product production would be significant).

Methane (CH_4). Four hydrogens at +1 each indicate that the carbon in this molecule has an oxidation state −4. That is the minimum for carbon, so additional reduction is not possible. This explains why methane persists in reactive zone applications that rely on chemical or biological reduction mechanisms.

Perchloroethene (Cl_2C-CCl_2). Each carbon has two chlorines at −2, so each carbon is assigned the oxidation state +2, and the mean oxidation number for the molecule is +2. Perchloroethene is the most oxidized of the chlorinated ethenes and is more susceptible to reduction and less susceptible to oxidation than the other chlorinated ethenes.

1,1,1-Trichloroethane (Cl_3C-CH_3). The chlorine-bearing carbon has a +3 oxidation state, balancing the three chlorines at −1 each, and the saturated ($-CH_3$) carbon has a −3 oxidation state, balancing the three hydrogens at +1 each. This gives the molecule a mean oxidation number 0. 1,1,1-TCA is susceptible to either reduction or oxidation.

For organic molecules, the addition of hydrogen or the loss of oxygen or halogen atoms yields a reduction in the mean oxidation number, and would be indicative of chemical or biological reduction mechanisms at work. The addition of oxygen or halogen molecules or the loss of hydrogen yield an increase in the mean oxidation number, indicating that a chemical or biological oxidation mechanism is at work.

3.2.2 Thermodynamic Feasibility of Reactions

Thermodynamics is part of the analysis of reactions, quantifying the energy that can be gained from a reaction. Thermodynamics does not reflect the rate at which a reaction will occur, and a large thermodynamic energy yield does not always lead to high reaction rates. In many cases, less energy-yielding reactions occur more quickly, so the most stable product is not always the first to form.

3.2.2.1 Bond Energies

Chemical reactions entail the breaking and making of bonds between atoms. The strength of these bonds determines the energy input required to start a reaction by the breakage of bonds, and the energy that will be gained when the reaction is completed by the formation of new bonds. Chemical reaction strategies for the destruction of soil and groundwater contaminants are most often directed at breaking bonds between carbon atoms and hetero-atoms, that is, atoms of elements other than carbon in organic molecules. Removal of hetero-atoms from toxic organic compounds normally reduces or eliminates their toxicity, and typically increases their mobility and susceptibility to natural attenuation mechanisms such as biodegradation.

Table 3.2 summarizes dissociation energies for chemical bonds between atoms in organic carbon molecules. From this data, we see that the hetero-atoms chlorine and bromine have relatively lower dissociation energies than the bonds between carbon atoms and oxygen or hydrogen. Bond energies between carbon and halogens decrease with increasing halogen atomic mass: $F > Cl > Br > I$. The decrease in

TABLE 3.2
Selected Bond Dissociation Energies

Bond	Bond Dissociation Energy (kJ mol^{-1})
Methyl bonds	
CH_3-H	435
CH_3-Cl	351
CH_3-Br	293
CH_3-OH	381
Primary carbon bonds	
CH_3CH_2-H	410
CH_3CH_2-Cl	339
CH_3CH_2-Br	285
CH_3CH_2-OH	381
Secondary carbon bonds	
$(CH_3)_2CH-H$	397
$(CH_3)_2CH-Cl$	335
$(CH_3)_2CH-Br$	285
$(CH_3)_2CH-OH$	381
Tertiary carbon bonds	
$(CH_3)_3C-H$	381
$(CH_3)_3C-Cl$	331
$(CH_3)_3C-Br$	272
$(CH_3)_3C-OH$	381
Carbon–carbon bonds	
CH_3-CH_3	368
$CH_3CH_2-CH_3$	356
C=C (second bond)	264[a]
Other bonds	
Ph–Cl	402[a]

[a] Data from Fox, M.A., and J.K. Whitesell, *Organic Chemistry*, Jones & Bartlett, Sudbury, MA, 1997.

Source: From Wade, L.G., Jr., *Organic Chemistry*, 4th ed., Prentice Hall, Upper Saddle River, NJ, 1999, unless otherwise noted.

bond strength is linked to the increase in atomic radius, which leads to greater bond lengths for the larger halogens.

3.2.2.2 Gibbs Free Energy

Gibbs free energy incorporates changes in enthalpy (ΔH) and entropy (ΔS) to reflect the total energy change entailed in any reaction.

When the Gibbs free energy associated with a reaction is positive, the reaction requires energy input to occur, and the reaction cannot occur spontaneously.

When the Gibbs free energy associated with a reaction is negative, the reaction releases energy, and can occur spontaneously. Spontaneous reactions are those that can occur without the input of energy into the system. Spontaneity does not imply speed.

Gibbs free energy is defined by the change in enthalpy (H) and entropy (S) in a reaction, and can be calculated by the van't Hoff equation (Equation (3.1)):

$$\Delta G = \Delta H - T\Delta S \qquad (3.1)$$

The change in Gibbs energy for a reaction is calculated by the difference between the Gibbs energy of products and reactants, as shown in Equation (3.2):

$$\Delta G_{rxn} = \sum G^0_{products} - \sum G^0_{reactants} \qquad (3.2)$$

where G^0 are values for the standard free energy of formation for individual compounds, referenced to a standard physical–chemical state. Values of ΔG for a reaction can be classified according to the following.

1. *Endergonic reactions.* Reactions that are energy-consuming as measured by an increase in Gibbs free energy are termed endergonic. Endergonic reactions cannot occur spontaneously. Metabolic reaction sequences may include endergonic steps, but the overall reaction must be exergonic to proceed.
2. *Exergonic reactions.* Reactions that are energy releasing, as measured by a decrease in Gibbs free energy, are termed exergonic. Exergonic reactions may occur spontaneously.

3.2.2.3 Nonstandard Conditions — The Nernst Equation

The Gibbs free energy for a reaction at nonstandard conditions is described by the Nernst equation. The free energy change is determined by the standard free energy of the reaction, the reaction temperature, reaction stoichiometries, and the concentrations of products and reactants, as shown in Equation (3.4), where A and B are reactants and C and D are products, as shown in Equation (3.3). For the reaction

$$aA + bB \rightarrow cC + dD \qquad (3.3)$$

$$\Delta G_{rxn} = \Delta G^0_{rxn} + RT \ln \left[\frac{[C]^c [D]^d \ldots}{[A]^a [B]^b \ldots} \right] \qquad (3.4)$$

where A and B indicate reactants, and C and D indicate products.

The free energy for a compound at nonstandard conditions can be determined by the partial molar free energy calculation, shown in Equation (3.5). The bar above the Gibbs energy symbol, G, indicates a molar quantity for compound A, in contrast to the Gibbs energy for a reaction, shown in Equations (3.2) and (3.4).

$$\overline{G}_A = \overline{G}^0_A + RT \ln [A] \qquad (3.5)$$

The thermodynamic properties of reactions can be used as a screening tool to identify reactions that are energetically feasible. Thermodynamics does not, however, determine whether a reaction can be deployed effectively in a chemical reactive zone. Many reactions that are thermodynamically feasible do not occur at rates that can be used for reactive zone processes. Rate predictions are the domain of chemical kinetics, which is discussed in detail in Section 3.2.3.

3.2.3 KINETIC CONTROL OF REACTION RATES

Thermodynamic analysis tells us the energy yield, ΔG_{rxn}, when a reaction occurs, but it does not give us information on the probability that the reaction will actually happen. The rate at which a reaction occurs, measured as mol L^{-1} s^{-1} (or M s^{-1}), is a reflection of the probability of the reaction. There are two major factors that determine the probability of completing a particular irreversible reaction in aqueous solution:[*]

1. *Frequency of collisions between reactants.* For a reaction to occur, the reactants must make contact. The frequency of contact events is controlled by two variables: the concentration of reactants and their effective velocities in solution. Reactions are more likely to occur at high reactant concentrations, and at higher molecular velocities (i.e., higher temperatures). Molecules that rapidly diffuse in aqueous solution would also be expected to achieve more contact events than reactants that migrate slowly in solution.
2. *Likelihood that a contact event will result in a completed reaction.* Many collisions occur between reactants, which do not result in a completed reaction. The rate at which contact events generate reactions is determined by the steric interaction of the reactants (the "fit" of the reacting molecules), the velocity (energy) at which the contact occurs, and the free energy of formation of the reaction intermediate (activated complex), $\Delta G^{\ddagger}$.

The effects of contact frequency, likelihood of reaction, and other variables are lumped into a kinetic constant that can be used to estimate the rate that a reaction occurs under specified conditions. Kinetic analysis of possible reactions is important in the development of chemical reactive zone strategies, because many reactions that are thermodynamically feasible occur at rates that are too slow or too fast to achieve remedial objectives. Kinetic analysis of reactions can also be used to determine when competition for reactants will affect *in situ* chemical reactive zone operation. This allows remedial system designers to manage reactive zone chemistry to minimize competitive limitations on reaction effectiveness.

3.2.3.1 Reaction Order and Rate Equations

There are four basic kinetic classifications for reactions that might be of interest to *in situ* reactive zone designers: zero-order, first-order, second-order, and a special case of second-order reactions, pseudo-first-order reactions. In the sections that follow, reaction mechanisms and associated mathematical descriptions are provided for each of the reaction classifications. For a remedial system designer, the value of

---

[*] This treatment of kinetics is not intended to be a comprehensive treatment of the subject but rather a summarization offered to assist the nonchemist in understanding the major factors that underlie reaction kinetics.

kinetic analysis lies in the ability to calculate the rate of reaction as a function of target compound and injected reagent concentrations, and to estimate the rate of loss of injected reagents to competing, unproductive reactions.

In the simplest reactions, there is one reactant that is changed to make one or more products, as shown in Equation (3.6). This reaction may follow either zero- or first-order kinetics.

$$A \rightarrow products \qquad (3.6)$$

Zero order. Zero-order reactions proceed at a constant rate over time. This may occur in a catalytic reaction with a single reactant, when the reactant is present in large quantities. Equation (3.7) shows the zero-order rate equation in differential form.

$$\frac{dA}{dt} = k \qquad (3.7)$$

Integrating to find a formula for the concentration of A as a function of time,

$$[A] = [A]_0 - k \cdot t \qquad (3.8)$$

where k is the kinetic rate constant ($M^{-1} s^{-1}$), and $[A]_0$ is the concentration of A when $t = 0$. A zero-order reaction generates a linear decrease of the reactant over time, as shown in Figure 3.2. Zero-order reactions are unlikely to be encountered in reactive zone applications.

First order. First-order reactions also follow Equation (3.6) and proceed at a rate that declines over time. The reaction rate at any time is determined by the concentration of a single reactant. Radioactive decay and many enzyme reactions follow first-order reaction kinetics.

$$\frac{dA}{dt} = k \cdot [A] \qquad (3.9)$$

where k is the kinetic rate constant of the reaction (s^{-1}), and $[A]$ is the molar concentration of the reactant, A, at any time, t. Integrating, we obtain

$$[A] = [A]_0 \cdot e^{-kt} \qquad (3.10)$$

where $[A]_0$ is the concentration of A when $t = 0$. A first-order reaction generates a log-linear decrease of concentration over time, as shown in Figure 3.2. True first-order reactions are unlikely to be encountered in reactive zone applications, although pseudo-first-order reactions are quite common. Pseudo-first-order reactions are a special case of second-order reactions, and are discussed later.

Second order. Second-order reactions proceed at a rate that is determined by the concentrations of two reactants. The rates of most reactions of interest in chemical reactive zones are described by second-order reaction kinetics. For the reaction:

$$A + B \rightarrow products \qquad (3.11)$$

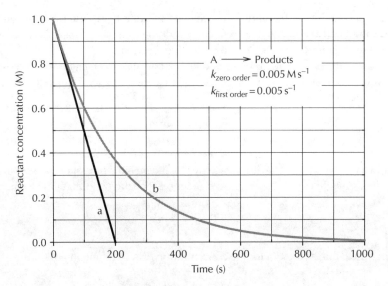

FIGURE 3.2 Concentration vs. time with zero-order (a) and first-order (b) kinetic rate constants. For these reactions, the initial reaction rates are identical. With zero-order kinetics, the reaction rate remains constant as the concentration of the reactant decreases. With first-order kinetics, the reaction rate declines as the reactant concentration decreases.

the rate of reaction is described by the following equation:

$$\frac{d[A]}{dt} = \frac{d[B]}{dt} = k \cdot [A] \cdot [B] \tag{3.12}$$

where the units for k are $M^{-1} s^{-1}$.

Notice that the rate of reaction for A and B are the same, because for each molecule of A reacted, a molecule of B also reacts. For simple reaction systems, solutions to the two-reactant second-order reaction differential equations (3.12) are provided by numerous authors. The following solution, given by Barrow,[3] is useful because it expresses the moles reacted per liter after t seconds, x, which can be isolated algebraically:

$$kt = \frac{1}{[B]_0 - [A]_0} \ln \frac{[B]_0([A]_0 - x)}{[A]_0([B]_0 - x)} \tag{3.13}$$

where x is the number of moles reacted per liter in time t. Solving for x (using the equation-solving facility of MathCad®*) at a specified time, t,

$$x = [B]_0 [A]_0 \frac{e^{(kt[A]_0 - kt[B]_0)} - 1}{[A]_0 \cdot e^{(kt[A]_0 - kt[B]_0)} - [B]_0} \quad [A]_0 \neq [B]_0 \tag{3.14}$$

---

*Expansion of the natural log term was required for MathCad to achieve a solution.

Chemical Reactive Zones

The concentration of each reactant remaining after t seconds is

$$[A] = [A]_0 - x \quad (\text{mol L}^{-1}) \tag{3.15}$$

$$[B] = [B]_0 - x \quad (\text{mol L}^{-1}) \tag{3.16}$$

Most reactions used for destruction of organic contaminant in chemical reactive zones are second order — the reaction rate depends on the concentration of the targeted contaminant as well as the injected reagent. Simplifying assumptions are sometimes used to collapse the second-order kinetics into a pseudo-first-order kinetics analysis, as described in the following. These assumptions must be made with care, to avoid overestimation of the expected reaction rates.

Pseudo-first order. Pseudo-first-order reaction kinetics are a special case of second-order kinetics, in which three conditions are met:

1. The reaction is first order relative to both the targeted contaminant and the injected reagent.
2. The injected reagent concentration is very high, relative to the targeted contaminant concentration.
3. The injected reagent is not significantly reactive with any other chemical species in the reactive zone.

These conditions are not always met, and the conservative approach is to use full second-order kinetic analysis of the expected reactions. In chemical reactive zone applications, pseudo-first-order reaction assumptions may be met at the point of injection; however, dilution of injected reagents by groundwater may invalidate the assumptions within a short distance of the injection point.

Cases in which the conditions are met allow a simplifying assumption: the concentration of the higher-concentration reactant is effectively constant relative to the lower-concentration contaminant, which allows us to collapse the kinetic rate equation to a pseudo-first-order calculation. This is accomplished by lumping the kinetic rate constant and the concentration of the excess reactant into a pseudo-first-order rate constant. For the reaction described in Equation (3.11), if A is present at a very large concentration, relative to the concentration of B, the terms are lumped as follows:

$$k' = k[A] \quad (\text{s}^{-1}) \tag{3.17}$$

where k' is the pseudo-first-order kinetic rate constant for the reaction.

The reaction can then be expressed in the same manner as a first-order reaction, as in Equation (3.10) (with A at a constant concentration).

$$[B] = [B]_0 \, e^{-k't} \tag{3.18}$$

The half-life of a reactant can be calculated for both first-order and pseudo-first-order reaction kinetics. The half-life is the time at which half of the reactant remains.

$$\frac{[B]}{[B]_0} = 0.5 = e^{-k't} \tag{3.19}$$

Solving for t,

$$\ln(0.5) = -k't \quad (s^{-1} \cdot s) \tag{3.20}$$

and

$$t_{1/2} = -\frac{\ln(0.5)}{k'} \quad (s) \tag{3.21}$$

Solving

$$t_{1/2} = -\frac{(-0.69)}{k'} \quad (s) \tag{3.22}$$

$$t_{1/2} = \frac{0.69}{k'} \quad (s) \tag{3.23}$$

Figure 3.3 shows the error that would be induced by pseudo-first-order analysis of a second-order reaction, when one of the reactants was initially present at double the concentration of the other reactant. The assumption of a constant concentration of reactant "B" (the pseudo-first-order assumption) overstates the rate of reaction, giving a lower concentration for reactant "A" than would actually be observed. As shown in Figure 3.3, the error is minimal during early stages of the reaction, and leads to the greatest error when the reaction is more than 50% complete. The magnitude of the error becomes quite large when the two reactants are at similar concentrations.

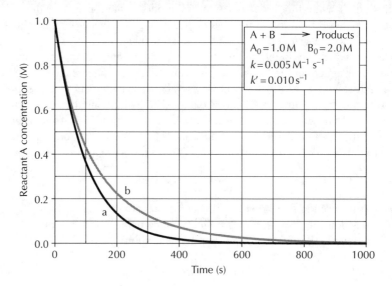

FIGURE 3.3 Comparison of second-order and pseudo-first-order kinetics for the reaction $A + B \rightarrow$ products. Curve (a) shows the concentration of reactant A, when the concentration of reactant B is assumed to remain constant (pseudo-first-order kinetics). Curve (b) shows the slower rate of reaction when the concentration of B declines through the reaction process. As the initial concentration of B is increased, curve (b) converges to match curve (a). See additional explanation in the text.

Table 3.3 summarizes reaction kinetic equations for zero-, first-, second-, and pseudo-second-order kinetics.

3.2.3.2 Kinetic Feasibility

Thermodynamic analysis determines whether a reaction is energy-yielding or energy-consuming. Energy-yielding reactions meet the criterion of spontaneity and are thus thermodynamically feasible. However, many reactions that are thermodynamically feasible do not occur at a rate that can be used cost-effectively in a remedial system. This leads to the second determination that must be made for candidate reaction systems: kinetic feasibility.

The determination of kinetic feasibility is more subjective than the determination of thermodynamic feasibility (refer to Section 3.2.2), because a judgment must be made on the economic viability of the reaction. The principle of kinetic feasibility can be stated as follows:

> A reaction is kinetically feasible if it proceeds at a rate, and at concentrations, that achieve remedial objectives in a cost-effective manner.

There are several potential kinetic limitations on the feasibility of an *in situ* chemical reactive zone strategy:

- It is possible for a reaction to occur too quickly. Many aquifer contaminants are present in two or more distinct phases. It is the aqueous-phase fraction of aquifer contamination that is typically quantified, and that is the phase in which most chemical reactive zone reactions occur. For hydrophobic organic

TABLE 3.3
Summary of Equations Describing Reaction Kinetics

Reaction Order	Differential	Solution to Differential	Units for k
A → products			
Zero order	$\dfrac{dA}{dt} = k$	$[A] = [A]_0 - k \cdot t$	M s^{-1}
First order	$\dfrac{dA}{dt} = k \cdot [A]$	$[A] = [A]_0 \cdot e^{-kt}$	s^{-1}
A + B → products			
Second order	$\dfrac{d[A]}{dt} = \dfrac{d[B]}{dt} = k \cdot [A] \cdot [B]$ $[A]_0 \neq [B]_0$	$kt = \dfrac{1}{[B]_0 - [A]_0} \ln \dfrac{[B]_0([A]_0 - x)}{[A]_0([B]_0 - x)}$ solving for x, the amount of A and B reacted after t seconds: $x = [B]_0 \, [A]_0 \dfrac{e^{(kt[A]_0 - kt[B]_0)} - 1}{[A]_0 \cdot e^{(kt[A]_0 - kt[B]_0)} - [B]_0}$	M^{-1} s^{-1}
Pseudo first order	$\dfrac{dB}{dt} = k' \cdot [B]$ $[A]_0 \gg [B]_0$	$[B] = [B]_0 e^{-k't}$	s^{-1}

contaminants, the nonaqueous fractions constitute a majority of the contaminant mass in many aquifers. During a reactive zone treatment for hydrophobic contaminants, the loss of the target compound from aqueous phase is balanced by desorption of contaminant from nonaqueous-phase. The half-life of highly reactive treatment agents, such as hydroxyl radical, are so short that the reagent may be spent before the desorption process is completed. This causes rebound of nonaqueous-phase contaminant concentrations, and under these conditions, a slower-acting reagent may be favored.

- *It is possible for a reaction to occur too slowly.* A reagent that acts too slowly may be washed out of the treatment area by groundwater movement, before reactions occur.
- *It is possible for a reactant to be consumed by competing chemical processes.* Most reagents used in chemical reactive zones can react with several chemical species that are likely to be present in the aquifer. It is not uncommon to observe significant reagent loss to reactions that have low kinetic rate constants. This occurs when the concentration of the competing reactant is high. The bicarbonate ion (HCO_3^-) is an example of a reactant that is only moderately reactive with the hydroxyl radical. Its second-order kinetic rate constant is approximately 1000-fold lower than the rate constant for hydroxyl radical reaction with trichloroethene. However, when the pH exceeds 7, a significant fraction of the dissolved-phase inorganic carbon is in the bicarbonate ion form. If the magnitude of the inorganic carbon pool is large, the molarity of the bicarbonate ion will be very high, relative to the trichloroethene concentration. As the oxidation reaction proceeds, and trichloroethene concentrations decline, an increasing portion of the hydroxyl radicals react with bicarbonate, rather than with the target, trichloroethene. The effect of carbonate competition on trichloroethene oxidation is quantified in Section 3.3.1.3.
- *The reactant concentration that must be sustained to fully treat a target contaminant can be uneconomical.* There are many cases in which the aquifer matrix consumption of a reagent is very large, and chemical reagent mass that must be supplied to achieve contaminant removal is economically unattractive.

All of these potential pitfalls should be examined in the determination of kinetic feasibility for a reaction strategy.

Kinetic feasibility generally must be determined on a site-specific basis. Here are elements of the kinetic analysis that should be considered for each *in situ* reaction scenario:

- Kinetic rate constants for the primary reactions.
- Kinetic rate constants for competing reactions that are reasonably anticipated. Examples of these are:
  - Decomposition reactions of the injected reagents
  - Reaction with dissolved inorganic species such as carbonates
  - Reaction with nontarget organic compounds

- Concentration patterns that will be achieved for the injected reagents: reagents are normally injected at a high concentration, in anticipation of dilution by groundwater. The injected reagent concentration may range over several orders of magnitude from a single injection point. The effectiveness of the planned reaction and the production of by-products will be dramatically different over the area of injection influence.
- Groundwater flushing rate. Will the injected reagent be swept out of the treatment area before the intended reactions have been completed?
- By-products review. As reagent strength increases, the likelihood of by-product formation increases. Reagents that are poorly matched to a target contaminant (slow reaction rates) require high reagent concentrations to achieve needed reductions in an acceptable time frame. With the high reagent strength comes an increased rate of by-product-forming reactions. Examples of by-product formation are given in Section 3.4.2.

3.2.4 Effect of Temperature on Reaction Rates

The temperature of a reaction solution is one of the variables controlling the frequency of contact events between reactants (another is concentration of reactants). Temperature also determines the energy at which contact occurs. Consequently, temperature can have a significant impact on reaction rates. The Arrhenius equation describes the relationship between the temperature and the kinetic rate constant for a reaction as follows:

$$k = A \cdot e^{-E_A/RT} \quad (3.24)$$

where A is the pre-exponential factor that describes the frequency of collisions with the correct geometry for a reaction to occur (L mol^{-1} s^{-1}), and the exponential term quantifies the fraction of collisions that occur with sufficient energy to complete a reaction. E_A is the activation energy, $G^\ddagger$ (J mol^{-1}), and R and T are the universal gas constant (8.3145 J mol^{-1} K) and reaction temperature (K), respectively.

In many cases, a kinetic rate constant is available for a reaction at one temperature, and we desire the rate of reaction at a different temperature. If the activation energy is known for the reaction of interest, the kinetic rate constant can be calculated for the reaction at an alternative temperature, using the following equation.

$$\ln\left(\frac{k_{T_1}}{k_{T_2}}\right) = \frac{E_A}{R}\left(\frac{1}{T_2} - \frac{1}{T_1}\right) \quad (3.25)$$

Many reactions that can be considered for deployment in chemical reactive zones are temperature-sensitive, and it is important to consider the expected reaction temperature when evaluating the use of these reactions in a reactive zone strategy. Nucleophilic substitutions and persulfate oxidation are among the especially temperature-sensitive reactions.

3.2.4.1 Estimating Temperature Effect on Thiosulfate-Mediated Destruction of Methyl Bromide

Methyl bromide is susceptible to nucleophilic substitution by thiosulfate, according to the reaction shown in Figure 3.4. From Wang et al.[4] we obtain the activation energy, $G^{\ddagger}$, 73.3 kJ mol^{-1}, and the second-order kinetic rate constant for the reaction at 20°C (293 K), 2.11×10^{-2} M^{-1} s^{-1}. The kinetic rate constant at 10°C (283 K) can be estimated using Equation (3.25) (notice that the numerical value of the gas constant, R, has been revised to 8.3145×10^{-3} reflecting the use of kJ units for the E_A term).

$$\ln\left(\frac{2.11 \times 10^{-2}}{k_{283}}\right) = \frac{73.3}{8.314 \times 10^{-3}}\left(\frac{1}{283} - \frac{1}{293}\right)$$

$$\ln(2.11 \times 10^{-2}) - \ln(k_{283}) = \frac{73.3}{8.314 \times 10^{-3}}\left(\frac{1}{283} - \frac{1}{293}\right)$$

$$\ln(k_{283}) = \ln(2.11 \times 10^{-2}) - \left(\frac{1}{283} - \frac{1}{293}\right)$$

$$\ln(k_{283}) = -3.858 - 2.896$$

$$k_{283} = e^{-6.75} \quad (\text{M}^{-1}\,\text{s}^{-1})$$

$$k_{283} = 1.16 \times 10^{-3} \quad (\text{M}^{-1}\,\text{s}^{-1})$$

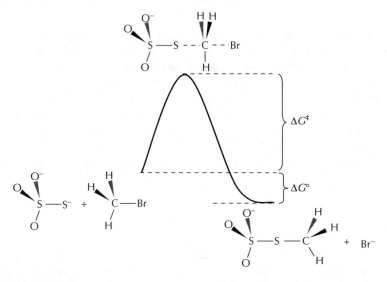

FIGURE 3.4 Reaction coordinate for nucleophilic substitution of methyl bromide by thiosulfate. The activation energy, $\Delta G^{\ddagger}$, was 73.3 kJ mol^{-1}, and the second-order rate constant for this reaction was 2.11×10^{-2} M^{-1} s^{-1} at 20°C, as measured by Wang et al.[11]

The 10°C decrease in temperature is expected to cause an 18-fold decrease in the reaction rate. Translation of published reaction data, which is commonly obtained at temperatures of 20 or 25°C, is essential to the process of selecting a chemical reactive zone strategy, where *in situ* temperatures may range between 10 and 15°C at locations in temperate and cool climates.

3.2.5 Effect of pH on Reactions

Changes in the solution pH for reactions that involve a hydrogen ion or hydroxyl anion directly affect the rate of reaction. The affect can be assessed by supplying alternative values for the hydrogen ion or hydroxyl anion, as appropriate for the reaction. The hydrogen ion concentration is calculated as the negative log of the pH, and the hydroxyl anion concentration for a specified pH value can be calculated from the dissociation of water, which is expressed by the following reaction equation.

$$H_2O \overset{k_w}{\rightleftarrows} H^+ + OH^-$$

The activity of water is defined as 1, and the dissociation rate is defined by the relationship

$$K_w = [H^+] \cdot [OH^-] \quad (mol^2 L^{-2} \text{ or } M^2)$$

Table 3.9 shows the dissociation constant for water as a function of temperatures ranging from 10 to 30°C. Equation (3.26) isolates the hydroxyl anion concentration as a function of pH and the dissociation constant for water.

$$[OH^-] = K_w \cdot 10^{pH} \quad (3.26)$$

Further, pH also effects reactions through the contribution to ionic strength from hydroxyl anion or proton concentration at extreme pH values. For example, at pH 11 and 20°C, $K_w = 0.68 \times 10^{-14}$ and

$$[OH^-] = 0.68 \times 10^{-14} \cdot 10^{11} = 6.8 \times 10^{-4} \quad (M)$$

The ionic strength that results from the hydroxyl anion concentration at pH 11 can reduce the activity of other ions in solution, as described in Section 3.2.6.

3.2.6 Effect of Ionic Strength on Reactions

Aqueous phase molar concentrations are utilized in calculations as an estimate of chemical activity. The activity of a chemical species, $\{A\}$, is related to its aqueous-phase concentration $[A]$ by its activity coefficient, γ_A, as shown in Equation (3.27).

$$\{A\} = \gamma_A [A] \quad (3.27)$$

The value for γ_A approaches 1 in very dilute aqueous solutions, and under those conditions, $[A]$ provides an accurate representation of chemical activity. However, as dissolved ionic solids' concentrations increase, the value of γ_A decreases significantly. Ionic solids' concentrations in many chemical reactive zone applications are very high, and calculations that use molar concentrations, in lieu of chemical activities, may overestimate reaction rates.

Equation (3.28) shows the calculation of ionic strength, I, which is a function of the concentration, C_i, and ionic charge, z_i, of the dissolved constituents.

$$I = \frac{1}{2}\sum C_i z_i^2 \qquad (3.28)$$

For solutions with relatively low ionic strength ($I < 10^{-2.3}$), the Debye–Huckel equation (3.29) provides an estimate of the activity coefficient at 25°C.

$$\log \gamma = -0.5 z^2 \sqrt{I} \qquad (3.29)$$

The extended Debye–Huckel equation (3.30) adds a correction factor, a, that accounts for the effects of the ionic radius of each species, and is valid for ionic strengths up to 0.1 M.

$$\log \gamma = -0.5 z^2 \left(\frac{\sqrt{I}}{1 + 0.33a\sqrt{I}} \right) \qquad (3.30)$$

Values for the coefficient, a, are given by Stumm and Morgan.[5] Figure 3.5 shows the effect of ionic strength on the chemical activity of selected ions in aqueous phase, using the extended Debye–Huckel equation (3.30).

The ionic strength of seawater can be calculated as 0.7 M, from its major ion composition, given in Stumm and Morgan.[5] This exceeds the range of ionic strengths for which Equations (3.29) or (3.30) are valid.

Another example of interest is the ionic strength of reagents at the solution working strength commonly used in chemical reactive zones. For example, the permanganate anion is likely to be applied in aquifers at a strength of 1% or greater. The molecular weight of the potassium permanganate that is often used to generate the oxidant solution is 158 g mol^{-1}; a 1% solution contains 10 g L^{-1} potassium permanganate, or 0.063 mol L^{-1}, of the potassium cation ($z = +1$) and the same concentration of the permanganate anion ($z = -1$). Using equation (3.28),

$$I = \frac{1}{2}\sum (0.063 \cdot (-1)^2 + 0.063 \cdot (+1)^2)$$

and

$$I = 0.063 \quad (M)$$

Figure 3.5 shows that a 1% permanganate solution can suppress the chemical activity of other ions in the solution.

Chemical Reactive Zones

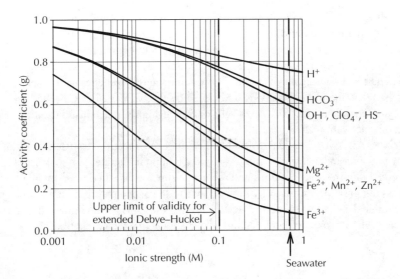

FIGURE 3.5 Activity coefficient as a function of ionic strength for selected ions. Values calculated by the extended Debye–Huckel equation (Equation (3.30)). Values for ion coefficients were taken from Stumm and Morgan.[5]

3.3 REACTION MECHANISMS

Most of the reaction systems considered for use in aquifer remedies were developed and tested in wastewater applications (e.g., ozonation) or industrial chemical synthesis (e.g., the Fenton's oxidation process). Useful remediation reactions have been drawn from four classes: electron transfers, nucleophilic substitutions and eliminations, precipitation reactions, and heterogeneous (liquid–solid) reactions. The general mode of action for each of these reaction classes is introduced in this section and detailed discussions of the most effective examples of each class of reactions is provided in the subsequent section.

3.3.1 Electron Transfer Reactions

The most commonly used chemical reactive zone strategies incorporate electron transfers. Even when the ultimate objective is to stimulate durable precipitation of a toxic inorganic compound, the reaction strategy will normally begin with modification of the chemical habitat through electron transfer reactions.

This section characterizes mechanisms that underlie oxidation–reduction reactions, the leading chemical reactive zone mechanism, and nucleophilic substitutions and eliminations. The special case of radical chain reactions is also discussed, due to its importance in all hydroxyl radical–based reaction sequences.

3.3.1.1 Electron Transfer (Oxidation–Reduction) Reactions

Electron transfer reactions, also known as oxidation–reduction or redox reactions, are the most common of the chemical reactive zone mechanisms. *In situ* chemical oxidation methods such as Fenton's reagent, permanganate or ozone injection, and chemical reduction methods using dithionite, vitamin B_{12}, or zero-valent iron all take advantage of electron-transfer reactions in the destruction of contaminants. The biological reactions described in Chapter 2 are predominantly electron-transfer reactions, as well. The most significant distinction between bacterial and chemical reactive zone chemistries is the use of catalysts (enzymes) by the bacteria. Lacking the benefit of catalysts, chemical reactive zones rely on much higher reactant concentrations to overcome activation energy limitations.

There are two chemical roles fulfilled in each electron transfer reaction — electron donor and electron acceptor — and a compound or element may serve as an electron donor in some reactions and as an electron acceptor in others. In the reaction process, an electron migrates from the donor to the acceptor, significantly changing the chemical behavior of both reactants. Equation (3.31) shows the reactants and products when two compounds, A and B, participate in a redox reaction.

$$A_{reduced} + B_{oxidized} \rightarrow A_{oxidized} + B_{reduced} \tag{3.31}$$

For each electron transferred in the reaction of Equation (3.31), the oxidation number of molecule A is increased by 1, and the oxidation number of molecule B is reduced by 1. A is said to be the reductant or reducing agent, and B is the oxidant or oxidizing agent.

The oxidation of methane by oxygen, shown in Equation (3.32), is an example of an electron transfer reaction.

$$CH_4 + 2O_2 \rightarrow CO_2 + 2H_2O \tag{3.32}$$

The oxidation potential of the carbon in methane was -4 and in carbon dioxide its oxidation potential increased to $+4$, which occurred through donation of electrons to the oxygen atoms of the carbon dioxide and water molecules. The oxidation potential of the diatomic oxygen molecule was 0, which decreased to -2 in each of four oxygen atoms in the products. The hydrogen atoms had an oxidation state of $+1$ in both the methane reactant and water product. In this reaction, the carbon atom in methane served as the electron donor (reductant) and oxygen served in the electron acceptor (oxidant) role.

Electron transfers utilized in chemical reactive zones include homogeneous aqueous-phase reactions (both reactants are aqueous phase) and heterogeneous reactions, in which one of the reactants is in the solid phase and the other is in the aqueous phase. The gaseous reagents that might be used in chemical reactive zones (e.g., the oxidant ozone) react in the dissolved phase.

Table 3.4 summarizes the second-order kinetic rate constants for reactions of hydroxyl radical, ozone, and permanganate with typical organic contaminant compounds, along with reaction rates for compounds that may act as scavengers, competing with target compounds for oxidant capacity. Table 3.4 also provides estimates of molarities that can be achieved in reactive zone applications for each of the oxidants. These values can be used in conjunction with kinetic rate constants and target contaminant concentrations to estimate treatment rates and competitive interactions that are likely to be observed in field applications.

TABLE 3.4
Summary of Kinetic Rate Constants for Oxidation Reactions with Common Organic Contaminant Compounds

Compound	Second-Order Kinetic Constants ($M^{-1} s^{-1}$)		
	O_3	$OH^{\bullet}$	MnO_4^{-}
PCE	$<10^{-1}$	3×10^9	4.5×10^{-2}
			6.5×10^{-1}
TCE	1.7×10^1	4×10^9	6.6×10^{-1}
			8.9×10^{-1}
cis-DCE	$<8 \times 10^2$		9.2×10^{-1}
1,1-DCE	1.1×10^2	7×10^9	2.4×10^0
1,1,1-TCA	$<1.2 \times 10^{-2}$	1×10^8	
1,2-DCP	$<4.0 \times 10^{-3}$	3.8×10^8	
MTBE	0.14^c	1.6×10^9	3.5×10^{-5a}
1,4-Dioxane	3.2×10^{-1}	2.4×10^9	Not reactive[b]
NDMA	10	3.3×10^8	
tert-Butyl alcohol	3×10^{-3}	5.9×10^8	
Methanol	2.4×10^{-2}	9.7×10^8	
Phenol	1.4×10^2	1.4×10^{10}	
Acetate	3×10^{-5}	7.9×10^7	
Benzene	2×10^0	7.8×10^9	
Fluorene		1.8×10^{10}	
Phenanthrene	1.5×10^4	2.3×10^{10}	
Naphthalene	3×10^3	9.4×10^9	
Nitrobenzene		4.0×10^9	
HCO_3^{-}		8.5×10^6	
CO_3^{2-}		4.0×10^8	
Fe^{2+}	5×10^5	1.7×10^{10}	
Formic acid	5×10^0		3.5×10^{-1}
Glyoxylic acid	1.7×10^{-1}		3.7×10^{-1}
Oxalic acid			1.1×10^{-1}
H_2O_2 (HO_2^{-})	(5.5×10^6)		

Oxidant	Feasible Oxidant Molarities (mol L^{-1})
O_3	1×10^{-3} (6% O_3 gas phase)
$OH^{\bullet}$	1×10^{-13} to 1×10^{-11}
MnO_4^{-}	2.5×10^{-1} (4%)

[a] Damm, J.H., C. Hardacre, R.M. Kalin, and K.P. Walsh, Kinetics of the oxidation of methyl-*tert*-butyl ether (MTBE) by potassium permanganate, *Water Res.*, 36, 3638–3646, 2002.
[b] ARCADIS bench-scale test results.
[c] Acero et al.[33]
Source: Ross et al.[7] unless otherwise noted.

3.3.1.2 Radical Reactions

The most aggressive of the chemical reactive zone mechanisms is the radical chain reaction. It is the basis for Fenton's reagent and much of the functionality of ozonation (ozone can act directly as an oxidant or serve as a feedstock reagent for hydroxyl radical formation). The kinetic rate constants of radical reactions are the highest of any of the chemical reactive zone mechanisms, with many of the reaction rates exceeding 10^{10} M^{-1} s^{-1}. Because radical reactions used in chemical reactive zones occur very quickly, concentrations of radicals in solution never reach high levels. The hydroxyl radical, for example, is expected to range from 1×10^{-13} to 1×10^{-11} M.[6]

3.3.1.2.1 Radical Chemistry

Radicals are atoms or compounds that have unpaired electrons, and are typically very reactive. Many chemical reaction mechanisms include formation and instantaneous reaction of radicals to form new compounds. To harness radical reactions in chemical reactive zones, it is necessary to generate useable radicals from a feedstock reagent *in situ*. The most sought-after of these highly reactive agents is the hydroxyl radical. Reflecting its high level of reactivity, the NDRL/NIST solution kinetics database[7] characterizes 1787 reactions in which the hydroxyl radical is a reactant, while it is a product of only 54 listed reactions.

Figure 3.6 shows the creation of a hydroxyl radical from hydrogen peroxide, through reaction with a ferrous iron cation (Fe^{2+}). The ferrous iron donates an electron to the hydrogen peroxide molecule, causing it to split into a hydroxyl anion (OH^-), and a hydroxyl radical ($OH^\bullet$). Notice that the oxygen atom in the hydroxyl anion has a total of eight valence electrons, including the pair shared in the covalent bond with a hydrogen atom. Hydrogen and oxygen in the hydroxyl anion, together, bring seven valence electrons. The eighth, donated by the ferrous iron atom, is an "extra," conferring a negative charge on the molecule. The hydroxyl anion is entirely soluble and acts as a moderately strong nucleophile ($n = 4.2$, as shown in Table 3.6). The fact that the electrons are all "paired" decreases their reactivity, relative to the unpaired electron of the hydroxyl radical.

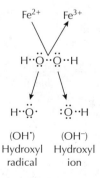

FIGURE 3.6 Creation of a hydroxyl radical from hydrogen peroxide, showing Lewis structures. Ferrous iron, Fe^{2+}, is the initiator, and is reduced to ferric iron, Fe^{3+}, by the donation of an electron to the hydrogen peroxide molecule.

The hydroxyl radical is electrically neutral, because there are seven valence electrons in the molecule — six from the oxygen and one from the hydrogen — so the protons and electrons are equal in number. Although the molecule is electrically neutral, the unpaired electron associated with the oxygen atom is highly reactive, readily capturing electrons from other atoms and molecules. The aggressive capture of electrons makes the hydroxyl radical a very strong oxidant. Many of its reactions are so fast that they are limited by the aqueous diffusion rates of its reactants (hence the term "diffusion-limited reaction rate"). There are many mechanisms of hydroxyl radical formation, but for *In situ* chemical reactive zones, the primary sources of hydroxyl radical are reactions with hydrogen peroxide in the Fenton's reagent reactions, and radical initiation reactions with the ozone molecule.

3.3.1.2.2 Chain Reactions

As a result of their high reactivity, radicals can enter into many reactions in any natural-water system. Hassan et al.[8] incorporated the kinetics of 72 reactions to effectively model the hydroxyl radical reaction network with ozone and hydrogen peroxide feedstock, in the presence of bromide, phosphate, and carbonate/bicarbonate ions.

Many of the potential reaction products are, themselves, radicals, and the formation of a radical chain reaction network is a feature of efficient use of radicals in chemical reactive zones. There are three basic processes occurring in a radical reaction network: initiation, promotion, and termination.

> *Initiation.* The first step in a radical chain reaction is the formation of a radical through reaction between an initiator and a feedstock reagent. A practical way to envision a radical-based chemical reactive zone is the injection of a compound that carries energy that can be used to drive reactions, but which is not sufficiently reactive to drive the process directly. Reaction with initiator cracks the feedstock reagent, forming a radical and thereby concentrating the available energy in the unpaired valence electron.
>
> *Promotion.* Many of the possible reactions of the first radical formed generate radicals, as products. The secondary radicals may react with target compounds in termination reactions, form other radicals that propagate the chain reaction, or serve as initiators. The secondary radicals might be more reactive with the target compound than the first radical formed.
>
> *Termination.* The final step in the radical chain reaction is termination of the radical. When radical chemistry is employed for oxidation of dissolved organic compounds in a chemical reactive zone, the radical propagation sequence generates a carbon radical. The carbon radical acts as an electron donor, reducing a susceptible molecule as the organic molecule bearing the carbon radical cracks and releases carbon dioxide. In an ideal radical chain reaction, the carbon radical reduces an oxidized initiator, recycling it to rejoin the radical chain reaction network.

Unproductive termination reactions also occur, in reactions that are often referred to as scavenging. Two of the prominent scavengers in chemical reactive zones are carbonate and bicarbonate ions.

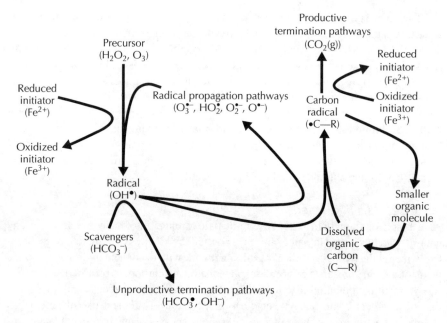

FIGURE 3.7 Radical chain reaction network (simplified), showing the network of reactions that develop when hydrogen peroxide or ozone are injected into an aqueous-phase solution containing dissolved organic carbon.

Figure 3.7 shows a conceptual radical chain reaction network, which is fed by either hydrogen peroxide or ozone as a feedstock reagent. Ferrous iron serves as an initiator, generating the first radical, a hydroxyl radical. The hydroxyl radical can then enter a promotion reaction, enter a target compound termination reaction sequence, or succumb unproductively to a scavenger reaction such as shown with the bicarbonate ion.

3.3.1.3 Carbonate Interference

Inorganic carbon in groundwater reacts readily with hydroxyl radical and other compounds in the radical chain reactions of ozonation, UV oxidation, and Fenton's reagent. For groundwater systems, the inorganic carbon pool is composed of three carbonates: carbonic acid (H_2CO_3), bicarbonate ion (HCO_3^-), and carbonate ion (CO_3^{2-}). To determine the extent of potential interference, it is first necessary to calculate the molarities of the two main interfering compounds, bicarbonate and carbonate. This is accomplished using alkalinity data, hydrogen ion concentration (from pH), and the acid–base dissociation constants for the two-step dissociation from carbonic acid to carbonate ion. After the carbonate species molarities have been calculated for a particular aquifer, the reaction rates between radical and target compound can be compared to the rates of radical chain reaction termination reactions that will be generated by dissolved carbonates. From that estimate, we can determine whether the expected losses of reagent to carbonate termination will be small, or alternatively, the level of pH adjustment that would be needed to achieve acceptable radical reaction rates with the target compounds.

Chemical Reactive Zones

3.3.1.3.1 Acid–Base Equilibrium

Aquifer carbonate concentrations are controlled by the two-step carbonic acid dissociation. The general form of the acid–base equilibrium is described by the following chemical equation,

$$HA \underset{}{\overset{K_A}{\rightleftarrows}} H^+ + A^- \tag{3.33}$$

where HA is the undissociated acid, H^+ is a proton, and A^- is the conjugate base. K_A is the kinetic rate constant for the acid dissociation, and is defined as follows:

$$K_A = \frac{[H^+] \cdot [A^-]}{[HA]} \tag{3.34}$$

In the special case when the acid is half-dissociated, the molarities of the undissociated acid and the conjugate base are equal, so the equation above reduces to

$$K_A = [H^+] \tag{3.35}$$

Taking the negative log of both sides,

$$pK_A = pH \tag{3.36}$$

This indicates that when the pH is at the same numeric value as the pK_A, molarities of the acid (HA) and base (A^-) are equal.

3.3.1.3.2 Two-Step Dissociations

Carbonic acid bears two hydrogen atoms (protons) that readily dissociate.

$$H_2CO_3 \underset{}{\overset{K_{A_1}}{\rightleftarrows}} H^+ + HCO_3^- \underset{}{\overset{K_{A_2}}{\rightleftarrows}} 2H^+ + CO_3^{2-} \tag{3.37}$$

In the generalized form,

$$H_2A \underset{}{\overset{K_{A_1}}{\rightleftarrows}} H^+ + A^- \underset{}{\overset{K_{A_2}}{\rightleftarrows}} 2H^+ + A^{2-} \tag{3.38}$$

The mole fraction of each constituent can be calculated by specifying a pH value, and solving the two kinetic equations simultaneously. The first step is to set the sum of mole fractions of the carbonate system constituents at 1.0:

$$H_2CO_3 + HCO_3 + CO_3 = 1.0 \tag{3.39}$$

The next step is to express H_2CO_3 and CO_3^{2-} in terms of the common ion, HCO_3^-, using the equations for the two dissociations. From the first dissociation,

$$k_{A_1} = \frac{[H^+] \cdot [HCO_3^-]}{[H_2CO_3]} \tag{3.40}$$

Rearranging,

$$[H_2CO_3] = \frac{[H^+] \cdot [HCO_3^-]}{k_{A_1}} \tag{3.41}$$

From the second dissociation,

$$k_{A_2} = \frac{[H^+]^2 \cdot [CO_3^{2-}]}{[H^+] \cdot [HCO_3^-]} = \frac{[H] \cdot [CO_3^{2-}]}{[HCO_3^-]} \tag{3.42}$$

Rearranging,

$$[CO_3^{2-}] = \frac{k_{A_2} \cdot [HCO_3^-]}{[H]} \tag{3.43}$$

Combining terms,

$$\frac{[H^+] \cdot [HCO_3^-]}{k_{A_1}} + [HCO_3^-] + \frac{k_{A_2} \cdot [HCO_3^-]}{[H^+]} = 1.0 \tag{3.44}$$

Solving for [HCO$_3^-$],

$$[HCO_3^-] = \frac{1}{\left(1 + \dfrac{[H^+]}{k_{A_1}} + \dfrac{k_{A_2}}{[H^+]}\right)} \tag{3.45}$$

Solving for [H$_2$CO$_3$],

$$[H_2CO_3] = \frac{[H^+]}{k_{A_1}} \cdot \frac{1}{\left(1 + \dfrac{[H^+]}{k_{A_1}} + \dfrac{k_{A_2}}{[H^+]}\right)} \tag{3.46}$$

Solving for CO$_3^{2-}$,

$$[CO_3^{2-}] = \frac{k_{A_2}}{[H^+]} \cdot \frac{1}{\left(1 + \dfrac{[H^+]}{k_{A_1}} + \dfrac{k_{A_2}}{[H^+]}\right)} \tag{3.47}$$

3.3.1.3.3 Dissociation Constants and Mole Fraction Diagrams
The two dissociation constants for carbonic acid are:

$$k_{A_1} = 4.467 \times 10^{-7} \quad pk_{A_1} = 6.33$$

$$k_{A_2} = 4.677 \times 10^{-11} \quad pk_{A_2} = 10.33$$

Chemical Reactive Zones

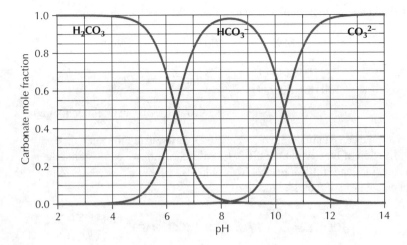

FIGURE 3.8 Mole fractions of three inorganic carbon species in aqueous phase, as a function of pH.

The simultaneous solution can be applied to calculate mole fractions for a range of pH values. The resulting graph is provided in Figure 3.8 for pH values ranging from 2 to 14. Notice that the mole fractions are 0.5 when the pH equals the values for the *pk* of each dissociation.

3.3.1.3.4 Calculation of Competitive Interaction

The loss of hydroxyl radical to carbonate/bicarbonate scavenging can be calculated as a function of target contaminant concentration at a specified aquifer pH and alkalinity. The calculations just shown are first used to determine the mole fraction of the bicarbonate and/or carbonate ion for the pH of interest. Then, the molarity is calculated by multiplication of the mole fraction and total carbonate alkalinity, expressed in molar concentration.

Consider a site at which hydroxyl radical oxidation is proposed for an *in situ* chemical reactive at a site that has a pH of 8.0 and a carbonate alkalinity of 300 mg L^{-1} as $CaCO_3$. The target compound is trichloroethene, and we would like to know whether the reaction can be run effectively without pH adjustment, using a chelating agent to maintain iron levels in solution. In this circumstance, competition from carbonate ions could be significant. The calculation strategy is as follows:

1. Estimate mole fraction of bicarbonate and use that value, combined with the carbonate alkalinity data, to calculate the bicarbonate molarity at the specified pH and alkalinity.
2. Develop a formula that expresses the reaction rate of hydroxyl radical with trichloroethene, divided by the rate of reaction between hydroxyl radical with bicarbonate ion. This gives a value for the fraction of productive, relative to total, reactions.
3. Prepare a graphic representation of the results.

First, calculate the bicarbonate molarity under the specified conditions.

$$MW_{CaCO_3} = 100 \ (mg \ mmol^{-1})$$

$$\frac{300 \ (mg \ L^{-1})}{100 \ (mg \ mmol^{-1})} = 3 \ (mmol \ L^{-1}) \ CaCO_3$$

$$= 3 \times 10^{-3} \ (M) \ CaCO_3 \ total \ alkalinity$$

Now, we can calculate the bicarbonate molarity, using the total alkalinity and the mole fraction of bicarbonate at pH 8. From Figure 3.8, we can see that the bicarbonate mole fraction is 0.95 at pH 8.0 (we could also use Equation (3.45) to make a more exact calculation of the mole fraction).

$$[HCO_3^-] = 3 \times 10^{-3} \ (M) \ CaCO_3 \times 0.95 \ \frac{mol \ HCO_3^-}{mol \ CaCO_3}$$

and

$$[HCO_3^-] = 2.85 \times 10^{-3} \ (M)$$

This is the value that will be used in the kinetic rate calculations that follow.

Next, we develop a formula that expresses the relative rates of reaction with hydroxyl radical. To do this, we work from the differential equation of the second-order rate constants for the two competing reactions.

$$TCE + OH^{\cdot} \rightarrow Products$$

and

$$HCO_3^- + OH^{\cdot} \rightarrow H_2O + CO_3^{\cdot -}$$

The first differential is based on the TCE oxidation reaction,

$$\frac{dTCE}{dt} = k_{TCE}[TCE][OH^{\cdot}] \quad \left(\frac{mol}{L \ s^{-1}}\right) \tag{3.48}$$

and the second differential is calculated for the bicarbonate anion reaction with hydroxyl radical, as follows.

$$\frac{dHCO_3^-}{dt} = k_{HCO_3}[HCO_3^-][OH^{\cdot}] \quad \left(\frac{mol}{L \ s^{-1}}\right) \tag{3.49}$$

To establish a comparison, we create a ratio of the two reaction rates.

$$\frac{\frac{dTCE}{dt}}{\frac{dHCO_3^-}{dt}} = \frac{k_{TCE}[TCE][OH^{\cdot}]}{k_{HCO_3}[HCO_3^-][OH^{\cdot}]} \tag{3.50}$$

Chemical Reactive Zones

Simplifying,

$$\frac{dTCE}{dHCO_3^-} = \frac{k_{TCE}[TCE]}{k_{HCO_3}[HCO_3^-]} \qquad (3.51)$$

The bicarbonate concentration is much larger than the TCE concentration for the concentrations of interest, so it can be assumed to be constant (this assumption is also supported by the fact that bicarbonate will be refreshed by conversion from carbonate soil minerals).

$$\frac{dTCE}{dHCO_3^-} = \frac{k_{TCE}}{k_{HCO_3}[HCO_3^-]} \times [TCE] \qquad (3.52)$$

The rate constants for hydroxyl radical reaction with trichloroethene and bicarbonate can be obtained from Table 3.4, or from Ross et al.[7]

$$k_{TCE} = 4 \times 10^9 \quad (M^{-1} s^{-1})$$
$$k_{HCO_3} = 8.5 \times 10^6 \quad (M^{-1} s^{-1})$$

Using the value of $[HCO_3^-]$ just inferred,

$$\frac{dTCE}{dHCO_3^-} = \frac{4 \times 10^9}{(8.5 \times 10^6)(2.85 \times 10^{-3})} \times [TCE] \qquad (3.53)$$

$$\frac{dTCE}{dHCO_3^-} = 1.65 \times 10^5 \times [TCE] \qquad (3.54)$$

The value of the function described by equation (3.54) can be graphed as a function of trichloroethene concentration values. Figure 3.9 provides results of the comparison for trichloroethene values ranging from 1 to 10,000 μg L^{-1}. At high trichloroethene levels, the ratio is approximately 10.5, indicating that more than 90% of the radical reactions occur with trichloroethene, fewer than 10% with bicarbonate. When the trichloroethene concentration declines to 1000 μg L^{-1}, only a little more than 10% of the reactions occur with trichloroethene, and when the trichloroethene concentration approaches the typical target value established for drinking water protection (5 μg L^{-1}), fewer than 1 reaction in 100 is occurring between trichloroethene and hydroxyl radical. From this exercise, we can conclude that bicarbonate ion is a significant competitor for trichloroethene in hydroxyl radical oxidation, and it may be necessary to lower aquifer pH to reduce trichloroethene levels to the 5 μg L^{-1} range.

3.3.2 Nucleophilic Substitution and Elimination Reactions

Most of the reactions used in chemical treatment strategies are very energetic, with reaction half-lives measured in minutes to hours. In this section, we introduce

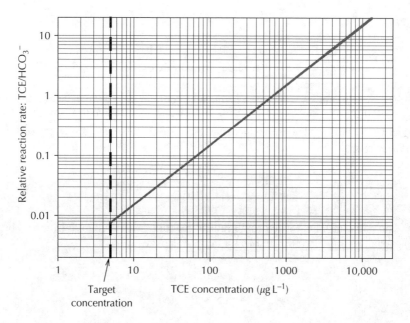

FIGURE 3.9 Relative reaction rates of TCE and bicarbonate ion (HCO_3^-) reactions with hydroxyl radical, as a function of pH. Total carbonate alkalinity is 100 mM (300 mg L^{-1} as $CaCO_3$).

nucleophilic substitutions and elimination reactions, which comprise a much quieter class of reactions, with half-lives measured in days to years in aquifers. With their low reaction rates, substitution and elimination reactions are most often associated with monitored natural attenuation processes — hydrolysis for 1,1,1-trichloroethane is an example. It may be possible to accelerate hydrolysis reactions through temperature increase or pH change. It is also possible to generate substitution reactions with other nucleophiles. That is the reason this relatively low-intensity reaction mechanism has become a subject of study for chemical reactive zone design.[9–11]

Hydrolysis is an example of a ubiquitous nucleophilic substitution reaction, in which a water molecule or hydroxyl ion is a reactant. The reaction is slow, but very sustainable. Hydrolysis is likely to be considered a mechanism supporting the natural attenuation of contamination, but most of the reactions are too slow to be deployed in a reactive zone strategy. However, recent work[9–11] shows that other nucleophilic reactions can be used effectively for treatment of some of the halogenated alkanes, and can be considered for deployment in chemical reactive zones. These studies centered on treatment of fumigants, including methyl bromide and 1,3-dichloropropene. In this section, we describe the nucleophilic substitution mechanism and provide a basis for estimating rates of proposed nucleophilic substitution reactions from known hydrolysis rates.

The substitution reaction we consider in greatest detail is termed a *concerted* reaction, because the substitution of one functional group for another occurs in a single action. The substituting group (the nucleophile) approaches the target carbon atom in the molecule from a position opposite the atom or group of atoms that will be displaced (the leaving group). In a single concerted reaction, electrons from the nucleophile begin migration into a bond with the target carbon atom, while electrons of the existing bond migrate to the leaving group. These reactions are also known by the acronym S_N2, denoting "substitution, nucleophilic, bimolecular." Other important modes of nucleophilic substitution are the S_N1 reaction, a unimolecular nucleophilic substitution, and the E2 reaction, a bimolecular elimination that generates an alkene. For further readings on S_N1, S_N2, and E2 reactions, see Schwarzenbach et al.[12]

Target sites for nucleophilic substitution reactions are electron-depleted due to a more electronegative bonding partner. Electrons in carbon atoms that are bonded to halogens or other strongly electronegative atoms are drawn away from the carbon, and this electron depletion endows the carbon atom with a slightly positive charge. An electron-depleted atom is a target for reaction with anions and molecules with electron-rich atoms that can be donated to a covalent bond. Because the substituting molecules are attracted to the positive charge centered in an atom's nucleus, they are termed nucleophiles. Stated more formally, a nucleophile is an electron-rich molecule or ion that can react with a locus of electron depletion in an organic molecule.*

Figure 3.10 shows reaction sequences for two nucleophilic substitution reactions with chloroethane: base-mediated elimination (E2) and neutral-pH hydrolysis (S_N2). In the base-mediated elimination reaction, the hydroxyl ion is the nucleophile, capturing a hydrogen nucleus (a proton, H^+) from the CH_3 end of the chloroethane molecule. The electron of the C–H bond migrates to form a second bond between the carbon atom, while the electrons of the C–Cl bond migrate to the chlorine atom, generating ethene and a chloride ion. In the neutral-pH hydrolysis, the electron-rich oxygen atom of the water molecule attacks the chlorine-bearing carbon atom from a position opposite the chlorine atom. An electron from one of the water's hydrogen bonds migrates to form an O–C bond, releasing the proton (H^+). At the same time, the electrons in the C–Cl bond migrate to the chlorine atom, releasing a chloride ion (Cl^-). The resulting products of the neutral-pH hydrolysis are ethanol, and chloride and hydrogen ions.

Activation energies for nucleophilic substitution reactions are typically high, so they are very temperature-sensitive. For hydrolysis of chlorinated ethanes and methanes, the activation energies (E_A) generally were estimated near 110 kJ mol^{-1}, with a minimum 92.4 kJ mol^{-1} for 1,1,2,2-tetrachloroethane and a maximum value of 122.9 kJ mol^{-1} for chloroform (based on data presented by Washington).[13] As an example, the neutral pH hydrolysis half-life for 1,1,1-trichloroethane increased from

---

*Electrons shared by atoms of different elements in covalent bonds are not evenly distributed between the atomic nuclei. The elements vary in their affinity for electrons, and the density distribution of the electron bonding cloud is tipped toward the more electronegative atom. Carbon atoms lie in the middle of the range between electronegative and electropositive atoms. The halogens, such as chloride and fluoride, are strongly electronegative elements. When these elements form a bond with other elements, the bonding electrons are drawn strongly toward the halogen nucleus. The alkalis, such as hydrogen and sodium, are strongly electropositive. Atoms of these elements readily release their bonding electrons, and a positive charge accumulates at these atoms.

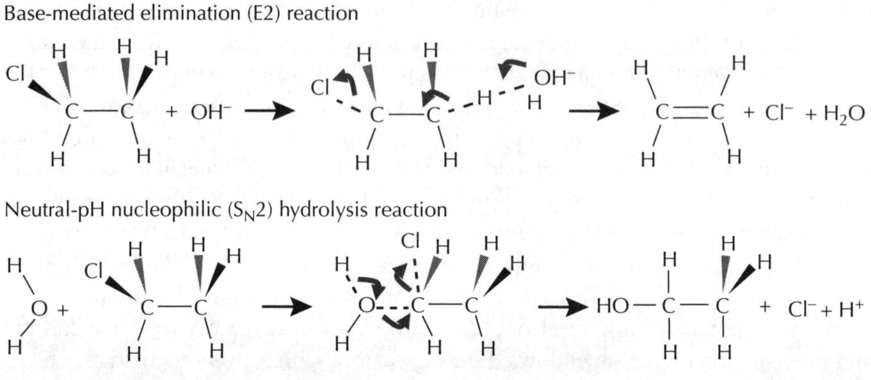

Base-mediated elimination (E2) reaction

[Reaction scheme showing chloroethane + OH⁻ → transition state → ethene + Cl⁻ + H₂O]

Neutral-pH nucleophilic (S_N2) hydrolysis reaction

[Reaction scheme showing H₂O + chloroethane → transition state → HO-CH₂-CH₃ + Cl⁻ + H⁺]

FIGURE 3.10 Potential reaction pathways for base-mediated elimination and nucleophilic substitution (water hydrolysis) of chloroethane. Curved arrows show electron migrations.

0.86 to 10.3 years, as the reaction temperature decreased from 25 to 10°C. The hydrolysis reaction activation energy for 1,1,1-trichloroethane is 119.4 kJ mol^{-1}.[13] Section 3.1.5 provides a discussion of temperature effects on reaction rates, along with methods for kinetic rate calculation.

There are three hydrolysis motifs that can occur: acid mediated, neutral pH, and base mediated. The observed hydrolysis rate for any compound is the sum of the rates of the three motifs.

$$k_{obs} = K_A + K_N + K_B \quad (3.55)$$

Washington (1995) summarized rate constant and Arrhenius parameters for the neutral-pH and base-mediated reaction mechanisms, pointing out that natural aquifers are normally buffered into pH ranges that do not support acid-mediated hydrolysis. Table 3.5 presents data excerpted and converted from Washington.[13]

3.3.2.1 Nucleophilicity

Nucleophilic substitution reactions can be driven by compounds other than water and hydroxyl anion. Table 3.6 provides a listing of common aqueous-phase nucleophiles. The strength of each compound's reactivity in nucleophilic substitution reactions, or nucleophilicity, is also provided in Table 3.6. The nucleophilicity values given are for each nucleophile's reaction with methyl bromide, scaled relative to the neutral-pH hydrolysis reaction for methyl bromide. Schwarzenbach et al.[12] provide Equation (3.56) developed by Swain and Scott (published in 1953) for the extrapolation of nucleophilic substitution reaction kinetics for methyl bromide, from its neutral-pH hydrolysis rate constant, k_{H_2O}.

$$\log\left(\frac{k_{nuc}}{k_{H_2O}}\right) = s \cdot n_{nuc,CH_3Br} \quad (3.56)$$

TABLE 3.5
Kinetic Rate Constants and Arrhenius Parameters for Base-Mediated (K_B) and Neutral-pH (K_N) Hydrolysis

Target compound	Product	Neutral-pH Reaction		Base-Mediated Action			Half-life
		A min^{-1}	E_A kJ mol^{-1}	A L mol^{-1}min^{-1}	E_A kJ mol^{-1}	k_{obs} min^{-1}	years
Carbon tetrachloride	$CO_2 + H_2O$	4.07×10^{12}	114.5			3.46×10^{-8}	38.1
Chloroethane	Ethene + HCl Acetaldehyde + HCl	1.37×10^{13}	110.8	3.72×10^{13}	101.2	5.18×10^{-7}	2.54
1,1-Dichloroethane	Chloroethene + HCl Acetaldehyde + HCl	3.54×10^{11}	109.5	9.95×10^{13}	114.8	2.26×10^{-8}	58.2
1,2-Dichloroethane	Chloroethene + HCl 2-Chloroethanol	2.83×10^{10}	103.7	1.75×10^{13}	97.4	1.88×10^{-8}	70.0
1,2-Dichloropropane	1-Chloropropene + HCl 2-Chloropropanol	1.6×10^{12}	110.1	1.4×10^{14}	110.1	7.93×10^{-8}	16.6
1,1,1-Trichloroethane	Acetic acid 1,1-Dichloroethene	4.06×10^{14} 3.7×10^{14}	117.2 119.4			1.6×10^{-6}	0.826

Note: All values are for reactions at 25°C and pH = 7.0.

Source: Excerpted from Washington, 1995.

TABLE 3.6
Nucleophilicity of Selected Ions, n_{nuc,CH_3Br}, Based on Reaction with Methyl Bromide

Nucleophile, nuc Compound Name	Chemical Formula	n_{nuc, CH_3Br}
Water	H_2O	0.0
Nitrate ion	NO_3^-	1.0
Fluoride ion	F^-	2.0
Sulfate ion	SO_4^{2-}	2.5
Acetate ion	CH_3COO^-	2.7
Chloride ion	Cl^-	3.0
Bicarbonate ion	HCO_3^-	3.8
Phosphate ion	HPO_4^{2-}	3.8
Bromide ion	Br^-	3.9
Hydroxyl ion	OH^-	4.2
Iodide ion	I^-	5.0
Cyanide ion	CN^-	5.1
Bisulfite ion	HS^-	5.1
Thiosulfate ion	$S_2O_3^{2-}$	6.1

Note: Values for n_{nuc,CH_3Br} can be used in Equation (3.57) to estimate kinetic rate constants for nucleophilic substitution reactions.

Source: Adapted from Schwarzenbach et al., 2003.

Simplifying,

$$\frac{k_{nuc}}{k_{H_2O}} = 10^{s \cdot n_{nuc,CH_3Br}} \quad (3.57)$$

Using Equation (3.57) and Table 3.5, the second-order kinetic rate constant for reaction of methyl bromide with other nucleophiles can also be estimated. For example, the nucleophilicity of the hydroxyl anion is 4.2. That value can be substituted into Equation (3.57) to determine the second-order reaction rate constant for base-mediated hydrolysis of methyl bromide.

$$k_{OH,CH_3Br} = k_{H_2O,CH_3Br} \cdot 10^{n_{OH,CH_3Br}} \quad (3.58)$$

Schwarzenbach et al. (2003) indicate that the value for $k_{H_2OCH_3Br}$ is approximately 5×10^{-9} M^{-1} s^{-1} at 298 K (25°C). That value can be used to project a value for k_{obs}, the combined rate constant for k_N and k_B. Using Equations (3.55) and (3.58), we can estimate k_{obs} as a function of hydroxyl anion concentration (note that k_{obs} is typically shown as a pseudo-first-order rate constant).

$$k_{obs} = (k_{H_2O} \cdot [H_2O]) + (k_{H_2O} \cdot 10^{4.2} \cdot [OH^-]) \quad (s^{-1}) \quad (3.59)$$

Solving,

$$k_{obs} = 2.8 \times 10^{-7} + 2.8 \times 10^{-7} \cdot 10^{4.2} \cdot 10^{-7}$$
$$k_{obs} = 2.8 \times 10^{-7} + 4.4 \times 10^{-10}$$
$$k_{obs} = 2.8 \times 10^{-7} \quad (s^{-1})$$

From experimental studies reported by Wang et al.,[11] the rate constant for overall hydrolysis, k_{obs}, was reported as 9.19×10^{-4} h^{-1} (2.6×10^{-7} s^{-1}) at 20°C. This indicates good agreement between theoretical and experimental observations for methyl bromide. Larger differences are likely to arise for nucleophilic reactions with target compounds other than methyl bromide, however. Some of these differences will be related to the susceptibility factor in Equation (3.56), and in other cases, unrecognized nucleophiles may be present in reaction solutions (e.g., phosphate buffers have nucleophilic potential). Pagan et al.[14] noted significant differences among reported hydrolysis rate constants for chlorinated ethanes and propanes.

There are many potential nucleophilic reaction pathways and a large number of potential nucleophiles that could be used to support destruction of aqueous-phase contaminants. Most of the reactions are very slow at aquifer temperatures and would only be considered as part of a natural attenuation calculation. Candidate reaction systems should be tested at the bench scale, with great care taken to run the tests at expected aquifer temperatures, due to the temperature sensitivity of the nucleophilic reactions.

3.3.3 Precipitation Reactions

The objective of chemical reactive zone precipitation reactions is to create insoluble compounds directly from the reaction of injected reagents with target compounds, or indirectly through modification of the physical/chemical habitat, reducing the solubility of the target compound in the aquifer. In an ideal precipitation reaction, the solid that forms is effectively insoluble in water and the target compound will be undetectable in aqueous phase. This ideal outcome sometimes cannot be achieved, but compliance criteria can still be met. There are two modes of successful precipitation:

Thermodynamic control. Durable precipitates are formed, for which the aqueous-phase equilibrium concentration, under foreseeable aquifer physical–chemical conditions (e.g., temperature, pH, E_H, ionic strength), is below the compliance criterion. This is the ideal result for precipitation reactions — compliance will be achieved, regardless of groundwater velocity.

Dissolution control. Limited-solubility precipitates are formed, for which the aqueous-phase equilibrium concentration is greater than the compliance criterion, but the rate of dissolution is slow, and groundwater flux is sufficient to maintain target compound concentrations below the compliance criterion.

Encapsulation is a special case of precipitation in which the target compound is captured in a coprecipitation reaction that would not meet either the thermodynamic or limited-dissolution definitions of successful precipitation. Coprecipitation reagents are added to generate non-target precipitation reactions at rates that far exceed the target precipitation reaction. This forms a combined precipitate mass in which the target element is only a minor component. When the combined precipitate dissolves, the aqueous-phase concentration of the target element remains at an acceptable level.

Application of precipitation reactions to contaminated aquifers is challenging because groundwater flux continuously refreshes the aqueous phase. Precipitation that may be achieved through pH adjustment or common ion effect, for example, might be reversed when groundwater flow modifies the aqueous-phase chemistry to favor dissolution of the precipitate. In the development of a precipitation treatment strategy, it is essential to identify processes that generate durable precipitates.

The following sections describe the solubility mechanisms that are employed in reactive zone precipitation strategies. These mechanisms are combined with chemical reduction strategies, described earlier, which create the physical/chemical habitat favoring formation of durable precipitates.

3.3.3.1 Solubility in Aqueous Phase

Solubility is a critical variable in all reactive zone systems, whether biological or chemical. Among the important solubility effects are:

- Solubility determines the level of exposure of hydrophobic organic species to aqueous-phase reactions. This controls the effectiveness of any method intended to control or destroy hydrophobic contaminants *in situ*. Remedial system designers typically seek to enhance the solubility of hydrophobic organics, making them available to aqueous-phase chemical or biological reactions.
- Precipitation reactions rely on low-solubility products to remove target compounds from aqueous phase. One of the important strategies applied in both chemical and biological *in situ* reactive zones is the formation of insoluble metal precipitates.
- Changes can be induced in the aqueous-phase chemical environment during remedial actions that solubilize natural soil minerals and contaminant precipitates in unintended side reactions. This can lead to the appearance of new contaminants in a reactive zone treatment.

3.3.3.2 The Solubility Product

Precipitation reactions of ionic solids (salts) are reversible, and the equilibrium concentrations of dissolved ions that form the salt can be calculated from published observations.

In the simplest cases, the dissolution of ionic solids is described by the following chemical equation.

$$A_mB_n(s) \underset{\text{precipitation}}{\overset{\text{dissolution}}{\rightleftarrows}} mA(aq) + nB(aq) \tag{3.60}$$

Chemical Reactive Zones

The equilibrium constant for the reaction is calculated from chemical activities of reactants and products, as for other chemical reactions. Because the activity coefficient of the solid-phase reactant, AB, equals 1, it does not contribute to the calculation of the equilibrium constant. Hence, the equilibrium constant is referred to as the solubility product, K_{sp}.

$$K_{sp} = [A]^m \cdot [B]^n \tag{3.61}$$

The dissolved-phase concentration of an ion due to dissolution of solid-phase can be calculated as follows:

$$[A] = \sqrt[m]{\frac{[B]^n}{K_{sp}}} \tag{3.62}$$

For the simple salt AB, the dissolved-phase equilibrium concentration calculation reduces to:

$$[A] = [B] = \sqrt{K_{sp}} \tag{3.63}$$

3.3.3.3 Calculation of Nickel Concentrations from K_{sp}

Nickel sulfide (NiS), and other metal sulfides such as cadmium or lead sulfide are relatively insoluble in the aqueous phase, when considered in isolation. The solubility product for amorphous nickel sulfide dissociation (α-NiS) is 3×10^{-21} and the dissociation equation is written as:

$$NiS_{(s)} \rightleftarrows Ni^{2+} + S^{2-} \quad K_{sp} = 3 \times 10^{-21} \tag{3.64}$$

The dissolved nickel, Ni^{2+}, can be isolated to show its predicted concentration as a function of sulfide ion (S^{2-}) and the K_{sp}.

$$Ni^{2+} = \frac{K_{sp}}{S^{2-}} \tag{3.65}$$

From examination of Equation (3.65), we can see that if a sulfide scavenger were present, the predicted dissolved nickel concentration becomes very high. Such a scavenger exists in the aqueous phase: the sulfide ion (S^{2-}) is a strong base that participates in a two-step acid–base equilibrium, drawing nickel from solid-phase nickel sulfide into solution by suppressing the sulfide ion concentration. Salts of strong bases, such as the sulfide anion, gain a boost in solubility due to the rapid hydrolysis of the base anion when it enters the aqueous solution.

The two-step hydrogen sulfide dissociation is as follows (K_a values for 18°C from Weast):[15]

$$H_2S \rightleftarrows H^+ + HS^- \quad K_{a_1} = 1 \times 10^{-7} \tag{3.66}$$

$$HS^- \rightleftarrows H^+ + S^{2-} \quad K_{a_2} = 1 \times 10^{-19} \tag{3.67}$$

These equilibria can be rewritten as base reactions, taking the inverse of the dissociation constants. We can also add the nickel sulfide dissolution reaction, to develop a net reaction for the nickel dissolution and sulfide scavenging process that enhances nickel solubility, particularly at low pH:

$$H^+ + HS^- \rightleftarrows H_2S \qquad K_{b_1} = 1 \times 10^7$$

$$H^+ + S^{2-} \rightleftarrows HS^- \qquad K_{b_2} = 1 \times 10^{19}$$

$$NiS_{(s)} \rightleftarrows Ni^{2+} + S^{2-} \qquad K_{sp} = 3 \times 10^{-21}$$

$$NiS_{(s)} + 2H^+ \rightleftarrows H_2S + Ni^{2+} \qquad K_{net} = K_{sp} \times K_{b_1} \times K_{b_2} = 3 \times 10^5 \quad (3.68)$$

Now, the concentration of nickel in solution, Ni^{2+}, can be calculated from the aqueous hydrogen sulfide concentration and pH:

$$K_{net} = 3 \times 10^5 = \frac{[H_2S] \cdot [Ni^{2+}]}{[H^+]^2} \qquad (3.69)$$

and

$$Ni^{2+} = 3 \times 10^{-5} \cdot \frac{[H^+]^2}{[H_2S]} \qquad (3.70)$$

The concentration of soluble nickel can be plotted as a function of the pH and hydrogen sulfide concentration. Sulfide concentrations are elevated above background levels during biological reducing zone operations, which can push dissolved nickel out of solution. Figure 3.11 shows predicted dissolved nickel concentrations as a function of pH, for sulfide concentrations of 0.1 to 0.01 M, to show the effects of each on nickel solubility. This shows that nickel precipitation can be promoted by the addition of sulfide ion and maintenance of near-neutral pH.

The long-term solubility of nickel, however, will be based on its solubility at "background" sulfide and pH levels. To develop a conservative estimate of nickel solubility, we must assume that nickel sulfide is the only source of sulfide in the aquifer. In that case, Equation (3.69) reduces to the following:

$$K_{net} = 3 \times 10^5 = \frac{[Ni^{2+}]^2}{[H^+]^2} \qquad (3.71)$$

and

$$[Ni^{2+}] = \sqrt{3 \times 10^5 \cdot [H^+]^2} \qquad (3.72)$$

In groundwater, other precipitation reactions are likely to intervene before nickel reaches the highest dissolved-phase concentrations shown by these calculations. The fact that many compounds interact in the dissolution and precipitation of metals may explain the wide range of solubility values available for any reaction. In groundwater,

Chemical Reactive Zones

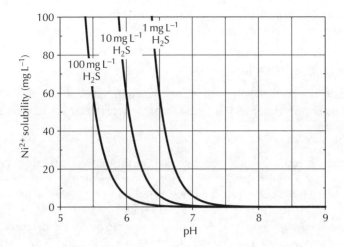

FIGURE 3.11 Solubility of nickel sulfide as a function of pH, at three aqueous-phase total sulfide concentrations.

where there are many interactions to account for, we can be confident of the direction in which we are pushing reactions (e.g., greater or lesser dissolved-phase concentrations), but we need to be cautious about predicting specific concentration outcomes without site-specific testing.

The solubility product, alone, does not adequately describe all of the precipitation processes that occur in aquifers. For many aquifer minerals, including sulfides, hydroxides, and carbonates, the hydrogen ion concentration (pH) and the reduction–oxidation potential are critical to predictions of solubilities. Transition metal precipitation strategies are discussed further in Section 5.4.2.1. Stumm and Morgan[5] provide an extended discussion of mineral solubilities.

3.3.3.4 Aging of Precipitated Solids

Precipitation reactions often do not lead to immediate formation of the most stable form of the precipitated compound. The first solids that form may be more chemically active than the stable precipitates that form over very long periods, and have higher solubilities. At least two factors control the rate and sequence of precipitation reactions.

Amorphous (finely divided) crystalline masses. The very high surface area-to-mass ratios of the very fine crystal masses that form in precipitation reactions are more available to aqueous-phase reactions than the larger, mature crystals that form over long time spans. It may be necessary to sustain the chemical conditions that generated the initial precipitate for an extended period, to allow the formation of a stable crystal structure. Failing that, rebound of the target compound may occur.

Reaction kinetic effects. Stability calculations and the resulting selection of precipitation strategies are based on analysis of the chemical thermodynamics of the target compounds and the aquifer. In many cases, the most stable precipitate is not the one that forms most quickly. Reaction kinetics dictate the sequence of product formation and control the rate at which conversions occur.

3.3.4 HETEROGENEOUS REACTIONS

Heterogeneous reactions involve two phases, the liquid and solid phases in the case of reactive zone chemistries. Solid surfaces can participate in reactive zone chemistry in several ways:

> *Catalysis.* Electrical charge concentrations on the mineral surfaces can attract ions or polar molecules to the surface. There, reaction geometries can be optimized for aqueous-phase molecules reacting with atoms on the mineral surfaces or with other aqueous-phase molecules.
> *Regenerable reactive centers.* Mineral surfaces can be activated, amended, or reactivated through injection of reagents such as dithionite ($S_2O_2^{2-}$) or iron sulfide (FeS). This is the basis for the creation of *in situ* reactive zones from iron minerals in aquifer soils.
> *Adsorption sites.* Aquifer minerals also can provide adsorption sites for contaminants such as the chlorinated alkenes, and for reactive metal ions such as ferrous iron (Fe^{2+}) and arsenic.

The reduction reactions mediated by reduced iron compounds are likely heterogeneous processes. Amonette et al.,[16] for example, studied the abiotic reductive dechlorination of carbon tetrachloride (CCl_4). They found that the mineral goethite (α-FeOOH) adsorbed Fe^{2+} from aqueous solution, and promoted dechlorination of carbon tetrachloride, while Fe^{2+} in the aqueous phase was unreactive. They postulated that the Fe^{2+} was adsorbed to the hydroxy groups of the goethite and that the charge concentration supported the multiple-electron-transfer reactions of reductive dechlorination.

3.4 CHEMICAL OXIDATION ZONES

Chemical oxidation zones employ very energetic reaction mechanisms in the destruction of organic contaminants. Costs for chemical oxidation zones are often much higher than for alternative remedies. Consequently, the choice to proceed with a chemical oxidation zone, rather than a biological or natural attenuation system, is most often driven by a requirement to achieve significant contaminant mass reduction in a short time interval.

Among the *in situ* oxidation alternatives, Fenton's reagent and permanganate remedies have generated the greatest interest, with only a few examples of ozone- and persulfate-based systems.

3.4.1 FENTON'S REAGENT

The Fenton's reaction is an advanced oxidation strategy, first described by H.J.H. Fenton in 1895. There are two reaction mechanisms at the core of the Fenton's process: radical initiation and ferrous iron regeneration. In the initiation reaction, ferrous iron (Fe^{2+}) reacts with hydrogen peroxide, as shown in Equation (3.73) and Figure 3.6:

$$Fe^{2+} + H_2O_2 \rightarrow Fe^{3+} + OH^\bullet + OH^- \qquad (3.73)$$

In ideal Fenton's chemistry, the ferric iron (Fe^{3+}) produced in the initiation reaction reacts with additional hydrogen peroxide to regenerate ferrous iron. The ferrous iron regeneration includes a series of steps, shown in Equations (3.74) through (3.76).[17]

$$Fe^{3+} + H_2O_2 \rightarrow FeOOH^{2+} + H^+ \qquad (3.74)$$

$$FeOOH^{2+} \rightarrow HOO^\bullet + Fe^{2+} \qquad (3.75)$$

$$Fe^{3+} + HOO^\bullet \rightarrow Fe^{2+} + O_2 + 2H^+ \qquad (3.76)$$

The net of the regeneration reactions is:

$$2Fe^{3+} + H_2O_2 \rightarrow 2Fe^{2+} + O_2 + 2H^+ \qquad (3.77)$$

The back-reaction of ferric iron (Fe^{3+}) consumes one hydrogen peroxide molecule for every two initiation reactions. In this idealized stoichiometry, one-third of the hydrogen peroxide is consumed in regeneration of ferrous iron, and 1 mol of oxygen is produced for every 3 mol of hydrogen peroxide injected into a formation. In real systems, a number of competing reactions may redirect the perhydroxyl radical ($HOO^\bullet$) from ferrous iron regeneration (Equation (3.76)), and ferric iron rapidly precipitates from solution if the pH rises above 3.0. These alternative reactions limit the efficiency of the Fenton's process in aquifers, and is discussed more completely in the following.

Ferric iron and other soil minerals may also catalyze a hydrogen peroxide disproportionation reaction that dramatically increases the oxygen yield of injected hydrogen peroxide.

$$2H_2O_2 \xrightarrow{\text{Metal}} 2H_2O + O_2 \qquad (3.78)$$

When the Fenton process is used in conventional industrial applications, the reaction mixture contains the target compound to be oxidized and the ferrous iron that will be required to quickly react with hydrogen peroxide. The reaction mixture is adjusted to a pH between 3 and 5. Hydrogen peroxide is then injected into the reaction mixture at a rate that is controlled to minimize loss of hydrogen peroxide to disproportionation.

In a chemical reactive zone application, it is not possible to maintain the reaction conditions within the narrow limits defined for industrial applications of the Fenton's process. There are several critical challenges for the field application of Fenton's reagent:

pH adjustment limitations. Aquifer minerals at most sites resist significant adjustments to pH. Injection of acids to push pH into the preferred range (between 3 and 5) may generate large quantities of gas (carbon dioxide) and heat. Moreover, pH levels typically rebound if acid loading is too small.

Large-scale gas production. Even under ideal stoichiometry, the Fenton's process generates large quantities of gas that displace water in the porous medium. At an injection rate of only 1 gal min^{-1} of 10% peroxide, more than 11 mol peroxide are being injected each minute. If one-third of the injected peroxide is converted to oxygen (ideal stoichiometry), oxygen is being generated at a rate of 83 L min^{-1} (nearly 3 cfm) at standard temperature and pressure. Under ideal stoichiometries, even this small liquid injection rate generates off-gas that is comparable to an aquifer sparge well. In the event that disproportionation reactions are significant, off-gas rates can climb even higher. The gas produced has at least three consequences:

Pore water displacement. The displacement of pore water by gas significantly reduces the hydraulic conductivity of porous media. Reagent injection backpressures can increase to very high levels as the aquifer formation becomes "gas-locked."

Oxygen gas buildup. Most of the gas generated in Fenton's reactions is oxygen, which pushes its way into the vadose zone overlying the treated volume. Oxygen partial pressures can build to very high levels in the vadose zone, creating significant safety hazards. Vadose zone soils should be monitored during Fenton's applications, and vented as needed to maintain safe conditions.

Volatile organic carbon stripping. The elevated temperatures associated with Fenton oxidation, combined with the sparge-like injection of gas into the treated aquifer, combine to cause significant mass transfer of volatiles from the aquifer into the overlying vadose zone. These fugitive vapors may constitute a significant contaminant removal mechanism as well as a safety hazard. As with the oxygen vapors, volatile organic carbon vapors that accumulate in the vadose zone should be monitored and vented, as needed, to maintain safe conditions.

Short-duration reactions. The speed of reagent consumption in the Fenton's process may cause reagents to be expended before the nonaqueous contaminant mass is desorbed or dissolved into solution. Contaminant rebound will occur if Fenton's reagents are exhausted before nonaqueous-phase target mass enters the aqueous phase where hydroxyl radical reactions occur.

3.4.1.1 Hydroxyl Radical Production Strategies

It is not possible to deploy "classic" Fenton's chemistry to *in situ* reactive zones — the presence of a porous medium constrains mixing and the aquifer matrix may, itself, participate in many of the oxidation or reduction reactions that accompany the process. One of the most important deviations from classic Fenton's chemistry is the natural aquifer's tendency to buffer pH through carbonate and other alkalinity mechanisms. This introduces competition for hydroxyl radical and foils the iron regeneration reactions outlined in Equations (3.74) through (3.76).

Chemical Reactive Zones

Three hydroxyl radical production approaches have been suggested for use in aquifers, and all are attempting to generate Fenton's or Fenton's-like reactions.

Conventional. The conventional Fenton's process relies on soluble ferrous iron as the radical initiator. This is normally achieved by preplacing iron in the reaction mixture and adjusting the pH to near 3. In a reactive zone, aquifer minerals may buffer pH into a range too high to sustain iron recycling reactions, so the conventional Fenton's strategy may only be operable in short episodes of acidification (with iron) followed immediately by peroxide injection. The peroxide volume injected in any pulse is limited to avoid waste of reagent. The acidification–oxidation cycle is repeated as needed to achieve remedial objectives. It is also possible to simply accept the ferric iron loss rate, and supply ferrous iron to match its consumption by initiation reactions. This strategy increases injection volume significantly and may reduce hydraulic conductivity through iron precipitation.

Iron chelation. The main limitation on radical production strategies is loss of iron from solution when pH exceeds 3. The ferrous iron that is injected to serve as radical initiator is sufficiently soluble, but the initiation reaction converts the ferrous iron to ferric iron, which has a very low solubility above pH 3. Effective Fenton's reactions require recycling of ferrous iron (Equations (3.74) through (3.76)), and this cannot be sustained above pH 3. Chelation of injected iron has been recommended[18] as a strategy to avoid loss of iron. Chelation solves the iron retention problem, and allows operation of the process at pH above 3. Competition for hydroxyl radical scavengers is another pH-related limitation on Fenton's efficiency, and it cannot be addressed by chelation.

Mineral iron strategies. Many aquifers contain iron minerals that can catalyze radical formation from hydrogen peroxide. In these cases, it has been suggested that the Fenton's reaction can be developed with injection of hydrogen peroxide, alone, relying on aquifer iron minerals in lieu of ferrous iron injections or pH adjustment.

Each of these radical production strategies is examined in more detail, in the following.

3.4.1.2 Conventional Fenton's Process

The conventional approach to the Fenton's oxidation process is to acidify groundwater to low pH (3 to 5) to maintain solubility of Fe^{3+}, supporting the iron regeneration reactions shown in Equations (3.74) through (3.76). This approach requires significant acid addition and, in high-carbonate aquifers, neutralization reactions may limit the duration of pH suppression. Bench-scale treatability testing can be performed to determine the neutralization capacities of aquifer soils. Repeated acidification suppresses the neutralization capacity of aquifer soils, likely due to occlusion of the carbonate mineral surface by reaction products.

When conventional, acidified Fenton's reagent is applied, a pre-oxidation injection with acid only can be used to gauge the carbon dioxide off-gas production due to carbonate neutralization. Carbon dioxide off-gas due to carbonate–acid reactions can add to carbon dioxide concentrations observed in off-gas monitoring during the oxidation process. Pretreatment with acid only may also suppress the carbonate reactivity, extending the duration of iron availability during the subsequent acidified iron/hydrogen peroxide injections.

3.4.1.3 Iron Chelation Strategies

Ferrous iron or a comparable reduced metal species is critical to initiation of the hydroxyl radical chain reaction. The initiation reaction oxidizes the ferrous ion (Fe^{2+}), and the resulting ferric ion (Fe^{3+}) rapidly precipitates from solution when the pH is greater than 3 (Wang and Brusseau).[17] It is possible to use chelating agents that maintain dissolved-phase iron concentrations for extended periods of high-redox potential and high pH.

Sun and Pignatello[18] tested numerous inorganic and organic chelators for effectiveness in supporting Fenton's oxidation of 2,4-dichlorophenoxyacetic acid (2,4-D). Their studies showed that nitrilotriacetic acid (NTA) was a highly effective chelator and several other organic chelators were highly effective, as well, including gallic and picolinic acids. In field applications, the acidified ferrous iron solution commonly applied as a component of Fenton's reagent would be replaced with a chelated iron solution. The chelator maintains soluble iron levels at high pH, allowing the ferrous iron regeneration sequence to proceed without significant precipitation losses of ferric iron.

Sun and Pignatello[18] also tested numerous phosphate compounds for their ability to support Fenton's oxidation of 2,4-D at pH 6.0, but none were rated as effective. Their phosphate findings contrast sharply with the later observations of Wang and Brusseau,[17] who showed that chelation of iron by the pyrophosphate ion (added to a Fenton's reaction mixture as sodium pyrophosphate decahydrate) generated significant increases in both the rate and extent of perchloroethene dechlorination, relative to addition of unchelated ferrous iron solution. The extent of the increase in dechlorination was related to the dose of chelator. In Borden soil* the initial dechlorination rate increased more than 40-fold in the presence of 30 mM pyrophosphate. The extent of reaction increased more than 60-fold with the pyrophosphate chelator.

Although the pyrophosphate chelation was highly effective in sustaining solution iron levels, organic chelators generated much greater acceleration of reaction rates over the first 24 to 48 h of reaction time. Picolinic and gallic acid solutions both generated immediate reactions that reached their full extent in less than 12 h. The oxidation reaction using pyrophosphate reached the extent of the picolinic-chelated treatment after 24 h and the gallic acid treatment after 48 h. It was presumed that the organic acids were oxidized by the Fenton's reagent system, limiting the extent of reaction. However, given the many variables that affect oxidant performance in an aquifer, it is often not possible to sustain energetic reactions for extended periods, and the inorganic chelators may not have as great an advantage in a field setting.

---

*Borden soil is a sandy loam commonly used in environmental soil chemistry studies. The Borden soil for the Wang and Brusseau study was collected at the Canadian Forces base at Borden, Ontario. It contained 0.29% organic carbon ($f_{oc} = 0.0029$), and was composed of 98% sand, 1% silt, and 1% clay, with a pH of 8.3.

3.4.1.4 Mineral Iron Strategies

A limited amount of heterogeneous initiation may occur through reaction with soil minerals, although this process is much slower than initiation by dissolved, reduced metal species.[17] Several other studies have also suggested that aquifer minerals can catalyze hydroxyl radical initiation. For example, Teel et al.[19] showed that the iron oxy-hydroxide mineral goethite (α-FeOOH) catalyzed initiation of hydroxyl radical formation from hydrogen peroxide. However, pH adjustment (to 3.0) was required to achieve complete dechlorination of TCE; very limited dechlorination was observed at pH 7. Pilot-scale testing would be required if mineral iron radical initiation is considered for a Fenton's reagent application. The following example describes a field trial for peroxide-only treatment at a site where moderate dissolved-phase iron levels (20 mg L^{-1}) indicated that Fenton's reactions might be sustained by iron minerals in the aquifer soils.

3.4.4.1.1 Field Trial for Peroxide-only
A field test was performed to determine whether hydroxyl radical formation could be sustained without acidification and without chelation. The selected aquifer segment had a pretreatment aquifer pH of 6.1 and dissolved iron concentration of 0.36 mM (20 mg L^{-1}). The pretreatment redox potentials were generally reducing, with values of -270 mV in the area of highest dissolved iron concentration. A 1150 gal batch of 5% hydrogen peroxide was injected steadily over a 30 h period, at a location 10 ft upgradient from the observation point with the highest pretreatment iron concentrations. The observation well was located outside the radius that could be reached by the liquid injection, however, gases generated by the reactions reached the observation location within 5 h of the start of injections. Dissolved oxygen concentrations immediately increased to three times atmospheric saturation from a pretreatment concentration of <1 mg L^{-1}, and the redox potential increased from -272 to $+135$ mV in the same period.

Losses of dissolved iron were observed beginning 24 h after the start of injections, a period during which redox potential increased, but only a trace of hydrogen peroxide was observed. On the eighth day following injection, hydrogen peroxide measurements climbed to more than 30 mM (1000 mg L^{-1}), and dissolved iron levels fell to less than 10^{-3} mM, as groundwater directly impacted by the injected reagents reached the observation well. Figure 3.12 shows the loss of soluble iron that was associated with the arrival of hydrogen peroxide residual from the injection well. Unreacted hydrogen peroxide is an indication that the Fenton's reaction sequence lacked the soluble iron needed for radical initiation.

3.4.1.5 Radical Consumption by Carbonates and Other Competitive Inhibitors

Hydroxyl radicals produced by the Fenton's reagent or other oxidation processes are highly reactive, with target compounds as well as other compounds and minerals that are commonly observed in aquifers. These reactions may compete significantly with target compounds, affecting the efficiency of hydroxyl radical oxidation. pH has a particularly strong impact on hydroxyl radical efficiency, due to reaction losses to the

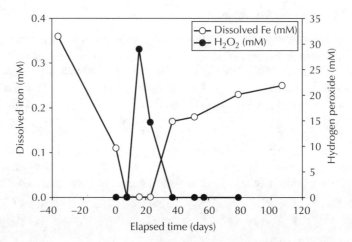

FIGURE 3.12 Loss of soluble iron from aqueous phase during a Fenton's reagent oxidation test to determine reaction rates with mineral iron radical initiation.

bicarbonate ion. This is one reason that, even when chelated iron is used, the Fenton's reactions are most productive at pH less than 7.

Second-order reaction kinetics can be used to determine the impact of competitive reactions on the efficiency of hydroxyl radical reactions. The efficiency is defined as the ratio of target compound reactions to nontarget reactions. Equation (3.52) was developed to describe competition between trichloroethene and bicarbonate ion. That equation can be generalized as follows:

$$\text{Efficiency} = \frac{d\text{Target}}{d\text{Competitors}} = \frac{k_{\text{Target}}}{\sum k_{\text{Comp}_i}[\text{Comp}_i]} \times [\text{Target}] \quad (3.79)$$

The competing compounds that are most important are those that are present at high concentrations, relative to target compound. The high concentration of bicarbonate that can be present at pH greater than 7 makes it a significant competitor, despite the fact that its kinetic rate constant is approximately 1000-fold lower than for most potential target compounds.

Figure 3.13 shows the results of efficiency calculations for phenol and dichloropropane as target molecules, with bicarbonate ion (HCO_3^-) as a competitor. For this analysis, the pH was assumed to be 8.0, and the total alkalinity was 100 mM as $CaCO_3$ (300 mg L^{-1}).

3.4.1.6 Common Reaction By-Products

Oxygen and carbon dioxide are the two most significant by-products of the Fenton oxidation processes. These gases can be produced at rates that match or exceed typical aquifer sparge injection well flow rates, and was discussed more completely earlier. Partial decomposition of soil organic matter and naturally occurring dissolved

Chemical Reactive Zones

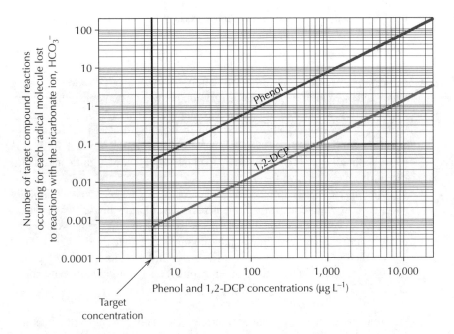

FIGURE 3.13 Comparison of radical reaction efficiencies for Fenton's oxidation of 1,2-dichloropropane (1,2-DCP) and phenol, in an aquifer at pH 8.0, with 300 mg L^{-1} carbonate alkalinity.

organic matter is likely to produce a wide array of products, including oxygenated molecules such as aldehydes, ketones, and carboxylic acids. Some such compounds are reactive with hydroxyl radical, and their presence in groundwater following Fenton's applications may indicate that the reagent was exhausted — not production of oxidant-resistant products.

3.4.1.7 Monitoring Fenton's Reagent Applications

The reactions that comprise the Fenton's oxidation process occur very rapidly, generating significant quantities of gas and heat. The gas mix that erupts from the aquifer surface during a Fenton's application provides a window on the reaction process. These gases can be monitored as the reaction proceeds, allowing adjustments to be made during the course of the oxidation treatment. At a minimum, monitoring should be conducted for the oxygen content of the off-gas, and the site-specific health and safety plan should include provisions for safely venting the vadose zone to prevent the development of explosive conditions.

Several processes contribute to the gas mixtures observed in the vadose zone soils overlying Fenton's reagent applications:

- *Carbon dioxide propagation.* The injected hydrogen peroxide that contributes to hydroxyl radical generation forms carbon dioxide gas in the radical chain reaction termination step. The presence of elevated carbon dioxide in

the off-gas is an indication that organic carbon oxidation is occurring. If carbon dioxide is not present in the off-gas, organic carbon oxidation is not occurring, which may indicate lack of radical formation, or may be an indication that no organic carbon remains in the formation.
- *Oxygen propagation.* Disproportionation of hydrogen peroxide, forming water and oxygen, is an alternative reaction pathway, often consuming a very large fraction of injected hydrogen peroxide. Reaction of hydrogen peroxide with Fe^{3+} is a major source of oxygen in the off-gas. Because productive (radical-forming) reactions of hydrogen peroxide with Fe^{2+} result in formation of Fe^{3+}, the disproportionation reaction and formation of significant quantities of oxygen are unavoidable.
- *Vaporization of target compounds.* Fenton's reagent is most commonly applied to organic contaminants that partition into gas phase, when one is present. Oxygen and carbon dioxide gases escaping the aquifer also carry gas-phase volatile and semivolatile organic compounds into the vadose zone. Gas-phase contaminant concentrations often surge in the vadose zone at the start of a Fenton oxidation treatment, and remain elevated until oxidation reactions are completed.

Vadose zone gas composition can be interpreted to gauge the status of a Fenton's treatment, in "real time." Table 3.7 provides a possible interpretation rationale for patterns of off-gas composition. It is important to note that the site-specific behavior of Fenton's reactions may invalidate the interpretations suggested by Table 3.7.

3.4.2 PERMANGANATE OXIDATION

Permanganate ion (MnO_4^-) is a limited-target oxidant that can be effective for treatment of many aquifer contaminants, particularly chlorinated alkenes. Chlorinated ethenes, phenols, sulfides, organosulfur compounds, and double-bonded, nonaromatic organics are suitable targets for permanganate oxidation, while fuels, aromatic organic compounds (e.g., benzene), chloromethanes (e.g., carbon tetrachloride), and methyl-*tert*-butyl ether (MTBE) are not suitable targets for permanganate oxidation.[20]

Permanganate reaction rates are much slower than those of the hydroxyl radical, which gives it a significant advantage in achieving contact with contaminants, especially those that reside in the static groundwater, nonaqueous-phase liquids and sorbed phase mass. Permanganate's slower reaction rates and more limited range of reactivity also decrease the violence of *in situ* reactions, which simplifies its application, relative to Fenton's reagent processes (refer to Section 3.4.1).

Permanganate is available as salts of potassium or sodium, and the determination of which to use is based on the required solution strength, purity, and ease of handling.

Potassium permanganate ($KMnO_4$, molecular weight = 158.03 g mol^{-1}). The potassium permanganate salt is sold as a fine crystalline solid that can be dissolved in the field, creating solutions up to 6% strength at 20°C. If the aquifer temperature is

TABLE 3.7
Possible Interpretations of Vadose Zone Gas Composition as a Reflection of the Status of Fenton's Reagent Application in an Underlying Aquifer

Relative Process Off-Gas Composition			
Oxygen	Carbon dioxide	VOCs	Interpretation
High	High	High	Hydroxyl radical is forming; organic carbon oxidation is occurring; VOCs remain to be treated
High	Low	High	Hydroxyl radical formation is not occurring at a sufficient rate — likely indicates shortage of Fe^{2+}; VOCs remain to be treated
Low	Low or high	Low or high	Process off-gas is not reaching the monitoring point — may indicate one of the following: either hydrogen peroxide is not reacting — possible lack of reactive iron in the formation or insufficient peroxide — or process off-gas is being intercepted
High	High	Low	Organic carbon oxidation is occurring, but VOC treatment may be complete; confirm through aqueous-phase sampling
High	Low	Low	Organic carbon oxidation is not occurring, and VOC treatment may be complete; confirm through aqueous-phase testing

significantly lower than ambient air temperature, it is possible to experience a significant decrease in solubility between the mixing station and the aquifer. It is important to limit the injected solution strength to less than the solubility at aquifer temperature, to avoid precipitation of permanganate from solution and the associated aquifer clogging that might occur.

The AWWA-grade potassium permanganate incorporates up to 5% silica sand component as an anticlumping agent. Although it is an inert ingredient, the silica sand must be settled or filtered out of the permanganate solution before it is injected into the aquifer to avoid plugging of the formation. Technical-grade potassium permanganate does not contain the silica sand; therefore, it is simpler to use.

Commercially available potassium permanganate contains trace levels of metals, including chromium, that may cause secondary water quality impacts following chemical reactive zone applications. To avoid potential metals contamination, "remediation grade" low-metals potassium permanganate may become available, to reduce potential injection permit limitations or posttreatment monitoring for metals. It is important to note that metals may be solubilized from aquifer soils, independent of metals content of the injected oxidant formulation.

Sodium permanganate (NaMnO$_4$, molecular weight = 141.93 g mol^{-1}). The sodium salt of the permanganate ion is much more soluble than the permanganate salt, and can be used to generate higher *in situ* solution strengths. It is commercially available as a 40% liquid solution, which can be highly reactive with permanganate-susceptible materials and should be handled with great caution. Sodium permanganate is normally diluted substantially prior to injection into a chemical reactive zone.

Dilution and neutralization of excess reagent can be very dangerous with sodium permanganate and improper handling has caused serious injury at a field application site.[21] It is critically important to follow all manufacturer's recommendations for safe management of oxidants.

3.4.2.1 Permanganate Demand Estimation and Injection Loading Requirements

Many compounds that occur naturally in aquifers can react with permanganate ion and, in most cases, the oxidant consumed by nontarget reactions exceeds that of reactions with target compounds. The nontarget demand is termed matrix demand, and it must be estimated during feasibility testing to assure that adequate reagent loading is anticipated during process design. Matrix demand is a function of aquifer soil type and reduction–oxidation status. Lower matrix demand is expected from coarse-grained sands and gravels while relatively higher demands are expected for fine-grained soils such as silts and clays. Table 3.8 shows results of oxidant demand testing for a range of soil types collected at a midwestern site. The permanganate demand ranged from 6.3 to 18.5 g oxidant per kilogram soil. For an aquifer with a bulk density of 1500 kg m^{-3}, that translates to 9.5 to 27.8 kg m^{-3} (27 to 79 lb yd^{-3}). The matrix demand values shown in Table 3.8 span most of the range observed at

TABLE 3.8
Permanganate Matrix Demand Results for Aquifer Soils at a Midwestern Site

Sample	Soil Description	Observed Permanganate Demand (g KMnO$_4$/kg dry wt)
1	Sand, gravel	7.53
1-rep	Sand, gravel — replicate	9.84
2	Sand, gravel, rock	6.34
3	Clay	18.53
4	Clay loam (wet)	7.91
5	Sand, clay (hydrocarbon odor)	8.21

Note: Batch testing was conducted over a 4-day exposure period on soil samples collected from an aquifer contaminated by trichloroethene. Total organic carbon concentrations were 550 and 580 mg kg^{-1} in two samples that were analyzed for that parameter.

sites studied by ARCADIS. For conceptual budget preparation (prior to bench demand testing) most sites can be expected to fall within the range of 5 to 20 grams permanganate per kilogram soil.

For projects that rely on potassium permanganate, the solubility limit has a significant effect on the aquifer pore volume displacement that will accompany oxidation treatment. If a solution concentration limit is set at 4% to prevent *in situ* precipitation of permanganate, the permanganate carrying capacity of injected solution is 40 g L^{-1}. To provide the required permanganate mass to an aquifer that exerts a 5 g kg^{-1} matrix demand, 9.5 kg KMnO$_4$ must be injected. At 40 g L^{-1}, that entails injection of 238 L of solution per cubic meter of aquifer matrix, which exceeds the migratory pore volume in many formations. To satisfy matrix demand at the 20 g kg^{-1} end of the range, nearly 700 L m^{-3} of 4% permanganate solution must be injected, which constitutes several pore volume changes. Equation (3.80) provides a method for calculating the reagent infusion volume per unit aquifer volume.

$$\text{Solution volume}\left(\frac{L_{\text{ox soln}}}{m^3_{\text{aquifer}}}\right) = \frac{\text{Demand}\left(\frac{g_{\text{ox}}}{kg_{\text{soil}}}\right) \times \rho_{\text{soil}}\left(\frac{kg_{\text{soil}}}{m^3_{\text{aquifer}}}\right)}{\text{Solution strength}\left(\frac{g_{\text{ox}}}{L_{\text{ox soln}}}\right)} \quad (3.80)$$

where ρ_{soil} is the aquifer matrix bulk density; a value of 1500 kg m^{-3} is commonly assumed.

These calculations highlight one of the significant issues that face any chemical reactive zone application — aqueous-phase liquid injections are limited in the reagent mass that can be delivered per unit volume, and multiple aquifer pore volumes of fluid injection are often required to deliver the specified reagent mass. Multiple injection episodes or days-long, slow-paced injections may be required to deliver oxidant at a rate that matches the rate at which the reactions occur in the aquifer. Dilution that occurs when the injected fluid reaches the formation and travels along preferred flow paths further complicates the delivery process. For permanganate treatments, practitioners can shift to sodium permanganate to achieve higher solution strength, however application of the full-strength (400 g L^{-1}) solution entails significant risk of violent reaction and is not recommended.

The hydraulic challenges associated with multipore-volume chemical floods are evaluated by Ibaraki and Schwartz.[22] They concluded that contact efficiencies of chemical floods will be low in heterogeneous media. Because their study did not include consideration of carbon dioxide gas production associated with permanganate oxidation (and the associated localized reduction of hydraulic conductivity), chemical flood efficiencies are likely to be even lower than their modeling projected.

3.4.2.2 Permanganate Reaction Mechanisms

The active ingredient in permanganate oxidation is the permanganate ion, MnO_4^-. The manganese atom has a "+7" valence and is shown as Mn(VII). During the oxidation process, the manganese is reduced to a "+4" valence, and is shown as Mn(IV). The main reduction product is manganese dioxide (MnO_2), which precipitates from solution.

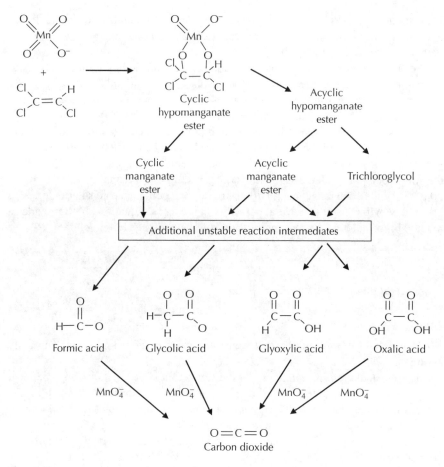

FIGURE 3.14 Reaction pathways for the permanganate oxidation of trichloroethene. (Redrawn from Yan and Schwartz, 2000.)

Yan and Schwartz[23] provided a reaction network for permanganate oxidation of trichloroethene that showed several alternative pathways, all of which began with a two-center reaction that forms a hypomanganate ester (see Figure 3.14). From this point, several potential unstable intermediates were postulated, leading to formation of four intermediate products that were observed in their experiments: formic acid, glycolic acid, glyoxylic acid, and oxalic acid. All of these products were susceptible to reaction with permanganate, forming the final product, carbon dioxide.

3.4.2.3 By-Product Formation

Like all energetic reagents, permanganate reacts with many compounds found in aquifers, accompanying the target contaminants. These naturally occurring compounds and cocontaminants may react only partially with injected permanganate, yielding undesirable by-products. The higher the oxidant solution strength, the more likely it is to form by-products. We have observed the many by-products during bench testing of permanganate at solution strengths up to 4% as $KMnO_4$, including acetone, carbon disulfide, chloroform, and other trihalomethanes.

Sodium permanganate was used in a bench study testing dechlorination of trichloroethene, *cis*-dichloroethene, and chlorobenzene. Kinetic rate constants for permanganate reaction with trichloroethene and *cis*-dichloroethene are available (Table 3.4), but values for dechlorination of chlorobenzene were not readily available. To determine whether chlorobenzene could be treated with permanganate, and to observe whether by-product formation may be a problem, a bench trial was conducted with a batch groundwater sample from a contaminated site. The large-volume sample was divided among treatments receiving sodium permanganate at four concentrations (1, 4, 7, and 10%), plus an untreated control.

Trichloroethene and *cis*-dichloroethene were both destroyed to below detectable levels in the 1% sodium permanganate solution, as shown in Figure 3.15. Treatment of chlorobenzene was less effective, with a 10-fold reduction observed at the 7% solution strength and reduction to below detectable levels observed only at the 10% permanganate solution strength. Chloroform was not observed in groundwater from the site, but appeared in treatments with permanganate concentrations of 4% and greater, as shown in Figure 3.15.

Trichloroethene and *cis*-dichloroethene were the remedial targets for the site under study, and the field application can be completed without achieving high

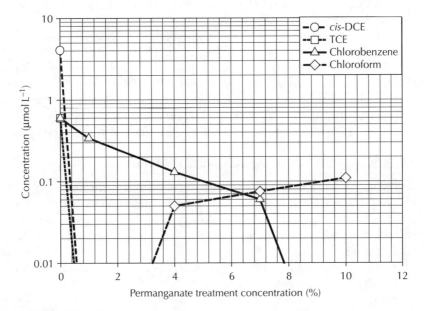

FIGURE 3.15 By-product formation during permanganate oxidation bench trial. Chloroform was observed as a by-product at concentrations that increased as a function of increasing permanganate solution strength.

permanganate levels throughout the formation. However, it is necessary to inject permanganate at concentrations in the 4% range to achieve efficient use of the reagent injection network, and limited chloroform formation can be expected to occur near the injection points. This type of by-product formation is a tradeoff that must be considered for any oxidant injection, and site-specific testing is useful in determining the upper acceptable permanganate concentration. It is important to note that some potential target compounds would require such a high oxidant strength that the benefit gained by initial target destruction may be outweighed by the formation of more troublesome by-products.

3.4.2.4 Permanganate Reaction Kinetics

Reaction kinetics for permanganate oxidation are several orders of magnitude lower than for reactions between similar target compounds and hydroxyl radical. Compare, for example, the second-order kinetic rate constants for oxidation of trichloroethene by permanganate and hydroxyl radical given in Table 3.4. For permanganate, the rate constant is approximately 0.7 M^{-1} s^{-1}, while the rate constant for hydroxyl radical is 4×10^9 M^{-1} s^{-1}. That 10-order-of-magnitude gap is largely closed by the concentrations of oxidant that can be achieved: more than 0.1 M for permanganate, and only 10^{-11} M for hydroxyl radical. Still, hydroxyl radical reactions are, relatively, quite fast.

For oxidation reactions between permanganate and potential target compounds, it is helpful to calculate expected concentrations as a function of time, at various permanganate concentrations. This was conducted for trichloroethene, perchloroethene, and MTBE. The results of calculations are shown in Figure 3.16(A)–(C). For each plot, the permanganate concentration was held constant at the level shown on the plot, and target compound concentrations were calculated by pseudo-first-order degradation [Equations (3.17) and (3.18)]. The starting concentration for each compound was 1 mg L^{-1}. Permanganate concentrations ranged from 5 to 50,000 mg L^{-1} as MnO_4. The visual detection limit for permanganate is 5 mg L^{-1} (Reference 20) and 50,000 mg L^{-1} is the maximum working strength of $KMnO_4$ that would be applied in a field application (higher concentrations can be achieved with $NaMnO_4$, but significantly higher concentrations are not recommended). These data can be interpreted to develop expectations for permanganate performance for each compound in a field application.

3.4.2.5 Sorption Effects on Permanganate Oxidation

Bench-scale studies were performed to determine whether permanganate oxidation could be applied successfully to a perchloroethene source area in a sandy aquifer at a site located in the eastern U.S. The project objective was reduction of perchloroethene soil concentrations to levels that were protective of groundwater for drinking water supply use. The study measured the reduction of perchloroethene soil concentrations, as well as the change in perchloroethene leaching characteristics that would result from permanganate treatment.

Previous experience indicated that between-sample variance was very high for pretreatment contaminant concentrations in samples collected from source area aquifers. The variance is typically also quite high for other characteristics affecting remedy performance — total organic carbon, for example. Samples collected in

Chemical Reactive Zones

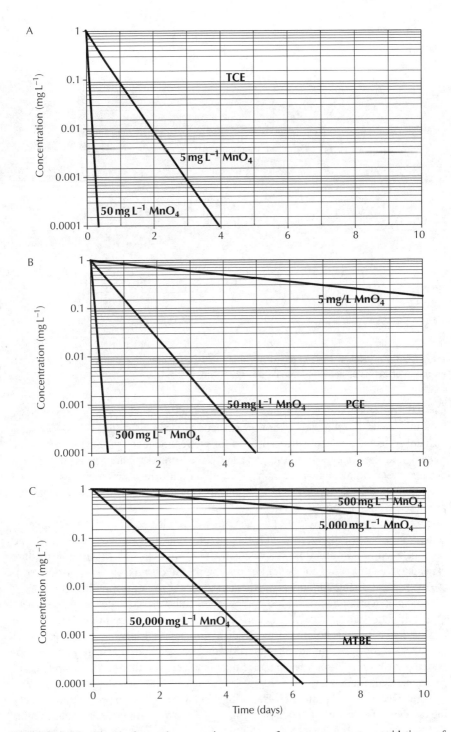

FIGURE 3.16 Pseudo-first-order reaction rates for permanganate oxidation of trichloroethene (curve A), perchloroethene (curve B), and methyl-*tert*-butyl ether (curve C). For each compound, reaction rates were calculated for multiple permanganate concentrations.

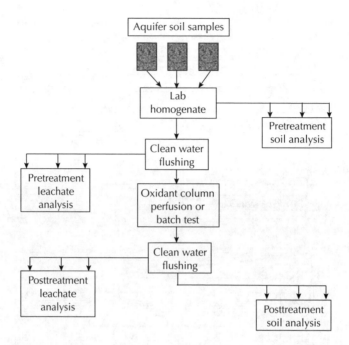

FIGURE 3.17 Bench-scale protocol for analysis of permanganate demand, pre- and post-leachate performance and contaminant mass removal.

parallel from a contamination source zone are often too variable to be represented by a sample mean (the standard error may exceed the mean estimate). If multiple samples are collected and some are sacrificed for pretreatment analysis, while others are treated, and then analyzed, the high variance may prevent comparison of pre- and posttreatment sample means. To gain the benefit of uniform pretreatment test material and to obtain useable replication in the bench-scale trials, a sample homogenization protocol was developed.

Four Shelby tube samples of aquifer soils were collected from the contamination source area. At the laboratory, contents of the split spoons were combined and homogenized in a minimal head space container. Although volatilization losses are expected from this setup, the benefit of starting bench testing from as uniform a sample mass as is possible far outweighs the importance of volatilization losses.

Samples were drawn from the homogenate, and were allocated to pretreatment characterization analysis or permanganate treatment, as described in Figure 3.17. In this case, the sample volume was sufficient to support triplicate sampling of the homogenate for pretreatment characterization, but the remaining volume only supported preparation of one permanganate treatment column. Replicated treatments are recommended, when possible.

The column was initially perfused with clean water to obtain pretreatment leaching analysis for perchloroethene. Then, a 3% w/w permanganate solution (as $KMnO_4$) was perfused through the column until the permanganate demand was satisfied, which was determined to be achieved when permanganate effluent matched

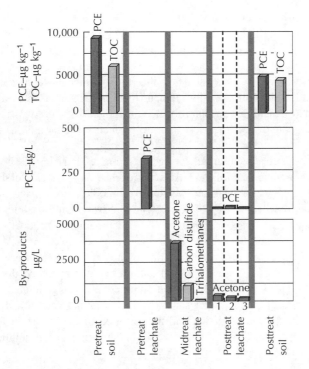

FIGURE 3.18 Results of bench-scale pilot studies on permanganate oxidation of perchloroethene-contaminated aquifer soil. Panel A, pre- and posttreatment analyses of perchloroethene (PCE) and total organic carbon (TOC) in aquifer soils. Panel B, results of pre- and postoxidation leachate analyses for PCE. Panel C, analysis of aqueous-phase byproducts. Posttreatment leachate samples were collected from the first, second, and third pore volumes following treatment, as noted on the graph.

influent concentrations. Clean water was once again perfused through the column, and postoxidation leachate samples were collected and analyzed. After leachate sampling was completed, the column was split into upper and lower halves for postoxidation characterization analyses. Because differences in the two results were relatively small, the results were averaged to represent the treated aquifer soils.

The results were very interesting, and are shown in Figure 3.18(A)–(C). The permanganate treatment yielded a very large reduction in perchloroethene leaching concentrations, from 300 to 10 μg L^{-1}. However, soil concentrations of PCE were only reduced from a pretreatment average 9400 to 4850 μg kg^{-1} in posttreatment sampling, despite the fact that permanganate consumption in the column had ceased. The total organic carbon (TOC) was only reduced from 6150 to 4000 mg kg^{-1}, indicating that a substantial fraction of the organic carbon mass resisted oxidation. The posttreatment leachate concentration and soil concentrations can be used to calculate an apparent organic carbon partition coefficient for the oxidant-resistant carbon fraction. According to this calculation, the apparent k_{oc} for perchloroethene was 104,000 L kg^{-1}. For the carbon fraction that was oxidized, the pretreatment leachate data suggested an apparent k_{oc} of 2600 L kg^{-1}. Both of these estimates far exceed the

U.S. EPA-published value of 265 L kg^{-1} (Reference 24) and highlight differences between partition coefficients based on absorption processes (which generate low k_{oc} values) and those observed for adsorption processes, which tightly bind hydrophobic organic compounds. Sorption processes are discussed in more detail in Chapter 5.

These results led to the conclusion that the permanganate oxidation process would not oxidize the entire perchloroethene mass in the source area of this site, but the reduction in leaching concentration would reduce the rate of perchloroethene supply to the groundwater to levels that are protective of the aquifer for drinking water use.

Byproducts were also observed during the testing, with the highest concentrations noted during permanganate flushing. Acetone reached 3400 μg L^{-1}, carbon disulfide rose to 650 μg L^{-1}, while the trihalomethanes, bromodichloromethane, and dibromochloromethane, reached 55 and 59 μg L^{-1}, respectively. Only acetone remained in the posttreatment leachate, at 95 μg L^{-1}. The organic by-products presented a concern, but they were expected to dissipate after completion of the treatment process. Hexavalent chromium concentrations, however, reached very high levels. Permanganate interfered with chromium analyses, but it appeared the Cr^{6+} levels exceeded 30,000 μg L^{-1} during the oxidation process, which generated an unacceptable risk to a nearby public water supply well. It was not determined whether the chromium resulted from oxidation and mobilization of chromium in the site soils (from plating waste) or from chromium metals in the permanganate solution.

3.4.2.6 Phase-Transfer Catalysts

One of the limitations of chemical reactive zone mechanisms is that they all depend on aqueous-phase reactants, while many of the target compounds are hydrophobic organic compounds. The oxidation or reduction reactions occur in the aqueous phase, and their scope of action is limited to the relatively small portion of the target compound that is found in the aqueous phase at any time, which may be a very small fraction of the total mass to be treated. If the aqueous-phase treatment reagent could be delivered into the nonaqueous phase mass, reaction productivity could be dramatically increased. Phase-transfer catalysts offer such a possibility.

A phase-transfer catalyst is a molecule that is composed of polar and nonpolar segments. They are sufficiently soluble to allow dispersal in aqueous phase. In the presence of a hydrophobic organic liquid interface, the nonpolar end of the phase-transfer catalyst enters the organic liquid, while the polar end protrudes into the aqueous liquid. The polar end of the phase-transfer molecule attracts ions of opposite polarity, forming a neutral complex that is essentially nonpolar. This draws the entire complex across the phase boundary. If the complexed ion reacts with the organic compound, the phase-transfer catalyst regains its polar end and it resurfaces from the interface and is available to draw another ion into the organic phase.

One reactant-transfer catalyst combination that has been considered is permanganate ion and quaternary amines. The quaternary ammonium compounds are composed of a pentavalent nitrogen, the polar locus, and four alkanes, the nonpolar bulk of the molecule. As shown in Figure 3.19, the alkane end is submerged in the

Chemical Reactive Zones

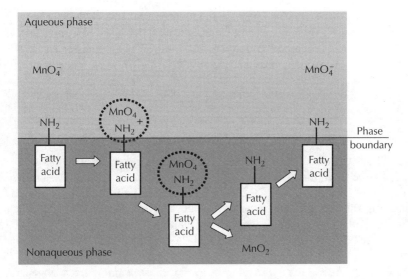

FIGURE 3.19 Phase-transfer catalysis to accelerate permanganate oxidation of liquid, non-aqueous-phase organic compounds.

hydrophobic organic liquid while the ammonium ion remains on the aqueous side of the phase boundary. The electrophilic ammonium attracts a permanganate ion from the aqueous-phase solution, forming a complex that suppresses the polarity of the ammonium end of the transfer catalyst. The complex is pulled across the phase boundary, where the permanganate ion reacts with a molecule of the organic liquid, releasing a manganese dioxide molecule (solid) and regenerating the polarity of the ammonium. The ammonium resurfaces from the organic phase and is available to transfer another permanganate ion across the interface. Chloride ions released in dechlorination reactions migrate into the aqueous phase.

Seol and Schwartz[25,26] tested the phase-transfer catalyst strategy for oxidation of trichloroethene, in bench-scale studies using quaternary ammonium salts as phase transfer catalysts. In their second study,[26] the authors tested three phase-transfer catalysts: tetra-n-ethylammonium bromide (TEA), $(C_2H_5)_4NBr$; tetra-n-butylammonium bromide, $(CH_3(CH_2)_3)_4NBr$; and pentyltriphenylphosphonium bromide (PTPP), $CH_3(CH_2)_4P(C_6H_5)_3Br$. For these compounds, the aqueous solubility ranged from 2.4×10^{-4} to 7.3×10^{-4} M, and their solubilities in methylene chloride ranged from 0.235 to 1.46 M (data on solubilities in trichloroethene were unavailable). In test tube experiments on separate-phase trichloroethene, overall dechlorination rates increased from 0 (for TEA) to approximately 25% (for PTPP), relative to permanganate solution controls. This indicates that three fourths or more of the oxidation reactions occurred in the aqueous phase, with replenishment of aqueous-phase trichloroethene exceeding the rate of phase-transfer catalysis. Seol and Schwartz[26] limited the concentration of phase-transfer catalysts, to avoid precipitation or coagulation of the catalyst–permanganate complex. It is possible that higher phase-transfer rates could be achieved at higher transfer catalyst application rates. However, enhancement of NAPL dissolution has been observed in association with unassisted permanganate oxidation in other studies,[27-29] and the incremental increase in productivity that might be gained through use of phase-transfer catalysis must be weighed against the added cost of materials and underground injection control permitting. Because the phase-transfer catalysts that would be used are oxidant resistant, the residual catalyst mass will remain in the aquifer.

3.4.3 OZONE

3.4.3.1 Ozone as the Sole Oxidant Source

Ozone (O_3) is a gas that can be generated from air or oxygen at the location where it is intended to be used. Ozone is a suitable oxidant for many compounds, and may be more effective than hydroxyl radical for some of the polyaromatic hydrocarbons. Comparing, for example, the pseudo-first-order rate constants for hydroxyl radical and ozone oxidation of naphthalene (rate constants and maximum feasible molarities can be obtained from Table 3.4; pseudo-first-order rate constants can be calculated using Equation (3.17)), ozone oxidation of naphthalene is slightly faster ($k' = 0.87$ s^{-1} for ozone and 0.12 s^{-1} for hydroxyl radical). For most *in situ* reactive zone targets, however, hydroxyl radical oxidation is dramatically faster.

Comparing reaction rates for trichloroethene, the hydroxyl radical rate constant is 4×10^9, and for ozone it is 1.7×10^1 M^{-1} s^{-1}. For perchloroethene, the hydroxyl radical rate constant is 3×10^9 M^{-1} s^{-1}, a slight reduction relative to the rate for trichloroethene, while ozone is effectively unreactive with perchloroethene, with a kinetic rate constant $<1 \times 10^{-1}$ M^{-1} s^{-1}. This is a striking illustration of the importance of reaction kinetics, which reflect steric effects, rather than oxidant "strength" as measured by the half-cell reaction potential (a thermodynamic quantity) for predicting reaction rates and selection of oxidants (or reductants) to achieve target reactions.

Figure 3.20 shows the direct ozonation reaction with an ethene. The ozone molecule (O_3) is electrically neutral, but carries opposing charge concentrations (Figure 3.20(A)). This structure is known as a zwitterion, and contributes to ozone's reactivity. Reaction with ozone requires bonding at two sites simultaneously, forming the first reaction intermediate, a molozonide (Figure 3.20(B)). Formation of the initial linkage is a relatively improbable event for an ozone–ethene reaction, and the probability decreases as the level of chlorination increases. This low reaction probability is reflected in the second-order kinetic rate constant for ozone reaction with ethenes, and by far the lowest rate with perchloroethene, the most highly chlorinated of the ethenes.

The most popular uses of ozone are in aboveground treatment systems, in which ozone is used for its direct oxidation capacity, or in combination with another oxidant, such as hydrogen peroxide, to generate hydroxyl radicals. Ozone has also been deployed *in situ*, through injection of an ozone–air combination in aquifer sparging. Both aboveground and *in situ* ozonation systems combine the effects of direct ozone oxidation of target compounds with ozone-driven formation of hydroxyl radical, which is discussed in the following.

Ozone solubility and the concentration of ozone in the injected gas stream is one of the key variables affecting the rate of treatment that can be established. Commercial

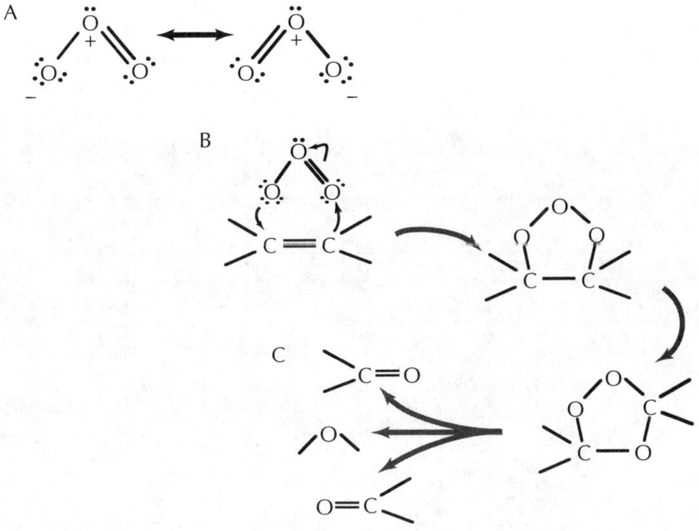

FIGURE 3.20 (A) Ozone is a zwitterion, an electrically neutral molecule that carries a resonating charge concentration that imparts reactivity. (B) Sequence of ozone reaction with an ethene. (C) Breakup of the reaction intermediate to form water and two aldehyde molecules.

ozone-generation equipment produces ozone from air or oxygen. Typical air-based systems can produce 30,000 ppmv* (3%), and oxygen-fed systems can produce up to 60,000 ppmv (6%), with advanced systems now available that may significantly exceed those production rates. Figure 3.21(A) shows the aqueous-phase solubility of ozone in balance with gas-phase ozone concentrations ranging from 0 to 1000 ppmv, and Figure 3.21(B) shows the solubility for gas-phase ozone from 0 to 60,000 ppmv (0 to 6%), at temperatures ranging from 10 to 35°C. The calculations for Figures 3.21(A) and (B) were made from Henry's law data for ozone given in Weast.[15] At 15°C, a 6% ozone injection gas is expected to generate approximately 1.2×10^{-3} M ozone in solution, if no reactions occur to scavenge the injected ozone. However, ozone is the target of numerous aqueous-phase reactions, and the sustainable ozone concentration in an aquifer is likely to be much lower.

Ozone reacts with hydroxyl anion in aqueous phase, to form oxygen and hydroperoxide anion, as follows:

$$O_3 + OH^- \rightarrow O_2 + HO_2^- \quad k = 4 \times 10^1 \quad (M^{-1} s^{-1}) \tag{3.81}$$

In a back-reaction, hydroperoxide anion reacts quickly with ozone to form hydroxyl radical, as follows:

$$O_3 + HO_2^- \rightarrow O_2 + OH^\bullet + O_2^{\bullet -} \tag{3.82}$$

---

*Parts-per-million by volume (μL ozone per L total gas phase).

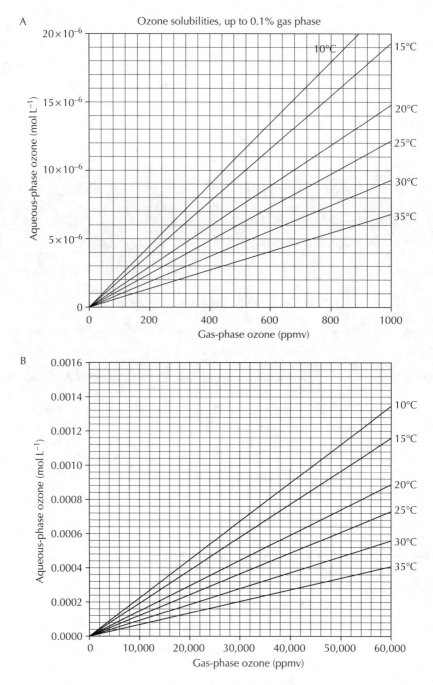

FIGURE 3.21 (A) Aqueous-phase zone solubility for gas-phase concentrations up to 0.1% ozone, by volume (0 to 1000 ppmv), at 1 atm total pressure. (B) Aqueous-phase zone solubility for gas-phase concentrations up to 6% ozone, by volume (0 to 60,000 ppmv), at 1 atm total pressure.

Chemical Reactive Zones

This reaction has a very high kinetic rate constant, relative to other ozone reactions ($k = 5.5 \times 10^6$ M^{-1} s^{-1}) (refer to Table 3.4). However, hydroperoxide anion is the conjugate base of hydrogen peroxide, a weak acid (Equation (3.83)).

$$H_2O_2 \leftrightarrow HO_2^- + H^+ \quad pK_a = 11.75 \tag{3.83}$$

At normal aquifer pH levels, hydrogen peroxide will be predominantly in its undissociated form. Ross et al.[7] show that ozone will also react with hydrogen peroxide, producing hydroxyl and perhydroxyl radicals, along with oxygen.

$$O_3 + H_2O_2 \rightarrow O_2 + OH^\bullet + HO_2^\bullet \tag{3.84}$$

The direct reaction between ozone and hydrogen peroxide is slow, compared to other potential reactions ($k = 3.6 \times 10^{-2}$ M^{-1} s^{-1}), and may not contribute significantly to hydroxyl radical formation in an aquifer system. Hydroxyl anion consumption of ozone at typical aquifer pH levels may be significant, but is unlikely to generate hydroxyl radicals at a useable rate.

Equation (3.81) can be used to estimate the half-life of ozone in reaction with hydroxyl anion, as a function of pH. For a well-buffered groundwater, we can assume the hydroxyl anion concentration is stable, allowing us to calculate a pseudo-first-order reaction rate constant for the reaction. Using Table 3.9 to obtain values for the ionization constant of water at various temperatures, we can see that at 15°C, K_w equals 0.45×10^{-14} M^2. We can then use the following equation to calculate [OH$^-$].

$$K_w = [H^+] \cdot [OH^-] = 10^{-pH} \cdot [OH^-] \tag{3.85}$$

and Equation (3.85) can be rearranged to isolate [OH$^-$] as a function of K_w and pH, as follows:

$$[OH^-] = \frac{K_w}{10^{-pH}} = K_w \cdot 10^{pH} \tag{3.86}$$

TABLE 3.9
Ionization Constants for Pure Water

Temperature (°C)	$K_w^a = [H^+] \cdot [OH^-]$
10	0.29×10^{-14}
15	0.45×10^{-14}
20	0.68×10^{-14}
25	1.01×10^{-14}
30	1.47×10^{-14}

[a] The units for K_w are mol^2 L^{-2}.
Source: From Kotz and Treichel.[1]

When the pH is 7.5, and the temperature is 15°C, the hydroxyl ion concentration is calculated by

$$[OH^-] = 0.45 \times 10^{-14} \times 10^{pH} = 0.45 \times 10^{pH-14} = 1.4 \times 10^{-7} \quad (M)$$

The second-order kinetic rate constant for ozone reaction with hydroxyl anion is $4 \times 10^1 \, M^{-1} \, s^{-1}$, which can be used to calculate a pseudo-first-order rate constant for 10°C and pH 7.5:

$$K' = K \cdot [OH^-] = 4 \times 10^1 \cdot 1.4 10^{-7} = 5.7 \times 10^{-6} \quad (s^{-1})$$

Using Equation (3.22) to solve for the half-life of ozone under these conditions,

$$t_{1/2} = -\frac{(-0.69)}{5.7 \times 10^{-6}} = 121{,}000 \quad (s)$$

If the temperature were increased to 30°C, and a different (higher) literature value for the kinetic rate constant were used, the estimate of ozone half-life would be reduced to approximately 20,000 s. These values far exceed the ozone life span observed in natural waters. Ternes et al.,[30] for example, observed a 12-min half-life for ozone in water treatment facilities (postflocculation), which indicates that other ozone-consuming reactions are occurring.

Among alternatives for ozone consumption in groundwater, hydroxyl radical initiation pathways are the most desirable for contaminant destruction because the hydroxyl radical is highly reactive with most contaminants. Ozone can be converted to hydroxyl radical or superoxide radical anion through initiation reactions with dissolved organic matter, and hydroxyl radical may be produced by reactions with naturally occurring reduced metal species. Pines and Reckhow[31] showed hydroxyl radical initiation by Co^{2+}, for example. The only reaction shown in the NIST database[7] for ozone and Fe^{2+}, however, produces a ferryl ion (Fe(IV)), as shown in Equation (3.87).

$$O_3 + Fe^{2+} \rightarrow FeO^{2+} + O_2 \quad k = 8.2 \times 10^5 \quad (3.87)$$

It is possible that the ferryl ion completes the radical initiation process through a subsequent reaction with ozone, however, kinetic rate data and documented reaction mechanisms are not available in peer-reviewed studies. Section 3.4.3.3 discusses the results of a bench-scale ozonation trial for 1,4-dioxane in groundwater from which we inferred significant radical initiation.

When ozone injection is deployed to create an *in situ* chemical oxidation zone, three modes of ozone demand are observed. When ozone is first injected, a large transient demand is observed as reduced inorganic species, aquifer matrix material, and target contaminants combine to rapidly consume injected ozone. After the transient ozone demand has been satisfied, the reactive zone is expected to settle into a lesser, constant demand composed of two compartments: target contaminant demand and ozone scavenger losses. Figure 3.22 shows the conceptual development of the

Chemical Reactive Zones

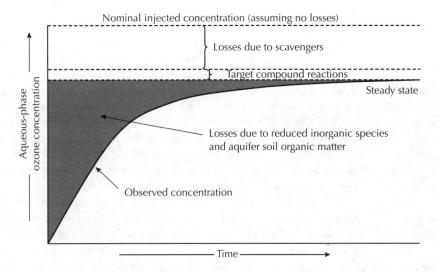

FIGURE 3.22 Conceptual ozone demand pattern, during continuous ozone sparge gas injection. The initial ozone consumption is high, due to interaction with reduced aquifer minerals and soil organic matter. Ozone losses decrease with time, as the matrix demand is satisfied. The steady-state ozone demand is determined by the rate at which groundwater carries scavenger and target contaminant molecules into the ozone sparge zone.

demand pattern in an aquifer ozone injection system. If the scavenging demand is sufficient to generate an ozone half-life in the range of 10 to 15 min, which may be typical for groundwater, the effective treatment zone that can be established around any injection point will be limited by time of travel of the injected gas through the formation.

3.4.3.2 Ozone–Peroxide Systems

The combined use of ozone and hydrogen peroxide is termed the peroxone process, which generates hydroxyl radical at rates sufficient to drive treatment reactors. The mechanism of reaction, however, is not well known.[32] The hydroxyl radical-generating reaction between hydroperoxide anion and ozone (Equation (3.82)) is unlikely to occur at typical aquifer pH levels, because the hydrogen peroxide is more likely to be in its undissociated form (Equation (3.83)). The ozone reaction with undissociated hydrogen peroxide (Equation (3.84)) generates a hydroxyl radical, but the reaction rate is too slow to be useful as a radical generator. Xu and Goddard[32] reported that a possible mechanism of the peroxone reaction involves formation of H_2O_3 and cyclical HOOHOOO intermediates.

Ozone–peroxide systems have been shown to generate bromate anions,[33] as has been shown for ozone-only oxidation systems.[34–38] Acero and coworkers[33] found that it was not possible to achieve more than 35 to 50% reduction of MTBE concentrations, in the presence of 50 $\mu g\ L^{-1}$ bromide anion, without exceeding regulatory limitations on bromate anion (10 $\mu g\ L^{-1}$). Bromate and other by-product formation by ozone and ozone–peroxide systems is discussed further in Section 3.4.3.4.

3.4.3.3 Bench-Scale Ozonation of 1,4-Dioxane

1,4-Dioxane (Figure 3.23) is a miscible organic compound that is persistent in aquifers, forming extended groundwater plumes. It is a cyclical ether compound and appears to be resistant to microbial degradation processes in aquifers. Although studies[39] have shown 1,4-dioxane can be cometabolically destroyed in the presence of a structural analog (tetrahydrofuran, Figure 3.23), the injection of tetrahydrofuran to support 1,4-dioxane degradation is not likely workable in an aquifer setting. Other studies have shown successful degradation of 1,4-dioxane through phytoremediation[40–42] and by aerobic bacteria.[43–45] Phytoremediation will likely be a viable alternative for instances of shallow contamination, but not in deeply embedded plumes, and microbial decomposition of 1,4-dioxane in deeper plumes has not yet been reported.

Lack of proven alternatives for treatment of deep, 1,4-dioxane contamination led us to test whether *in situ* oxidation may be a viable option for this compound. Ross et al.[7] provided kinetic rate constants for reactions of hydroxyl radical and ozone with 1,4-dioxane, and other bench trials we have conducted indicated that 1,4-dioxane was unreactive with the permanganate ion. That led us to consider direct ozonation and ozone-derived hydroxyl radical for *in situ* chemical oxidation of 1,4-dioxane.

The kinetic rate constants for hydroxyl radical and ozone oxidation of 1,4-dioxane were 2.4×10^9 and $3.2 \times 10^{-1}\,M^{-1}\,s^{-1}$, respectively.[7] If a 6% ozone gas stream were injected into a 1,4-dioxane-contaminated aquifer formation, both direct ozonation and hydroxyl radical oxidation would be expected to occur. Masten and Davies[6] indicate that feasible molarities in groundwater systems are $3 \times 10^{-5}\,M$ for ozone and range from 10^{-13} to $10^{-11}\,M$ for the ozone-derived hydroxyl radical. These concentration values can be used to estimate reaction rates for 1,4-dioxane in an ozone sparging system, using a pseudo-first-order approximation (Equations (3.17) and (3.18)). The two oxidation reactions are

$$1,4\text{-Dioxane} + O_3 \rightarrow \text{products} \quad k_{O_3} = 0.32 \quad (M^{-1}s^{-1})$$

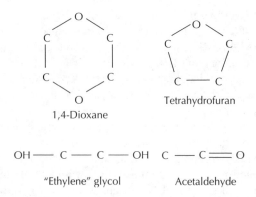

FIGURE 3.23 The structures of 1,4-dioxane and its potential analogs.

and

$$\text{1,4-Dioxane} + \text{OH}^\bullet \rightarrow \text{products} \quad k_{\text{OH}^\bullet} = 2.4 \times 10^9 \quad (\text{M}^{-1}\,\text{s}^{-1})$$

For the ozonation of 1,4-dioxane

$$k'_{O_3} = [O_3] \cdot k_{O_3} = 3 \times 10^{-5} \cdot 0.32 \quad (\text{M} \cdot \text{M}^{-1}\text{s}^{-1})$$

$$k'_{O_3} = 1 \times 10^{-5} \quad (\text{s}^{-1})$$

and for hydroxyl radical oxidation of 1,4-dioxane, the minimum-pseudo-first-order rate constant is defined by the minimum feasible hydroxyl radical molarity.

$$k'_{\text{OH}^\bullet} = [\text{OH}^\bullet] \cdot k_{\text{OH}^\bullet} = 10^{-13} \cdot 2.4 \times 10^9 \quad (\text{M} \cdot \text{M}^{-1}\text{s}^{-1})$$

$$k'_{\text{OH}^\bullet} = 2.4 \times 10^{-4} \quad (\text{s}^{-1})$$

At the maximum potential hydroxyl radical molarity, 10^{-11} M,

$$k'_{\text{OH}^\bullet} = 2.4 \times 10^{-2} \quad (\text{s}^{-1})$$

The oxidation reactions driven by ozone and hydroxyl radical can be composited into a single kinetic rate equation,

$$k_{\text{composite}} = k'_{O_3} + k'_{\text{OH}^\bullet} \quad (\text{s}^{-1})$$

To account for the range of feasible molarities for hydroxyl radical, the composite kinetic rate constant, k'_{comp}, is also expressed as a range variable,

$$2.5 \times 10^{-4} \leq k'_{\text{comp}} \leq 2.4 \times 10^{-2} \quad (\text{s}^{-1})$$

For 1,4-dioxane, the composite rate constant is not significantly affected by the participation of ozone, due to the assumed effective molarity of ozone (which is less than its solubility, due to ongoing scavenging losses that will occur in natural groundwater), and the high kinetic rate constant for the hydroxyl radical reaction. In many other cases, the hydroxyl radical rate constant is much lower, and the participation of ozone in the reaction is very important (as is the case for n-nitrosodimethylamine, examined in Section 5.4.2.2).

A bench-scale test was performed to confirm the ozonation of 1,4-dioxane from a contaminated groundwater source in the U.S. midwest. The pretreatment 1,4-dioxane concentration was 9000 μg L^{-1}, and the initial pH was 7.8, with a carbonate alkalinity of approximately 100 mM as $CaCO_3$. A 1-L sample of the groundwater was placed in a 1-L ozonation chamber, through which a 300 mL min^{-1} gas sparge injection was maintained. An oxygen cylinder provided the sparge gas stream, which was passed through a two-channel UV spectrophotometer enroute to the sparge chamber and upon exit, to monitor ozone gas-phase concentrations. To account for possible 1,4-dioxane losses due to sparging, the chamber was sparged with oxygen only (no ozone) for a 40-min control period, then the ozone generator was placed online, supplying a 6% ozone-in-oxygen gas feed for an additional 40-min period, to measure the oxidation of 1,4-dioxane.

Figure 3.24 shows the results of the ozonation trial for 1,4-dioxane. There was no loss of 1,4-dioxane during the 40-min sparge period with oxygen and no ozone added, confirming the contention that 1,4-dioxane is not spargeable. A pseudo-first-order exponential decay line was drawn on Figure 3.24, using a rate constant at the middle of the calculated range of values, $0.0024\ s^{-1}$. Data from the first 10 min outperformed the pseudo-first-order projection, then the kinetic-calculated line overestimated 1,4-dioxane decomposition for the balance of the experiment. The average 1,4-dioxane remaining after 40 min was $145 \pm 7.1\ \mu g\ L^{-1}$, compared to a kinetic-predicted concentration of $28\ \mu g\ L^{-1}$. Other trials with a 3000 to 4000 $\mu g\ L^{-1}$ initial concentration decomposed 1,4-dioxane to less than $1\ \mu g\ L^{-1}$ in 40 min of ozonation.

In these studies, the 1,4-dioxane decomposition appeared to be bimodal, with the early-phase data representing a much faster decomposition process. This may be due to radical initiation reactions associated with naturally occurring dissolved organic matter or dissolved ferrous iron (Fe^{2+}) that was exhausted before the 1,4-dioxane decomposition was completed in the samples with high starting concentrations. These results indicate the importance of site-specific testing for ozonation rates, to account for effects of scavengers and initiators that may vary significantly from site to site.

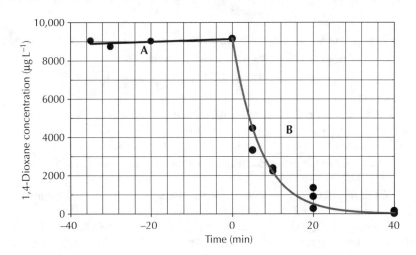

FIGURE 3.24 Results of bench-scale ozone sparging for destruction of 1,4-dioxane. During part A of the trial, an oxygen-only gas stream was sparged into the 1 L chamber at 300 ml min^{-1}. During the second phase (B), the gas mix was shifted to 6% ozone, 94% oxygen, and the flow of 300 ml min^{-1} was maintained. The individual data points are shown as solid circles. A linear regression line was drawn through the pretreatment sparge data, and the mid-range pseudo-first-order kinetic rate constant, $k' = 0.0024\ s^{-1}$, was used to generate the exponential decay line drawn through the ozonation phase data.

3.4.3.4 By-Products

Ozonation generates numerous classes of partial decomposition products and reaction end-products that may be of concern for drinking water protection. Reaction intermediates include ketones, aldehydes, and carboxylic acids. Many of these intermediates are less reactive with ozone or hydroxyl radical than the original organic in the reaction, and are susceptible to accumulation when there is kinetic competition from more reactive substrates. Formaldehyde is an example of a potential product of ozonation reactions and the second-order kinetic rate constant for its decomposition with ozone is 0.1 $M^{-1} s^{-1}$ (Acero et al.),[33] lower than for many potential ozonation targets. In most cases, these highly soluble reaction intermediates and by-products are very biodegradable in the aerobic aquifer conditions that follow application of ozone.

Ozone also has the capacity to react with bromide anion, Br^-, forming the bromate anion,[33] BrO_3^-, which often has a very low regulatory threshold concentration (e.g., 10 $\mu g\ L^{-1}$, or lower). The rate of bromate formation from bromide ion increases with increasing pH, and with increasing ozonation intensity. Von Gunten provided a detailed review of ozone oxidation chemistry[37] and by-product formation and minimization,[38] indicating that bromate formation can be limited somewhat by operation of ozonation systems at reduced pH and through the addition of ammonia, which suppresses formation of a key intermediate in the bromate formation reaction sequence.

3.4.4 PERSULFATE

The peroxidisulfate anion, more commonly known as persulfate, has recently gained attention as a potential oxidant in chemical reactive zone applications. At high temperatures (e.g., 50°C), persulfate anion ($S_2O_8^{2-}$) dissociates to form two sulfate radical anion molecules ($SO_4^{\cdot -}$).

$$S_2O_8^{2-} \xrightarrow{50°C} SO_4^{\cdot -} + SO_4^{\cdot -}$$

Sulfate radical anion is reactive with many organic aquifer contaminant compounds, although high activation energies for many potential target compounds will require thermal enhancement to generate workable reaction rates *in situ*. Thermal decomposition of persulfate will be uneconomical for most reactive zone applications, due to the high cost of heating an aquifer formation.

Reactions with reduced metal ions can generate sulfate radical anion from persulfate at ambient aquifer temperatures, although the radical yield is only half of the disproportionation approach. Ross et al.[7] cataloged 17 reactions of the persulfate anion, all of which are between persulfate and a metal cation or alkyl radical anion, producing sulfate radical anion. The reaction of persulfate anion with zinc is an example.

$$S_2O_8^{2-} + Zn^+ \rightarrow Zn^{+2} + SO_4^{2-} + SO_4^{\cdot -} \qquad (3.88)$$

The second-order reaction rate constant for Equation (3.88) is $1.3 \times 10^9\ M^{-1} s^{-1}$ at 25°C, according to Ross et al.[7] Similarly, rapid disproportionation of persulfate has been observed for cadmium, nickel, and cobalt,[7] and reactions with other reduced metal ions or minerals may be possible.

Ferrous iron ion (Fe^{2+}) reacts with persulfate anion according to the following reaction equation, although rates may be low at ambient aquifer temperatures.[46]

$$Fe^{2+} + S_2O_8^{2-} \rightarrow Fe^{3+} + SO_4^{2-} + SO_4^{\cdot-}$$

Cost-effectiveness of the metal-driven production of sulfate radical anion is likely to be greater than for the thermally driven disproportionation reaction. However, activation energies for reactions with the sulfate radical anion are high for many compounds, and reaction rates at ambient aquifer temperatures may be too slow to be of use for many potential target compounds.

Ross et al.[7] reported 340 reactions of the sulfate radical anion, including many with compounds that are notably unreactive with other oxidants. Low-molecular-weight alkanes, for example, are very reactive near 20°C, with second-order kinetic rate constants ranging from 1×10^6 through 4×10^7 for oxidation of methane through propane, respectively. Benzene and toluene kinetic rate constants were reported as 3×10^9 and 3.1×10^9, respectively. Reaction rates with common aquifer anions are also relatively high: for hydroxyl anion, the kinetic rate constant is $1.4 \times 10^7 \, M^{-1} \, s^{-1}$ at 25°C, and produces a hydroxyl radical.

$$SO_4^{\cdot-} + OH^- \rightarrow OH^{\cdot} + SO_4^{2-} \tag{3.89}$$

Hydroxyl radical production at high pH might generate increased oxidation rates for compounds that are otherwise resistant to sulfate radical anion, such as 1,1,1-trichloroethane or trichloroethene.

Reactivity with typical reactive zone target compounds limits the applicability of sulfate radical anion (i.e., persulfate). The activation energy for reaction of sulfate radical anion with trichloroethene is 98 kJ mol^{-1} and for 1,1,1-trichloroethane the value is 164 kJ mol^{-1}, according to studies by Liang et al.[47] Consequently, these compounds are essentially unreactive with persulfate at normal aquifer temperatures. In aqueous-phase oxidation trials the half-life for 1,1,1-trichloroethane decreased from 54.6 to 1.3 h as the temperature was increased from 40 to 60°C.

In field applications, many other compounds would be expected to compete for sulfate radical anion, and this was borne out by soil slurry studies reported by Liang et al.[47] In 1:5 w/w soil:water slurries, the half-lives of trichloroethene and 1,1,1-trichloroethane increased dramatically. Competitors may have included soil organic matter, as well as aquifer matrix minerals. Costs for aquifer heating that will be required to establish productive reaction rates for sulfate radical anion, as well as high dosing rates that will be required to overcome competitive losses, will reduce the utility of this oxidant for chlorinated alkene oxidation.

The reactivity of sulfate radical anion with many aliphatics, alcohols, and ethers may offer a suitable oxidant for those compounds in metal-catalyzed persulfate-based chemical oxidation zones. Although many of these compounds are reactive with hydroxyl radical, metal-catalyzed persulfate may provide an advantage, relative to metal-catalyzed hydrogen peroxide (Fenton's reagent), which generates large amounts of oxygen. For example, reactions between sulfate radical anion and *tert*-butyl alcohol, 1,4-dioxane and tetrahydrofuran have been cataloged by Ross et al.[7] Sulfate radical anion also reacts with reduced metals, including Fe(II), Mn(II), and As(II), and it may affect solubility of aquifer minerals.

3.5 CHEMICAL REDUCING ZONES

Reductive dechlorination and precipitation processes that were described in Chapter 2 can also be driven abiotically, through reagent injections. Lacking the catalytic function of microbially driven reduction processes, direct chemical reduction approaches require injection of energetic reagents at high concentrations, in lieu of the carbohydrates or organic acids that are used to stimulate bacterial reductions. The speed of reactive zone development is much greater for chemical reducing zones than for biological reducing zones, and that is most often the basis for selecting a chemical reducing zone over the less expensive, but slower, biological systems.

3.5.1 HYDROGEN

Molecular hydrogen is an electron donor (reducing agent) for microbial dechlorination reactions, and has been injected into aquifers to support dechlorination (e.g., Newell et al.).[48] The only applications of molecular hydrogen for direct chemical reduction in aquifers have used metal catalysts to overcome activation energy limitations. Studies reported by Lowry and Reinhard[49] and McNab et al.[50] indicate that palladium (Pd) and palladium-on-aluminum (Pd-on-Al) catalysts can dechlorinate a wide range of aromatic and aliphatic hydrocarbons. Lowry and Reinhard[49] pointed out that rapid and complete dechlorination can be achieved with catalytic hydrogenation, while zero-valent iron reactive walls react more slowly, and can only partially dechlorinate certain compounds, such as carbon tetrachloride.

Field deployment of catalytic hydrogenation systems has been limited to in-well systems, and the cost and maintenance requirements of the catalyst are likely to prevent *in situ* reactive zone deployment of these systems. There is not yet any indication in the literature suggesting it will be possible to develop catalytic hydrogenation systems with native aquifer minerals.

3.5.2 DITHIONITE

3.5.2.1 Dithionite Chemistry

The dithionite anion ($S_2O_4^{2-}$) is a moderately strong reducing agent that is used in paper-making processes to bleach pulp (e.g., Svensson et al.).[51] Dithionite is available as a sodium salt, and has recently been used to generate strong reducing conditions in aquifers, supporting processes that precipitate metals and dechlorinate ethene solvents. The dithionite anion dissociates to form two sulfoxyl radical anions, as shown in Figure 3.25 and Equation (3.90).

$$S_2O_4^{2-} \leftrightarrow SO_2^{\cdot-} + SO_2^{\cdot-} \qquad (3.90)$$

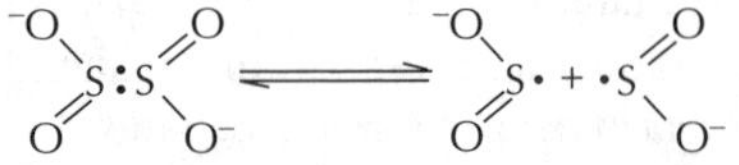

FIGURE 3.25 Dissociation of dithionite anion, forming two sulfoxyl radical anions.

The back reaction to re-form dithionite is quite fast ($k = 1.7 \times 10^9$ M^{-1} s^{-1} at 25°C, pH 6.5 according to Ross et al.),[7] so the molarity of the sulfoxyl radical anion is likely to remain low.

Dithionite may also react with water, as shown in Equation (3.91), to form sulfite (SO_3^{2-}) and thiosulfate ($S_2O_3^{2-}$) anions.

$$2S_2O_4^{2-} + H_2O \rightarrow 2SO_3^{2-} + S_2O_3^{2-} + 2H^+ \quad (3.91)$$

The hydrolysis reaction contributes to loss of dithionite in solution, particularly at elevated temperatures and low pH. The sulfite ion is the second dissociation product of sulfurous acid (H_2SO_3), a weak acid with a second dissociation constant of 1.23×10^{-7} (p$K_2 = 6.91$). In lower pH ranges, the sulfite ion produced by dissociation of dithionite is consumed by protonation, drawing more dithionite into the hydrolysis reaction.

$$H^+ + SO_3^{2-} \rightarrow HSO_3^- \quad pK_a = 6.91 \quad (3.92)$$

To minimize hydrolysis losses, dithionite solutions are prepared shortly before their planned application, and adjusted to high pH. Chilakapati et al.[52] buffered dithionite to pH 11, using a potassium carbonate/bicarbonate buffer, when they used dithionite to generate an *in situ* reducing zone for chromium reduction. The high pH suppressed the hydrolysis reaction of dithionite, shown in Equation (3.91).

The sulfoxyl radical anion ($SO_2^{\cdot-}$) is very reactive with oxygen in solution, producing superoxide radical anion ($O_2^{\cdot-}$) and sulfur dioxide gas (SO_2). The kinetic rate constant for the reaction is 2.4×10^9 M^{-1} s^{-1}. Dithionite-generated sulfoxyl radical anion should, therefore, be a very effective oxygen scavenger.

$$SO_2^{\cdot-} + O_2 \rightarrow SO_2 + O_2^{\cdot-} \quad (3.93)$$

The dithionite reactions that are valuable in the formation of *in situ* chemical reducing zones are heterogeneous reactions that occur with soil minerals. The sulfoxyl radical anion ($SO_2^{\cdot-}$), formed by dissociation of dithionite (Equation (3.90)), reacts with solid-phase iron(III) and water, to form iron(II) solid and sulfite anion, as shown in Equation (3.94).[53]

$$SO_2^{\cdot-} + Fe^{3+}(s) + H_2O \rightarrow Fe^{2+}(s) + SO_3^{2-} + 2H^+ \quad (3.94)$$

Dithionite-generated sulfoxyl radical does not always react quickly with dissolved metals or solvents. Larson and Cervini-Silva[54] studied reduction of substituted trihalomethanes (including trichloroacetic acid, pentachloroethane, trichloroacetonitrile, and carbon tetrachloride) by the reduced iron(II) of heme molecules.* Second-order dechlorination rate constants varied widely, from approximately 5×10^{-5} to 10^1 min^{-1}, with higher rate constants for compounds with more electronegative substitutions to the trihalomethanes molecule. They tested dithionite-only control solutions and found that no dechlorination products were formed,** although excess dithionite, relative to heme, increased observed dechlorination. This led the authors to conclude that the reduced iron(II) center of the heme molecule served as an electron transfer agent for dithionite, which supplied reducing power via sulfoxyl radical.

Nzengung et al.[55] found that homogeneous solutions of dithionite and heterogeneous solutions of dithionite and clay minerals slowly reduced perchloroethene in aqueous solution. Solution concentrations ranged from 11.5 to 137.8 mM and the cumulative dechlorination was greatest at 34 mM dithionite, for both homogeneous and heterogeneous reactions (55% reduction after 19 days in treated sodium-montmorillonite, 34% reduction in treated ferruginous smectite, and 33% reduction in the homogeneous dithionite solution). Dechlorination rates were much higher in heterogeneous solutions than in homogeneous dithionite, and more complete dechlorination was observed in solutions buffered at ph 8.5, compared to unbuffered solutions in which the pH dropped to 3.65 (consistent with expected loss of dithionite to hydrolysis at the lower pH). Minor amounts of TCE were formed by hydrogenation reactions, with higher rates of TCE formation at the lower pH. The authors observed that elimination reactions probably accounted for most of the dithionite-driven dechlorination, generating acetylene and related products. Dithionite-treated clays that were washed to remove dithionite prior to testing showed no perchloroethene dechlorination capacity. This suggests that, as with the heme structure studied by Larson and Cervini-Silva,[54] the clay mineral provided an electron transfer agent that increased dechlorination in the heterogeneous solution, relative to the homogeneous dithionite.

Both the Nzengung et al.[55] and Larson and Cervini-Silva[54] studies indicated that sulfoxyl radical-driven (dithionite) reduction was most effective in the presence of mineral or heme-stabilized metals that could serve as an electron-transfer agent. There was also an indication from these studies that coordination of the iron(II) in the heme and attached to solid mineral surfaces draws "inner sphere" electrons into the reaction, significantly increasing effectiveness of the reduction process.

Although homogeneous reactions may occur between dithionite and chlorinated organic compounds, it will be difficult to maintain high dithionite concentrations in flowing aquifers. *In situ* application of dithionite will be most effective when it generates reduced minerals that support heterogeneous reactions that dechlorinate perchloroethene.

An *in situ* dithionite-driven chromium reduction/precipitation strategy has been described by Chilakapati,[56] Istok et al.,[57] Fruchter et al.,[58] and Khan and Puls.[59]

---

*A heme group is a tetrapyrrole ring in which four nitrogens form a symmetrical pocket with four coordination bonding sites, similar to the structure of the cobalt-bearing vitamin B_{12} molecule, discussed later.
**It is possible that direct reduction occurred, but was too slow to have been observed during their study. Note the very slow homogeneous dithionite reduction of perchloroethene, reported by Nzengung et al.[55]

Dithionite is injected into aquifers to reduce structural ferric iron (Fe^{3+}) to ferrous iron (Fe^{2+}), or to generate other reduced metallic compounds that can serve in dechlorination reactions. The reduction of soil minerals generates an *in situ* chemical reducing zone that supports heterogeneous reduction of redox-sensitive metals, such as chromium, and reducible hydrophobic organics such as chlorinated ethenes and ethanes.

In field trials reported by Fruchter et al.,[58] 77,000 L (20,500 gal) of a buffered dithionite solution (average 0.066 M) was injected into a permeable formation that consisted of sandy gravel overlying sandy silt/clay. The injection fluid was pumped into the 8-in. diameter injection well at 20 gpm for 17.1 h. The fluid remained in place for the next 18.5 h. The spent dithionite solution was then recovered from the aquifer by withdrawal of 375,000 L (99,600 gal) of groundwater and dithionite solution, 4.9 times the initial injection volume. Estimates of the recovery mass balance were not provided in the cited publication.

The reduction of solid-phase mineral iron that was achieved by the dithionite injection was estimated by a coring study and subsequent estimation of oxygen reduction capacities of the samples as received, relative to maximum oxygen reduction capacity that could be induced by dithionite treatment in the laboratory. A small iron reduction effect (8% reduction) was detectable 27 ft from the injection well, while most of the dithionite impact was observed within a 10-ft radius, where seven samples, collected at various depths, at radial distances of 4 ft (four samples) and 10 ft (three samples), averaged 53% conversion of aquifer mineral iron content (calculated from data in Fruchter et al.).[58] The total aquifer volume to the greatest radius where iron reduction was observed was approximately 148,000 L (20% porosity, 8.2 m radius, 3.5 m average aquifer thickness), indicating that the injected fluid reached a greater aquifer volume than would have been achieved through plug-flow displacement. The pattern of iron reduction achieved was shown by Fruchter et al.[58] in cross-sectional figures, indicating heterogeneity of dithionite distribution. Results of the chemical reducing zone are discussed in the following section.

3.5.2.2 Dithionite-Driven Reduction of Chromium

Chromium VI minerals are highly soluble in groundwater, across the range of naturally occurring pH and redox conditions. Chromate (CrO_4^{2-}) and bichromate ($HCrO_4^-$) are among the commonly observed ions of Cr(VI). The dichromate ion ($Cr_2O_7^{2-}$) may also be observed at low pH. The solubilities Cr(VI) compounds in aerobic aquifers are very high: potassium chromate (K_2CrO_4) – $k_{sp} = 1.87 \times 10^2$; potassium dichromate ($K_2Cr_2O_7$) – $k_{sp} = 2.52 \times 10^{-1}$; calcium chromate ($CaCrO_4$) – $k_{sp} = 2 \times 10^{-2}$. Accordingly, removal of Cr(VI) ions from aquifers must be accomplished indirectly, through conversion to Cr(III) and subsequent precipitation. The potential Cr(III) precipitates are highly insoluble. Chromium(III) hydroxide [$Cr(OH)_3$], for example, is highly insoluble ($k_{sp} = 6.7 \times 10^{-31}$). Note that in plating fluid spill areas, there are often redox stabilizers, in addition to extremely high levels of dissolved chromium and other reducible metals such as nickel or zinc.

Precipitation strategies for chromium remediation entail two principal elements: reduction of Cr^{6+} to Cr^{3+}, and availability of a reactive anion that will form a durable

TABLE 3.10
Solubility Product Data for Common Soil Minerals

Chemical Compound Name	Mineral Name	Chemical Formula	Solubility Product K_{sp}	Temp. (°C)	Data Source
Barium carbonate	Witherite	$BaCO_3$	7×10^{-9}	16	
Barium sulfate	Barite	$BaSO_4$	8.7×10^{-11}	18	
Cadmium sulfide	Greenockite	CdS	3.6×10^{-29}	18	
Calcium carbonate	Calcite	$CaCO_3$	9.9×10^{-9}	15	
Calcium sulfate	Anhydrite	$CaSO_4$	1.95×10^{-4}	10	
Chromium fluoride		$Cr(F)_3$	6.6×10^{-11}	25	Internet
Chromium hydroxide		$Cr(OH)_3$	6.3×10^{-31}	25	Internet
Copper(II) sulfide	Covellite	CuS	8.5×10^{-45}	18	
Copper(I) sulfide	Chalcoctite	Cu_2S	2×10^{-47}	16–18	
Ferric oxyhydroxide	Goethite	$FeOOH$	3.16×10^{-42}		Internet
Iron(III) hydroxide		$Fe(OH)_3$	1.1×10^{-36}	18	
Iron(II) hydroxide		$Fe(OH)_2$	1.64×10^{-14}	18	
Iron(II) sulfide	Troilite	FeS	3.7×10^{-19}	18	
Iron(II) disulfide	Pyrite	FeS_2			
Lead(II) carbonate	Cerussite	$PbCO_3$	3.3×10^{-14}	18	
Lead(II) sulfate	Anglesite	$PbSO_4$	1.06×10^{-8}	18	
Lead(II) sulfide	Galena	PbS	3.4×10^{-28}	18	
Magnesium carbonate	Magnesite	$MgCO_3$	2.6×10^{-5}	12	
Magnesium hydroxide	Brucite	$Mg(OH)_2$	1.2×10^{-11}	18	
Manganese hydroxide	Pyrochroite	$Mn(OH)_2$	4×10^{-14}	18	
Manganese sulfide	Alabandite	MnS	1.4×10^{-15}	18	
Nickel sulfide	Millerite	NiS	1.4×10^{-24}	18	
Zinc hydroxide		$Zn(OH)_2$	1.8×10^{-14}	18–20	
Zinc sulfide	Sphaelerite	ZnS	1.2×10^{-23}	18	

Note: Data source is Weast[15] unless otherwise noted.

precipitate with Cr^{3+}. Table 3.10 provides solubility data for several chromium minerals. The solubility products (K_{sp} for chromium hydroxide, e.g., is 6.6×10^{-31}) indicate that Cr^{3+} minerals will be generally insoluble. The selection of an anion to drive the precipitation process is based on the kinetics of the precipitation process. Ultra-stable minerals that only form at high temperatures, or that form over extended periods (tens or hundreds of years) do not form the basis of an effective *in situ* reactive zone strategy. Kinetic rate constants for the precipitation reactions must form durable precipitates within minutes to hours (days at the most), because it is often not desirable (from a cost perspective as well as a secondary water quality impact perspective) to sustain the pH and redox conditions necessary to induce precipitation for extended periods.

The reduction of chromium VI may be achieved by reductants that can be expected in groundwater of low redox potential. Hydrogen sulfide is one such example, and has received attention as one of the major reduced inorganic species that is associated with reduction of aquifer matrices.

Patterson et al.[60] studied the reduction of chromium(VI) by ferrous iron in solution, with pH ranging from 5 to 8. They generated dissolved ferrous iron using a solid-phase ferrous sulfide (FeS) suspension in water, under a nitrogen atmosphere that limited oxidation losses of the dissolved ferrous iron. The mineral mackinawite, FeS, is slightly soluble in water.

$$H^+ + FeS(s) \rightarrow Fe^{2+} + HS^- \quad k_{sp} = 2.86 \times 10^{-4} \quad (3.95)$$

The solubility is pH sensitive, as seen by the participation of a proton, H^+, in the solubilization reaction. Patterson et al.[60] observed increasing Fe^{2+} concentrations as pH decreased: 0.050 mM at pH 7, and 0.77 mM at pH 5.

Two reactions between ferrous iron and chromium were observed.

$$3CrO_4^{2-} + FeS(s) + 9H_2O \rightarrow 4[Cr_{0.75}Fe_{0.25}](OH)_3 + Fe^{2+} + S_2O_3 + 6OH^- \quad (3.96)$$

and

$$3Fe^{2+} + HCrO_4^- + 8H_2O \rightarrow 4[Cr_{0.25}Fe_{0.75}](OH)_3 + 5H^+ \quad (3.97)$$

The authors observed increasing pH during the chromium reduction and concluded that the former reaction was the dominant mechanism of chromium removal. Chromium removal was observed over the pH range studied (5 to 8); however, a small amount of chromium(VI) remained in solution at pH 8, due to the limited solubility of FeS at high pH.[60] The initial aqueous-phase chromium concentration in their studies ranged from 50 to 5000 μM (2.6 to 260 mg L^{-1} Cr^{6+}).

The occurrence of reduced iron minerals such as FeS in the oxidized groundwater normally associated with high Cr^{6+} concentrations is unlikely, so the ferrous iron–chromium precipitation strategy requires a method to solubilize Fe^{2+} from soil minerals, or to deliver and sustain Fe^{2+} to the aquifer directly through reactive zone injections. One very effective strategy for dissolution of Fe^{2+} and precipitation of Cr^{6+} is to utilize aquifer microbial communities for the depression of aquifer redox potentials, driven by injection of degradable organic carbon compounds at rates that significantly exceed recharge of electron acceptors such as oxygen or nitrate (Suthersan).[61] Microbially driven chromium and nickel precipitation is discussed in detail in Chapters 2 and 5.

It is also possible to reduce Fe(III) aquifer minerals, through injection of reducing agents that attack Fe(III) minerals in the aquifer soil. This enables solid-phase chromium reduction (Equation (3.96)) and solubilizes Fe^{2+} from the newly formed Fe(II) minerals, providing an opportunity for the aqueous-phase chromium reduction reaction (Equation (3.97)). The dithionite ion, $S_2O_4^{2-}$, is one such reductant.

Chilakapati et al.[52] described a process for creation of an *in situ* reduced-iron barrier, using periodic sodium dithionite injections. The main reaction of the dithionite ion with Fe(III) soil minerals is:

$$S_2O_4^{2-} + 2Fe(III)(s) + 2H_2O \xrightarrow{k_1} 2SO_3^{2-} + 2Fe(II)(s) + 4H^+ \quad (3.98)$$

Both reactions are near first order with respect to the dithionite concentration.

Chilakapati et al.[52] noted that the disproportionation reaction is minimized, and the iron reduction rate is maximized, at high pH. To take advantage of those reaction characteristics, they delivered dithionite in a potassium carbonate/bicarbonate buffer, at four times the dithionite concentration. This produced an injection fluid of pH 11. The tested aquifer was poorly buffered by the soil matrix, and the high pH condition was likely maintained while the dithionite reagent remained active.

While the pH remains high, the chromium precipitation reactions are expected to be limited to the solid-phase reaction, Equation (3.96). As groundwater flushes the reactive barrier zone of the aquifer, greater Fe^{2+} solubilization is likely to occur and aqueous-phase reaction with Cr^{6+} is likely to occur.

The resulting Fe(II) minerals react with dissolved chromium arriving in the reactive zone, gradually consuming the active sites. Reducing capacity is also lost to other electron acceptors such as dissolved oxygen and nitrates. These are reversible reactions, and the dithionite reductant must be applied periodically to reestablish Fe(II) minerals in the reactive zone, at a frequency that is determined by the total electron acceptor flux (which determines the rate of consumption of the Fe(II) minerals). The number of regenerations that can be accomplished, reestablishing an effective *in situ* reactive zone is determined by the accumulated irreversible mineralization of the Fe(II) over time. Cr(III)/Fe(III) and any other irreversible nonchromium minerals formation permanently removes reducing capacity from the reactive zone. The dimensions of the reactive zone must be sufficient to accept all the irreversible mineralization that will occur over the life of the project.

3.5.3 Zero-Valent Iron

Zero-valent iron metal, Fe^0, can be used as a solid-phase reducing agent in chemical *in situ* reactive zones. A number of factors determine the reactivity of the metal, including surface area per volume, processing and handling protocols (affecting corrosion reactivity properties at deployment), and alloys and impurities (catalytic metals such as palladium, Pd, have been alloyed to achieve desired catalytic properties in some applications).

To achieve effective reduction, the iron surface area concentration[*] must be significant and the contact time must be long, relative to other *in situ* chemical reactive processes. To achieve these requirements, zero-valent iron is generally deployed in a trench, perpendicular to groundwater flow, that has been backfilled with a mixture of zero-valent iron and sand, or other porous supporting medium. This approach is suitable for creating a containment barrier, but has not been adapted to use in contaminant source zones.

An alternative method of zero-valent iron deployment that can achieve high surface area concentrations via direct injection into aquifer formations is the use of nanoscale iron. Conventional zero-valent iron particles are millimeter sized. Nanoscale iron particles (10 to 200 μm) have very high surface area to volume ratios, and therefore can achieve high surface area concentrations with a much smaller mass of metal than is required with millimeter-sized iron particles employed in conventional zero-valent iron reactive zones.

---

[*]The surface area of solid per volume of liquid, expressed as $m^2 L^{-1}$, is the surface area concentration.

Numerous efforts are under way to commercialize nanoscale iron technologies, which, if successful, may allow direct injection of zero-valent iron metals and alloys into contaminant source zones, to achieve reductive destruction or precipitation of target compounds. The very high surface area to mass ratio of nanoscale iron makes it extremely reactive and difficult to handle safely and high rate of hydrogen production can occur in aquifer formations. Particle size selection is also critical — if too small, electrostatic interaction with aquifer mineral surfaces can immobilize and passivate the material; if too large, the particles sediment. These are all issues under study in the commercialization efforts.

3.5.3.1 Zero-Valent Iron Reaction Mechanisms

Zero-valent iron reactive zones initiate a complex network of homogeneous and heterogeneous reactions that may directly or indirectly destroy or precipitate contaminants. Some of the reactions may be catalyzed by mineral surfaces or by metal-reducing microbial populations that develop *in situ*. Many reactive zone processes consume the zero-valent iron, and any zero-valent iron application has a finite life span that is directly related to the flux and chemical composition of groundwater passing through the reactive zone. The most common applications of zero-valent iron reactive walls have targeted hexavalent chromium and other dissolved metals (e.g., Puls et al.).[62] Many examples of zero-valent iron applications for chlorinated solvents such as perchloroethene and trichloroethene have also been published (e.g., Roberts et al.).[63]

Blowes et al.[64] provide a summary of theory and field experience for conventional zero-valent iron applications. They identified two key reactions that drive chromium removal in zero-valent iron reactive walls: a heterogeneous reduction reaction that is followed by precipitation of the reduced chromium as shown in Equations (3.99) and (3.100), respectively.

$$CrO_{4(aq)}^{2-} + Fe_{(s)}^{0} + 8H_{(aq)}^{+} \rightarrow Fe_{(aq)}^{3+} + Cr_{(aq)}^{3+} + 4H_2O_{(l)} \qquad (3.99)$$

$$(1-x)Fe_{(aq)}^{3+} + (x)Cr_{(aq)}^{3+} + 2H_2O_{(l)} \rightarrow Fe_{(1-x)}Cr_{(x)}OOH_{(s)} + 3H_{(aq)}^{+} \qquad (3.100)$$

The net consumption of protons would be expected to raise pH during the reaction series.

Numerous competing reactions are expected in aquifer formations, and the longevity of zero-valent iron systems is more likely to be determined by consumption of nontarget compounds. Furakawa et al.[65] provided two examples of competing reactions.

$$2Fe_{(s)}^{0} + O_{2(g)} + 2H_2O_{(l)} \rightarrow 2Fe_{(aq)}^{2+} + 4OH_{(aq)}^{-} \qquad (3.101)$$

$$Fe_{(s)}^{0} + 2H_2O_{(l)} \rightarrow Fe_{(aq)}^{2+} + H_{2(g)}^{+} + 2OH_{(aq)}^{-} \qquad (3.102)$$

The oxic reaction shown in Equation (3.101) is likely to occur in the upgradient interface between the zero-valent iron mass and influent groundwater. Further along the migration pathway, the anoxic reaction shown in Equation (3.102) is likely to occur. Competitive reactions are also expected with oxyanions such as nitrate and sulfate, and Puls et al. (1999) observed significant decreases of sulfate concentrations in groundwater passing through a reactive barrier constructed for chromium reduction.

Reactions with zero-valent iron lead to significant pH increases, through consumption of protons (e.g., Equation (3.99)) as well as release of hydroxyl anions (e.g., Equations (3.101) and (3.102)). Puls et al. (1999) observed pH increases across a zero-valent iron reactive barrier. On the influent side pH ranged from 5.5 to 6.1, while values on the downgradient side ranged from 7.5 to 9.9.

The longevity of zero-valent iron reactive zones is determined by the rate of decrease of the Fe^0 surface area concentration. There are two main modes of surface area concentration loss:

Consumption. The reactions between Fe^0 and target compounds, as well as reactions with nontarget oxidants, consume the Fe^0 reagent. Nontarget reactions (e.g., with O_2, NO_3^-, and SO_4^-) are expected to dominate reactive iron consumption in most systems.

Passivation. The Fe^0 surfaces may become coated with minerals that block the heterogeneous reactivity, or passivate, the Fe^0. Formation of magnetite, for example, passivates the zero-valent iron surfaces and has been observed in field application sites.[65]

Mineral precipitate formation accompanies many of the iron-consuming reactions, so that dissolved iron concentrations on the effluent side of a reactive barrier may be relatively low, considering the large mass of iron deployed. Puls et al.,[62] for example, observed Fe^{2+} concentrations ranging from 2 to 20 mg L^{-1} in the effluent of a zero-valent iron reactive barrier wall in the eastern U.S.

Furakawa et al.[65] studied the mineralogy of the barrier described by Puls et al.,[62] and an additional iron reactive barrier near Denver, Colorado. Both walls had been in place for 4 years at the time of mineral sample collections. Ferrihydrite formation was significant, but the mineral did not accumulate on the Fe^0 surfaces and therefore did not contribute to passivation. The authors pointed out that ferrihydrite may be beneficial, as the site of arsenate and Cr(III) adsorption. They also observed numerous other minerals in the reactive wall samples, and concluded that the geochemical environment within the reactive wall was highly heterogeneous. They also pointed out the possible role of metal-reducing bacteria in the formation of many of the observed minerals.

3.5.3.2 Zero-Valent Iron Applicability

Zero-valent iron, through direct or indirect reactions, has the potential to treat numerous inorganic and organic contaminants.

Chlorinated alkenes. The chlorinated alkenes are not equally reactive with zero-valent iron. Perchloroethene, trichloroethene, and *trans*-dichloroethene react more quickly than *cis*-dichloroethene, 1,1-dichloroethene, and vinyl chloride. Roberts et al.[63] studied dechlorinating reactions between Fe^0 and chlorinated ethenes, and postulated that a two-electron concerted reaction process drives most of the dechlorination reactions with Fe^0, and this elimination reaction process is much more rapid for *trans*-chlorides.

Wang and Farrell[66] point out that multiple reaction mechanisms are probably at work, which contributes to differences in relative reactivities that have been observed for perchloroethene vs. trichloroethene, and for trichloroethene vs. 1,2-dichloroethene reaction rates. Potential dechlorination reaction mechanisms include β-elimination and catalytic hydrogenation (note that molecular hydrogen forms in the anoxic corrosion of Fe^0, as shown in Equation (3.102)).

Chlorinated methanes. Johnson et al.[67] studied the degradation of carbon tetrachloride by zero-valent iron and found that reaction rates were controlled by heterogeneous reaction with Fe^0.

$$Fe^0_{(s)} + CCl_{4(aq)} + H^+_{(aq)} \rightarrow Fe^{2+}_{(aq)} + CHCl_{3(aq)} + Cl^-_{(aq)} \quad (3.103)$$

A homogeneous reaction, Equation (3.104), may be possible with the ferrous ion (Fe^{2+}), but the authors pointed out that it is likely to be very slow.

$$2Fe^{2+} + CCl_4 + H^+ \rightarrow 2Fe^{3+} + CHCl_3 + Cl^- \quad (3.104)$$

Johnson et al.[67] did not discuss degradation of the chloroform produced by reaction of carbon tetrachloride with zero-valent iron. The fate of chloroform should be investigated further if zero-valent iron reactive barriers are considered for the treatment of carbon tetrachloride.

Arsenic. Arsenic is a semimetallic element that occurs most commonly in two oxidation states in groundwater, As(V), arsenate, and As(III), arsenite. Under highly reducing conditions, elemental arsenic, As (0), and its gaseous form, arsine, As (–III) may also be possible. Arsenite forms arsenous acid, H_3AsO_3, which dissociates to $H_2AsO_3^-$ above pH 9, and arsenate forms arsenic acid, H_3AsO_4, which can undergo three dissociations from pH 2 to 12, yielding AsO_4^-. Bissen and Frimmel[68] and Smedley and Kinniburgh[69] provide extended reviews of arsenic geochemistry and mobility.

Arsenic solubility is dominated by the interaction of its oxyanions with iron and manganese minerals, particularly through coprecipitation and adsorption into the hydrous structure of amorphous iron minerals.[68] Aquifer pH or redox potential decreases solubilize arsenic bound in the iron and manganese minerals, as those compounds are released into the aqueous phase (the ferric–ferrous switch occurs as the redox potential falls below +200 mV). Arsenic solubility decreases at very low redox potentials and low pH, due to the formation of arsenic sulfides (AsS and As_2S_3) and arsenopyrite (FeAsS). The sulfides are readily oxidized and solubilized if redox potentials or pH increase.[68]

Zero-valent iron reactive barriers entail large-scale corrosion of zero-valent iron surfaces, and many iron mineral precipitates form that may serve to trap arsenic. Su and Puls[70,71] studied arsenic immobilization in zero-valent iron batch and column studies, and concluded that arsenic precipitation may be a viable mechanism in field-scale iron reactive barriers. However, arsenic sorption processes will be significantly impacted by local variations in the pH and reduction–oxidation conditions that develop in these reactive barriers, and arsenic precipitation will not occur in all sectors of the wall. Site-specific aquifer chemistry will control the ability of a zero-valent iron to achieve desired arsenic treatment, and pilot-scale testing is essential to support feasibility determinations for this application.

Chromium. The small-scale field trial reported by Puls et al.[62] indicates that zero-valent iron reactive walls can generate significant reduction of dissolved chromium concentrations. In their test of auger-installed Fe^0, the authors observed a reduction of Cr^{6+} from 1 to 3 mg L^{-1} to less than 0.01 mg L^{-1} on passage through the reactive barrier. The reactive barrier in this case was constructed from equal parts of site surficial sand, aquifer sediments, low-grade steel lathe turnings, and cast iron particles. The iron particles ranged from 0.1 to 2 mm and the steel turnings ranged from 1 to 10 mm size; the specific surface areas were 1.1 m^2 g^{-1} for the iron and 8.3 m^2 g^{-1} for the steel turnings. For comparison, the specific surface area of the surficial sand was <1 and 5.8 m^2 g^{-1} for the aquifer sediments. Bulk densities were 2.1 g cm^{-3} for the iron particles, 2.3 g cm^{-3} for the steel turnings, 1.8 g cm^{-3} for the aquifer sediments, and 1.4 g cm^{-3} for the surficial sand. A total of 21 auger "tubes" of the mixture were installed over a 5-m-thick portion of the aquifer, in a pattern that measured approximately 3 m perpendicular to groundwater flow and 1.5 m along the groundwater flow path. The authors estimated the in-place diameters of the 21 mixture tubes was 20 cm.

Pesticides. Comfort et al.[72] reported on the use of zero-valent iron, combined with aluminum sulfate, $Al_2(SO_4)_3$, and acetic acid for treatment of the pesticide metolachlor in surface windrow stockpiles. Metolachlor soil concentrations were reduced from 1402 to 13 mg kg^{-1} during a 90-day incubation with the combined reagents. Fe^0 alone achieved much less metolachlor destruction, from 1789 to 504 mg kg^{-1}. The authors indicated that the aluminum sulfate generated pH reduction, and sulfates may have enhanced iron corrosion. Their visual observations of the soil stockpiles suggested that green rust was formed, which may have enhanced metolachlor decomposition rates. The reagent dosages were as follows (based on soil oven-dry weight): Fe^0, 5% w/w; $Al_2(SO_4)_3$, 2% w/w; CH_3COOH, 0.5% v/w. Based on their assumption that the soil bulk density was 1400 kg m^{-3}, the mass dosing rates were 70 kg m^{-3} for Fe^0, 28 kg m^{-3} for $Al_2(SO_4)_3$, and 7 L m^{-3} for acetic acid. The zero-valent iron was unannealed iron with a specific surface area of 2.55 m^2 g^{-1}. The reagents were applied to the soils in windrows by mechanical mixing, with water and diluted acetic acid applied to raise the soil water content to between 0.35 and 0.40 kg H_2O per kilogram soil. The windrows were rewetted five times during the 90-day treatment period, and soils were mixed at the outset and prior to the final sampling at 90 days.

3.5.4 REDUCED VITAMIN B_{12}

The reduced form of the cobalt-bearing vitamin B_{12} is a strong reducing agent that can dechlorinate alkenes and methanes. In its normal biological function, B_{12} is a coenzyme that transfers reducing and oxidizing potential as it cycles between its reduced, Co(II), and its oxidized, Co(III), states. According to Voet and Voet,[73] B_{12} (also known as cobalamin or cyanocobalamin) is "a reversible free radical generator," and provides the only known example of a carbon–metal bond in biological systems.

The B_{12} molecule contains a cobalt atom, coordinated with four nitrogen atoms at the center of a tetrapyrrole ring.[74] The fifth bonding position of the cobalt atom is taken by a nitrogen bond with dimethylbenzimidazole, a heterocyclic base, and the sixth coordination position of the cobalt atom is the "active" carbon–cobalt bond. In organisms, the active site is bonded to a hydroxyl anion or a water molecule, while in the isolated form, the sixth position is occupied by a cyanide group (hence the molecule's name, cyanocobalamin).[74]

In its application in chemical reactive zones, B_{12} is purchased in its cyanocobalamin form, hydrated to aquocob(III)alamin, then reduced to its Co(II) form, B_{12r}, and its fully reduced Co(I) form, B_{12s}. Effective reduction of chlorinated hydrocarbons depends on formation of the fully reduced B_{12s} form, and this is accomplished by providing excess titanium(III) citrate. In laboratory studies[75,76] the B_{12} remained viable through multiple reduction–oxidation cycles. Reduced B_{12} and titanium(III) citrate are supplied in aqueous phase for *in situ* applications, and several researchers have studied immobilization of B_{12} for fixed-bed applications.

Fully reduced vitamin B_{12} has been shown to dechlorinate alkenes, including perchloroethene and trichloroethene,[76] and carbon tetrachloride.[77,78] Field applications have shown effective dechlorination of alkenes,[79] although publication of peer-reviewed data has not been completed. Woods et al.[75] studied the dechlorination of penta-chlorinated biphenyls (penta-PCBs), in aqueous phase and sorbed to sediments. They prepared solutions of cyanocobalamin (0.5 mM in aqueous tests and 0.33 mM in sediment suspension tests), with excess titanium(III) citrate (7 mM in aqueous tests and 80 mM in sediment tests). They found that fully reduced B_{12} could achieve between 20 and 30% reduction of the penta-PCB over a 20-day period, after which the remaining penta-PCBs were not susceptible to reduction. Dechlorination of the tetra- and less-chlorinated PCBs continued for up to 200 days, after which titanium(III) citrate was exhausted.

Zou et al.[80] investigated intracellular vitamin B_{12} levels in microbial cultures that dechlorinated carbon tetrachloride, and found dechlorination rates were linked to vitamin B_{12} levels. However, they observed that kinetic rate constants from their "live culture" testing were lower than those observed in vitamin B_{12} extracts maintained under highly reducing conditions by the addition of titanium(III) citrate, reported by other researchers.[78] This suggests that more rapid reactions could be achieved with the injection of reduced vitamin B_{12} than can be achieved through stimulation of B_{12}-producing microbial communities.

Application of chemical reduction through vitamin B_{12} injection is likely to be limited because the cost of application of reduced vitamin B_{12} is quite high, relative to reductive dechlorination through microbial stimulation. For systems in which B_{12} is selected, the loss of B_{12} in an injection-only mode of application would be very costly, so it may be more likely to be applied in a recirculation mode, with regeneration of fully reduced B_{12} through aboveground addition of titanium(III) citrate and augmentation of B_{12}, only as necessary to maintain acceptable reduction rates.

3.6 CHEMICAL PRECIPITATION STRATEGIES

Chemical precipitation strategies are built on modification of an aquifer chemical environment to promote conversion of soluble compounds to insoluble compounds. Development of a chemical precipitation strategy entails two main elements: (1) characterization of the chemical environment that will persist after the physical and chemical impact of the injected reagent has dissipated, and (2) selection of a precipitation reaction that will maintain acceptable aqueous-phase contaminant concentrations in the persistent chemical environment. It is important to note that the chemical environment that develops during reagent injection (e.g., pH, ionic strength, E_H) may be very different for the persistent chemical environment that emerges after reagent injections are completed.

Precipitation of dissolved inorganic contaminants such as arsenic, chromium, or nickel can be driven by the creation of oxidative or reductive *in situ* reactive zones, such as those described earlier. Precipitation strategies are built on pH and E_H effects on the stabilities of target elements. The status of the aqueous-phase chemical environment can be summarized by two master variables:* pH, which measures the tendency of the solution to accept protons, and pϵ (the negative of the log of electron activity), which measures the tendency of the solution to accept electrons. The concept of pH and pϵ as master variables is described in Stumm and Morgan,[5] which is recommended for further reading. Stability diagrams can be used to identify the compounds of each element that are most stable for any pH–E_H condition. There are three important limitations on the use of stability diagrams in reactive zone design:

Availability of chemical equilibrium data. Stability diagrams are built from chemical equilibrium data, and published data may not adequately describe the reactions that will occur in a treatment zone.

Equilibria are thermodynamic properties. Chemical equilibria, and the determination of the "most stable" compound, are based on thermodynamic properties of the compounds under study. The rate at which reactions occur is also critical — it is possible to identify "most stable" compounds from stability diagrams that are formed only over very long time scales at aquifer temperatures. The compounds that form in reactive zones are determined by kinetic as well as thermodynamic properties of potential reactions. Kinetic rate constants often control the outcome of reactive zone endeavors.

Reactive zones violate chemical activity assumptions. The chemical activity of charged elements and compounds in solution (ions) is controlled by the solution's ionic strength, a measure of the effect of charged species on solution chemistry (refer to Section 3.2.6). Many chemical reactive zones are operated at very high ionic strength, and chemical properties of potential reactants should be corrected for ionic strength effects, when necessary.

---

* For *in situ* chemical reactive zones, the conventional assumption of a dilute aqueous-phase solution is often violated. Consequently, ionic strength may also be considered as a critical variable in characterization of aqueous-phase chemical environment. Ionic strength influences both homogeneous and heterogeneous reactions.

REFERENCES

1. Kotz, J.C. and P. Treichel, Jr., *Chemistry and Chemical Reactivity*, 3rd ed., Saunders College Publishing, Philadelphia, 1996.
2. Vogel, F., J. Harf, A. Hug, and P.R. von Rohr, The mean oxidation number of carbon (MOC) — a useful concept for describing oxidation processes, *Water Res.*, 34(10), 2689–2702, 2000.
3. Barrow, G.M., *Physical Chemistry*, 4th ed., McGraw-Hill, New York, 1979.
4. Wang, Q., J. Gan, S.K. Papiernik, and S.R. Yates, Transformation and detoxification of halogenated fumigants by ammonium thiosulfate, *Environ. Sci. Technol.*, 34, 3717–3721, 2000.
5. Stumm, W. and J.J. Morgan, *Aquatic Chemistry*, John Wiley & Sons, New York, 1996.
6. Masten, S. and S. Davies, Michigan State University, personal communication.
7. Ross, A.B., W.G. Mallard, W.P. Helman, G.V. Buxton, R.E. Huie, and P. Neta, NDRL-NIST Solution Kinetics Database — Version 3, Notre Dame Radiation Laboratory, Notre Dame, IN, and NIST Standard Reference Data, Gaithersburg, MD, 1998.
8. Hassan, K.Z.A., K.C. Bower, and C.M. Miller, Numerical simulation of bromate, formation during ozonation of bromide, *J. Environ. Eng.*, 129(11), 991–998, 2003.
9. Gan, J., S.R. Yates, J.O. Becker, and D. Wang, Surface amendment of fertilizer ammonium thiosulfate to reduce methyl bromide emission from soil, *Environ. Sci. Technol.*, 32, 2438–2441, 1998.
10. Gan, J., Q. Wang, S.R. Yates, W.C. Koskinen, and W.A. Jury, Dechlorination of chloroacetanilide herbicides by thiosulfate salts. *Proc. Nat. Acad. Sci.* 99(8): 5189–5194, 2002.
11. Wang, Q., J. Gan, S.K. Papiernik, and S.R. Yates, Transformation and detoxification of halogenated fumigants by ammonium thiosulfate, *Environ. Sci. Technol.*, 34(17), 3717–3721, 2000.
12. Schwarzenbach, R.P., P.M. Gschwend, and D.M. Imboden, *Environmental Organic Chemistry*, 2nd ed., John Wiley & Sons, New York, 2003.
13. Washington, J.W., Hydrolysis rates of dissolved volatile organic compounds: principles, temperature effects and literature review, *Ground Water*, 33, 415–424, 1995.
14. Pagan, M., W.J. Cooper, and J.A. Joens, Kinetic studies of the homogeneous abiotic reactions of several chlorinated aliphatic compounds in aqueous solution, *Appl. Geochem.*, 13(6), 779–785, 1998.
15. Weast, R.C., *Handbook of Chemistry and Physics*, 50th ed., The Chemical Rubber Company, Cleveland, 1969.
16. Amonette, J.E., D.J. Workman, D.W. Kennedy, J.S. Fruchter, and Y.A. Gorby, Dechlorination of carbon tetrachloride by Fe(II) associated with goethite, *Environ. Sci. Technol.*, 34(21), 4606–4613, 2000.
17. Wang, X. and M.L. Brusseau, Effect of pyrophosphate on the dechlorination of tetrachloroethene by the Fenton reaction, *Environ. Toxicol. Chem.*, 17(9), 1689–1694, 1998.
18. Sun, Y. and J.J. Pignatello, Chemical treatment of pesticide wastes. Evaluation of Fe(III) chelates for catalytic hydrogen peroxide oxidation of 2,4-D at circumneutral pH, *J. Agric. Food Chem.*, 40(2), 322–327, 1992.
19. Teel, A.L., C.R. Warberg, D.A. Atkinson, and R.J. Watts, Comparison of mineral and soluble iron Fenton's catalysts for the treatment of trichloroethene, *Water Res.*, 35(4), 977–984, 2001.
20. Carus Chemical Corporation, Permanganate product documentation, 2003.
21. U.S. Department of Energy, Type B Accident Investigation. Injury resulting from violent exothermic chemical reaction at X-701B Site, Portsmouth Gaseous Diffusion Plant, DOE/ORO-2103, 2000.

22. Ibaraki, M. and F.W. Schwartz, Influence of natural heterogeneity on the efficiency of chemical floods in source zones, *Ground Water*, 39(5), 660–666, 2001.
23. Yan, Y.E. and F.W. Schwartz, Kinetics and mechanisms for TCE oxidation by permanganate, *Environ. Sci. Technol.*, 34(12), 2535–2541, 2000.
24. U.S. EPA, Soil Screening Guidance: Technical Background Document, EPA/540/R95/128, May, 1996.
25. Seol, Y., and F.W. Schwartz, Phase-transfer catalysis applied to the oxidation of non-aqueous phase trichloroethylene by potassium permanganate, *J. Contam. Hydrol.*, 44, 185–201, 2000.
26. Seol, Y., and F.W. Schwartz, Oxidation of binary DNAPL mixtures using potassium permanganate with a phase transfer catalyst, *Ground Water Monitoring Remediation*, 21(2), 124–132, 2001.
27. MacKinnon, L.K. and N.R. Thomson, Laboratory-scale *in situ* chemical oxidation of a perchloroethylene pool using permanganate, *J. Contam. Hydrol.*, 56(1–2), 49–74, 2002.
28. Schnarr, M., C. Truax, G. Farquhar, E. Hood, T. Gonullu, and B. Stickney, Laboratory and controlled field experiments using potassium permanganate to remediate trichloroethylene and perchloroethylene DNAPLs in porous media, *J. Contam. Hydrol.*, 29(3), 205–224, 1998.
29. Schroth, M.H., M. Oostrum, T.W. Wietsma, and J.D. Istok, *In-situ* oxidation of trichloroethene by permanganate: effects on porous medium hydraulic properties, *J. Contam. Hydrol.*, 50(1–2), 79–98, 2001.
30. Ternes, T.A., M. Meisenheimer, D. McDowell, F. Sacher, H-J. Brauch, B. Haist-Gulde, G. Preuss, U. Wilme, and N. Zulei-Seibert, Removal of pharmaceuticals during drinking water treatment, *Environ. Sci. Technol.*, 36(17), 3855–3863, 2002.
31. Pines, D.S., and D.A. Reckhow, Effects of dissolved cobalt(II) on the ozonation of oxalic acid, *Environ. Sci. Technol.*, 36(19), 4046–4051, 2002.
32. Xu, X. and W.A. Goddard III, Peroxone chemistry: formation of H_2O_3 and ring-$(HO_2)(HO_3)$ from O_3/H_2O_2, *Proc. Nat. Acad. Sci.*, 99(24), 15308–15312, 2002.
33. Acero, J.L., S.B. Haderlein, T.C. Schmidt, M.J.F. Suter, and U. Von Gunten, MTBE oxidation by conventional ozonation and the combination ozone/hydrogen peroxide: efficiency of the processes and bromate formation, *Environ. Sci. Technol.*, 35(21), 4252–4259, 2001.
34. Von Gunten, U. and Y. Oliveras, Kinetics of the reaction between hydrogen peroxide and hypobromous acid: implication on water treatment and natural waters, *Water Res.*, 31(4), 900–906, 1997.
35. Von Gunten, U. and Y. Oliveras, Advanced oxidation of bromide-containing waters: bromate formation mechanisms, *Environ. Sci. Technol.*, 32(1), 63–70, 1998.
36. Pinkernell, U. and U. Von Gunten, Bromate minimization during ozonation: mechanistic considerations, *Environ. Sci. Technol.*, 35(12), 2525–2531, 2001.
37. Von Gunten, U., Ozonation of drinking water: Part I. Oxidation kinetics and product formation, *Water Res.*, 37, 1443–1467, 2003.
38. Von Gunten, U., Ozonation of drinking water: Part II. Disinfection and by-product formation in the presence of bromide, iodide or chlorine, *Water Res.*, 37, 1469–1487, 2003.
39. Zenker, M.J., R.C. Borden, and M.A. Barlaz, Mineralization of 1,4-dioxane in the presence of a structural analog, *Biodegradation*, 11(4), 239–246, 2000.
40. Aitchison, E.W., S.L. Kelley, P.J.J. Alvarez, and J.L. Schnoor, Phytoremediation of 1,4-dioxane by hybrid poplar trees, *Water Environ. Res.*, 72(3), 313–321, 2002.
41. Dietz, A.C. and J.L. Schnoor, Advances in phytoremediation, *Environ. Health Perspect.*, 109, 163–168, 2001.

42. Ying, O.Y., Phytoremediation: modeling plant uptake and contaminant transport in the soil–plant–atmosphere continuum, *J. Hydrol.*, 266(1–2), 66–82, 2002.
43. Burback, B.L. and J.J. Perry, Biodegradation and biotransformation of groundwater pollutant mixtures by *Mycobacterium vaccae*, *Appl. Environ. Microbiol.*, 59(4), 1025–1029, 1993.
44. Parales, R.E., J.E. Adamus, N. White, and H.D. May, Degradation of 1,4-dioxane by an actinomycete in pure culture, *Appl. Environ. Microbiol.*, 60(12), 4527–4530, 1994.
45. Kelley, S.L., E.W. Aitchison, M. Desphpande, J.L. Schnoor, and P.J.J Alvarez, Biodegradation of 1,4-dioxane in planted and unplanted soil: effect of bioaugmentation with *Amycolata* sp. CB1190, *Water Res.*, 35(16), 3791–3800, 2001.
46. FMC Corporation, Oxidation product literature.
47. Liang, C., C.J. Bruell, M.C. Marley, and K.L. Sperry, Thermally activated persulfate oxidation of trichloroethylene (TCE) and 1,1,1-trichloroethane (TCA) in aqueous systems and soil slurries, *Soil & Sediment Contamination*, 12(2), 207–228, 2003.
48. Newell, C.J., P.E. Haas, J.B. Hughes, and T. Khan, Results from two direct hydrogen delivery field tests for enhanced dechlorination, Proc. Second International Conference on Remediation of Chlorinated and Recalcitrant Compounds, Monterey, California, May, 2000.
49. Lowry, G.V. and M. Reinhard, Hydrohalogenation of 1- to 3-carbon halogenated organic compounds in water using a palladium catalyst and hydrogen gas, *Environ. Sci. Technol.*, 33(11), 1905–1910, 1999.
50. McNab, W.W., Jr., R. Ruiz, and M. Reinhard., *In-situ* destruction of chlorinated hydrocarbons in groundwater using catalytic reductive dehalogenation in a reactive well: testing and operational experiences, *Environ. Sci. Technol.*, 34(1), 149–153, 2000.
51. Svensson, E., H. Lennholm, and T. Iverson, Pulp bleaching with dithionite: brightening and darkening reactions, *J. Pulp and Pap. Sci.*, 24(8), 254–259, 1998.
52. Chilakapati, A., M. Williams, S. Yabusaki, C. Cole, and J. Szecsody, Optimal design of an in situ Fe(II) barrier: transport limited reoxidation, *Environ. Sci. Technol.*, 34(24), 5215–5221, 2000.
53. U.S. Department of Energy, Innovative technology summary report — in situ redox manipulation, OST/TMS ID 15, 2000.
54. Larson, R.A. and J. Cervini-Silva, Dechlorination of substituted trihalomethanes by iron(II) porphyrin, *Environ. Toxicol. Chem.*, 19(3), 543–548, 2000.
55. Nzengung, V.A., R.M. Castillo, W.P. Gates, and G.L. Mills, Abiotic transformation of perchloroethylene in homogeneous dithionite solution and in suspensions of dithionite-treated clay minerals, *Environ. Sci. Technol.*, 35(11), 2244–2251, 2001.
56. Chilakapati, A., Optimal design of a subsurface redox barrier, *AIChE J.*, 45(6), 1342–1350, 1999.
57. Istok, J.D., J.E. Amonette, C.R. Cole, J.S. Fruchter, M.D. Humphrey, J.E. Szecsody, S.S. Teel, V.R. Vermeul, M.D. Williams, and S.B. Yabusaki, *In situ* redox manipulations by dithionite injections: intermediate-scale laboratory experiments, *Ground Water*, 37(6), 884–889, 1999.
58. Fruchter, J.S., C.R. Cole, M.D. Williams, V.R. Vermeul, J.E. Amonette, J.E. Szecsody, J.D. Istok, and M.D. Humphrey, Creation of a subsurface permeable treatment zone for aqueous chromate contamination using *in situ* redox manipulation, *Ground Water Monit. Rem.*, 20(2), 66–77, 2000.
59. Khan, F.A., and R.W. Puls, In situ abiotic detoxification and immobilization of hexavalent chromium, *Ground Water Ground Water Monit. Rem.*, 23(1), 77–84, 2003.
60. Patterson, R.R., S. Fendorf, and M. Fendorf, Reduction of hexavalent chromium by amorphous iron sulfide, *Environ. Sci. Technol.*, 31(7), 2039–2044, 1997.

61. Suthersan, S.S., *Natural and Enhanced Remediation Systems*. Lewis Publishers, Boca Raton, FL, 2002.
62. Puls, R.W., C.J. Paul, and R.M. Powell, The application of an in situ permeable reactive (zero-valent iron) technology for the remediation of chromate-contaminated groundwater: a field test, *Appl. Geochem.*, 14, 989–1000, 1999.
63. Roberts, A.L., L.A. Totten, W.A. Arnold, D.R. Burris, and T.J. Campbell, Reductive elimination of chlorinated ethylenes by zero-valent metals, *Environ. Sci. Technol.*, 30(8), 2654–2659, 1996.
64. Blowes, D.W., C.J. Ptacek, S.G. Benner, C.W.T. McRae, T.A. Bennett, and R.W. Puls, Treatment of inorganic contaminants using permeable reactive barriers, *J. Contam. Hydrol.*, 45, 123–137, 2000.
65. Furakawa, Y., J.-W. Kim, J. Watkins, and R.T. Wilkin, Formation of ferrihydrite and associated iron corrosion products in permeable reactive barriers of zero-valent iron, *Environ. Sci. Technol.*, 36(24), 5469–5475, 2002.
66. Wang, J. and J. Farrell, Investigating the role of atomic hydrogen on chloroethane reactions with iron using Tafel analysis and electrochemical impedence spectroscopy, *Environ. Sci. Technol.*, 37(17), 3891–3896, 2003.
67. Johnson, T.L., W. Fish, Y.A. Gorby, and P.G. Tratnyek, Degradation of carbon tetrachloride by iron metal: complexation effects on the oxide surface, *J. Contam. Hydrol.*, 29, 379–398, 1998.
68. Bissen, M. and F.H. Frimmel, Arsenic — a review. Part I: Occurrence, toxicity, speciation and mobility, *Acta Hydrochim. Hydrobiol.*, 31(1), 9–18, 2003.
69. Smedley, P.L. and D.G. Kinniburgh, A review of the source, behavior and distribution of arsenic in natural waters, *Appl. Geochem.*, 17(5), 517–568, 2002.
70. Su, C. and R.W. Puls, Arsenate and arsenite removal by zerovalent iron: kinetics, redox transformation, and implications for in situ groundwater remediation, *Environ. Sci. Technol.*, 35(7), 1487–1492, 2001.
71. Su, C. and R.W. Puls, Arsenate and arsenite removal by zerovalent iron: effects of phosphate, silicate, carbonate, borate, sulfate, chromate, molybdate, and nitrate, relative to chloride, *Environ. Sci. Technol.*, 35(22), 4562–4568, 2001.
72. Comfort, S.D., P.J. Shea, T.A. Machacek, H. Gaber, and B.-T. Ho, Field-scale remediation of a metolachlor-contaminated spill site using zerovalent iron, *J. Environ. Qual.*, 30, 1636–1643, 2001.
73. Voet, D. and J.G. Voet, *Biochemistry*, 2nd ed., John Wiley & Sons, New York, 1995.
74. Matthews, C.K., K.E. van Holde, and K.G. Ahern, *Biochemistry*, 3rd ed., Addison Wesley Longman, San Francisco, 2000.
75. Woods, S.L., D.J. Trobaugh, and K.J. Carter, Polychlorinated biphenyl reductive dechlorination by vitamin B12s: thermodynamics and regiospecificity, *Environ. Sci. Technol.*, 33(6), 857–863, 1999.
76. Burris, D.R., C.A. Delcomyn, M.H. Smith, and A.L. Roberts, Reductive dechlorination of tetrachloroethylene and trichloroethylene catalyzed by vitamin B12 in homogeneous and heterogeneous systems, *Environ. Sci. Technol.*, 30(10), 3047–3052, 1996.
77. Lewis, T.A., M.J. Morra, and P.D. Brown, Comparative product analysis of carbon tetrachloride dehalogenation catalyzed by cobalt corrins in the presence of thiol or titanium(III) reducing agents, *Environ. Sci. Technol.*, 30(1), 292–300, 1996.
78. Chiu, P.C. and M. Reinhard, Transformation of carbon tetrachloride by reduced vitamin B12 in aqueous cysteine solution, *Environ. Sci. Technol.*, 30(6), 1882–1889, 1996.
79. Mowder, C., ARCADIS, personal communication.
80. Zou, S., H.D. Stensel, and J.H. Ferguson, Carbon tetrachloride degradation: effect of microbial growth substrate and vitamin B12 content, *Environ. Sci. Technol.*, 34(9), 1751–1757, 2000.

4 Components of an *in Situ* Reactive Zone

The realization, during the last decade, that contaminant mass removal efficiencies, in addition to containment, can be significantly enhanced via various *in situ* technologies has led to the necessity of refining and improving these reactive processes. While it can be argued that the initial motive for developing the various modes of *in situ* applications has been one of saving money, the end result is much quicker cleanup times to more acceptable cleanup levels (Figure 4.1). This win–win situation for the entire remediation industry fostered continuous innovation, which led to (1) faster, cheaper solutions, (2) less-invasive *in situ* processes that can be applied within *in situ* reactive zones, and (3) processes that are also complementary to the natural environment, which took advantage of nature's capacity to degrade the contaminants. Thus holistically, environmentally, and economically sound and sustainable solutions were provided.

Figure 4.2 illustrates the evolutionary reduction in remediation costs from the late 1970s to the present time. These cost reductions were possible due to the reduction in capital costs and operation and maintenance (O&M) costs achieved during the various phases of this evolution.[1] *Ex situ* extractive techniques such as excavation and pump and treat systems were replaced by *in situ* extractive techniques, namely, soil vapor extraction and *in situ* air sparging during the first phase. Cost reductions were possible mainly due to the significant improvement in efficiencies of mass removal and time frames leading to substantial reductions in O&M costs. Subsequently these *in situ* extractive techniques gave way to *in situ* nonextractive techniques such as funnel and gate systems and, eventually, to *in situ* mass destruction techniques that can be implemented within *in situ* reactive zones (IRZs). These developments led to reductions in both capital and O&M costs. Recently, this evolutionary pattern has focused toward more natural solutions and/or enhancing existing subsurface biogeochemical conditions that contribute to remediation.

The most recent shift occurred with the recognition and demonstrated value of natural mechanisms that contributed toward containment, control, and mass reduction of contaminants in soil and groundwater. Under a host of names — including natural attenuation, bioattenuation, natural remediation, and monitored natural attenuation (MNA) — this remediation approach has taken root as a viable remediation method at appropriate sites and under the right biogeochemical conditions. Used in conjunction with ongoing remediation systems or as a stand-alone remedy,

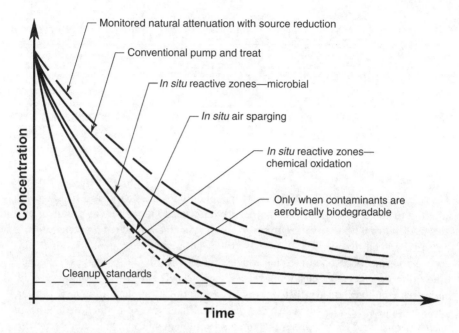

FIGURE 4.1 Evolution of *in situ* remediation technologies and improvements in efficiencies when applied under the right conditions at the appropriate site.

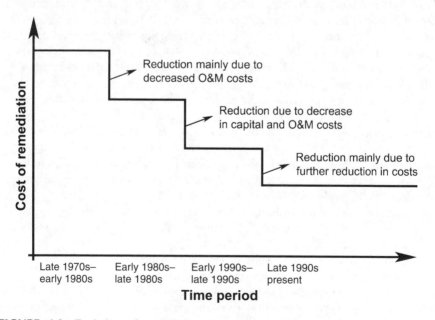

FIGURE 4.2 Evolution of remediation technologies leading to reduction in cost of remediation.

MNA can increase the probability of a successful, cost-effective, and well-documented restoration of a contaminated site. However, based on recent experience and data collected from many sites, questions have been raised regarding the validity of MNA as a stand-alone remedy for sites with certain types of contaminants.

The development of IRZs is essentially an outgrowth of the efforts to implement the same proven reactions and processes, successfully applied aboveground, in an *in situ* environment (Figure 4.3). These systems can be designed to implement microbial reactions, whether oxidizing or reducing, and similarly oxidizing or reducing chemical reactions.

The effectiveness of the reactive zone is determined largely by the relationship between the kinetics of the target reactions and the rate at which the mass flux of contaminants passes through it with the moving groundwater.[1] Creation of a spatially fixed reactive zone in an aquifer requires not only the proper selection of the reagents, but also a clear understanding of the initial biogeochemical regime and the selection of the most optimum reaction mechanisms. Furthermore, such reagents and ensuing reactions must cause negligible or minimal side reactions and be relatively nontoxic in both the original and treated forms.

In the recent past, only the *proven* processes and reactions have been selected to be implemented in an IRZ for the remediation of contaminated groundwater (Figure 4.3). These processes can also be termed as *active treatment* processes. This *active treatment* in a way can be considered as "wastewater engineering" applied *in situ* to groundwater systems. *Active treatment* differs from *passive treatment* and emphasizes the improvement of groundwater quality by methods and reactions that require ongoing inputs of (bio) chemical reagents and/or added energy. The added energy referred to can be in the form of aeration, mixing, or heat to enhance reaction rates, or pressure to control injection rates.

Because of the varying types of contaminants and biochemical regimes encountered in groundwater systems, and the familiarity with the physical and (bio) chemical processes necessary to remediate the contamination, there is a wide range of conventional and innovative techniques available to be implemented within IRZs. Selection of the most appropriate technique/reaction for a given contaminated groundwater system should be based on consideration of the "raw" groundwater quality, its biogeochemical regime, and the desired cleanup goals.

The most widespread processes and reactions that are implemented within engineered IRZs include physical mass transfer mechanisms, such as air sparging; biochemical mechanisms, such as microbial oxidation and reduction; chemical mechanisms, such as chemical oxidation, reduction, and nucleophilic substitutions; and passive mechanisms, such as monitored natural attenuation (MNA).

There is no technical limit to the quality of the water that can be achieved using the various reactions within an IRZ. However, the complexity of implementation is significantly more for some reactions and the cost may be astronomical based on the nature of some reaction mechanisms (Figure 4.4). At the same time, inclusion of the time requirements as a critical evaluation parameter to achieve the cleanup objectives, specifically within complex hydrogeological settings, changes the scenario, as presented in Figure 4.5. For instance, demanding a very fast cleanup time regardless of cost and complexity may lead to choosing chemical oxidation as the reaction

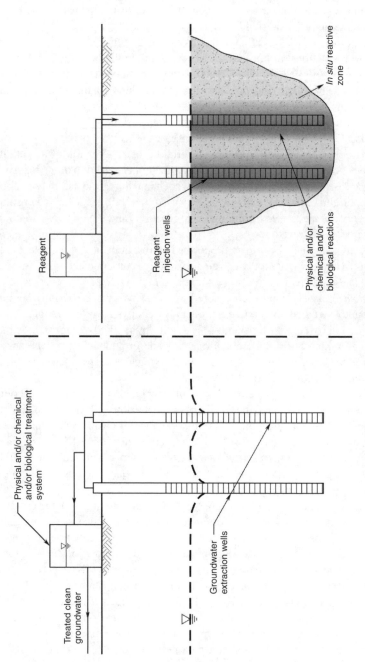

FIGURE 4.3 Conceptual description of an *in situ* reaction zone.

Components of an *in Situ* Reactive Zone

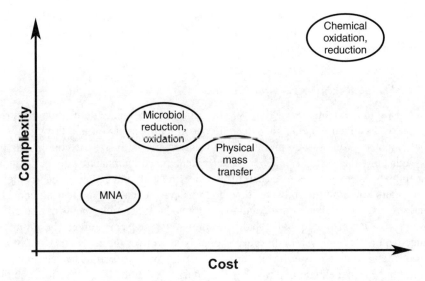

FIGURE 4.4 Description of the combination of cost and complexity for various reaction mechanisms that can be applied within IRZs.

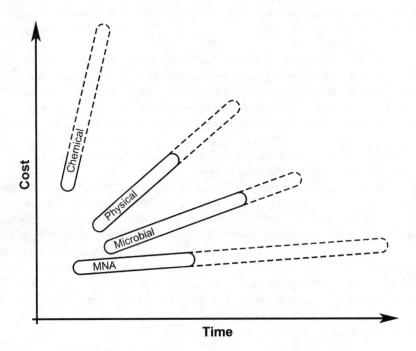

FIGURE 4.5 Variability of cost and time frames to achieve the cleanup objectives for different reaction mechanisms that can be applied within IRZs.

mechanism to be implemented. Again, this will be possible only if the target contaminants can be chemically oxidized and the selected reagents can be injected under controlled conditions.

4.1 SITE SCREENING

Successful implementation of IRZ systems relies mainly on the delivery and distribution of dissolved reagents, at selected concentrations and volume, throughout a contaminated groundwater plume. Administering the delivery of those reagents through both the vertical and horizontal extent of contaminated plumes sounds deceptively easy, but it requires careful engineering and knowledge of the geologic parameters affecting groundwater flow and transport. Creative engineering considerations have to be taken into account to accommodate the requirements of a smaller plume vs. a larger plume and a shallower plume vs. a deeper plume. The ultimate objective of the IRZ system designer should be to deliver the reagent(s) as fast as possible and to create a uniformly mixed reactive zone, as well as to maintain the optimum biogeochemical conditions for the target reactions to occur. Site-specific conditions will obviously influence the total time required for treatment. For example, to implement an IRZ for enhanced reductive dechlorination, a contaminant plume with already existing reducing conditions will require less time for complete transformation of contaminants than sites with initial aerobic conditions and no evidence of even partial reductive dechlorination. Similarly, preexisting reducing anaerobic conditions will influence the time required for chemical oxidation reactions to be established.

Other factors influencing the time required to treat a site include aquifers with low groundwater velocities, which will require more time for distributing the reagents throughout the plume, and the presence of NAPL or high concentrations of the contaminant in the sorbed phase, which would require considerably longer treatment times due to the rate limitation imposed by dissolution of the contaminant. The assessment of a particular site for IRZ application, regardless of the type of reaction to be selected, should include the development of a contaminant profile, a hydrogeological profile, and a biogeochemical mapping/profile.

An inventory of contaminant(s), their concentrations, and distribution throughout the plume will be the first step in assessing the feasibility of an IRZ. The presence, relative concentration, and distribution of daughter products is particularly important when assessing the potential for enhanced reductive dechlorination. Cocontaminant impacts may be either beneficial or detrimental, so it is important to assess before selecting the target reactions. Development of an overall site conceptual model is discussed in a later section. A summary of site screening criteria to evaluate the feasibility of microbial and/or chemical IRZs is presented in Tables 4.1 and 4.2.

4.1.1 SITE ACCESS

Contaminated sites can have various degrees of convenience or difficulties in terms of successfully implementing an IRZ. These sites may include an operating manufacturing facility where certain portion(s) of the plume may never be accessible for

TABLE 4.1
Suitability of Site Screening Characteristics for an Anaerobic, Reducing, Microbial IRZ Implementation[2]

Site Characteristics	Suitable for IRZ	Marginally Suitable for IRZ	Unsuitable for IRZ
Hydraulic conductivity	$10^{-1} > K \geq 10^{-4}$ cm s^{-1}	$10^{-6} \leq K \leq 10^{-4}$ cm s^{-1}	$K < 10^{-6}$ cm s^{-1}
Groundwater velocity	Greater than 30 ft year^{-1} and less than 2.5 ft day^{-1} Organic carbon demand will be reasonable to high	2 ft year^{-1} to 3 ft day^{-1} Organic carbon demand will be very high	Less than 20 ft year^{-1}, hard to deliver organic carbon Greater than 4 ft day^{-1}, TOC demand is excessive
pH	5.5–8.0	4.0 < pH < 5.5 8.0 < pH < 9.0	pH < 3.0 pH > 9.5
Natural degradation prior to injection of organic substrate	Slow, complete, or partial degradation of parent compound(s)	No or very little degradation but the aquifer system is mildly anaerobic or transient anaerobic	No degradation at all and the system is highly aerobic; TOC demand will be excessive and injection costs will be high
DNAPL presence	Presence of dissolved, sorbed, and emulsified contaminants	Most of the mass is still in emulsified NAPL form	Recoverable separate phase contaminant within the targeted zone of remediation
Presence of electron acceptors O_2, NO_3^-, SO_4^{2-}	Low enough to enable reasonable consumption of TOC to create sufficiently reducing conditions	Moderate flux of incoming electron acceptors will demand continuous injection of high levels of organic carbon	Excessively high flux of incoming electron acceptors will make it difficult for the target biogeochemical regime(s) to be established
Contaminant level	Technology has advanced toward implementing IRZs from 100s of ppb concentration to 100s of ppm concentrations (of dissolved contaminants) within contaminated plumes	Dissolved contaminants with a large fraction of adsorbed or emulsified mass of LNAPL or DNAPL	Very large volumes of NAPL mass, either floating or sinking, present within a large area
Presence of metals	Most dissolved metals at nontoxic concentrations less than 100s of ppm each	Metal concentrations at inhibitory levels	Metal concentrations at toxic levels, 1000s of ppm
Oxidation state of contaminants	Oxidized	Neutral	Highly reduced

TABLE 4.2
Suitability of Site Screening Characteristics for Chemical IRZ Implementation

Site Characteristics	Suitable for IRZ	Unsuitable for IRZ
Hydraulic conductivity	10^{-1}–10^{-5} cm s^{-1}	Less than 10^{-6} cm s^{-1}
Groundwater velocity	Greater than 30 ft year^{-1} and less than 3 ft day^{-1}	Less than 20 ft year^{-1}, hard to deliver reagents
pH	Depends on the reagent(s) chosen — extreme pH values unsuitable	Less than 3.0 and greater than 11.0
Pre-injection biogeochemical conditions	Neutral redox values not very reduced, low natural organic carbon in the system	Very reduced redox conditions Very high hard and soft organic carbon
Oxidation state of contaminants	Reduced or neutral	Very oxidized
Contaminant level	10s–100s of ppm	Metals concentration in the groundwater and soils
DNAPL presence	Dissolved, adsorbed, mass 10s of ppb–100s of ppm	Flowing, recoverable NAPL free product

direct injections; an inactive, closed site with significant portions of the plume having migrated off site and thus may be inaccessible for installing off-site injection wells; a closed landfill with an expensive, regulated cover in place, thus precluding any injections within the footprint of this installed cover. These examples are described only to show the difficulties that can be caused by site access issues.

Since the success of an IRZ application primarily depends upon the effective delivery and distribution of the reagents throughout the contaminated subsurface, complete access to the surface and subsurface locations is critical for the proper design and implementation of an IRZ system. The ability to control and select the frequency and dosage of the injections as well as the locations of the injection wells and performance monitoring wells is an important factor for the optimum design. Different sites, with varying initial site conditions, prior to an IRZ system implementation are shown in Figure 4.6(A)–(I).

Once site access issues have been sorted out, the designer has to consider many other tangible and peripheral site conditions to be included in the IRZ system design evaluation. These conditions include:

- Precise and thorough knowledge with respect to the health and safety effects of the contaminants of concern, if and when field workers may be exposed via inhalation, absorption, ingestion, or dermal contact pathways
- Knowledge of potential pathways for chemical vapor(s) migration and the required detection, monitoring, and mitigation systems
- Similarly, the potential for the formation and buildup of explosive gases and the required detection, monitoring, and mitigation systems

Components of an *in Situ* Reactive Zone

FIGURE 4.6 (A) Site conditions at a contaminated site (former strip mall) showing easy access to install the injection points. (B) Successful remediation and successful property redevelopment of the same property shown above — near Milwaukee, WI. (C) Injection well installation inside former plating building using a drill rig that does not require a lot of height.

FIGURE 4.6 (continued) (D) Injection trenches within an active facility covered by metal plates. (E) Implementation of an IRZ system during site development and building activities. (F) Implementation of an IRZ system in the vicinity of a residential area.

FIGURE 4.6 (continued) (G) Implementation of an IRZ system where access is not a problem. (H) Implementation of an IRZ system where injections have to be done in wells placed along a highway. (I) Injection of reagents under increased level of health and safety protection.

- Steep terrain and unstable surface conditions for the installation of injection and monitoring wells
- Safe and secure locations for short-term and long-term storage of reagents — some of them may be explosive if not handled properly
- Knowledge and identification of underground and overhead utilities, in addition to aboveground and underground tanks
- Access and pathways for the movement of vehicles and heavy equipment.

4.1.2 Safety Considerations

Every remediation project, whether consisting of an *in situ* or aboveground system, should have a comprehensive safety plan that includes the various aspects of contaminant migration, exposure, reagent storage and handling, and system implementation. However, this section discusses only the safety aspects that need to be considered when designing an IRZ where injection of reagents is the key factor.

The injection reagents to be used should be evaluated during the design phase of the project for the following factors:

- Toxicity and safety of the reagent to humans and the environment; generally, the preferred reagent will be the material with lowest toxicity to humans and the environment while providing a reasonable time frame for achieving project cleanup goals
- Methods of transport, storage, mixing, and application
- Volume of a reagent required to effectively achieve the contaminant destruction effects vs. use of an alternative reagent requiring less volume of injection
- Materials of construction that will come in contact with the reagents should be evaluated for chemical compatibility with the selected reagents
- A pressure relief valve should be provided with all vessels of storage of the reagents; the release of the gases that build up should be evaluated for air quality requirements and safety — potential for buildup of CO_2, CH_4, and H_2S within closed molasses storage tanks, and high concentrations of O_2 within closed hydrogen peroxide tanks are examples of such conditions
- Gravity feed injection systems should consider the requirements of injection pressures necessitated by the hydrogeologic conditions at the site.

The substrates that are generally used for the creation of anaerobic microbial IRZs include molasses, corn syrup, cheese whey, lactate, acetic acid, methanol, vegetable oil, and similar organic compounds. The authors prefer to use molasses, corn syrup, or cheese whey mainly because of the fact that these are substrates with multiple compounds and the perceived advantage of using food grade substances. Using substrates with multiple compounds including complex organics helps in maintaining a multispecies mixed microbial ecology within the reactive zone.

Reagents that are commonly used in the creation of oxidizing microbial IRZs include air, oxygen, magnesium peroxide, persulfate, ammonium nitrate, nitrate containing fertilizer, etc. Magnesium peroxide is a mild oxidizer and is considered a low chemical hazard to humans and the environment. Nitrate-containing fertilizer products are commonly used materials and considered to be harmless if handled properly — particularly when not mixed with materials that produce harmful effects. Proper storage and handling procedures should be followed when pressurized cylinders are used for the delivery of O_2 gas.

Reagents used to create chemically oxidizing IRZs include potassium permanganate ($KMnO_4$), sodium permanganate ($NaMnO_4$), Fenton's reagent, hydrogen peroxide, ozone, etc. The permanganates are strong oxidizing agents and are considered a low to moderate chemical hazard if used and stored appropriately, particularly $KMnO_4$. However, if used improperly, permanganates can quickly produce a violent exothermic reaction or produce toxic fumes that can cause serious injury or death. The sodium permanganate solution poses serious dangers, which must be carefully avoided when handling the product during dilution and application. During mixing of these materials, use of a high-efficiency particulate air (HEPA) filter dust mask is a must. Permanganates can also cause severe eye irritation and/or permanent eye damage; therefore, safety goggles and face shields are also required during mixing. Enclosed trailer or building mounted systems require ventilation during mixing.

Hydrogen peroxide, which is also the key ingredient in the application of Fenton's reagent, is a strong oxidizer and considered to be a moderate or high chemical hazard to humans and the environment, depending on dilution factors. Extra caution should be observed in handling and storing of hydrogen peroxide, completely avoiding any introduction of impurities. Hydrogen peroxide and Fenton's reagent produce violent, exothermic reactions when handled improperly. The potential for the groundwater around the injection zone to boil has to be taken into consideration. Improper mixing, use, or neutralization of this product may cause serious injury or death. The requirements of maintaining a low pH also necessitates the use of concentrated acids (commonly H_2SO_4) during the application of Fenton's reagent. Handling of all these chemicals require high levels of caution and attention to safety details.

Sodium bicarbonate and orthophosphate, used as buffering agents, are mixed with other injection reagents and are not considered to be a serious hazard when handled properly. Lime [$Ca(OH)_2$] and other concentrated alkaline reagents are used when there is a need to raise the pH within the engineered reactive zone. Storage and handling of concentrated alkaline solutions require extreme caution and proper attention to the health and safety guidelines.

4.1.3 Vapor Migration

Given the planned site activities that would take place during the implementation of an IRZ system and physical properties of the contaminants of concern as well the intermediate products, the potential exists for chemical exposure via migration of chemical vapors. In addition, the formation of certain by-products — for example, O_2 gas during chemical oxidation by Fenton's reagent, and methane during enhanced reductive dechlorination — may exacerbate the potential safety concerns during the implementation of IRZ systems. The safety concerns stemming from vapor migration have to be dealt with seriously when the entire IRZ or portions of it is implemented under the floor slab of an active facility, where the potential for human exposure is significant.

The potential exposure pathways, as a result of vapor migration, are inhalation of volatilized contaminants of concern and dermal contact with chemicals and particulates. Exposure limits for various contaminants of concern have been developed by the American Conference of Governmental Industrial Hygienists (ACGIH) and the National Institute of Occupational Safety and Health (NIOSH).[2] Exposure limits, based on toxicological and epidemiological data, have been developed at various levels: Threshold Limit Value (TLV) as a Time Weighted Average (TWA); Short Term Exposure Limit (STEL); and Immediately Dangerous to Life and Health Limit (IDLH).[3]

Another concern resulting from vapor migration that should be taken into consideration during the design of an IRZ system is the potential for the creation of a hazardous atmosphere. When the designed IRZ has to be implemented under a building, this concern must be taken seriously. An atmosphere is said to be *hazardous* by reason of being explosive, flammable, poisonous, corrosive, oxidizing, oxygen deficient, toxic, or otherwise harmful, and may cause injury, illness, or death.

Intrusive activities or cracks along impermeable surfaces (such as floor slabs or parking lots) increase the potential for the occurrence of elevated concentrations of "dangerous" vapors. Explosive concentrations of some of these compounds — such as methane and O_2 — could develop in small and confined spaces. Confined spaces include excavations, storage tanks, sewers, in-ground vaults, tunnels, manholes, and even buildings. These enclosures, because of inadequate ventilation and/or the introduction of hazardous vapors, may present conditions that could produce injury or explosions. In the event of any concern for vapor migration arising during the design phase of the IRZ system, a proper ventilation and exhaust system has to be incorporated as part of the IRZ system implementation.

Both microbial and chemical IRZ systems have the potential to produce/liberate "gases of concern" during system implementation. Oxygen gas is liberated as the hydrogen peroxide undergoes decomposition during Fenton's reaction. This gas evolution can become violent if the hydrogen peroxide is added too rapidly. The gas evolved during the addition of hydrogen peroxide can potentially entrain volatile organic contaminants and should be scrubbed appropriately.

Similarly, under the very reducing conditions required for complete reductive dechlorination within a microbial IRZ, formation and evolution of methane is a real possibility. Similar potential exists for the formation of hydrogen sulfide. If microbially reducing IRZ systems are implemented under building floors, a low-flow ventilation system has to be incorporated to remove the "gases of concern," and the removed gases have to be scrubbed according to the air quality treatment requirements.

A combustible gas indicator (CGI) equipped to monitor percentages of Lower Explosive Limits (LEL) of percentages of O_2 must be used for monitoring inside the buildings. If certain target parameters are exceeded (e.g., % LEL $>$ 10%; or $19.5 <$ % $O_2 >$ 23.5%), IRZ implementation should be immediately stopped and should not be allowed to resume until the mechanical ventilation system eliminates the potential for the accumulation of the "gases of concern."

4.1.4 TRENDS IN CONTAMINANT CONCENTRATION AND ELECTRON ACCEPTOR PROCESSES

Numerous processes occurring under natural conditions in groundwater systems can transform contaminants into other compounds. In addition to the biotic and abiotic transformations of the contaminant(s), transport processes such as dilution, advection, volatilization, dispersion, and phase transfer participate in the natural attenuation process. However, all these processes may not yield the desirable end-product, and hence the need for engineered systems.

The process of performing a detailed analysis of a contaminated site in order to assess the potential of the site for the application of an IRZ has much in common with the assessment performed for MNA. Such an assessment bases decision making on a review of data on the contaminants themselves, electron donors, electron acceptors, partial-degradation by-products, geochemical parameters, and hydrogeology. In particular, the trend analysis should help determine whether an ongoing source exists at the site that is still contributing dissolved phase contaminants at a rate that natural attenuation processes cannot overcome (i.e., increasing trends). In such a situation, *in situ* treatment may accelerate contaminant destruction significantly. In extreme cases a more aggressive source removal approach may need to be initiated first.

If decreasing trends in contaminant concentrations are observed, then the contribution of destructive degradation processes should be discerned from nondestructive attenuation processes described earlier, before concluding that biotic or abiotic processes are present that can be enhanced. This can be accomplished by deriving bulk attenuation rates for individual contaminants and then subtracting the effects of dilution using the synoptic trends of conservative tracers, such as sodium (a cation), or chloride (an anion). More expensive carbon, oxygen, and hydrogen isotope chemistry (e.g., tritium) can also be used in tracer studies. The cost of laboratory analysis precludes these techniques at most contaminated sites and historical data from a site will generally not contain a record of these more costly analyses. In general, if the trends for source and daughter products are identical for many wells located throughout the plume, then dilution may be playing a large part in the overall attenuation that is being observed. At a minimum, a qualitative trend analysis should be considered using the concepts discussed earlier. Quantitative trend analysis using statistical techniques can also be helpful.

Many trends and observations based on conventional screening techniques used for MNA evaluation, however, might disqualify a site from consideration for implementation of a microbial IRZ. For example, the use of traditional screening tools that use a scoring system with indicators of aerobic conditions, such as high dissolved oxygen, high redox potential, and the absence of daughter products, will disqualify that site for further evaluation of an anaerobic microbial IRZ. However, recent experience has shown that implementation of a microbial IRZ by design can drive the conditions more anaerobic so that initial aerobic conditions need not be a barrier to processes like enhanced reductive dechlorination. It should also be noted that the use of inappropriate field purging and sampling methods to collect sensitive screening data may lead to aerobically biased false negative data and can result in a preference for aerobic bioremediation processes.

Many of the concepts described here are also applicable for interpreting data from a pilot study and a full-scale operating system. The development of a site conceptual model to summarize contaminant migration pathways, if possible in three dimensions, and links to potential receptors is highly recommended. The exercise of developing a conceptual model can also ensure that the IRZ system designer identifies the ongoing biogeochemical processes, if any, that are controlling contaminant movement at the site. The use of contaminant isoconcentration maps and biogeochemical maps to portray the distribution of oxidation–reduction processes and other biogeochemical parameters across the site is helpful. These maps can be used to track the biogeochemical changes induced by the designed reactions. The use of time-series plots to understand changes in contaminant concentrations over time is vital. The use of vertical contour plots oriented along the axis of the plume to understand vertical migration is also important. The use of multiple lines of evidence for continuous decision making can keep critical conclusions from being influenced by a few spurious, false negative data points.

It is important to stress that scientific documentation of the *mechanism claimed as responsible for contaminant destruction or control is scientifically feasible* and that *the proposed mechanism is actually occurring at the site*. Thus, in addition to documenting contaminant removal in a pilot test or full-scale treatment process, an IRZ system designer should also be able to demonstrate the *footprints* of the contaminant(s) transformation processes. For example, these would include the appearance of intermediate degradation products (e.g., *cis*-DCE and ethene for TCE/PCE), the appearance of indicators that the desired alternate electron acceptor processes are operative (e.g., reduced dissolved oxygen (DO) and ORP, increased methane and hydrogen sulfide), and the utilization of appropriate electron donors during the implementation of a microbial IRZ system for enhanced reductive dechlorination.

4.2 OXIDATION STATE OF CONTAMINANTS

When designing contaminant transformation reactions to be implemented with an IRZ, it is important to evaluate the oxidation state of the contaminant(s) and decide whether the oxidative or reductive pathway is the right strategy to follow. We can determine if the contaminant will have a sensitivity to oxidative or reductive reactions by considering the oxidation state of the subject molecule.

When dealing with molecular transformations of contaminants, it is important to know whether or not electrons can be transferred between the reactants. For determining the number of electrons that can be transferred, it is necessary to examine the oxidation states of all atoms involved in the reaction.[4] Of particular interest to the designer will be the (average) oxidation state of carbon, nitrogen, and sulfur in a given contaminant molecule, since these are the elements most frequently involved in redox reactions.

Many chemical reactions result in the transfer of electrons from one atom or molecule to another; this transfer may or may not accompany the formation of new chemical bonds. The loss of electrons from an atom or a molecule is called *oxidation*, and the gain of electrons by an atom or molecule is called *reduction*. Because electrons are neither created nor destroyed in a chemical reaction, if one atom or molecule is oxidized, another must be reduced.

When evaluating a contaminant molecule an oxidation state of zero is assigned to the uncharged element; a loss of n electrons is then an oxidation state of $+n$. Similarly, a gain of electrons leads to an oxidation state lower by an amount equal to the number of gained electrons.[1] A simple example is the oxidation of sodium by chlorine, resulting in the formation of sodium chloride:

$$Na^0 \xrightarrow{oxidation} Na^{+1} + e^- \quad (4.1)$$

$$Cl + e^- \xrightarrow{reduction} Cl^{-1} \quad (4.2)$$

Unlike the ionic redox reactions, like the ones shown earlier, electrons are shared in covalent bonds. For any atom in a molecule with covalent bonds the oxidation state may be computed by adding 0 for each bond to an identical atom, -1 for each bond to a less electronegative atom or for each negative charge on the atom, and $+1$ for each bond to a more electronegative atom or for each positive charge. Once a value for each bond is determined, the assigned values are summed to provide the oxidation state of the specific atom. As shown in Figure 4.7, each carbon in tetrachloroethene (PCE) has an oxidation state of $+2$. Similarly, the two carbon atoms in the trichloroethene molecule have oxidation states of $+2$, and 0, with the average oxidation state for the TCE molecule being $+1$. Hence, it can be seen that the formation of TCE from PCE by the reductive dechlorination reaction involves the transfer of one electron.[5] Similarly, each step of the sequential transformation of PCE → TCE → cis-DCE → VC → Ethene → Ethane takes place as a result of one electron transfer (Figures 4.8 and 4.9).

For larger, more complex molecules, we must only consider the atoms directly involved in the reaction process to determine if a change in oxidation state can take place. Although our understanding of oxidative or reductive transformations in the environment has progressed to the point that we can identify the types of functional groups that will be susceptible to redox reactions within engineered IRZs, our limited

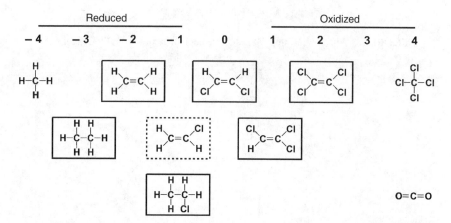

FIGURE 4.7 Oxidation states of various chlorinated alkanes, alkenes, and CO_2.

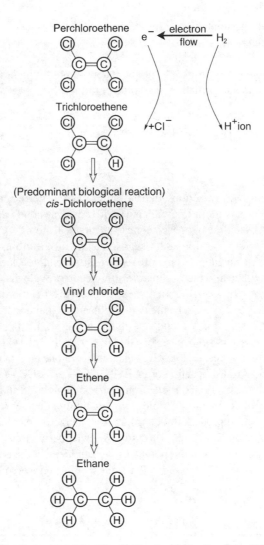

FIGURE 4.8 Sequential steps of reductive dechlorination from PCE to ethene.

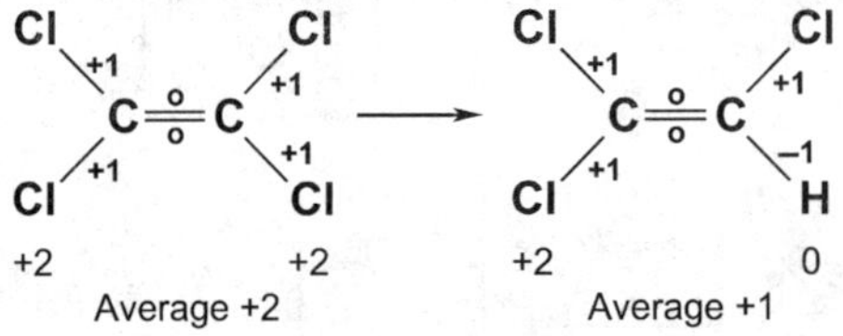

FIGURE 4.9 Oxidation states of PCE and TCE.

understanding of precise reaction mechanisms for such transformations currently is a barrier to the prediction of absolute transformation rates and how reaction rates will vary from one engineered system to the next.

Based on the discussion presented in this section and Figure 4.7, it becomes abundantly clear that determination of the oxidation state(s) of the contaminant(s) of concern is the critical factor in selecting the preferred contaminant transformation reactions to be implemented within the IRZs. Figure 4.10 presents a simplified notion in terms of which reactive strategies will be preferred depending upon the oxidation state of the contaminant. Hence, it is very clear that for oxidized compounds such as carbon tetrachloride and PCE, a reductive strategy must be chosen. Similarly, if the target contaminants are reduced compounds such as vinyl chloride an oxidative strategy will be the preferred one. However, it should be noted that vinyl chloride can still be reductively transformed to ethene, albeit at a considerably higher expenditure of biochemical energy. Compounds with neutral oxidation states can be transformed under both oxidative and reductive conditions, however, only at moderate rates (Figure 4.11).

Now, let us look at some *nonchlorinated* organic compounds which are, at times, encountered as contaminants in the subsurface. The C=O double bond (*carbonyl group*) is the most common group in organic chemistry. Many tomes have been devoted entirely to this functional group. Figure 4.11 shows only the oxidation–reduction reactions of compounds containing this functional group. Aldehydes prefer to go to acids rather than to alcohols: aldehydes are reducing agents, ketones are not.

In addition to the oxidation states of the contaminants, the baseline (preengineered) biogeochemical condition of the contaminated groundwater system also has to be evaluated before making the final decision on the reactive strategy. For example, an environment that has a high background concentration of electron acceptors has a high electrical potential and is considered oxidizing. Hence, if a reductive strategy is preferred, based on the oxidation states of the contaminants, a significant level of reducing agents will be required to overcome the elevated oxidative poise of the system to create the optimum biochemical regime necessary for the contaminant transformations to take place.

4.3 SITE CONCEPTUAL MODEL

4.3.1 Hydrogeologic Model

In selecting the reactive strategy to be implemented within a reactive zone, it is often important to postulate a conceptual model or hypotheses for the contaminated site. This effort can begin on the basis of publicly available geologic and geographic information about its surroundings. Additional conceptualization can be done on the basis of generic data about similar regions and the characteristics and properties of similar contamination and geologic materials. Yet each site is unique and so virtually guaranteed to reveal additional features, properties, and behaviors when characterized in detail at the local level. Hence the strategy should consider a proper site characterization as essential for the postulation of robust site conceptual models. It should be noted that regional and site-level hydrogeologic data tend to represent a wide range of measurement scales.

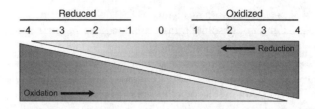

FIGURE 4.10 Carbon oxidation state and remedy selection.

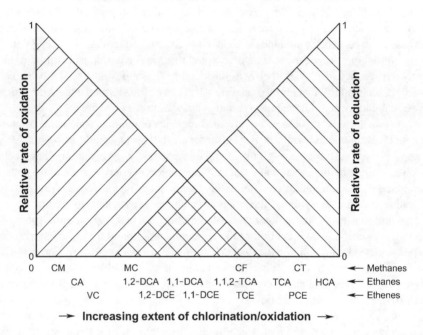

FIGURE 4.11 Relative rates of oxidation and reduction from a range of C1 and C2 chlorinated compounds.[1,6]

While hydrogeologic characterization of a site and its surroundings makes conceptualization possible, it does not provide the means to test the conceptual model or compare them with other alternatives. For this, it is necessary to have monitoring data that constitute observations of actual hydrologic behavior at and around the site.

Hydrogeologic systems are inherently heterogenous on a multiplicity of scales arising from surface and subsurface flow conditions; complexity of flow pathways, and interactions between the geometry of the flow domain and the prevailing hydraulic gradients and flow pathways (including both "dead zones" and "preferential" flows). The problem in defining heterogeneity is that detailed characterization of a site may be technically impractical or infeasible due to expense. Even if this was practical and feasible, it would not necessarily lead to an improved representation of the system on larger scales; interactions among smaller-scale features and processes often lead to effects that are not completely captured by effective parameters in larger-scale models.[7]

In practice one has to represent the unique heterogeneous and scale-dependent characteristics of a site and its system complexities in a simplified and reduced form so that the conceptual model provides an acceptably realistic representation consistent with available site data. To ensure that relevant aspects of hydrogeologic complexity are reflected in the conceptualized system behavior and performance, it is important that the process of simplification be done systematically and objectively. This can be done by filtering out undesirable details through formal averaging in a way that retains the overall system behavior. Only with such data can one evaluate the ability of models to mimic and calibrate real system behavior and determine the required refinement and/or optimization. Conceptual models should include the representation of hydrogeologic heterogeneity and scale on the basis of field observation data.[7]

For the purpose of developing site remediation strategy, the hydrogeologic model is a conceptual (and mathematical) construct that serves to analyze, qualitatively and quantitatively, subsurface flow and transport at the site in a way that is useful for the IRZ system designer and subsequent performance evaluation. The tools that are commonly used to develop hydrogeologic models include hand-drawn geologic cross-sections and/or software tools for reconstructions and visualization of site hydrostratigraphy and structure; statistical and geostatistical analysis of site data; interpretation of pumping and/or slug tests; and, simulation of flow and transport processes. The evaluation typically consists of simulating (reconstructing or predicting) and analyzing space–time variations in quantities such as hydraulic head or pressure, solute concentration, fluid and solute flux and velocity, solute travel time, and associated performance measures for the IRZ system implementation. A comprehensive hydrogeologic model for a complex site, thus, should consist of a conceptual and a mathematical component.

Much has been written about the mathematical component of hydrogeologic models, but relatively little attention has been devoted to the conceptual component. In most mathematical models of subsurface flow and transport, the conceptual framework is tacitly assumed to be accurate and unique.

It was stated earlier that a hydrogeologic model should consist of a description of the site and circumstances, a set of regional and site-specific data, and the cohesive evaluation of system behavior and performance. A conceptual model embodies the descriptive component of the model, which may be both qualitative and quantitative. To assimilate the many separate pieces of information for the design of an IRZ system, the *conceptual model* describes, correlates, connects, systematizes, interprets, and integrates all available information into a "body" of knowledge (Figure 4.12(A) and (B)).

Eventually, the site conceptual model, or the "body of knowledge," is considered to be a hypothesis that describes the main features of site geology, hydrology, geochemistry, and relationships between geologic structure and fluid flow and contaminant fate and transport. The conceptual model for an IRZ system design is a pictorial, qualitative description of the groundwater system in terms of hydrogeologic units, system boundaries including time-varying inputs and outputs, and hydraulic as well as transport properties including their spatial variability. A mathematical model is a quantitative representation of such a conceptual model and a process of testing this hypothesis.

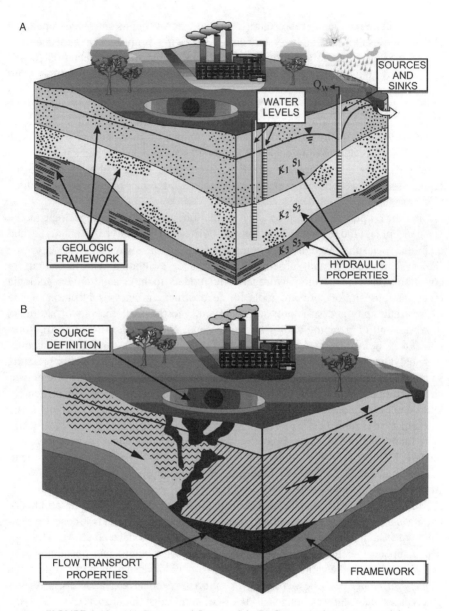

FIGURE 4.12 (A) Conceptual flow model. (B) Conceptual transport model.

A contaminated groundwater plume is a hydrogeochemical system and hence the site conceptual model should also include a qualitative assessment of how chemicals enter, move through, or are retained in and leave the system. It also should include a qualitative description of the source, transport, fate, and distribution of each targeted contaminant or, in the case of unknown sources, a hypothesis concerning source locations and strengths based on conceptualized transport and fate processes and known distribution of contaminants. The *envirochemical* system should be described by the previously discussed tools but also by iso-concentration contour maps in both horizontal and vertical directions.

4.3.2 Contaminant Distribution

For the development of a comprehensive site conceptual model, the first step prior to designing an IRZ system, it is important to understand the characteristics of the source zones that create dissolved contaminant plumes and discuss the processes that control the partitioning of contaminants into groundwater. It should be emphasized that only a few key relationships and simple models should be utilized for evaluation and development of site conceptual models.

At most contaminated sites, the bulk of the contaminant mass is in what remediation professionals call "source zones." Examples of source zones include landfills, areas of chemical spills, buried tanks that contain residual chemicals, deposits of tars, and so forth. Some of these sources can be easily located and complete or partial removal of contaminant may be possible. However, other common types of sources often are extremely difficult to locate and remove or contain. One example of a source in this category is contaminants that have sorbed to soil particles but have the potential to later dissolve into groundwater that contacts the soil. Another extremely important example is the class of organic contaminants known as "nonaqueous-phase liquids" (NAPLs). There are two types of NAPLs: those that are denser than water (dense nonaqueous-phase liquids, or DNAPLs) and those that are less dense than water (light nonaqueous-phase liquids, or LNAPLs). When released to the ground, these types of fluids move through the subsurface in a pattern that varies significantly from that of the water flow, because NAPLs have different physical properties than water. As shown in Figure 4.13(A)–(C), LNAPLs can accumulate near the water table; DNAPLs can penetrate the water table and form pools along geologic layers; and both types of NAPLs can become entrapped in soil pores. These NAPL accumulations contaminate the groundwater that flows by them by slow dissolution. Common LNAPLs include fuels (gasoline, kerosene, and jet fuel), and common DNAPLs include industrial solvents (trichloroethene, tetrachloroethene, and carbon tetrachloride) and coal tar. Once they have migrated into the subsurface, NAPLs are often difficult or impossible to locate in their entirety.

Normally, the total mass of a contaminant within source zones is significantly larger compared to the mass dissolved in the plume. Therefore, the source usually persists for a very long time. The rate at which contaminants dissolve from a typical NAPL pool is so slow that many decades to centuries may be needed to dissolve the NAPL completely by dissolution without any intervention. Given the persistent nature of contaminant sources, removing them would seem like a practical way to speed the cleanup of the contaminant plume. In many cases, environmental regulators require source removal of contaminant as part of a natural attenuation remedy. Although requiring source control or removal is good policy for many sites, expert opinions conflict on whether source removal is advisable when using natural attenuation as the preferred remedy, even when such removal is technically feasible.

In theory, if one can delineate the source completely and succeed in removing most of the mass, then a significant benefit may be achieved. There are many case

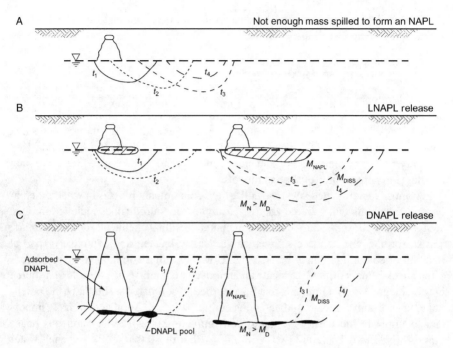

FIGURE 4.13 (A), (B), and (C): Various possibilities of source zone contamination.

studies available in the literature indicating that even for compounds like polycyclic aromatic hydrocarbons (PAHs), after removal of the source, the plumes attenuated rapidly. However encouraging this example might be, this kind of success may not always be realized. Particularly, DNAPL sources in fractured bedrock environments cannot be delineated completely and/or cannot be removed to any significant degree at a reasonable cost. Hence, source removal options may be rejected because none is anticipated to remove all of the source mass without leaving a significant level of residual mass; thus, the expense and risks of the removal efforts are unwarranted.

The manner in which natural attenuation and active remediation measures implemented within an IRZ are combined depends on the *natural attenuation capacity* (NAC) of the system. If the NAC is small, for example, active remediation measures will need to remove or degrade a high proportion of the contaminant flux via more closely spaced IRZ curtains to achieve faster site closure. Conversely, if the NAC is large, lesser mass flux of the contaminants has to be destroyed within the actual reactive zones. In either case, it is necessary to quantify the NAC of the biogeochemical system to develop the site conceptual model as the first step in the design of the IRZ system. NAC is a concept that refers to the capacity of a hydrobiogeochemical system to degrade/transform mass along aquifer flow paths.

Once a thorough evaluation has been made of the impacts of the sources (of contamination) on the design, implementation, and performance of a full-scale IRZ system, educated screening tools can be used to observe the patterns of contaminant behavior within the dissolved plume. Contaminated groundwater plumes can take a variety of

Components of an *in Situ* Reactive Zone 339

forms: they might be expanding, stable, or shrinking, depending on the spatial variations of contaminant concentrations with time (Figure 4.14(A)–(C)). A common pattern in all naturally attenuating plumes is a decline in the dissolved contaminant mass with time.

Once these patterns of contaminant distribution are observed initially, the following list of questions should be answered to develop the site conceptual model and also to evaluate whether additional data need to be collected to complete the design of the IRZ system:

- What physical, chemical, and biological processes are in effect within the defined contaminated groundwater plume?
- What biogeochemical conditions are present to support these processes?
- What level of information is needed to characterize and delineate the footprint of the contaminant plume(s)?
- Is there one single plume emanating from a single source of contamination or multiple plumes extending from multiple source locations?
- Are the plumes comingled or can the plumes be physically separated?
- How does one prove the known or "speculative" contaminant transformation pathways?
- How long is it reasonable to monitor to ensure the permanence of the treatment achieved via the IRZ system, particularly beyond the active period of reagent injections?
- What kinds of specific monitoring and testing methods are needed to confirm that the designed target reactions are in fact taking place within the IRZs?
- How viable are institutional controls and can they be enforced?

4.3.3 BIOGEOCHEMICAL CHARACTERIZATION

Biogeochemical characterization is a key element in understanding site conditions so that an IRZ system can be designed. Biogeochemical characterization is intended to confirm or refute the applicability of a set of target reactions to the specific contaminants present at the site, help establish the baseline along with historical data, and provide data that help guide system design. Biogeochemical characterization data can be used to modify the design in numerous ways. Some illustrative examples are presented below.

- Existing reducing conditions and the presence of higher concentrations of natural organic carbon, both in the dissolved and adsorbed forms, will create a high demand for oxidation reagents, if chemical oxidation is the selected remediation strategy. In addition to getting the footprint of the baseline biogeochemical conditions, it is important to estimate the continuous flux of the reducing compounds that will flow into the target zone.
- High ORP and dissolved oxygen measurements indicate that higher amounts of electron donor are required to create and maintain the desired reducing environment during the implementation of an anaerobic microbial IRZ. Higher

amounts of carbon may be delivered by increasing the amount and/or frequency of reagent injections. If the groundwater velocity and/or the potential for oxygen recharge were also high, a more dense injection well network and more frequent injection would be a likely response.
- Low pH in the groundwater system and geology suggests that the use of buffers would be appropriate. In some extreme cases the use of an acid or an alkali agent may be necessary depending on the preinjection pH values.
- In systems with high levels of alternate electron acceptors, such as sulfate, nitrate, or iron, it is likely that the amount of injected electron donor and the required time to reach the target equilibrium conditions will be high during the implementation of a microbially reducing IRZ.

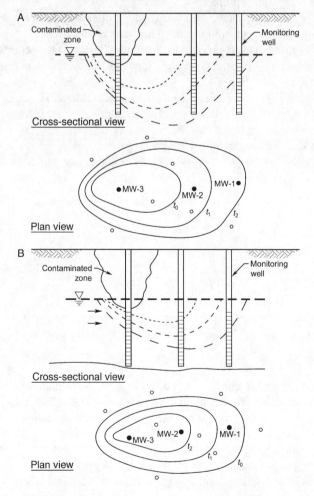

FIGURE 4.14 (A) Contaminant plume is continuing to grow and move downgradient from the source area. (B) Contaminant plume is almost stationary over time and concentrations at points within the plume are relatively constant over time with a slight declining trend.

Components of an *in Situ* Reactive Zone

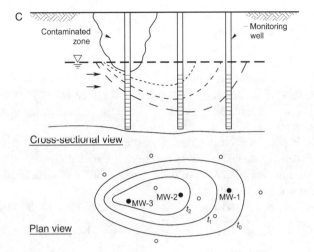

FIGURE 4.14 (continued) (C) Contaminant plume is receding back toward the source area over time and the concentrations at points within the plume are declining over time.

- A groundwater system with very low pH values and high dissolved metals contamination may not be conducive to apply the *in situ* metals precipitation technology, unless sustainable, long-term pH adjustments can be achieved.

Often biogeochemical characterization data are available from site characterization studies, especially if natural attenuation alternatives have been evaluated; however, where not available, a sampling program needs to be implemented before the final design of an IRZ system, and initiation of any pilot studies. Regular and frequent sampling also must occur during IRZ implementation, both to control the formation of the desired biogeochemical conditions critical in process monitoring and to monitor its effectiveness (performance monitoring).

Furthermore, ongoing monitoring of key biogeochemical characteristics of the site is critical to the proper control, operation, and maintenance of the system. Monitoring should be conducted more frequently (e.g., monthly or biweekly) during the initial operation of the system and less often (e.g., quarterly or semiannually) as desired conditions are established. In the beginning, it is often desirable to schedule monitoring rounds to occur between injection events, so that the results of field measurements can be used to refine injection volumes and frequencies.

4.4 PERFORMANCE MEASURES

Ongoing monitoring of IRZ system implementation is done for two purposes: to help control the system (process monitoring) and to evaluate its performance (performance monitoring). Each of these is discussed here.

Performance measures developed must consider many factors that should include, but are not limited to:

- The spatial and temporal scales of the IRZ system(s) being implemented
- The type of hydrogeologic environment — whether saturated, unsaturated, confined or unconfined, porous or fractured

- The driving hydraulic mechanisms for the groundwater flow regime; infiltration or pumping; recharge or discharge
- Three-dimensional and generally transient nature of groundwater flow regimes
- The uncertain nature and distribution of any contaminants and their sources above, at, and below the groundwater surface
- The extent of concentration decline and mass destruction of the contaminant(s) within the implemented IRZ(s) and the rates of those specific decline in concentrations
- Specific testing methods and techniques to verify the designed-target reactions that are designed to take place within the engineered reactive zones
- The duration and frequency of the collection of the biogeochemical parameters and water quality data
- Limited access to and possibility to extensively explore the interior of the engineered IRZ system

The designer of an IRZ system must decide early on the degree of reliability, certainty, and accuracy with which given performance measures need to be developed. Their importance in the decision process, the amount of time, effort, and resources spent collecting and assessing them, and the specified levels of reliability, certainty, and accuracy have to be justified in each specific case. These questions are of a regulatory nature, in most cases, and cannot be addressed by IRZ system designers alone.

4.4.1 Process Monitoring

Groundwater sampling for process monitoring should be performed both in select injection wells and associated monitoring wells, ones within and immediately downgradient of the injection zone. The intent is not to verify or quantify the effectiveness of the technology, but rather to provide almost real-time feedback to control the selected IRZ reactive processes and maintain it within the specified ranges of the biogeochemical parameters to optimize the treatment efficiency. The frequency for these sampling events should be selected and fine tuned as the process is ongoing. However, it is likely that these events will be more frequent near the beginning of the injection program, when the optimum reagent-dosing plan for each site is being established. A typical frequency is weekly to biweekly during the first month of injections, biweekly or monthly during the next 2 to 3 months of implementation, and bimonthly to quarterly for the remainder of the active implementation period. Long-term process monitoring is generally quarterly and the required postinjection compliance monitoring to ensure the permanence of treatment can be quarterly or semiannual. The parameter list for these events may include temperature, pH, DO, ORP, and specific conductance (measured in the field) as well as TOC (laboratory analysis) and the appropriate list of electron acceptors. In the case of a chemical IRZ, concentrations of the injected reagent and their decomposition products also have to be measured frequently. These parameters provide information on the efficacy of reagent delivery and distribution within the reactive zone and the redox conditions of the zone. From this information, reagent injection regimes can be fine tuned and more involved monitoring events can be effectively scheduled.

In selected monitoring rounds including the baseline, the contaminants of concern and their known degradation products should be measured to determine treatment effectiveness. Some information about initial transformation products can be obtained from the initial analyses but concentrations of final end-products will need to be measured to determine if complete transformation of contaminants has been achieved. The incremental cost of measuring the additional parameters that are used primarily to describe the general biogeochemical processes ongoing in the system is low, but can be extremely useful.

Monitoring frequency should always remain flexible during the life cycle of the reactive zone. This concept is critical to the performance of the reactive zone being implemented and the success of the process. Acceptance of the need to be flexible is critical to the regulatory agency's perceptions of the monitoring effort and the client's understanding of the budget and schedule.

4.4.2 Performance Monitoring

Performance monitoring is required to determine and, if needed, enhance the treatment's effectiveness. Performance monitoring generally includes a baseline sampling event, which occurs during the biogeochemical characterization phase previously discussed, and periodic monitoring events during the period of IRZ system implementation. All samples should be collected using low-flow or passive diffusion techniques and appropriate QA/QC procedures discussed in later sections.

Performance monitoring should be initiated only after process monitoring has demonstrated that the target reactions within the IRZ have been achieved. The list of parameters measured during performance monitoring will likely include the concentration of contaminants, their intermediate by-products, and complete transformation end-products. The frequency of monitoring will vary between the monitoring wells immediately associated with the reactive zones and other monitoring wells located within the plume. Generally, the monitoring wells associated with the reactive zones will be measured more frequently than the site-wide monitoring wells that are monitored as part of the required compliance monitoring.

Many of the same tools that are used for data evaluation and interpretation prior to and during IRZ system design can be used during IRZ implementation as well. However some additional tools become valuable during and after implementation. For example, comparison of rate(s) of change of contaminant(s) concentration before and after treatment from time series curves are often helpful in clearly demonstrating the effectiveness of the remedy. However, since the reaction processes that take place within a microbial IRZ proceed through several mechanisms and are controlled by microbial population dynamics and desorption processes, the changes in contaminant concentration(s) will not necessarily fit a simple kinetic model. Time series concentration curves should be done on a molar rather than mass basis so that the effects of sequential degradation processes can be more clearly perceived.

Usually, two units to represent concentrations are used, amount concentration and mass concentration. *Mass concentrations* (e.g., mg L^{-1}) can significantly distort the actual performance derived from a target reaction within a reactive zone. It is necessary to prefer *amount concentrations* (m moles L^{-1}) or "equivalent concentration" when evaluating the mass balance and total mass of contaminants transformed. The authors like to represent the stoichiometry and efficiency of sequential reductive dechlorination by "amount of substance concentrations" on a molar basis (m or μmol L^{-1}) rather than the conventional and widely used mass concentrations (ppm). This makes sense and is a necessity due to significant decreases in molecular weights with every step of the reductive dechlorination process, PCE (166) → TCE (131.5) → *cis*-DCE (97) → VC (62.5) → Ethene (28).

A mass budgeting type approach could be useful in understanding the overall system performance. If a DNAPL is present, estimation of the mass flux from the DNAPL before and after implementation of the engineered IRZ becomes important. The mass flux from the DNAPL would then need to be compared to the estimated DNAPL mass and the degradation rate in the dissolved phase to assess the degree of treatment. However, estimating the mass of contamination in the subsurface is one of the most difficult tasks involved in environmental investigations. Numerical solute transport models incorporating a degradation term are not usually needed but can be helpful in evaluating the performance of complex systems.

4.4.3 GROUNDWATER SAMPLING AND ANALYSIS

Groundwater samples from active IRZs, in which contaminants are being continuously transformed, are often in dramatic nonequilibrium with ambient conditions. Furthermore, contact of these samples with the atmosphere can cause significant shifts in aqueous biochemical equilibrium. The key to minimizing or avoiding shifts in the chemical equilibrium of collected samples is minimizing contact with atmospheric air (Figure 4.15).

Hence, groundwater sampling methods during *in situ* remediation projects should utilize low-flow or micropurge procedures, consistent with published protocols. The basic tenet of the micropurge technique is to collect groundwater from a dis-

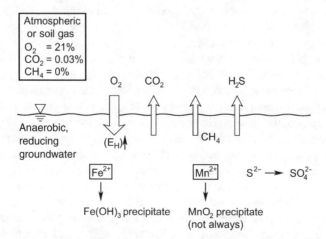

FIGURE 4.15 Possible changes, for example, within an anaerobic microbial IRZ, on the biogeochemical parameters when the samples are in contact with ambient conditions.

crete portion of the well screen at a rate that most closely replicates the natural recharge of groundwater from the formation into the well screen (Figure 4.16(A)). This is accomplished by removing groundwater at low-flow rates (typically between 100 and 500 mL min^{-1}) while monitoring the water level within the well to ensure minimal (or preferably no) draw down. While the well is being purged, field parameters are monitored at the wellhead using a flow-through cell. DO, ORP, temperature, pH, and conductivity are monitored and recorded at (typically) 10-min intervals while the well is purged. When these readings stabilize within 10%, the groundwater is considered to be representative of the aquifer (as opposed to stagnant water within the well) and groundwater samples for laboratory analysis are collected directly from the pump discharge at the surface.

Depending on the depth to water and diameter of the existing wells at the site, different pumps should be utilized. A suitable pump, for example, is a submersible Grundfoss Rediflow 2 pump, which can be inserted into wells of 2-in. diameter or larger and can purge at low rates with a constant flow (important for measuring parameters that are influenced by atmospheric conditions). These pumps can be used for depths up to approximately 125 ft bgs. At extreme depths, the pumps tend to get hot, which can also volatilize constituents in the sample. For deeper depths to groundwater, a bladder pump will be the preferred pump. These pumps require a compressed gas supply and an appropriate cycling control mechanism at the surface to achieve relatively continuous sample delivery. The operating pressure should only limit the depth capability of a bladder pump, as it affects the burst strength of the various pump components. Bladder pumps have only a minimal potential for chemical alteration of the sample, and can be used to sample groundwater up to 400 ft bgs. However, these pumps are generally more expensive than low-flow submersible or peristaltic systems. Sites that have small diameter wells and shallow depth to water may require the use of a peristaltic pump. A peristaltic pump acts by lifting water using a small vacuum (suction) created at the surface. These pumps cost very little and are useful for sampling shallow groundwater to water table depths of 26 ft bgs or less. At any site, efforts should be made to use the same pump and purging method, so that any variability associated with the purging method can be minimized in the generated data set.

While reliable, the sampling techniques described here have certain drawbacks. These techniques require a significant amount of equipment, supplies, and labor. Furthermore, three to five volume purging can produce large amounts of impacted water, which must then be handled appropriately.

Efforts to simplify and reduce the cost of groundwater sampling without compromising data quality have resulted in the development of Passive Diffusion Bag (PDB) sampling devices. The underlying principle behind PDB sampling devices is that contaminants in the dissolved phase are naturally driven from areas of high concentration to areas of low concentration.[8] Polyethylene bags are utilized as semi-permeable diffusion membranes and filled with deionized water as a reference medium. These bags were installed in the screened interval of monitoring wells and left in contact with the groundwater for varying periods of time. The distilled water in the bag contains none of the volatile organic compounds (VOCs) present in the groundwater and the resulting concentration gradient causes VOC molecules to diffuse through the membrane until the concentrations in the bag are equilibrated with the surrounding

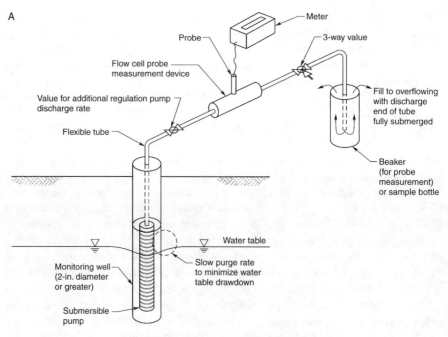

FIGURE 4.16 (A) Schematic of minimal aeration, low-flow groundwater sampling technique.[1] (B) Deployment of a single passive diffusion bag in a monitoring well.

groundwater. This process also works in reverse, such that if VOC concentrations in groundwater drop relative to the water in the bag, the VOC concentrations in the bag would equilibrate accordingly.

Groundwater containing most VOCs can be sampled with this method, and laboratory testing indicated close correlation (within 10%) between diffusion sampler water and test vessel water. In general, the sampling devices should be deployed approximately 2 weeks prior to collection to allow sufficient time for equilibration. However, after initial equilibration, there is no specified time for sample recovery. PDB sampling devices have been left in monitoring wells for up to 1 year without a reported loss of sampler integrity. After collection, the used PDB is simply discarded and a new device is deployed (Figure 4.16(B)).

The primary benefits of using PDB sampling devices include the following: disposable, inexpensive, and easy to deploy/collect (minutes per location, with limited amount of equipment); no purge water to handle; practical for use where access is a problem or where discretion is desirable; multiple PDB samplers distributed vertically along a screened or open interval can provide insight into vertical contaminant distribution; the bags prevent sediment from entering the sample, mitigating sample interference from turbidity.

The primary limitations include the following: not appropriate for all compounds, including MTBE, acetone, humic acids, inorganic ions, and phthalates (bags for metallic contaminants are currently under development); fluctuations in contaminant concentrations not immediately reflected in bags (time delay to accommodate equilibration); and rely on the free movement of groundwater through the well, and thus are representative of conditions in the discrete interval immediately adjacent to the sampling device. Vertical transport in the well would compromise sample representativeness.

4.4.3.1 Dissolved Gases

Remediation processes implemented within an IRZ produce gases that can provide useful information about the process. Additionally, in some cases, the gases produced may need to be managed for health and safety reasons. The production of methane in a microbial IRZ can help determine if desirable alternate electron acceptors are being utilized, while the increased production of ethene and/or ethane indicates that the microbial treatment is proceeding to completion. Evidence from our case studies and peer-reviewed literature suggests that the production of ethane correlates well with the complete dechlorination of CAHs in an ERD system implemented within an IRZ.[9] Indeed, like ethene, naturally occurring concentrations of ethane are generally quite low and the appearance of significant concentrations of both are generally not observed until complete dechlorination is achieved. On the other hand, hydrogen sulfide and methane are potential process by-products that must be controlled if generated and/or released near confined structures.

Monitoring and measuring different types of gases is an important requirement while implementing a chemical IRZ system. Whether it is a permanganate or Fenton's reagent or ozone-based injection, collection and measurement of gases present within the reactive zone is an important process control step. Carbon dioxide, dissolved oxygen, oxygen gas, and ozone are some of the gases that have to be monitored depending on the reaction mechanism being implemented.

Health and safety issues with respect to gas generation should be considered during the engineering design of an IRZ system. The depth to the zone of interest, likely paths for vapor migration, proximity of structures and other receptors, and potential volumes of gases produced must be assessed in this context. When the potential risk of vapor exposure is significant, the engineering design should be modified accordingly to prevent or negate such exposures.

Analytical methods for the dissolved gases including light hydrocarbons and CO_2 typically rely on gas chromatographic techniques similar to those reported in the literature, using the SW3810 modified method, which is a static headspace technique for extracting volatile organic compounds from samples.[10] Hydrogen is usually measured in the groundwater using the "bubble strip" method.[11] It is important that sample collection is performed to ensure steady-state hydrogen levels representative of the undisturbed conditions.

4.4.3.2 Dissolved Oxygen Field Measurement

While dissolved oxygen (DO) probes are usually quite reliable when used in conjunction with a down-hole sonde or flow-through cell, they tend to be relatively expensive, require maintenance and calibration, and may be subject to interferences by temperature, salinity, and other dissolved gases. As such, various simple test kits offered by field instrument vendors (e.g., Hach, Chemetrics, LaMotte) are available as an alternate approach to measuring DO with a probe. These test kits are typically based on one of several methods, including the Winkler (iodometric) titration method; the indigo carmine method, whereby the reduced form of indigo carmine reacts with DO to form a blue product; and methods developed using proprietary reagents such as Chemetrics' Rhodazine D method for measuring trace levels of DO. The choice of which kit to use should be based primarily on its specified range, detection limit, and sensitivity, while considering the kit's ease of use and applicability. For field-testing DO at microbial IRZ systems, a range of 0 to 10 mg L^{-1}, and detection limit and sensitivity of 0.2 mg L^{-1} would be sufficient. It is very important to follow strict low-flow purging and sampling procedures when sampling for DO measurement. Furthermore, note that DO samples may not be diluted to get the measurement within the range of the analytical method. Samples must be tested immediately after sampling.

4.4.3.3 Analysis of Biogeochemical Parameters

Most of the sampling and analyses required for biogeochemical characterization are routine analyses for evaluating soil and groundwater systems. However, even routine analyses need to be carried out with a high level of quality assurance and quality control. The dissolved gas analyses are less often performed and only a select group of laboratories are fully competent to conduct these analyses. Table 4.3 lists the methods used for most biogeochemical analyses needed to support system design and monitoring. Most of the parameters listed in Table 4.3 are typically analyzed offsite, but DO, pH, ORP, conductivity, temperature, sulfide, and iron should be done in the field. There are few standard EPA methods for the analysis of dissolved gases.

TABLE 4.3
Analytical Methods for Groundwater Parameters[2]

Parameter	Analytical Method	Concentration Reporting Units	Location of Test
Temperature	EPA 170.1	°C	Field
ORP	U.S. Geological Survey, National Field Manual for the Collection of Water-Quality Data, 1997	MV	Field
Dissolved oxygen	EPA 360.1	mg L^{-1}	Field
pH	EPA 150.1	S.U.	Field
Conductance	*Standard Methods for Examination of Water and Wastewater*, 15th ed., Method 205 or EPA Method 1201.1	mS (milliSiemen)	Field
Alkalinity	EPA Method 310.1	mg L^{-1}	Fixed lab
Nitrate	EPA Method 300.0A	mg L^{-1}	Fixed lab
Nitrite	EPA Method 300.0A	mg L^{-1}	Fixed lab
Sulfate	EPA Method 300.0A	mg L^{-1}	Fixed lab
Chloride	EPA Method 300.0A	mg L^{-1}	Fixed lab
Methane, ethane, ethene	See text	μg L^{-1}	Fixed specialty gas lab
Carbon dioxide	See text	mg L^{-1}	Fixed specialty gas lab
Chemical oxygen demand	EPA Method 410.4 or 410.1	mg L^{-1}	Fixed lab
Biochemical oxygen demand	EPA Method 405.1	mg L^{-1}	Fixed lab
Total organic carbon (TOC)	EPA Method 415.1	mg L^{-1}	Fixed lab
	EPA Method 415.1	mg L^{-1}	Fixed lab
Ammonia	EPA Method 350.1	mg L^{-1}	Fixed lab
Sulfide	Color chart/effervescence of H_2S (Hach Kit 25378-00)	mg L^{-1}	Field
Total iron	6010B and/or CHEMetrics kit in field	μg L^{-1}	Fixed lab or field
Total manganese	6010B and/or CHEMetrics kit in field based on APHA 314C and CHEMetrics kit in field	μg L^{-1}	Fixed lab or field
Dissolved iron	EPA Method 6010B and/or CHEMetrics kit in field	μg L^{-1}	Fixed lab or field
Dissolved manganese	EPA Method 6010B and/or CHEMetrics kit in field (APHA 314C)	μg L^{-1}	Fixed lab or field
CAHs	EPA Method 8260	μg L^{-1}	Fixed lab
Hydrogen	RSK-196	μM L^{-1}	Fixed specialty gas lab
Bromide	EPA Method 300.0	mg L^{-1}	Fixed lab

Key groundwater parameters that should be sampled in most monitoring rounds include the following basic *process monitoring* parameters:

- pH
- Temperature
- Redox potential
- Dissolved oxygen
- Conductivity
- Organic carbon

While these parameters are of utmost importance in assessing performance of an IRZ system, a number of other parameters may also provide important and useful information. These other parameters generally fall into one of two categories: (1) potentially mobilized metals that may be present at a specific site depending on geology and/or geochemistry, and (2) additional biogeochemical parameters to assist in the better understanding of conditions and/or reaction pathways within the reactive zone. This more complete list of parameters may need to be monitored only once or twice after active treatment begins at sites with simple conceptual models. At more complex sites they may be monitored more frequently to ensure that the desired biogeochemical changes have occurred for the target reactions to take place.

The typical list of biogeochemical parameters includes the following:

- Temperature
- Total dissolved solids
- Total suspended solids
- NO_3^-
- NO_2^-
- SO_4^{2-}
- S^{2-}
- Fe (total and dissolved)
- Mn (total and dissolved)
- Carbonate content
- Alkalinity
- Dissolved hydrocarbons
- Any other organic or inorganic parameters that have the potential to interfere with the target reactions
- Metals: target metals (filtered and unfiltered)

Characterization of the redox chemistry should be a main objective during the predesign studies, since the biogeochemical conditions are a crucial factor in controlling the potential and rate of the target reactions within the reactive zone. In addition to the methods discussed so far to determine the distribution of aqueous (and solid) redox species there are advanced techniques available to quantify the redox processes in the aquifer better. These techniques include measurement of stable isotopes $\delta^{13}C$

(of dissolved inorganic carbon, dissolved organic carbon, and methane), δ^2H of methane, $\delta^{34}S$ of sulfate, $\delta^{15}N$ of nitrate, and dissolved gases (N_2, Ar, H_2).

Degradation of certain organic compounds favors the ^{12}C with respect to the ^{13}C isotopes, which results in an enrichment of $\delta^{13}C$ in the residual fraction. Sulfur isotopes of sulfate can give clues with respect to the origins of the sulfate and confirm whether sulfate reduction is occurring. Similarly, $^{15}\delta N\text{-}NO_3$ and partial pressure of N_2 exceeding atmospheric equilibrium will help in determining the precise nature of denitrifying reductions. Isotope chemistry analysis can also help in shedding light on some of the precipitation reactions taking place within the reactive zone.

4.4.3.4 Oxidation–Reduction Potential and E_H

The oxidation–reduction potential (ORP) of groundwater is a measure of electron activity and is an indicator of the relative tendency of a biogeochemical system to accept or transfer electrons. In a very oxidizing environment the activity of electrons is low and in a very reducing environment the activity of electrons is high. The electron activity is characterized by using the lowercase "p" notation (just as in pH). The pϵ can be calculated from the measured concentrations of reactants and products in a REDOX half-reaction.

A scale equivalent to the pϵ is the E scale, which is expressed in volts and is based on the determination of electron activity using electrochemical methods. E_H is often confused with the closely related REDOX potential or ORP measurement. The ORP or REDOX is measured by placing a REDOX electrode into the water sample; the REDOX electrode is a piece of metallic platinum, which acquired a more negative potential with respect to its reference electrode under reducing conditions where electron activities are higher (Figure 4.17).

In electrochemistry, the *standard-state* potential of an oxidation–reduction reaction, expressed in volts (V), is typically determined per the following equation:

$$E°(\text{reaction}) = E°(\text{electrode 1}) + E°(\text{electrode 2}) \quad (4.3)$$

Standard-state conditions are indicated by the superscript "°", which means that during the reaction, the oxidized and reduced species (often referred to as a redox couple) are both present at unit activity (1 mol L^{-1}), under one atmosphere of pressure, and at 25°C. In order to determine the standard potential for one of the electrodes in Equation (4.3) (known as a half-cell reaction), a reference electrode of

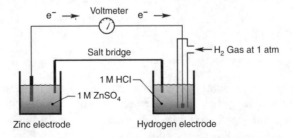

FIGURE 4.17 Description of the measurement of redox.

known potential is needed. This need is filled by the standard hydrogen electrode, which has a potential of V.

Figure 4.17 depicts the half-cell reactions for the oxidation of Zn to Zn^{2+}, measured relative to the standard hydrogen electrodes. In the lab, the above reaction yields the following:

$$E° = +0.76V = E°(Zn/Zn^{2+}) + 0V \qquad (4.4)$$

Because the conversion of Zn to Zn^{2+} is an oxidation reaction, Equation (4.4) confirms that the standard oxidation potential of zinc is +0.76 V. Conversely, its standard reduction potential would be −0.76 V (simple reversal of the reaction). Because standard-state conditions are difficult to maintain, an equation is needed to allow the calculation of potential at other temperatures, pressures, and when the oxidized and reduced species are present at different activities. The Nernst equation provides such a solution for systems at *thermodynamic equilibrium*:[12–14]

$$E_H = E° + \frac{2.300RT}{nF} \times \log\frac{\text{(activity product of oxidized species)}}{\text{(activity product of reduced species)}} \qquad (4.5)$$

E_H can also be translated into a measure of electron activity (the negative log of electron activity expressed in moles per liter, typically represented as pϵ). Assuming standard temperature and pressure, E_H can be converted to pϵ using the following equation:

$$p\epsilon = 16.9\,(E_H) \qquad (4.6)$$

In the past, it was theorized that E_H (pϵ) could be used with pH as a master variable for characterizing groundwater systems.[12–14] This led to the development of methods and equipments for measuring E_H in the field as a tool for characterizing oxidation–reduction processes. Because standard hydrogen reference electrodes are not practical tools, ORP measurements are typically collected using a platinum reference electrode. Consequently, ORP measurements collected in the field are not the same as E_H. To determine the E_H from an ORP measurement, the ORP measurement must be corrected using a reference solution of known E_H. The correction can then be made using Equation (4.7):

$$E_{H(observed)} - E_{H(reference)} = ORP_{observed} - ORP_{reference} \qquad (4.7)$$

Therefore:

$$E_{H(observed)} = ORP_{observed} - ORP_{reference} - E_{H(reference)} \qquad (4.8)$$

If it is reasonable to assume constant pressure, this correction must be made at the same temperature as the system being observed. Common reference solutions include Zobell's solution and Light's solution. ORP electrode manufacturers will often provide a generic correction that can be added to the observed ORP to approximate E_H. This correction is typically in the range of +200 millivolts (mV).

In recent years it has become more and more clear that the interpretation and use of groundwater E_H measurements is problematic. Two key reasons for this are as follows:

- The determination of E_H assumes that the system for which the determination is being completed is at *thermodynamic equilibrium*. In natural systems (i.e., groundwater), the presence of active, respiring microorganisms, which require available free energy to drive their metabolisms, precludes the ability of redox couples to achieve equilibrium.
- ORP electrodes respond to all reactions at the platinum surface, not just those of a particular redox couple. Consequently ORP measurements can drift due to mixed potentials, impurities at the electrode surface, or electrokinetic phenomena.

Based on this, it is apparent that groundwater systems do not have a true E_H. Consequently, E_H/ORP measurements should be used with care when evaluating geochemical environments. Appropriate use may include the evaluation of general trends in redox activity (such as the environment becoming more or less oxidizing or reducing), and the interpretation of other site-specific groundwater data, such as the appearance/buildup of an electron acceptor's reduced species.[12–14]

The measurement of ORP, obtained by using the REDOX electrode, provides a useful, approximate characterization of REDOX conditions in the aquatic environment, although it lacks a precise theoretical definition. It should be remembered that even though ORP and E_H are both measured in volts (millivolts) and do show a rough correlation, they are defined quite differently and should not be treated as synonymous.

4.4.3.5 Microbial Assessments

Microbial communities differ in both qualitative and quantitative composition. The relative proportion of their community members is subject to biogeochemical changes of the environment as well as changes caused by the physiological and metabolic changes caused by the organisms. Organisms that are abundant and culturable under some conditions may develop into dormant and possibly uncultured forms under other conditions. What has been learned from recent studies on microbiological analyses of environmental samples can be summarized as follows: (1) the cultured microorganisms represent only a small fraction of natural microbial communities and hence the microbial diversity in terms of species richness and species abundance is grossly underestimated, and (2) our understanding of microbial diversity is not represented by the cultured fraction of the diversity.

A recent report published by the American Society of Microbiology points to two critical concerns regarding studies on microbial diversity. First, those few microbial species that are grown and studied in culture poorly represent the diversity of physiology, biochemistry, and morphologies that can be found in nature. "Although the general patterns of macroorganismal diversity are relatively well known, spatial patterns of microorganismal diversity are completely unknown."[15] Constraints on knowledge are caused largely by methodologies that do not contend well with the complexity of field sites and the scale differential between field habitats and laboratory microorganisms. It is easy to understand this difficulty when we realize how small a sample we take to represent a typical field site. A contaminated aquifer that measures 500 ft in length, 100 ft in average width, and 20 ft in thickness has a total volume of 1,000,000 ft^3. Each split spoon sample we collect has a volume of 0.025 ft^3, so there are more than 40,000,000 possible split spoons in the aquifer soil population. Even with advanced, phenotype-free methodologies such as nucleic acid extraction followed by cloning and separation of phylogenetically diagnostic 16S rRNA sequences, the results of lab assays on a few split spoons cannot be representative of an entire field site.

As a result of our inability to cope with microbial diversity, efforts have been spent in the last decade to develop methodologies that are considered to be a quantum leap from the previous ones. These methodologies include molecular cloning, polymerase chain reaction (PCR), DNA probing, and the like. A whole new discipline, *molecular microbial ecology*, enables us to use the semantic molecules for phylogenetic studies.

However, in the view of the authors, evaluation of hydrogeologic and baseline biogeochemical data generally is sufficient to determine whether or not a microbial IRZ application is feasible at a site. Pilot tests are conducted to collect predesign data on injection well spacings, reagent requirements, and injection frequency. Laboratory-scale studies generally are not required and will be considered to be redundant when a field pilot test is already being performed. The costs in time and dollars are going to be only slightly different for laboratory and pilot-scale efforts for a microbial IRZ system. However, a laboratory treatability study using groundwater and core samples obtained from the subject site is highly recommended for evaluating a chemical IRZ design.

If there is a reason in the biogeochemical data to raise the doubt as to whether a microbial IRZ system will be successful, a laboratory study may be warranted. Laboratory study costs can vary significantly. For instance, Denaturing Gradient Gel Electrophoresis (DGGE) and Phospholipid Fatty Acid (PLFA) techniques used for microbial screening have lower costs per sample and can thus be used to screen for the presence of microorganisms at multiple areas of the site. PCR analysis used for the identification of dehalorespiring bacteria can cost reasonably higher per sample (Table 4.4). On the other hand, cost of microcosm studies can be very high, and may take 4 to 12 months to complete.

There are many reports where designers of IRZs preparing to implement ERD systems have performed microcosm studies coupled with microbial identification techniques to determine whether or not complete dechlorination is likely to occur at a site. However, the authors of a recent study stated that the heterogeneous distribution of dechlorinating activity . . . points to potential weaknesses in using microcosms to predict responses at a given site.[16] In addition they stated that the time, trouble, and expense involved in running microcosm studies clearly dictate that the locations for testing must be carefully chosen according to the best and most current site data. When looking at a site and the alternatives available to determine the feasibility of using reductive dechlorination as a treatment option, the costs may be similar for

TABLE 4.4
Summary of Microbial Assessment Techniques

Test Method	Description	Data Required	Method of Data Collection	Pros	Cons
High complexity microcosm tests	Microcosm tests including various easily utilized pure carbon sources coupled with small highly instrumented field pilot tests	Laboratory data include COC data from both liquid and headspace and donor taken from liquid; field data included COCs from groundwater	Lab: gas-tight glass and Teflon syringe for GC analysis Field: low-flow/ micropurge methods	Detailed, definitive information on contaminant removal, info for substrate and nutrient selection and dose is generated	High cost and extended time needed; pilot testing must be used to define design parameters; primarily uses pure substrates that have a higher cost associated with them
Lower-cost, lower complexity microcosm Testing	Microcosm tests using single complex low-cost carbon sources such as molasses, vegetable oil, etc.	Lab data includes COCs from liquid	Gastight glass and Teflon syringe	Directly yields information on contaminant removal and completeness of treatment; guidance to substrate dose selection	Moderate cost and time; must be coupled with an engineering assessment or pilot test to evaluate reagent distribution
Phospholipid fatty acid (PFA) and denaturing gradient gel electrophoresis (DGGE)	Analysis provides a determination of total viable biomass and characterizes the types of organisms present and their general physiological status	Soil or groundwater sample	Soil coring from shelby tube or split spoon Low-flow/micropurge procedures	Provides detailed information on microbial community; can identify known degraders; low cost per sample; can screen multiple areas of site	Specific only for known degraders; excludes other species that have not yet been identified
Dehalorespiration genetic screening tool (PCR analysis)	DNA-based screening technique to detect the presence of *Dehalococcoides*	Soil or groundwater sample	Soil coring from shelby tube or split spoon Low-flow/micropurge procedures	High correlation between complete degradation of chlorinated compounds and presence of *Dehalococcoides*; can screen multiple areas of site	Specific only for *Dehalococcoides*; excludes other species or consortia known to have the same capability

either a complex microcosm study or for installation and completion of a simple field pilot test. Note that a pilot test will also allow microorganisms that may be initially present in only a minority of subsurface locations to flourish and become more widely distributed within the reactive zone once the optimum biogeochemical conditions have been established.

PCR analysis can yield a positive for the bacterial species *Dehalococcoides*, but overlooks many other species of bacteria that are capable of reductive dechlorination. *Dehalococcoides* has been chosen as an indicator microorganism as it is capable of completely degrading PCE to ethene. However, a study found that at least two populations of dehalorespirers were responsible for the sequential dechlorination of PCE to ethene in one mixed anaerobic culture.[17] Other studies have been performed to evaluate how widely distributed *Dehalococcoides* strains were in the environment and to determine their association with dechlorination at chloroethene-contaminated sites.[18] One study determined that *Dehalococcoides* organisms are widely distributed in the environment and are associated with full dechlorinating processes, which suggest there may be little need for routine microbial assessments. This study indicates that the organism *Dehalococcoides* had a widespread distribution and also stresses that other organisms may be capable of converting PCE to ethene.

Thus currently there is no universally applicable answer regarding what techniques should be used and when they should be applied in determining whether or not enhanced reductive dechlorination will be successful at a site. The benefits associated with bench-scale microbial assessment generally do not outweigh the costs of performing them when the biogeochemical data are favorable. When bench-scale work is needed, simple microcosm studies coupled with PCR identification appear to be the most helpful bench-scale tools in predicting to what extent reductive dechlorination will occur.

4.4.3.5.1 What Is Meant by 16S rRNA?
Some new terminology has entered the groundwater microbiology literature, marketing materials, and questions posed by many regulators. One of those terms, 16S rRNA, is being used in association with the examination of aquifer soils for the presence of *Dehalococcoides ethenogenes*. Here is a brief description of the "16S" part of the term 16S rRNA.

RNA is the abbreviation for ribonucleic acid, a part of every organism's system for translating genetic instructions carried in their DNA (deoxyribonucleic acid). One of the very important functions of RNA is the synthesis of proteins, a process associated with subcellular bodies called ribosomes. Ribosomal RNA molecules (rRNA) are built from thousands of nucleotides, each composed of a ribose sugar residue, a phosphoryl group, and one of four bases (adenine, uracil, guanine, or cytosine). Using a high-speed centrifuge, rRNA molecules can be extracted, separated, and classified according to their sedimentation rate in a density-graded sucrose solution.

The RNA extract to be examined is placed in a centrifuge tube at the top of a sucrose solution that has been constructed with a gradation of densities, low density at the top and high density at the bottom. Under high-speed centrifugation, individual molecules from the RNA extract settle into the sucrose gradient at levels where the

Components of an *in Situ* Reactive Zone

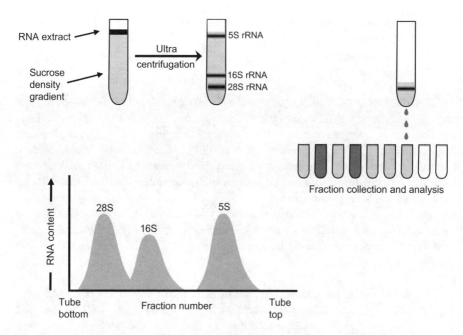

FIGURE 4.18 Ribosomal RNA (rRNA) is separated by molecular weight using density gradient ultracentrifugation. The rRNA extract is placed at the top of a sucrose gradient and the rRNA molecules migrate during centrifugation to sucrose density zones corresponding to their respective molecule sizes. Following centrifugation, the density tube is punctured and a series of fraction samples is collected for analysis of rRNA content. The 16S fraction is used for characterization of microbial community composition. (Redrawn from Voet and Voet.[19])

sucrose and RNA densities match. After centrifugation, the tube is drained from the bottom and collected in a series of fractions with decreasing molecular size. Figure 4.18 describes the process.

The capital S stands for Svedberg, the standard unit of the sedimentation coefficient. The 16S fraction contains bacterial rRNA, composed of approximately 1500 nucleotides, that can be analyzed for biochemical markers unique to a particular species of bacteria, such as *Dehalococcoides ethenogenes*.

4.5 SYSTEM DESIGN

4.5.1 Design Considerations

A number of design considerations need to be understood to successfully design and implement an IRZ system. The goal is to create an optimum hydro-biogeochemical environment in the subsurface that provides the optimum conditions for the selected (bio)chemical reactions to accelerate the remediation of target contaminants. Among the critical design considerations for an IRZ system are:

- Geology
- Groundwater chemistry

- Baseline definition
- Microbiology
- IRZ layout options
- Reagent selection

Selection of drilling techniques would also be considered as part of the system design for an *in situ* remediation system. Since there is so much information available on this subject, we are not discussing drilling techniques in this book. Another important factor in design consideration is cost of implementation — including both capital and operation and maintenance costs. Budgetary cost limitations can often directly or indirectly affect design decisions such as source reduction and/or plume-wide treatment and/or containment.

Based on the authors' experience and analysis, the two largest cost factors for the implementation of an IRZ system are the injection well(s) installation and reagent delivery. The three site-specific factors that contribute to the cost of an IRZ system implementation are as follows:

- *Plume size to be treated.* This is a primary factor driving the cost of the system as the larger the plume area to be treated the more wells are needed (drilling costs) and the more time it takes for each reagent injection (more details are discussed in Chapter 5).
- *Depth of target zone.* Drilling costs are another primary factor affecting overall system costs. Deep contaminant settings and/or those requiring specialized drilling techniques (bedrock drilling, multiple conductor casings, etc.) can significantly increase system costs.
- *Groundwater flux through zone of treatment.* Reagent injections also play a large role in overall costs. At sites where there is a high groundwater flux, more reagent injections will be required at a higher volume and frequent intervals, thus increasing costs.

These factors need to be given special consideration during design in order to develop the most cost-effective approach for site remediation.

4.5.1.1 Geology

It is important to obtain specific hydrogeologic data in order to design a delivery system to deliver the reagents and other additives such as buffering solutions, tracers, and the like at the desired concentrations and distribution to the target zone. While a complicated lithology can place constraints on the use of the IRZ technology at a given site, in most cases it will not completely eliminate it as a remediation option. Complex lithologies are likely to be problematic for most remediation technologies. By properly placing injection well screen zones or other delivery mechanisms to target specific impacted groundwater bearing layers, the technology can be effectively applied in most environments. From a design perspective, delivery in complex geologic settings will more often be dictated by remedial goals and timeframes, economic considerations, or nontechnical factors such as regulatory and public perception. Thus,

TABLE 4.5
Specific Geologic/Hydrogeologic Parameters Required for the Design of an IRZ[2]

Geologic/Hydrogeologic Parameter	Design Impact
Depth to impacted groundwater	Injection well depth and screen locations
Width of contaminant plume	Number of injection wells
Thickness of contaminant plume	Number of injection points within a well cluster
	Pressure injection vs. gravity feed
Groundwater velocity	Injection volume and frequency, residence time for the target reactions, dilution of end-products
Hydraulic conductivity (horizontal and vertical)	Mixing zones of reagents, extent of reactive zone
	Number of injection points within a well cluster
Geologic variations, layering of various soil sediments	Location of well screens at injection points
Soil porosity and grain size distribution	Removal of end-products resulting from immobilization reactions (such as heavy metals precipitation)

understanding the complexity and defining the lithologic variability as it relates to the groundwater impacts is an important first step in design (Table 4.5).

A tool that can be part of the evaluation of a site's suitability for IRZ application is aquifer testing such as pumping tests. This can be important as a means to refine the knowledge of hydrogeology and thus predict the performance of a full-scale delivery system. It can help reduce the cost of pilot testing and the full-scale system by allowing an optimum monitoring strategy, including the locations of the injection and monitoring wells, to be well defined.

4.5.1.2 Hydraulic Conductivity

Understanding a formation's hydraulic conductivity is critical to designing IRZ systems. A formation's hydraulic conductivity is used, along with the hydraulic gradient, to determine the groundwater velocity (and thus the flux) and the amount of reagent to be injected. In addition, this information also helps determine injection well spacing and distances between injection well arrays. The higher the hydraulic conductivity of the formation the easier it is to deliver the reagent into the subsurface and the more effective a single delivery point can be. As the hydraulic conductivity increases — all other factors remaining equal — the distribution of reagent from a single injection point along the direction of advective flow increases but the radius of influence perpendicular to the flow direction decreases. In addition, lower permeability will generally contribute to lower groundwater flow velocities and advective transport of the reagent. This, in turn, must be carefully considered when evaluating the design of full-scale treatment systems (i.e., spacing of injection points in the direction of groundwater flow and across in order to treat the impacts in a timeframe consistent with remediation goals). This is particularly true for pilot-scale or demonstration systems with respect to the placement of observation points/wells to see desirable results within the short timeframe of the study.

4.5.1.3 Groundwater Flow Characteristics

Groundwater flow characteristics are another important consideration in the design of reactive zones. The groundwater velocity, flow direction, and the horizontal and vertical gradients impact the effectiveness of reagent injections and the speed with which the reagent will spread and mix with the groundwater. Low velocity systems typically require lower reagent mass feed rates since the groundwater flux is reduced — all other conditions being equal. This is an important criterion for the concentration and amount of additive that is needed to create the reactive zone.

While the composition of interstitial water is the most sensitive indicator of the types and the extent of reactions that will take place between contaminants and the injected reagents in the aqueous phase, groundwater flow direction and gradients are also important to consider. It is critical to understand the dynamics of groundwater flow to ensure that the reagent injections will form a reactive zone in the target area. Horizontal and vertical gradients are used to define the lateral location (well point or injection device) and vertical location (screen or delivery zone) of the injection and monitoring points. It is also important to understand the heterogeneity of the aquifer. The groundwater flux is the carrier that moves the reagents downgradient of the delivery system. This advective transport accounts for the majority of the transport mechanisms that contribute toward the creation of the IRZ. It should be noted that during implementation of a microbial IRZ, injected reagents are also consumed by the microorganisms as they are transported advectively from the injection zone.

The authors have seen evidence that suggests that some influences of the reactive zone will propagate faster than would be predicted by advective transport, although this contribution is likely immaterial for design consideration. If there are pockets within the IRZ where groundwater movement is very slow, then the reagents will have difficulty reaching those areas and the environment will not be appreciably changed in the short term. It will not be possible to create an optimum IRZ in the low-flow areas except through the slow process of diffusion.

Advection is the main process that moves the reagents downgradient of the delivery system and dispersion moves the reagents in directions perpendicular to groundwater flow (this is referred to as transverse dispersion). Advection is movement by bulk motion, and is quantified by the value of the groundwater velocity. Under most conditions, groundwater is constantly moving, although this movement is usually slow. Groundwater flow velocities may be calculated using Darcy's law and calculating an initial estimate of the groundwater velocity is the basic first step in the design of IRZ systems at any site.

4.5.1.4 Saturated Thickness and Depth to Water

The depth to groundwater will define well design and contribute significantly to the capital cost of a full-scale system. The saturated thickness can also have an influence on cost, since there are practical limits on the maximum screened interval that can effectively be used in a single injection well. Based on our experience, a 25-ft screened interval represents a practical limit for an injection point (Figure 4.19). Of course, this limit will be impacted by the heterogeneity of the subsurface lithology,

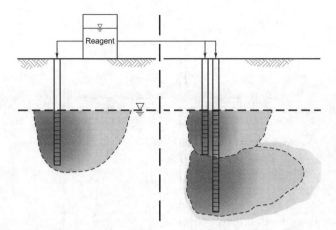

FIGURE 4.19 Single injection well point for shallow depths and injection well cluster for deeper *in situ* reactive zones.

hydraulic conductivity, and the resulting effects on permeability and groundwater flow characteristics. If the lithology and resultant groundwater flow characteristics are such that there are variations in the flow characteristics within the target saturated interval, the use of multiple screened zones or multiple well points should be considered — even if the interval is less than 25 ft.

In faster groundwater flow systems, very limited transverse dispersion in groundwater can limit the lateral extent of the reactive zone created by an individual injection point. This is of particular importance in settings where drilling costs may be high (i.e., deep settings). In such cases, an *in situ* recirculation well can yield considerable cost savings over use of direct injection wells. This *in situ* recirculation well concept aims primarily at delivering reagents in a cost-effective manner while remediating larger, deeper contaminant plumes at sites with relatively high groundwater velocities (Figure 4.20(A) and (B)).

4.5.1.5 Geochemistry

Organic carbon fraction (f_{oc}) and buffering capacity are two geochemical characteristics that are important to consider during design of an IRZ system. The fraction of organic carbon (f_{oc}) will impact the reductive poise of the system, as well as the sorptive capacity of the soil matrix for the contaminants. A site characterized by high f_{oc} soils will have a high contaminant sorptive capacity. Accordingly, these sites will have considerably higher sorbed, solid-phase mass of contaminants. Thus, as the target reactions proceed in a microbial IRZ, there is the potential to release a larger mass of contaminant(s) by desorption due to the formation of microbial surfactants and cosolvents. As such, the layout of the injection wells and planned treatment time should consider these desorption effects. In addition, caution should be exercised in installing systems in areas close (measured in groundwater travel time) to risk receptors. A careful "outside–in" approach would be needed in those cases to ensure that

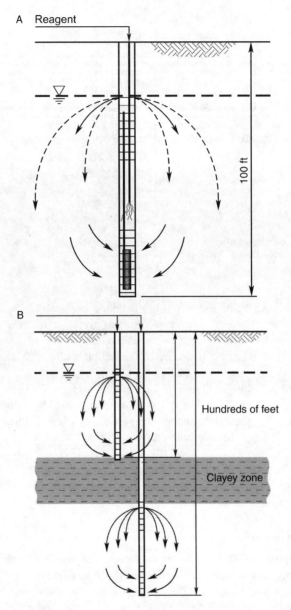

FIGURE 4.20 (A) In-well mixing systems with submersible pumps to create a reactive zone with wider lateral distribution per injection well. (B) Multiple cluster in-well mixing systems to create reactive zones up to hundreds of feet in depth.

desorbed contaminants or partial degradation products are contained by a downgradient IRZ and do not impact the receptor.

Optimal use of the *in situ* chemical oxidation technology is very much dependent on understanding oxidant demand from contaminant oxidation as well as the matrix demand. Matrix oxidant demand refers to the oxidant consumption that can be attributed to background soil and groundwater conditions. Matrix demand can be derived from oxidation of natural organic matter, reduced metals, carbonates, sulfides, and other sources of reductive poise. Matrix demand can be highly variable and will be influenced by background geochemistry. The oxidant demand caused by nontarget compounds can range from 10 to 100 times (or even higher) of the stoichiometric demand caused by the target contaminants. Hence it is very important to look at the oxidant demand of the entire (bio)geochemical system.

It should be noted that destruction of the natural organic matter can also release adsorbed contaminants into the dissolved phase. Even though mechanisms of contaminant mass desorption are different within microbial and chemical IRZs the end result is similar — higher concentrations of dissolved contaminants. Another important evaluation of the geochemistry should include the background concentrations of the naturally attenuated metallic contaminants such as chromium. The reduced Cr^{3+}, adsorbed to the soil matrix, can be mobilized as Cr^{6+} during chemical oxidation.

Other geochemical impacts on chemical oxidation systems include the effect of background iron on the efficiency of $OH^{\cdot}$ formation during Fenton's reactions. Redox cycling of manganese between the Mn(II) and Mn(IV) oxidation states consumes H_2O_2, and also does not yield $OH^{\cdot}$ ions.

The generation of low-pH groundwater zones during IRZ implementation needs to be minimized in order to maintain high reaction rates. The pH of an aquifer system is a function of the buffering capacity of the aquifer, a characteristic primarily imparted by the aquifer solids. Aquifer systems with lower buffering capacities are more susceptible to pH drops. Measuring the alkalinity of groundwater can generically assess the buffering capacity of an aquifer system. However, because of the importance of the aquifer solids in establishing buffering capacity, groundwater alkalinity will present only a partial picture of the true buffering capacity and is likely to underestimate it. Groundwater alkalinity samples are fairly stable and can thus be analyzed offsite, though alkalinity may also be readily measured in the field. Field test kits are readily available and should be sufficient to quickly field-screen the buffering capacity of an aquifer system.

Bicarbonates, carbonates, and hydroxides in the aquifer solids usually dominate alkalinity, although borates, silicates, and phosphates may also contribute. Because of the importance of carbonates in establishing the buffering capacity of a groundwater system, groundwater systems containing limestone minerals such as calcite ($CaCO_3$) and dolomite ($CaMgCO_3$) tend to yield the highest alkalinities, and therefore buffering capacity. In aquifers where buffering capacity is low, buffering reagents may need to be incorporated into the design of the reagent solution.

4.5.2 GROUNDWATER CHEMISTRY

Groundwater chemistry includes an understanding of the target contaminants, their daughter products, and the biogeochemical parameters. Understanding the conditions present in the groundwater system will make the selection and application of reactive zones more likely to succeed. Lacking that understanding, one may end up trying to undo nature and find that it is necessary to spend significantly higher effort to bring about the desired result. A key factor to keep in mind is that no single measurement or result should be relied on to define the predominant biochemical processes ongoing at the site. Redox processes in natural systems are rarely in equilibrium. Moreover, the predominant electron acceptor being utilized often varies in zones across the site. Instead, the full list of parameters analyzed should be reviewed both on a well-by-well and a site-wide basis, to determine which of the electron acceptors are primarily being utilized.

As discussed previously, the enhancement of natural conditions takes advantage of natural processes that are already contributing to the degradation of the target compounds. Of particular importance is the presence of degradation products, the presence and nature of electron acceptors, a definition of the redox conditions (ORP), and the presence of electron donors. The presence of degradation products which indicate that a particular environment has established itself is typically easy to verify. For many sites historical data are available that may date back years and that can be used to establish the presence of degradation products, as well as to evaluate trends in source and daughter products over time. These data can also provide information regarding historical impacts of variable organic substances that may have served as electron donors. Reducing reactive zones rely on the presence of an adequate source of electron donors to establish and maintain a bacterial population that can maintain the optimum biogeochemical environment until the contaminant destruction/transformation is complete.

In the case of chemical IRZs, certain anions present in the groundwater can impact oxidation efficiencies. Common groundwater anions such as NO_3^-, SO_4^{2-}, Cl^-, HCO_3^-, and CO_3^{2-} could scavenge hydroxyl radicals and may be a source of treatment inefficiency. The oxidation chemistry of contaminants has to be well understood. Oxidants must attack the C–C bonds in these molecules. The double bonds that characterize chlorinated ethenes are far more reactive than the single bonds of chlorinated ethanes. Hence, in the evaluation of a plume with mixed contaminants, equal concentrations of chlorinated ethanes and ethenes will cause us to question the feasibility of chemical oxidation.

4.5.2.1 pH

While microbial populations can endure a wide range of pH, a pH close to neutral (5 to 9) is the most conducive to the proliferation of healthy, diverse microbial populations necessary for implementation of ERD systems. In particular, low groundwater pH may indicate and/or encourage fermentative biochemical reactions unfavorable to ERD systems. In such cases, pH buffering, typically using common basic salts, may be required during implementation to raise pH and/or neutralize pH against further decreases. Sites with pH outside of the 5 to 9 range may require more thorough microbial screening to evaluate the effect of pH manipulation on the efficacy of existing microbial populations. Sites characterized by large areas having very high or very low pH as a result of their past contamination history may be poor choices for ERD implementation unless large-scale pH manipulation is feasible. The natural buffering capacity of a site against such pH changes can be generically assessed via the measurement of groundwater alkalinity and the evaluation of bulk mineralogy.

For a chemical IRZ, permanganate, though expensive, is a more stable and effective chemical oxidant over a broad pH. The optimum pH range for permanganate oxidation is in the range of 7 to 8. The stability of H_2O_2 increases with decreasing pH in Fenton's systems, and oxidation efficiency is optimum under acidic conditions.

4.5.2.2 Role of Sulfur in Enhanced Reductive Dechlorination Systems

The role of sulfur in the effectiveness of the reductive dechlorination mechanisms is complex and multifaceted. Existing literature reports tend to suggest that anaerobic bioremediation of CAHs proceeds best under methanogenic conditions. Even though they do document that CAH degradation under sulfate reducing conditions is feasible, it appears anecdotally to us that the majority opinion of field practitioners is that sulfate is inhibitory for reductive dechlorination. It has been stated that concentrations of sulfate greater than 20 mg L^{-1} may cause competitive exclusion of dechlorination. However, in many plumes with high concentrations of sulfate, reductive dechlorination still occurs.[20] Existing screening protocols and other reports score a site more poorly if sulfate exceeds that level.[21] Other guidelines state that abundant electron acceptors, such as sulfate, may inhibit reductive biodegradation. It has been stated that high sulfate concentrations may prevent methanogenic conditions from developing, and the documentation of high sulfate mineral abundance can be used to explain slow rates of reductive dechlorination. Sulfate, whether contributed in the injected reagent or already present, indeed must be reduced in order to reach methanogenic conditions, but there is ample evidence in the literature for dechlorination of a wide variety of CAHs under sulfate reducing conditions.[22,23]

It has also been postulated that the presence of abundant electron acceptors, including sulfate, can interfere with the enhancement of dehalorespirators and that dehalorespiring organisms compete for electron donors such as hydrogen with both methanogenic and sulfate reducing organisms.[21] It is further stated that depletion of electron acceptors should effectively eliminate competition between dechlorinators and such groups as nitrate reducers, iron reducers, and sulfate reducers. Competition from methanogens, on the other hand, may never be eliminated and must be managed by choice and delivery of the electron donor. The authors, however, from experience and review of the most recent literature, argue that halorespirators can coexist with methanogens and sulfate reducers at high hydrogen concentrations, and, thus, the strategies that limit the generation of hydrogen to favor dehalorespirators are not necessary.[1,24]

Sulfur, whether contained in the formulation of a donor such as molasses or present in the system already, offers important advantages, including sulfide production for metals precipitation and potentially aiding biological degradation of CAHs. Sulfur and sulfur containing compounds can aid the degradation of CAHs through a variety of mechanisms:

- Involvement (as reductants and/or intermediates) in the degradation mechanisms of CAHs[25]
- Through the stimulation of dehalogenation by sulfate reducers[26]
- Through abiotic degradation of CAHs by FeS[27]

The authors have successfully applied enhanced anaerobic bioremediation at sites with up to 500 to 700 ppm of sulfate. We have reviewed data that suggest natural attenuation at even higher levels (up to 2000 ppm). Thus, it is clear that CAHs can be effectively treated under sulfate reducing conditions and that sulfur can be directly involved in several processes that enhance this degradation. Practitioners should recognize that multiple complex processes occur in these systems, that research on them is ongoing, and that the presence of substantial sulfate concentrations will not necessarily preclude enhancement of CAH bioremediation.

The amount of sulfur already in the groundwater system cannot be controlled. However, the amount of sulfur added to the system can be controlled by selection of the donor reagent including low sulfur (i.e., corn syrup less than 1 mg L^{-1} sulfate) and medium sulfur (edible-grade molasses, 250 ppm sulfate) reagents. Sulfate concentrations are given for diluted 10% reagent solutions, which is the typical injected concentration, and substantial further dilution should be expected in the aquifer. Thus, the amount of additional sulfur provided to the system can be controlled to some extent by the selection of product and grade among those electron donors commercially available in the food industry.

4.5.3 IRZ Layout Options

There are a number of options available for delivery of reagents to form an IRZ. The most commonly used method is the use of injection wells or direct push well points to inject the reagent into the target zone (Figure 4.21). However, other delivery mechanisms would include gravity flooding via trenches/infiltration galleries for shallow plumes, *in situ* recirculation wells for deeper/thicker plumes, horizontal wells for shallow/thin plumes and plumes beneath buildings or other structures, and recirculation-well systems consisting of a closed system of extraction and injection wells oriented

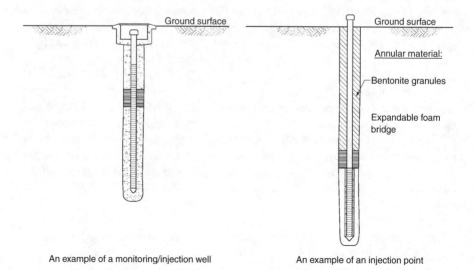

FIGURE 4.21 Schematic of an injection well and a monitoring well.

Components of an *in Situ* Reactive Zone

perpendicular to groundwater flow. For purposes of demonstrating options for applying an IRZ system we focus on injection wells as the delivery mechanism.

The injection systems may be mounted on a truck (Figure 4.22(A)–(C)) or may be placed in a (semi-) permanent structure. Figure 4.22(D)–(F) show the interior of such a structure as well as the physical layout of the injection distribution system. As shown in Figure 4.22(D)–(F), systems can either be manually controlled or automated using Programmable Logic Controllers (PLC).

4.5.3.1 Injection Well/Point Placement

Reactive zones can be implemented in a number of different configurations and using a variety of approaches. The variety and combinations used are limited only by the variety of potential scenarios that may be encountered in the field and the goals of the project. For the purposes of this text three basic layouts are discussed: cutoff/barrier, plume-wide, and hot spots.

Cutoff/barriers or containment curtains (or "fences") consist of a series of reagent injection wells or points typically established in a row perpendicular to the groundwater flow direction along a line that represents a critical boundary for remediation (Figure 4.23(A)). This layout is commonly employed along or near a property line, or other boundaries established for the purpose of site closure or regulatory closure. The location of this layout can also be selected based on practical means (e.g., located near a road for drilling and injection access) or in an available open area in developed settings (Figure 4.24(A) and (B)). In most cases, from a capital cost perspective, the cutoff layout is less expensive to deploy. Since the entire plume is not being remediated, fewer injection points are required. However, life cycle costs could be higher if the source of the CAHs upgradient of the cutoff barrier is not being addressed.

Plume-wide reactive zones target a large portion of the impacted groundwater for much more aggressive and short-term treatment. Typically the injection points will be spaced throughout the target-impacted groundwater (Figure 4.23(B)). By applying the reactive zone across the entire plume, site closure can be achieved more readily. Obviously, there are cost implications with such an application: higher capital costs are traded for shorter remedial time frames and commensurate reduction in total O&M costs.

Hot-spot reactive zones target the source area (Figure 4.23(C)). This layout is often employed in situations where the natural remediation process or a barrier method is successfully controlling the movement of the contaminant plume, but there is a need — regulatory or other — to speed up the overall remediation. In this case the source area is targeted for an IRZ, in order to reduce contaminant mass quickly. Once the IRZ has brought down concentrations in the source area to target concentrations, injections in the source area can cease. An IRZ in the source area is likely to result in a release of daughter compounds as well as the source constituent due to desorption from the soil matrix. Thus it is important to design the IRZ so that the reaction has been established on the downgradient "edge" before beginning the treatment within the "heart" of the source area.

In a number of cases, the remediation strategy includes establishing an IRZ in the source area along with installing one or more IRZs (injection well arrays)

FIGURE 4.22 (A) A trailer-mounted injection for batch injections of reagent. (B) Injection of molasses premixed with the amendments at selected dilution rates. (C) Injection of lower volumes of reagents.

Components of an *in Situ* Reactive Zone

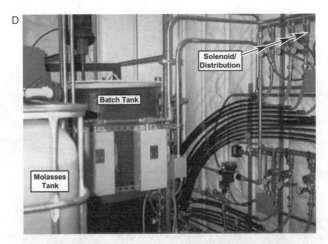

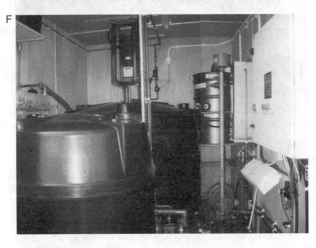

FIGURE 4.22 (continued) (D) Automated injection system. (E) Automated injection system for a large-scale IRZ with about 400 injection points. (F) Automated injection system for multiple reagents.

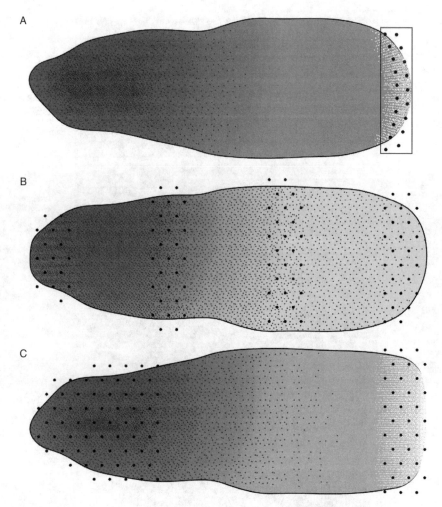

FIGURE 4.23 (A) Injection well placements to provide just a plume cutoff barrier. (B) More injection wells placed across the entire plume to achieve much faster cleanup time frames. (C) Injection wells placed to provide source area treatment accompanied by a cutoff barrier.

downgradient of the source area. The most appropriate application of the reactive strategy is site specific and is based on a site strategy that combines, among other things, costs, regulatory issues, and future use of the site.

Upon selection of the injection well layout or layouts for a given site, the specific reagent injection well spacing must be addressed. Spacing consideration will include both parallel and perpendicular to the direction of groundwater flow. Proper well spacing in the direction perpendicular to the direction of groundwater flow is more critical to the successful application of the IRZ technology as it will have a greater bearing on the ability to create a complete (i.e., overlapping) IRZ downgradient of the injection area thus ensuring complete treatment. In the absence of site-specific well spacing information obtained via a pilot study, this spacing must be based on evaluation

FIGURE 4.24 (A) Trailer-based premixed injection system. (B) Premixed molasses injected from a tanker.

of expected transverse dispersion of reagents following injection as well as experience obtained for field applications. Well spacing on the lower end of the range is recommended for more permeable, higher groundwater flow sites, or in cases where aggressive treatment of high concentration or source areas is desired.

Proper well spacing in the direction parallel to groundwater flow in terms of overall IRZ success is affected more by the balance between budget and desired cost of treatment and is discussed extensively in Chapter 5. Empirical application experience suggests the length of an individual zone where we can expect aggressive treatment of both dissolved and adsorbed phase impacts is generally 100 days of groundwater travel time from the point of injection at most sites. The authors have seen effective treatment within an individual IRZ even up to distances of 200 days of travel time, albeit in a very small number of sites. Beyond this distance the IRZ application will result in treatment of groundwater impacts in a more indirect manner (e.g., due to the flow of "clean" water into the downgradient area and enhanced desorption).

4.5.3.2 Monitoring Well Placement

The selection of monitoring or observation well placement for an IRZ application will be dictated by the degree of interest in monitoring the results from the regulatory agencies and customers. In terms of monitoring well placement and/or utilization of monitoring wells it is first advisable to utilize one or more previously existing monitoring wells for performance evaluation. This is because existing wells most often have useful (sometime extensive) historical data regarding the trends in constituent concentrations and can often offer clues on seasonal variability of the monitoring data. This is important as it often provides clear proof of increase in constituent degradation rates related to the implementation of the IRZ.

New process monitoring wells are typically placed within a reactive zone at monthly intervals of groundwater travel time predicted by a conservative tracer. Wells could be placed at distances of 1, 2, and 3 months travel time from the location of the injection wells. This ensures that the monitoring well system will be able to observe various stages of reagent utilization. This also makes the success of the monitoring program less vulnerable to variations in reaction rates or groundwater velocity. However, it should be recognized that on low groundwater flow settings, 1 or even 2 months travel time may be unrealistically close to the injection wells. It is advised that the minimum spacing between injection and observation wells be 10 ft, unless other constraints dictate closer placement.

Since most real-world systems display preferential flow paths and temporal variation in groundwater flow direction it is also advised to use transects of multiple monitoring wells perpendicular to the direction and/or to stagger the lateral placement of wells as you move downgradient from the injection area. When there is significant vertical migration of the contaminants within the reactive zone, monitoring wells have to be placed as clusters to monitor the performance in that direction.

4.6 REAGENTS

For an IRZ system, specifically for a microbial IRZ, the cost of the reagent material itself is relatively insignificant. The majority of the costs related to reagent injection include the labor associated with preparing the reagent mixture and injecting the material into the wells/points along with related costs (mobilization to the site, record keeping, preparation, etc.). However, for the implementation of a chemical IRZ the cost of reagent(s) itself will constitute a significant portion of the overall cost of system implementation.

4.6.1 MICROBIAL IRZ SYSTEMS

There are a variety of reagents that can be used for the creation of an IRZ to implement anaerobic microbial reactions (Figure 4.25). Table 4.6 identifies a number

Components of an *in Situ* Reactive Zone

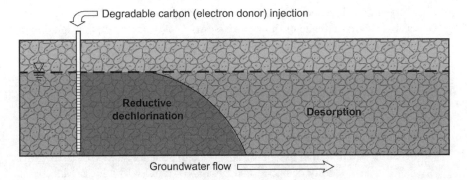

FIGURE 4.25 Creation of an anaerobic microbial IRZ by injecting an easily degradable carbon source for enhanced reductive dechlorination (ERD).

TABLE 4.6
Relative Cost of Various Electron Donors That Have Been Used To Enhance Reductive Dechlorination[2]

Electron Donor	Bulk Price $/lb of TOC
Molasses	0.05–0.20
Sugar (corn syrup)	0.15–0.30
Sodium lactate	1.25–1.46
Whey (powdered, dry)	1.17
Whey (fresh)	0.05
Edible oils	0.20–0.50
Flour (starch)	0.30
Cellulose	0.40–0.80
Chitin	2.25–3.00
Methyl cellulose	4.00–5.00
HRC™ (regenesis commercial material)	5.00–6.00

of soluble carbohydrate and other easily degradable organic carbon reagents reported in the literature.[2] As previously discussed, the goal of injecting a reagent is to create a reactive zone in the subsurface that is sustainable, easily maintained, cost effective, and appropriate to the target compounds. As such, molasses, lactate, high fructose corn syrup (HFCS), and cheese whey have been successfully applied for the treatment of CAHs in groundwater. Other substances that are not soluble carbohydrates have also been used for the same purpose — including vegetable oil, hydrogen release compound (HRC, a proprietary polyacetate ester), methanol, bark mulch, and hydrogen gas. The substrates used by the authors in their projects only are mentioned in the following.

- Molasses is injected in a water solution of 10% molasses or less and moves readily with groundwater. It can be injected using direct push technology, but

most often is injected via injection wells because of the number of injections necessary to reach site closure. Molasses is readily degraded, thus leading to the rapid formation of anaerobic conditions. Several grades of molasses are available and are based on the sulfur content.
- Lactate also can be utilized as a means of inducing ERD conditions in an aquifer. Many of the other, more chemically complex reagents are fermented to lactate. Thus, lactate is known to be a reasonable selection for supporting bacteria capable of reductive dechlorination.
- HFCS can be utilized in place of molasses in situations where the addition of sulfur to an aquifer is undesirable. Our experience with HFCS suggests that it is also readily degraded and capable of rapidly forming an anaerobic zone.
- Cheese whey is perhaps the most chemically complex of the soluble carbohydrates. Its increased chemical complexity makes it longer acting in the aquifer. Fresh whey can be obtained at a very low cost — typically the only cost is transportation. Powdered whey is substantially more costly, but it is easier to obtain, ship, and store.

As the efficacy of ERD has been demonstrated in the field by numerous practitioners, the number of soluble substrates applied has grown. Table 4.6 lists only a few of the many reagents tested in the laboratory or in the field. Other substrates include methanol, ethanol, sucrose, cellulose, acetic acid, acetate, pure hydrogen, and proprietary blends of these and other sources of soluble organic carbon. The selection of the substrate must take into account reagent unit cost, reagent availability, expected variations in rate of anaerobic zone generation, rate at which the substrate is utilized, and target aquifer inorganic water quality goals. Many reagents are being selected from the wide variety of available food-grade organic carbon sources (e.g., molasses, sucrose, and vegetable oils). Reagent selection as discussed later should also be driven by engineering considerations chiefly regarding its physical state, speed of utilization, and relative cost.

4.6.1.1 Suitability of Different Electron Donors

All commercially available carbon substrates have some characteristics that are similar, including some degree of degradability and solubility. They differ in the speed with which the material becomes bioavailable and is degraded, in the complexity of their composition, and in their cost. Complexity in composition is viewed as a desirable substrate feature because it thus stimulates a more diverse microbial community. As we discussed earlier, substrates range from those that slowly release soluble degradable carbon at lower concentrations to those that immediately produce soluble, degradable carbon at higher concentrations.

The effectiveness of ERD systems is governed by many site-specific conditions. The geochemical character of the matrix and groundwater and hydrologic conditions such as groundwater velocity influence the efficacy and areal extent of the created IRZ. In cases where extensive contaminant plumes are being treated, it is desirable that carbon supplements are consumed at a rate high enough to lower the redox conditions rapidly but low enough to propagate the maximum area of desired treatment from a given injection point. The rate of the carbon supplement release and consumption will influence the volume of aquifer being treated with each injection point and should be considered based on site-specific conditions. Excessive application can result in the production of increased levels of unwanted by-products such as methane or organic acids. Also, an excessive consumption rate can result in inadequate temporal distribution of the carbon substrate, resulting in an increase in frequency of injections and/or an increase in the number of injection points required to cover a given treatment area. It has been argued at length in the literature that a slow steady release of hydrogen from degradation of the electron donor is desirable to optimize the biological conditions for CAH degradation. The authors strongly feel and have demonstrated that this may not be necessary and may even be counterproductive.

The economic application of soluble carbon substrates thus requires the ability to match the biogeochemical and hydrodynamic character of the aquifer to the biogeochemical character of one or more sources of soluble carbon. The selection of a carbon substrate(s) will be primarily driven by overall reaction rates, which are, in turn, controlled by the site conditions. The primary goal should be to minimize overall project cost by minimizing the number of required injection points, the number of injection events, and reagent cost.

Physical characteristics of the substrate (i.e., phase and solubility) may also make certain substrates more suitable than others in particular applications. Examples of candidate electron donor and carbon substrate products for widespread field application include:

- Hydrogen (gas, very rapidly consumed, very low solubility)
- Butyrate, lactate, acetate, and other carboxylates (soluble, pure compounds, rapidly consumed)
- Corn syrup (soluble, readily consumed)
- Molasses (soluble, readily consumed, provides sulfate as a supplement, and is very suitable when dealing with a mixed plume contaminant)
- Vegetable oils (partially soluble, slowly consumed, and may solubilize the contaminants themselves in the emulsion that may be subject to advective transport)
- Yeast extract (partially soluble, readily consumed)
- Whey or other milk solids (solid, which can be dissolved in water and then is readily consumed or liquid, slowly-to-readily consumed)
- Soluble humates (soluble, slowly consumed)
- Chitin (partially soluble, slow release)
- Organic mulches (partially soluble, slowly-to-readily consumed)
- HRC™ (solid, slow releasing)
- Combination of various products

Where groundwater velocities are relatively high, the effect of the carbon injections may be reduced by dilution into a large volume of oxygenated groundwater, and thus a high consumption of substrate may be necessary to induce anaerobic conditions required for treatment. In systems that are naturally aerobic, it is necessary to use a highly soluble, rapidly acting carbon substrate to quickly drive the redox potential down. Additionally, a highly degradable substrate may aid in overcoming the microbial lag phase attributed to the facultative, anaerobic microbial populations. Once reducing conditions are achieved this substrate can be injected at low dosages and frequencies to minimize the cost of maintaining the optimum equilibrium conditions. Thus, using fast- and slow-acting products will be site specific.

A cost comparison for a variety of substrates is presented in Table 4.6. This makes clear that there are dramatic price differences on a cost per pound basis for various substrates. However, as discussed previously, cost per pound should not be the sole criterion for substrate selection. Based on experience, loading rates for differing scenarios are expected to be on the order of 0.001 to 0.01 lb of TOC per gal of groundwater flux per day.

4.6.2 Chemical IRZ Systems

Several oxidants have been employed in the recent past for *in situ* chemical oxidation applications. The most common oxidants used have been hydrogen peroxide, potassium permanganate, and ozone. Ozone is the strongest oxidant available with an oxidation potential ($E°$) of 2.07 V. However, ozone is a gas and therefore has to be applied differently than the liquid reagents. Persulfate ($S_2O_8^{2-}$) salts are also available with an $E°$ of 2.01 V, but this oxidant is relatively expensive and requires thermal activation.

Hydrogen peroxide works through two mechanisms: free radical generation and direct oxidation. The direct oxidation has an $E°$ of 1.76 V, and free radical formation has an $E°$ of 2.76 V. The latter relies on the so-called Fenton's chemistry in which iron acts as a catalyst. Therefore, iron in the form of $FeSO_4$ is added with hydrogen peroxide to achieve the Fenton's reactions. In addition, pH adjustment is common because oxidation efficiencies are higher under acidic conditions.

Permanganate has an $E°$ of 1.70 V and yields MnO_2 as an insoluble precipitate under most conditions. Potassium permanganate, $KMnO_4$, is a dry crystalline material that turns bright purple when dissolved in water. The purple color acts as a built-in indicator for unreacted chemical. Reacted $KMnO_4$ is black or brown, indicating the presence of the MNO_2 precipitate. Limitations of $KMnO_4$ include its low solubility and its inability to oxidize petroleum compounds effectively.

Sodium permanganate ($NaMnO_4$) is an oxidant that performs very similarly to $KMnO_4$. Its attributes and limitations are much the same as $KMnO_4$. However, $NaMnO_4$ has a much higher solubility in water, allowing it to be used at much higher concentrations to achieve the target reactions during *in situ* chemical oxidation. $NaMnO_4$ is more expensive than $KMnO_4$ on a pound per pound basis and users have to be concerned about safety during handling and storage of $NaMnO_4$.

Discussions about chemical reactants are provided in extensive detail in Chapter 3. The relative reaction kinetics of the different oxidants are shown in Figure 4.26(A) and (B).

4.7 DELIVERY SYSTEM DESIGN

As discussed previously, there are a number of injection or delivery systems used in IRZ implementation including injection wells, direct-push well points, *in situ*

Components of an in Situ Reactive Zone

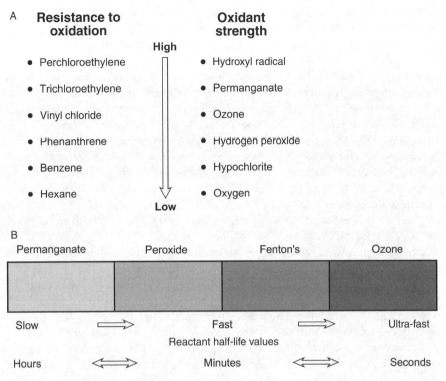

FIGURE 4.26 (A) Relative strength of oxidants and relative resistance of some common contaminants to chemical oxidation. (B) Relative reaction kinetics of various oxidants.

recirculation wells, recirculation well systems, and infiltration galleries. Permanent wells are constructed with materials appropriate for the geological formation, groundwater quality, and selected reagent. A typical injection well is shown in Figure 4.21. Wells in an unconsolidated formation are typically 2- to 4-in.-diameter stainless steel or polyvinyl chloride (PVC) construction with wire-wound screens sized for the formation. In bedrock, an open borehole in the target-saturated zone is acceptable with a PVC or steel casing in the overburden. However, conversion of existing open-hole wells in bedrock to screened wells has been performed in order to focus delivery of reagents into the fracture zones requiring treatment. Wells constructed in this fashion provide a permanent and repeatable means of delivering reagent to the subsurface. They are commonly applied in situations where readily soluble and degradable organic substrates — such as molasses — are applied. Permanent wells allow for multiple injections to establish and maintain the reactive zone. Permanent delivery systems are also necessary in situations where depths or soil strata make direct-push techniques impractical. Permanent delivery systems can also be implemented in situations where existing wells or remediation wells from a previous remediation system are reused for reagent delivery.

Direct-push techniques are also used for the creation of reactive zones in select applications. This type of delivery point is limited to shallow, unconsolidated formations at depths typically limited to 50 ft. This technique is also constrained by the soil characteristics, particularly grain size. In some cases, where direct-push wells are used, a permanent or temporary well point is placed using a direct-push drilling rig such as a cone penetrometer (CPT). This type of well is a small diameter point and is commonly applied where the number of injections will be limited and the need for well maintenance is minimal. This design is recommended only when the groundwater flow is relatively slow (less than 0.1 to 0.2 ft day^{-1}) and therefore the direct-push deployment of the reagent can be made at intervals that make sense economically (6 to 12 months). Depending on the delivery layout this technique may require a large number of injection points and the ability to repeat injections through the temporary point must be carefully considered in this layout. Another type of direct-push well point is a temporary geoprobe or hydro-punch well. In this case, the well drilling process itself is an injection process. Thus as the well point is placed, the reagent is injected. When the injection process is complete, no well points remain. This approach is applicable for slow-release materials such as polylactate ester or whey, which will dissolve slowly and typically require injection points on 3- to 10-ft centers to treat the target zone. As with CPT, this technique is limited by depth and geology.

IRZ wells must be designed to target the impacted groundwater. Thus the depth and screened interval will be determined by the vertical delineation of the groundwater impacts — with all the limitations this implies. In addition, the lithology can have an impact on the well design and screened interval since the injected reagent will flow with groundwater, following the path of least resistance. As a result, it is important to understand both the geology and the contaminant distribution when designing the wells. Where the saturated thickness of contamination exceeds 25 to 30 ft, multiple well points are recommended. As an alternative a single well point can be used in which multiple screened intervals are present and separated by packers during injections.

The number of injection points and the spatial distribution of these points are a function of the contaminant distribution, the hydrogeology of the impacted zone, the type of injection point selected, and the type of reagent being used. The injection wells need to cover the entire area targeted by the reactive zone. The geology and groundwater velocity will control how wide an area a single point can impact. For example, in a tight geologic unit, groundwater is likely to move relatively slowly, and the ability to inject is limited by the permeability of the formation. As a result, the reactive zone developed from a single point will have a limited impact laterally from the injection point and in the direction of groundwater flow. Therefore more points will be needed on a closer spacing.

The type of reagent used also can affect the spatial distribution of the injection points. If a water-soluble reagent, such as molasses or sucrose is used, the reagent will have flow characteristics very nearly identical to that of water and thus will move readily with the groundwater. As the reagent becomes more viscous, the ability to inject and achieve good lateral distribution will decline. As a result, in the latter case, more closely spaced well points will be required.

As discussed previously, there are a number of IRZ delivery options available. By far the most common is the batch injection system (Figure 4.27(A)–(E)). In this case a given batch of reagent is generated manually and injected manually into an

Components of an *in Situ* Reactive Zone

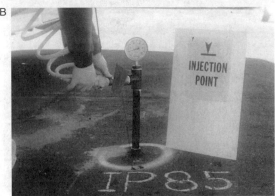

FIGURE 4.27 (A) Injection equipment. (B) An injection point being connected to the reagent tank. (C) Field mixing of reagent.

FIGURE 4.27 (continued) (D) Injection of reagent under pressure. (E) Injection of reagents in a well cluster.

injection well using a portable system. The main components of a batch delivery system consist of a mixing vessel, a centrifugal pump, a mixing device, and associated piping, fittings, pressure gauges, and flow indicators. A suitable mixing vessel is a polypropylene tank (mixing tank), the size of which can be selected based on the desired volume of injection and/or the availability of transport equipment. The most common application is the use of a plastic tank that can be temporarily deployed in a standard pick-up truck bed and is large enough for most individual well batch injections. Mobile systems can also use larger trailer-mounted mixing tanks ranging as high as 2000 gal in volume.

Once mixed, the reagent batch is injected into the injection well under pressure via a power-operated, general purpose centrifugal pump. Pressure readings will vary depending on site-specific characteristics and the type of soil within the reactive zone matrix. The injection equipment will also typically include lengths of 0.75- to 1-in. diameter heavy-duty rubber hose to connect the mixing tank to the pump inlet and the pump outlet to the injection well. The tank, pump, wellhead, and hoses should be fitted with Cam-Lok type fittings for ease of connection.

A uniform reagent mixture for a batch injection can be prepared using a variety of methods. An applicable method with proven success involves the operation of two

TABLE 4.7
Engineering Characteristics of Reagents Applied in ERD

Reagent	Method of Delivery	Common Form of Reagent
Molasses	Injection wells, direct push	Dissolved in water
Lactate	Injection wells	Dissolved in water
Cheese whey	Dissolved, powdered form can be in injection well or direct push; slurry, fresh form, can be injected by direct push or into a borehole	Material can be obtained in dry powdered form and dissolved in water (filtration is recommended before injection) or, in some cases, as a liquid slurry, which is more slowly released
High-fructose corn syrup	Injection wells, direct push	Dissolved in water

processes: power-operated submersible pump and powered mixer. The submersible pump is connected to a section of perforated PVC piping affixed to the interior of the mixing tank that circulates and mixes the reagent mix with water while filling the tank. At the same time a power-operated mixer can be utilized to agitate the solution while the mixing tank continues to fill (Figure 4.27(C)). Adequate mixing can be achieved when applying the two processes simultaneously. In conditions where the batch solution is being prepared in cold weather, it is suggested that the duration of mixing be prolonged after batch solution preparation to ensure a coherent mixture between the reagent and water.

4.7.1 REAGENT INJECTION STRATEGY

Based on the preliminary evaluation of the existing subsurface environment, appropriate reagents have to be selected to optimize the environment as well as to achieve the target reactions. Design of the reagent injection system requires an evaluation and understanding of the hydrogeologic conditions at the site and specifically within the plume and the location of the reactive zones. This understanding has to include both a macroscopic site-wide pattern and microscopic levels between layers of varying geologic sediments. Specific geologic hydrogeologic parameters required for the design of an IRZ have been discussed previously.

Injection of reagents can be implemented in two ways: gravity-feed and pressure injection (Figure 4.27(C) and (D)). Under gravity-feed conditions, injected reagents will tend to spread over the water table as a sheet flow and the mixing within the reactive zone will be dominated by diffusion rather than advective flow. Consequently, injection under pressure is generally the preferred approach and is usually more economical because individual injection events are shortened. When injecting under pressure there are two things to keep in mind. First, the injection well needs to be properly constructed to prevent short-circuiting between the borehole and the well casing. Second, particularly for shallow plumes, injection pressures must be controlled to prevent formation of vertical migration pathways to the surface.

When the depth of contamination is greater, multiple injection points may be required within a well cluster at each injection point. The reagent solution will have to be injected under pressure into each injection well. Concentration of the injected feed solution should be dilute enough to avoid downward migration due to density differences between the reagent and groundwater, unless such density-driven migration is intended.

Once the injection strategy has been determined and there is an understanding of the number of injection events that are needed, a decision needs to be made on whether to do manual or automatic injections. Normally that decision is made based on costs. However, other factors may need to be considered. In some cases, although manual injections may be the selected methodology, site conditions dictate that a central distribution system be used to deliver the reagents to individual injection wells. For example, where injection wells are in a high-traffic area, such as a parking lot, a central distribution system may be desirable, even though it adds to the cost of the remediation program.

The successful application of an IRZ system is first and foremost reliant on the timely, cost-effective, and consistent delivery of the reagent to the treatment zone in quantities carefully tuned to produce the desired reactions in the subsurface. Based on the application experiences with IRZ technology to date, reagent delivery becomes most complicated in low permeability geologic environments (10^{-5} cm s^{-1} (3×10^{-2} ft day^{-1}) or less) or those with low groundwater flow velocities (less than 30 ft year^{-1}). These settings can limit the area of influence of individual injection points due to the absence of sufficient reagent dispersion.

Poor reagent delivery and distribution can also result in other potentially undesirable impacts. These impacts include uneven application of reagent resulting in inconsistent treatment within the reactive zone, and/or requirement of too many injection points for the full-scale application. In low permeability (i.e., 10^{-5} cm s^{-1} or less) and/or low groundwater velocity environments, the reagent can accumulate in the vicinity of the injection point and result in the formation of excess undesirable by-products. As a result of the formation of organic acids, the ambient pH in the microbial treatment zone can be lowered and conditions not conducive to the target reactions (pH $<$ 5) may be created. When fermentation conditions are created, other by-products such as acetone and 2-butanone have been observed at sites where reagent dosing was high and pH decreased near the injection wells. The occurrences of these by-products are generally limited in extent and often sporadic in nature. It is expected that microbes in the IRZ can also utilize these ketones. The possibility of formation of these by-products needs to be accounted for in the project planning and operational stages, particularly when dealing with very low groundwater velocity sites. Monitoring of groundwater within the treatment zone should thus be provided in order to ensure optimum process conditions.

The implementation plan for an IRZ system should be flexible enough to allow for modification of both the delivery frequency and mass of the reagent delivered preventing the buildup of reagent or by-products. Modifications in reagent delivery should be tied to the process control parameters, especially pH, which can be measured in the field. For any IRZ system, the reagent feed characteristics such as rate, strength, and frequency need to be optimized to deliver adequate reagent mass, in order to create and maintain optimum conditions for the target reactions in the subsurface.

In a microbial IRZ, the goal is to maintain dissolved organic carbon above background levels in the reactive zone at a distance of 100 days of travel time. Based on experience, this translates to one or two orders of magnitude higher target DOC concentration in the injection wells. The reason higher concentrations must be fed in the injection wells relates to the fact that the organic carbon will be metabolized as it flows with groundwater. Therefore it is necessary to establish a DOC gradient between the injection points and the rest of the reactive zone.

The aforementioned criteria are simply guidelines for a preliminary calculation of the reagent feed rate. Experience has proven them to be adequate means to define a reasonable reagent feed rate to begin the reactive zone. Field data collected after the reactive zone has been established are the true measure of the adequacy of the reagent feed. Field analytical data (in particular redox potential, pH, and TOC for a microbial IRZ) from the injection wells and monitoring wells within the reactive zone should be used to confirm that the reactive zone has been established and is expanding with groundwater flow. In the long term, once the reactive zone is established, reagent feed concentrations can be reduced to a sustainable *equilibrium maintenance* feed rate.

Finally, in designing an injection strategy, the advantages of frequent, low-dose reagent injections, which can give the practitioner the opportunity to carefully control the dosing on the basis of feedback from field parameters, must be weighed against less-frequent, higher-dose buffered injections that can provide important economic advantages at many sites.

4.8 PILOT TESTING

A pilot or demonstration test is used to gather critical design data — well spacing and reagent feed rate, strength, and frequency — as well as to demonstrate the efficacy of the target reactions and thus the technology and satisfy regulatory agency concerns regarding the technique. The pilot test should always focus on a significant area of concern within the plume and at times can be used as an interim remedial measure.

The following design and location selection criteria should be considered while evaluating the feasibility of an IRZ pilot test.

- Aquifer permeability and thickness in the targeted test area
- Distribution and magnitude of horizontal and vertical target contaminant(s)
- Potential concerns with respect to the formation of any by-products even during the short duration of the pilot test
- Technical specifications and equipment requirements for reagent delivery
- Implementation costs
- Monitoring and injection well(s) matrix
- Reagent injection methodology and design parameters, such as dosage, frequency, injection volumes, and duration
- Pilot test groundwater monitoring protocol

Prior to initiating the drilling activities, drilling permits must be obtained from the local environmental regulatory entity. An underground utility locating service must be contracted to investigate each of the proposed drilling locations to ensure that they are free of subsurface utilities prior to the initiation of field activities. Each of the installed wells must be completed with a locking cap and traffic-rated vault or riser, where appropriate. The location and top-of-casing elevation of each injection and monitoring well must be surveyed by a certified, licensed land surveyor.

4.8.1 Pilot Test Wells — Number and Location

To properly evaluate the IRZ technology in the field, a network of injection wells and monitoring wells is required. The injection wells need to be located in an area of the site where sufficient impacts are present, and should be installed in a manner similar to wells that would be employed as part of the full-scale system. The combination of injection and monitoring wells must be sufficient to create an IRZ that is of sufficient size for a realistic field test and that includes the locations of more than one monitoring well. The monitoring wells need to be located downgradient in the direction of groundwater flow in order to compute reagent utilization, constituent degradation reaction rates, local groundwater velocities, and the zone of influence of the IRZ. It is also preferable to have at least one monitoring well with some historical contaminant data. The monitoring wells should be located in a manner to evaluate both the performance of the degradation process and the extent, both parallel and perpendicular to the direction of groundwater flow, of the IRZ. These objectives can be achieved with different configurations of monitoring and injection wells.

The field test must employ a minimum number of injection wells that can be used to properly deliver the reagent to the target zone. The screened interval should intercept the impacted zone, with consideration given to the lithology and groundwater flow conditions. The injection wells should be constructed using appropriate drilling techniques. If there is little, or poor quality, geologic data available, consideration should be given to the need to gather supplemental geological data using split spoon sampling techniques during installation or other appropriate means. Following installation, the well should be developed to remove fine material and ensure hydraulic communication with the surrounding aquifer.

It is preferable for monitoring wells to be located at variable distances from the injection wells, both parallel to the direction of groundwater flow and perpendicular to the direction of groundwater flow. The wells or transects are typically spaced at 1, 2, and 3 months groundwater flow time downgradient (based on predicted travel times derived from existing site data). This will allow the reactive zone to be defined parallel and perpendicular to groundwater flow. Once again, consideration should be given to the variability of the site geology in locating screened intervals. It may be necessary to monitor multiple intervals if the geology dictates. Heterogeneity can also lead to channeling into preferential flow pathways, thus requiring more pilot test wells to ensure adequate evaluation. When possible, existing monitoring wells may be used as injection or observation wells in order to control costs.

A potentially difficult and time-consuming endeavor in the pilot phase is determining the optimal spacing of the injection wells and positioning of the monitoring wells. This can be accomplished using analytical solutions as a starting point, using computer modeling to predict injectant travel and behavior, using a tracer in the field, or by utilizing monitoring wells to view the effects over time. The first two methods require some knowledge of the longitudinal (α_L, length) and transverse (α_T, width) dispersivity of the aquifer. The ratio of these two parameters α_L/α_T, will affect the shape of the plume aerially. From field studies it has been shown that α_L/α_T ratios are typically in the range of 6 to 20, with 6 being a relatively broad shaped plume, and 20 being a thin elongated plume. If the plume from a site was created by a point source, the shape of the delineated plume can be used to determine this ratio, and the transverse dispersion estimated at a smaller scale for determining of pilot well spacing. Dispersivity can also be estimated using a one-dimensional dispersion equation and placing fluids through soil columns.[2,28] These data can be input into a groundwater flow model to predict plume shape, given the known hydraulic characteristics of the aquifer and the estimate of dispersivity. The effects of injection can also be predicted using numerical groundwater flow models in conjunction with a transport model. By manipulating the injection rates and durations, the longitudinal and transverse movement of the injected substrates can be evaluated and used to develop a pilot program. The accuracy of this approach is consistent with the available site data (it is only as good as the parameters input into the model) and is intended to form the basis for quantitative assessments of various pilot test configurations.

It is fiscally prudent to collect data at the field scale and use this information to design the larger-scale program. A common approach is to inject a conservative tracer mixed with the substrate and monitor for this tracer under natural or induced gradients at downgradient wells. A tracer such as sodium bromide is typically injected at a concentration approximately 100 to 1,000 times its method detection limit. This will allow for a more realistic derivation of the actual dispersivities and groundwater velocities that could be expected in the field during treatment. Downgradient concentrations will also indicate whether or not there are preferential flow paths related to subtle changes related to deposition (i.e., grain size and lithology).

However, injecting a reagent is much different from a conservative tracer since the reagent is reactive, sorbs to the geologic matrix, and the utilization of the reagent is transient. Monitoring for electrical conductivity and the reagent itself at monitoring wells placed downgradient of the injection well is a good first step to determine the dispersivity (both length and width) of the reagent. A typical pilot test would consist of the minimum number of injection wells required with four or five monitoring wells placed at monthly average flow distances downgradient and deviated from the downgradient center line by 10 to 50% of the monthly flow velocity. Actual reagent travel will be less than the average Darcian velocity, and by conducting a pilot test over 6 to 9 months, the actual field-scale longitudinal and transverse dispersion can be determined based on water quality monitoring. A full-scale system can then be designed so that both the transverse and longitudinal dispersion of an individual injection well overlaps with that of other injection wells.

The frequency of injections will vary with the geologic, biogeochemical, and hydrogeologic conditions of each site. As a result the frequency of injections can vary from once a week to once every 6 months. However, initially monthly injections are typical with less-frequent injections after the IRZ has formed and the zone reaches the optimum conditions for the target reactors. An initial loading rate per injection well per month is generally proposed for the initial injection, but these rates will have to be adjusted based on the results of field monitoring.

4.8.2 DURATION OF PILOT STUDY

Typical microbial IRZ pilot studies last between 6 and 12 months while the chemical IRZ pilot tests may be completed in a much shorter time frame. The rate of groundwater flow, biogeochemical considerations, and the proposed observation well locations will determine the site-specific duration of the test — the closer the observation wells are to the injection well(s) and the faster groundwater moves, the sooner results can be expected and the shorter the pilot test needs to be. Once the reagent has been delivered to an area within the aquifer, a period of several additional months is often required for the successive consumption of various electron acceptors, which in turn requires successive changes in the microbial community. The "testing" of a microbial IRZ is complete when redox conditions downgradient of the injection well(s) are substantially reduced, the source contaminant is degrading, and the amount of final end-product (e.g., ethene) has increased.

These results may sometimes, but rarely, be achieved within 1 or 2 months of implementation of the pilot test program; longer duration is generally needed to collect design data on the zone of influence of the injection well(s). Additional information may be necessary to satisfy regulators or clients and in order to optimize the full-scale design. For this reason the pilot is often extended beyond a simple demonstration of the microbial transformation process(es) itself.

Many times the "testing" ends, but the injections continue. Once the reactive zone is established within the plume, maintenance dosages of the reagent will allow the zone to continue to serve as an interim remedial measure, until such time as the system is expanded or the cleanup goals are achieved.

Regulators or other stakeholders often seek assurance that the technology is effective. One of the criteria to be evaluated is that observed treatment results are real and not due to dilution, dispersion, or natural attenuation. Matched, side-by-side, untreated controls rarely exist in the real world. Essentially, two types of "controls" are feasible in pilot test design: upgradient wells and wells with good historical data trends in the treatment zone. Upgradient wells can help control for the effect of natural attenuation. Similarly, sharp changes in the historical trends of contaminant concentration at the site after the implementation of IRZ can help rule out natural attenuation as an explanation for the observed treatment. Clear trends in the concentrations of transformation products such as *cis*-DCE and ethene or ethane can also be used to determine the effectiveness of treatment.

The water level should be routinely recorded during all sampling events and volume injected during all injection events. The potential for significant contaminant spreading through injection displacement can be evaluated first with relatively simple hydrogeologic computations. At many sites, calculations will indicate that the expected volume of solution added over the test period is equal to less than 10% of the volume of groundwater expected to move through the subject zone. In addition, at many sites the pilot test treatment zone is well within the existing plume. Thus there is not likely to be a detectable spreading effect at these sites.

Components of an *in Situ* Reactive Zone

Including a nonreactive bromide tracer in the pilot study enhances the ability to track the dispersion of the injected reagents. The bromide tracer also allows a simple computation of the magnitude of the potential observed "dilution effect" by comparing the concentration of bromide in the monitoring wells to the injected concentration (assuming a minimal initial groundwater concentration of bromide and true conservative behavior). The observed concentrations of contaminants can then be adjusted for the dilution effect. We should note, however, that these computations sometimes underpredict treatment because the continuous desorption effects are often ignored.

When bromide is used as a conservative tracer it flows along with and at the same rate as the injected liquids. Thus, the amount of dilution in a given well should be directly proportional to the amount of bromide that shows up in that well. As an extreme example, if 100 mg L^{-1} of bromide were injected into the injection well and then 100 mg L^{-1} bromide were withdrawn a week later at a monitoring well located 1 week downgradient of the injection well, we would surmise that the monitoring well contains only injection fluid and is 100% diluted. If 0 mg L^{-1} bromide were measured in the monitoring well, none of the water from the injection well has made it to the monitoring well and there is thus 0% dilution.

Based on this logic, concentrations of constituents of concern determined through chemical analysis can be corrected by calculating a dilution factor per the following equation:

$$DF = 1 - \frac{[Br]_{MW}}{[Br]_{inj}} \quad (4.9)$$

where DF is the dilution factor (i.e., the percentage of monitoring well sample that is not dilution water), $[Br]_{MW}$ is the bromide concentration measured on a sample taken from a monitoring well on a given date, and $[Br]_{inj}$ is injection fluid bromide concentration (an approximate running average of injection fluid bromide concentrations made around the assumed travel time of the injection well to the monitoring well). This takes into account both the molasses solution and the water "push" fluids.

Then the concentration of the constituent of concern for a given monitoring well sample can be corrected by calculating its actual concentration per the following equation:

$$VOC_{actual} = \frac{VOC_{measured}}{DF} \quad (4.10)$$

4.8.3 Scaleup Issues

Following demonstration of the preferred IRZ technology via a pilot study, site-specific scaleup to a larger or full-scale system is, typically, fairly simple. The main issue facing scaleup will be the addition of more injection wells to create a larger IRZ and perhaps multiple IRZs. The authors and others already have extensive full-scale commercial experience in applying this technology.

With this in mind, if pilot testing indicates that the effective area of influence of a given injection well does not propagate far from the injection well itself, many additional injection wells may be required for the ultimate system if the system is being used to treat the entire plume rather than to form a barrier or treat a source area. This could be a scaleup issue of concern if a large number of wells are required. If drilling costs are high due to the depth or to methodology required to install the well, the cost to implement the full-scale system may become prohibitive. However, this is generally understood prior to the pilot study and the preferred implementation strategy uses one or more arrays of injection wells to "segment" the plume vs. trying to remediate the entire plume all at once.

The only other scaleup issue regarding the technology is that of reagent injection methodology. In many cases even with a larger number of injection wells, the frequency and volume of the injection is such that it is still advantageous to use manual batch injections. However, in some cases, scaleup to a full-scale system will require the implementation of an automatic reagent feed system (Figure 4.28(A) and (B)). This type of system would be equipped with a source of bulk concentrated reagent, a source of potable water, metering and mixing equipment, and a network of injection piping that would allow for the metered injection of the solution to each well automatically. Generally, these systems are easy to construct, operate, and maintain given they are made up of commonly used equipment and technology (i.e., tanks, valves, piping, wells, and automation controls).

Given the *in situ* nature of the technology, possible interference during scaleup is expected to be minimal. The main interference concern would be if a portion of the contaminant plume needing to be treated was located beneath a building or other permanent structure. Hence, installation of reagent injection wells and performing injections (either batch or automatic) could be an issue. Gas generation under buildings or parking lots could also require installation of passive or active venting systems. However, given the nature of the technology these potential interference issues could be overcome through strategic design of the full-scale remedy (i.e., creating IRZs before and/or after the groundwater move past the interference). Alternately, horizontal wells could be used for installation at some additional cost.

4.8.4 Sustainability and Reliability

The implementation of an IRZ system is a dynamic process that requires a detailed understanding of the site geochemistry and hydrogeologic conditions before implementation and as it changes as a result of pilot or full-scale implementation. IRZ implementation can be successful when there is considerable process monitoring during the initial deployment of the pilot test that allows for adjustment of reagent deliverability (strength and frequency). Where the target reactions have failed to establish, or has required longer than expected treatment periods, it is usually the result of improper monitoring (the wrong parameters or the wrong frequency) or data evaluation in the early stages of the pilot test. Reagent loading and induced gradients must be reviewed early in the pilot process to allow delivery rates to be increased, for greater spreading and increased reagent concentrations within the treatment area. In some cases the reagent concentrations may have to be decreased and a buffering agent required to control the pH fluctuations.

Components of an *in Situ* Reactive Zone 389

FIGURE 4.28 (A) An automated injection system installed to handle injection of reagents into approximately 400 injection wells. (B) Rows of injection wells placed perpendicular to groundwater flow in a large IRZ system.

Similarly, the effects of reagent injections must be reviewed in the context of how the addition of aqueous solutions affect hydraulic gradients (i.e., mounding) and flow directions. Groundwater flow directions and gradients should be viewed both in a macro and micro scale before and during the demonstrations.

For either the manual solution injections or the more permanent system approach, the overall treatment system is expected to be reliable and easy to reproduce from one injection event to the next. The types of equipment being employed (transfer pumps, tanks, mixers, and controls) are commonly used for similar applications and can be expected to perform as designed and intended for the duration of system implementation. It is expected that some routine preventative maintenance will be required for all equipment. However, this work will be performed at planned intervals according to manufacturer's recommendations, and should not affect system reliability.

Other controls on the reliability of an IRZ system are related to how the geochemical data are collected, and eliminating as much variability as possible. Consistent equipment should be used for each facility so that variability related to field measurements (e.g., as a result of using different meters) can be eliminated. Similarly, consistent low-flow pumps and backup pumps should be assigned for each site. In some cases tubing or pumps dedicated and left in place in a given well can be advantageous.

All field monitoring should be conducted before initiating any planned injection event for that week. This will eliminate temporal variations in water elevations and chemistry that would be most pronounced immediately following an injection event. Efforts should be made to conduct field-monitoring events during fair weather conditions. In-field meters are very susceptible to moisture-induced electronic circuitry problems that can generate anomalous readings. Field instruments should be calibrated at the beginning and end of each field day, with a calibration check completed at midday. This will also help in producing reliable data and eliminate variability associated with poorly calibrated instruments.

In many cases IRZ systems are applied in plume treatment or source treatment modes that aim to complete treatment in a short number of years or even months. However, there are barrier-type applications where long-term sustainability and operability would be of greater importance.

4.8.5 BIOFILM DEVELOPMENTS

When injecting an electron donor such as molasses (and electron acceptors) into an aquifer via injection wells, biofilm development around the injection wells should be anticipated. Biofilms are large aggregations of bacteria and other microorganisms bound together in a sticky mass of tangled polysaccharide fibers that connect cells together and tie them to a surface. Aerobic and anaerobic bacteria not only can thrive side by side within biofilms when biogeochemical conditions permit, but also actually seem to collaborate to make themselves more powerful. The polysaccharide coating acts like armor, giving the microorganisms protection beyond their usual defense mechanisms.

While the typical average diameter of a bacterium in established biofilms is about 0.5 to 1 μm, biofilm bacteria rarely adhere directly to solid surfaces. Instead, at distances shorter than 1 nm, short-range forces such as hydrogen bonding and dipole formation tend to be the dominant adhesion effects. As bacteria are held in place and fed by the organic and inorganic molecules trapped by these short-range forces, they form slime that anchors them to solid surfaces. This slime becomes a home for additional bacterial growth. If the biofilm becomes too thick to permit adequate oxygen penetration, under aerobic conditions any additional biofilm growth may actually decrease biofilm adherence due to shearing. The thickness of the biofilm under anaerobic conditions is significantly smaller due to the above-mentioned shearing effects and the fact that the rate of biomass growth is substantially lower under anaerobic conditions.

Under unaerobic conditions, typical of an IRZ, reduction in porosity within the saturated zone due to biofilm growth will not be significant enough to impact the hydrogeologic conditions for reagent transport. The amount of biomass produced per unit mass of organic substrate under anaerobic conditions is significantly lower than under comparable aerobic conditions. However, well clogging around the injection wells is an issue to be taken into consideration. Electron donor solutions, such as dilute molasses, are injected at reasonably high concentrations of TOC, which are then further diluted by mixing with the groundwater within the IRZ. As a result of the higher concentrations of TOC present around the injection wells, the amount of biomass and biofilm growth will be significant.

Since the electron donor solutions are injected in a batch mode at most of the IRZ applications, resistance to injection due to clogging may be an operation issue only during the injection events. In all the sites in the authors' experience, there were only two sites where injection under pressure was difficult due to significant head buildup. Manual cleaning of the well screens will be required under those conditions. Since the anaerobic ERD systems are operated under high TOC levels, the pH around the injection wells are generally under acidic conditions. This may be another reason for the absence of biofouling in systems designed and operated under high organic substrate concentrations.

4.8.6 SITE CLOSURE

In any remediation project a fundamental choice needs to be made between containment and aggressive treatment aimed at closure. Currently, it is accepted that not all heavily contaminated source zones can be economically and feasibly treated. Of course, this depends on the type of contamination. These issues have recently been carefully discussed in the literature and at conference panel discussions and will not be discussed at length here.[29,30] A key point is that rapid closure cannot be achieved without some form of source area treatment — biological, chemical, or otherwise; however, IRZ may be applied either in a curtain/barrier mode, a source zone treatment mode, or a dissolved plume treatment mode.

The area and depth of the contaminated plume, the hydraulic characteristics of the aquifer and the costs associated with installing and maintaining different injection well layouts will ultimately dictate the potential remediation scenarios and cleanup time frames for a given site. For smaller sites (less than an acre), pilot testing can be used to define achievable field scale degradation rates and radius of influence from injection points. A grid of injection wells can then be set across the entire footprint of the plume and periodic injections completed over 2 to 4 years. For relatively thin, lower permeability or low velocity settings this may only require a few injection events during this time period.

At sites where larger plumes are present (>2 acres) or the depth of the plume makes installing injection wells difficult and expensive, multiple treatment lines can be established, located perpendicular to flow directions. The spacing of the lines can be determined from pilot data, or to be conservative, separated by 6 to 12 months of travel time. At some larger sites, treatment lines have been spaced at 1.5- to 3.0-year travel time intervals to be cost-effective.

In summary, capital cost can often be traded off against time. The better the injection well coverage, the quicker closure can be achieved. At any site there will be a ramp-up period where the reagent needs to be delivered, distributed, and mixed into the aquifer to create the optimum environment. The systems discussed in this section are generally passive systems that rely on aquifer advection, dispersion, and diffusion for delivery. In deeper or thicker settings, recirculation can also be used to achieve mixing more rapidly. These systems also require more maintenance and tend to foul, due to the anaerobic/aerobic interfaces in the system that cause rapid accumulation of floc. It generally takes 1 to 3 months to create a reducing environment across the site. Desorption of source material from the aquifer sediments is observed after the reagents have been delivered and a different target environment is created. The desorbed mass can then be treated within the reactive zone. Lag times on treatment of the intermediates are sometimes observed after the reagent delivery and desorption periods. Once the lag phase is complete in a microbial IRZ, degradation rates increase and periodic maintenance dosing of the system is required to maintain the reducing conditions.

When conservative placement of injection points and dosing is utilized, it has been our experience that levels of contaminants of concern near the detection limits and/or MCLs can be realized. Given the desorption that occurs within the treatment area, the effects of rebound are minimal compared to other technologies that address only the dissolved phase. More details of the desorption phenomena and contaminant rebound are discussed in Chapter 5.

REFERENCES

1. Suthersan, S.S., *Natural and Enhanced Remediation Systems*, Lewis Publishers, Boca Raton, FL, 2001.
2. Suthersan, S.S. et al., *Technical Protocol for Using Soluble Carbohydrates to Enhance Reductive Dechlorination of Chlorinated Aliphatic Hydrocarbons*, AFCEE, 2002.
3. Clarkson, R., ARCADIS, personal communication, 2003.
4. Schwarzenbach, R.P., P.M. Gschwend, and D.M. Imboden, *Environmental Organic Chemistry*, 2nd ed., Wiley-Interscience, New York, 2003.
5. Larson, R.A. and E.J. Weber, *Reaction Mechanisms in Environmental Organic Chemistry*, Lewis Publishers, Boca Raton, FL, 1994.
6. Semprini, L. et al., Anaerobic transformation of chlorinated aliphatic hydrocarbons in a sand aquifer based on spatial chemical distributions, *Water Resour. Res.*, 31(4), 1051–1062, 1995.
7. Newman, S.P. and P.J. Weingart, *A Comprehensive Strategy of Hydrogeologic Modeling and Uncertainty Analysis for Nuclear Facilities*, University of Arizona, Tucson, 2003.
8. Vrobleski, D., U.S. Geological Service, personal communication, 2003.
9. Debruin, W.P. et al., Complete biological reductive transformation of tetrachloroethene to ethane, *J. Appl. Environ. Microbiol.*, 58, 1996–2000, 1992.
10. Kampbell, D.H. et al., Dissolved oxygen and methane in water by a GC headspace equilibrium technique, *Int. J. Environ. Anal. Chem.*, 36, 249–257, 1980.
11. Pirkle, R., Microseeps, personal communication, 2003.
12. ASTM, Standard Guide for Remediation of Groundwater by Natural Attenuation at Petroleum Release Sites, Draft, Philadelphia, 1997.
13. Horst, J., ARCADIS, personal communication, 2003.

14. Hamond, H.F. and E.J. Fechner, *Chemical Fate and Transport in the Environment*, Academic Press, New York, 1994.
15. American Society of Microbiology, *ASM News*, September 2002, pp. 428–429.
16. Fennell, D., A. Carroll, J. Gossett, and S. Zinder, Assessment of indigenous reductive dechlorination potential at a TCE contaminated site using microcosms, polymerase chain reaction analysis, and site data, *Environ. Sci. Technol.*, 35(9), 1830–1839, 2001.
17. Flynn, S., F. Loeffler, and J. Tiedje, Microbial community changes associated with a shift from reductive dechlorination of PCE to reductive dechlorination of *cis*-DCE and VC, *Environ. Sci. Technol.*, 34(b), 1056–1061, 2000.
18. Hendrickson, E.R., J.A. Payne, R.M. Young, M.G. Starr, M.P. Perry, S. Fahnestock, D.E. Ellis, and R.C. Ebersole, Molecular analysis of *Dehalococcoides* 16S ribosomal DNA from chloroethene-contaminated sites throughout North America and Europe, *Appl. Environ. Microbiol.*, 68(2): 485–495, 2002.
19. Voet, D. and J.G. Voet, *Biochemistry*, 2nd ed., John Wiley & Sons, New York, 1995.
20. Wiedermeier, T.H. et al., *Technical Protocol for Evaluating Natural Attenuation of Chlorinated Solvents in Groundwater*, EPA/600/R-98/128, 1998.
21. Morse, J.J. et al., *Draft Technical Protocol: A Treatability Test for Evaluating the Potential Applicability of the Reductive Anaerobic in Situ Treatment Technology (RABITT) to Remediate Chloroethenes*, ESTCP, February 23, 1998.
22. ITRC, *Technical and Regulatory Requirements for Enhanced in Situ Bioremediation of Chlorinated Solvents in Groundwater*, 1998.
23. Devlin, J.F. and D. Muller, Field and laboratory studies of carbon tetrachloride transformation in a sandy aquifer under sulfate reducing conditions, *Environ. Sci. Technol.*, 33, 1021–1027, 1999.
24. Drzyzza, O. and J.C. Gottschal, Tetrochloroethene dehalorespiration and growth of *Desulfitobacterium frappieri* TCEI in strict dependence on the activity of *Desulfovibrio fructosivorans*, *J. Appl. Environ. Microbiol.*, 642–649, 2002.
25. Bushmann, J. et al., Iron porphyrin and cysteine mediated reduction of ten polyhalogenated methanes in homogeneous aqueous solution: product analyses and mechanistic considerations, *Environ. Sci. Technol.*, 33(7), 1015–1020, 1999.
26. Zwiernik, M.J. et al., $FeSO_4$ amendments stimulate extensive anaerobic PCB dechlorination, *Environ. Sci. Technol.*, 32(21), 3360–3365, 1998.
27. Butler, E.C. and K.F. Hayes, Kinetics of the transformations of trichloroethylene and tetrachloroethylene by iron sulfide, *Environ. Sci. Technol.*, 33(12), 2021–2027, 1999.
28. Bigham, W.E., Mixing equations in short laboratory columns, *J. Soc. Pet. Eng.*, 14, 91–99, 1974.
29. ITRC Work Group, *DNAPL Source Reduction: Facing the Challenge*, 2002.
30. Hinchee, R.E. (Moderator), DNAPL Source Zone Remediation, Panel Discussion, SERDP/ESTCP Partners in Environmental Technology Conference, November 29, 2001, Washington, D.C.

5 Building Reactive Zone Strategies

5.1 INTRODUCTION TO REACTIVE ZONE STRATEGIES

In Chapters 2 and 3, we introduced the mechanisms of chemical and biological *in situ* reactive zone (IRZ) technologies, which can be applied to a wide array of aquifer contaminants. Each of those methods has been shown to achieve remedial objectives at some scale — bench, pilot, or full field scale — and, from that base of information, remedy screening can be performed, as shown in Chapter 4, to match technology candidates to a particular contamination problem. Identification of technically viable methods is the starting point for development of the larger remedial strategy, which includes decisions on the scope of the remedy, the desired time to completion, and structuring of the reactive zone remedy to achieve those scope and timing objectives. In the transition from remedy screening to remedial strategy development, impacts of contaminant distribution and aquifer hydraulics are considered for each potential remedy, and a plan is constructed that includes placement and operation of reactive zones in source areas and along the contaminant plume at an intensity* that can achieve the project objectives.

The starting point for discussion of reactive zone strategies is an examination of the physical and chemical compartmentalization of contaminants and the migratory processes that extend the distribution of contaminants beyond their source areas, generating the large-scale, persistent distributions of contaminants we observe in aquifers. The mechanisms that determine contaminant distribution and migration are also at work in reactive zone systems, and reactive system designs must anticipate and, where possible, exploit those forces to be effective.

In this chapter, reactive zone structures are defined according to the physical, chemical, and biological processes that occur within, and downgradient from, the locations at which the managed reactions occur. Each of the identified zones— injection, reaction, desorption, and recovery — is a working component of an IRZ strategy. Reactive zones are deployed in several possible patterns, including source zone,

---

*The strength of reagents applied, the spacing of reagent injection points along and across the groundwater flow gradient, the frequency of reagent injections and the volume of reagent injection, relative to the aquifer pore volume and groundwater pore volume exchange rate, all are characteristics of the reactive zone intensity.

barrier, and whole-plume treatments, according to the project scope and timing objectives.

The chapter concludes with discussions of limitations on reactive zone effectiveness in contaminated aquifers. The fundamental problems of reactive zone design and operation include the challenge of achieving contact between reagents and target compounds, the variable distribution of reagent concentrations, and the associated impacts of reaction kinetics and by-product formation. Contaminant distribution factors such as the presence of nonaqueous-phase liquids (NAPLs) and large-scale sorbed mass add further to the difficulty of reactive zone design and operation.

5.2 A CONCEPTUAL MODEL FOR CONTAMINANT DISTRIBUTION

When contamination is discharged into the subsurface, it is immediately distributed among several physical and chemical locales that are determined by the physical and chemical nature of the contaminant, the mass released, and the physical and chemical nature of the subsurface itself. The point of release becomes a source of contaminant migration through the subsurface, and migratory losses from the source area often feed the development of a contaminant "plume" in the underlying groundwater. IRZ remedies are most often applied at sites where a contaminant release is feeding a significant groundwater contaminant plume. A conceptual framework that defines the physical and chemical compartmentalization and migratory potential of contaminants and injected reagents in the subsurface is a necessary precursor for discussion of reactive zone strategy development.

5.2.1 REACTIVE ZONE HYDROGEOLOGY

An aquifer is a geologic unit composed of porous granular or fractured massive solids, its interstitial spaces filled with water. In most aquifers of interest, the water is a transient component, migrating through the aquifer matrix, along an elevation or pressure gradient. The cross-sectional dimensions of the interstitial spaces, their volume (relative to the total aquifer volume), and their interconnectedness are key characteristics that determine the behavior of water and other fluids moving through the matrix.

The movement of water through an aquifer matrix at the microscopic scale follows tortuous paths, through pore apertures of varying dimensions, among particles of varying surface textures and charge distributions. Quantitative description of the complex water movement at a pore scale probably exceeds our theoretical and computational capacities. Darcy's law and related descriptions of groundwater flow through porous media reflect the emergent behavior[*] of water movement at a scale that is extremely large, relative to pore scale. Our ability to model and predict *net*

---

[*]The phrase "emergent behavior" is used here to note the fact that, at relatively large scales, the behavior of water in porous media can be described and useable predictive tools can be formulated. These emergent behaviors represent the composite of many small-scale events that cannot be described individually and composited, due to their complexity.

Building Reactive Zone Strategies

groundwater flows over large aquifer volumes (e.g., at the 1000-m scale) is quite good, while our ability to predict details of groundwater flow in the same formation at a 1-m scale is comparatively quite poor.

To support the design of IRZs, the behavior of injected fluids and the development of overlapping injection fields must be described at the 1- to 10-m scale, where we have the most limited predictive power. Collection of data on aquifer characteristics at the 1-m scale may not improve our predictive abilities.

5.2.1.1 Darcy's Law, Permeability, and Groundwater Velocity

Darcy's law provides a quantitative representation of the large-scale hydraulic behavior that emerges from the collective outcomes of countless pore-space — water interactions that occur at the ultra-fine scale. Darcy's law states that the volumetric flux of water through a defined cross-section, Q, is proportional to the difference in hydraulic head per unit path length, L, where path length refers to the macroscopic, not microscopic, path,

$$Q = K \cdot \frac{dh}{dL} \cdot A \quad (cm^3 \, s^{-1}) \tag{5.1}$$

where the constant of proportionality, K, is the hydraulic conductivity (cm s^{-1}), and h is the hydraulic head difference per unit net path length, L (cm cm^{-1}). A is the cross-sectional area (cm^2) of the flow, measured on a plane perpendicular to the longitudinal flow axis.

Groundwater flow direction is often inferred from water table surface elevation or piezometric surface maps. These inferences overlook the importance of hydraulic conductivity in the determination of groundwater flux and often reach incorrect conclusions. Remedial system designers should anticipate heterogeneities in the distribution of hydraulic conductivity throughout an aquifer and the associated meandering of natural groundwater flows and injected fluids.

Hydraulic conductivity, K, is a function of the porous matrix, as well as the fluid flowing through it. The porous medium is represented by the intrinsic permeability, k, and the fluid is represented by its viscosity, μ, and its density, ρ. These variables are related by the following equation:

$$K = \frac{k \cdot \rho \cdot g}{\mu} \left(\frac{cm^2 \cdot g \, cm^{-3} \cdot cm \, s^{-1}}{g \, cm^{-1} \, s^{-1}} = cm \, s^{-1} \right) \tag{5.2}$$

where g is the acceleration due to gravity (980 cm s^{-2}). Because the viscosity and density of fluids change as a function of temperature, we can see that the conductivity will also change as a function of temperature. The following values for viscosity and density of water provide a useful comparison between aquifer temperatures and the elevated temperature that might be encountered during chemical reactive zone operations.

$$\mu_{10°C} = 0.01307 \quad (g \, cm^{-1} \, s^{-1}) \, (poise)$$

$$\mu_{50°C} = 0.00547 \quad (g \, cm^{-1} \, s^{-1}) \, (poise)$$

$$\rho_{10°C} = 0.999728 \, (\text{g cm}^{-3})$$

$$\rho_{50°C} = 0.988066 \, (\text{g cm}^{-3})$$

For water, the decrease in density is outweighed by a greater decrease in viscosity as temperature increases from 10 to 50°C, so the permeability and hydraulic conductivity increase 2.4-fold over that 40° temperature interval.

The permeability for any fluid is at its maximum when the matrix pore space is entirely saturated with that fluid. When a second fluid is introduced to the porous matrix, permeabilities can change significantly if the two fluids are immiscible. For example, in a porous matrix that has been fully saturated with water, the introduction of a gas displaces a portion of the water, reducing the water-filled fraction of the pore volume. The following equation provides an approximation of the water permeability reduction that occurs due to the displacement:

$$k_{H_2O} = k \cdot \left(\frac{S_w}{n}\right)^3 \tag{5.3}$$

where n is the drainable porosity of the porous medium (L m^3), and S_w is the water-filled fraction of the porosity (L m^3). When the porous medium is fully saturated with water, k_{H_2O} equals k. The gas permeability is given by

$$k_{gas} = k \cdot \left(\frac{n - S_w}{n}\right)^3 \tag{5.4}$$

Equations (5.3) and (5.4) provide a rough approximation of the relationship between intrinsic permeability and the reduced permeabilities that occur when immiscible fluids interact in porous media. The hydraulic conductivity impact of gas generated during *in situ* oxidation reactions can be approximated using Equation (5.3); this is discussed in Section 5.6.5. For a more complete treatment of multi-fluid permeabilities, refer to the works of Brooks and Corey.[*]

The longitudinal velocity of groundwater (and fluids injected for an IRZ) can be calculated from Darcy's law and the porosity of the aquifer matrix,

$$\frac{Q}{A}\left(\text{cm}^3 \, \text{s}^{-1} \text{cm}^{-2}\right) = v_x(\text{cm s}^{-1}) = \frac{K \cdot \frac{dh}{dL}}{n}$$

where n is the effective porosity (cm^3 pore space/cm^3 matrix). The effective porosity refers to the portion of the aquifer matrix porosity that permits bulk movement of water or injected fluids. In many formations, the effective porosity is a small fraction of the total porosity, and the velocity of migrating fluids can be quite high. This concept is developed more completely later.

---

[*]Brooks, R.H. and A.T. Corey. Hydraulic Properties of Porous Media, Hydrology Papers, Colorado State University, Fort Collins, No. 3, March 1964.

Building Reactive Zone Strategies

When the migration of a cohort of contaminant or other tracer molecules is tracked through an aquifer, randomness of water movement at the pore scale causes the leading edge of the tracer fluid to become incoherent. This phenomenon is termed dispersion, and the dispersion that occurs along the flow axis is termed longitudinal dispersion. Dispersion effects are cumulative along a flow path, so the incoherence of the leading edge of a plume or tracer increases with distance traveled.

Field observations of plume migration combine the effects of dispersivity, the random walk effect, with the effects of imperfect estimates of hydraulic conductivity and contaminant sorption. Dispersion is measured in the field by the difference between calculated time of arrival for a tracer or contaminant front and its observed time of arrival. We term the dispersion measured in the field as *apparent* dispersion, to reflect the fact that it combines the effects of true dispersion (the random walk effect) with the natural variabilities of hydraulic conductivity and geosorbent distributions that are difficult to capture in large-scale estimates, such as those made from pump tests.

Figure 5.1 shows the dispersion that occurs at the leading edge of a contaminant plume. The concentration profile was constructed using the fate and transport equation developed by Domenico[1] (Equation (5.5)), iterated along the longitudinal axis of a plume, 1 year after initiation of a continuous source at $x = 0$. The contaminant was assumed to be unaffected by sorption (it will act as a tracer) and the first-order degradation rate constant was assumed to be 1.89×10^{-3} day^{-1}, which gave a half-life of 365 days. Groundwater velocity along the plume axis (v_x) was assumed to be 1 ft day^{-1}. Without dispersion, the contaminant plume in Figure 5.1 would have arrived 365 ft downgradient, at one half the source concentration (curve A). Instead, the

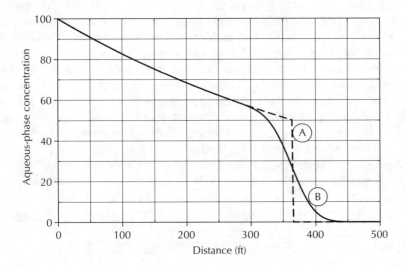

FIGURE 5.1 Concentration profile of a hypothetical contaminant with a 365-day half-life. The source release began at day 0, and both graphs represent a concentration profile along the longitudinal core of the plume. Curve A shows the concentration profile that would be expected if dispersion ≈ 0 ft (plug flow), and curve B shows the profile when dispersion = 1.0 ft. Calculations were made using the equation provided by Domenico.[1]

contaminant front had reached more than 400 ft downgradient, and the concentration at the 365-ft distance had not reached 50 mg L^{-1} after 365 days (curve B). That is caused by dispersion at the leading edge of the plume. In curve B of Figure 5.1, the center of the leading-edge cohort is at the 365-ft distance, after 365 days travel from the source, traveling at an average 1 ft day^{-1} with a portion of the cohort leading and an equal portion lagging. Note that Equation (5.5) does not incorporate organic carbon partitioning, since it is used primarily to describe steady-state conditions.

$$C = \frac{C_0}{8} \cdot e^{\left[\left(\frac{x}{2 \cdot \alpha_x}\right)\left(1-\sqrt{1+4\cdot\lambda\cdot\frac{\alpha_x}{v_x}}\right)\right]} \times \text{erfc}\left[\frac{x - v_x \cdot t \cdot \sqrt{1 + 4 \cdot \lambda \cdot \frac{\alpha_x}{v_x}}}{2 \cdot \sqrt{\alpha_x \cdot v_x \cdot t}}\right]$$

$$\times \left[\text{erf}\left(\frac{y + \frac{Y}{2}}{2 \cdot \sqrt{\alpha_y \cdot x}}\right) - \text{erf}\left(\frac{y - \frac{Y}{2}}{2 \cdot \sqrt{\alpha_y \cdot x}}\right)\right]$$

$$\times \left[\text{erf}\left(\frac{z + \frac{Z}{2}}{2 \cdot \sqrt{\alpha_z \cdot x}}\right) - \text{erf}\left(\frac{z - \frac{Z}{2}}{2 \cdot \sqrt{\alpha_z \cdot x}}\right)\right] \quad (5.5)$$

The magnitude of apparent longitudinal dispersion can be very large if hydraulic conductivity tests do not capture the effects of smaller-scale, high-conductivity segments of the aquifer matrix. For example, Rivett et al.[2] reported on the migration of three chlorinated solvents through the Borden aquifer, downgradient from an emplaced nonaqueous-phase contaminant mass. The three solvents tested were perchloroethene, trichloroethene, and trichloromethane (chloroform). Concentration mapping of groundwater at several times following the source placement suggested that trichloromethane traveled through the aquifer at nearly double the calculated groundwater velocity (15 cm day^{-1} for trichloromethane and 8.5 cm day^{-1} average groundwater velocity). In this case, the apparent longitudinal dispersion was very large, which was an indication that field estimates of hydraulic conductivity (from which groundwater velocity estimates were derived) did not represent the flow path for a significant portion of the trichloromethane.

Injection system designs for IRZ treatments must anticipate the variability of aquifer conductivities at the small scale. Hydraulic conductivity and groundwater velocity estimates drawn from larger scale observations can provide a basis for initial system design. However, because the areal distribution and mixing of injected reagents can be so critical to the success of reactive zone technologies, testing of injected fluid behaviors through pilot-scale injections is strongly recommended.

5.2.1.2 A Dual-Porosity Pore-Scale Model

Conceptually, we can envision a network of interconnected pore spaces, the larger and better connected of which provide less resistance to groundwater movement and

Building Reactive Zone Strategies

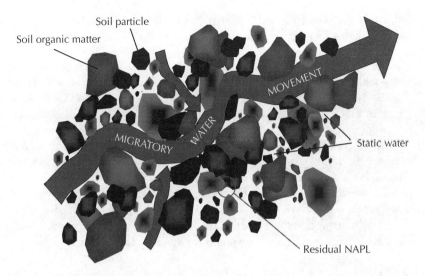

FIGURE 5.2 Pore-scale structure of an aquifer, showing the relative portions of migratory and static pore space.

serve as groundwater migration pathways. Figure 5.2 shows a pore-scale depiction of an aquifer matrix. Smaller and more poorly connected pore spaces present greater resistance to groundwater movement and harbor a static groundwater mass that lies astride the migratory pathways. One of the important elements of this construct is the idea that a relatively small fraction of the aquifer pore spaces may be "migratory." This concept is especially important when we consider the distribution of contaminants and design methods for distribution of reagents in aquifers.

Hydrogeologists have addressed the complexity of fine-scale flow regimes with dual-porosity models that represent the migratory and static pore water masses separately. Julian et al.[3] reported on the application of a dual-porosity mass transport model to support tracer studies in a heterogeneous aquifer in the U.S. midwest. They modeled the site at three scales, with the small-scale model covering an area 60 × 60 m^2 in the horizontal plane and 11 m in the vertical. The model grid spacing was 1 m^2 in the horizontal plane with 22 layers at 0.5-m intervals in the vertical. Bromide tracer was added to a trench located in the small-scale model study area, and mass transport was measured over a network of monitoring locations. A plan view map of bromide observations shows that the tracer moved in the general direction that was predicted, although the plume was much more variable in width and localized flow direction than predicted. Scatter diagrams that compared observed and predicted bromide tracer concentrations showed only order-of-magnitude agreement. To the extent their observations are representative of other sites, there will be a significant constraint on the precision at which we prescribe specific reagent concentrations in reactive zones. Their studies also indicated that advective transport at their study site was limited to between one eighth and one seventh (12.5 to 14.3%) of the drainable aquifer matrix pore volume.

Julian et al.[3] and similar studies help explain two problems related to hydrogeology that are encountered in reactive zone remedies.

Contaminant/reagent contact. Injected reagents tend to flow along the migratory pathways, while the contaminant mass may reside in the static portion of the aquifer matrix. This increases the difficulty of establishing contact between injected reagent and contaminants, and leads to a preference for more persistent reagents that can diffuse into static pore spaces while still "active." Contaminant mass in the static pore water fraction that is not treated during reagent injections may be a persistent source of contamination. This is discussed further in the following sections.

Reagent strength. The limited ability to predict bromide tracer concentrations suggests it would be very difficult to closely match a prescribed reagent concentration at any point in a reactive zone. Effective reagent distribution may require closely spaced injection points to overcome aquifer heterogeneities.

5.2.2 Contaminant Classes and Their Behaviors

The distribution of contaminants in the subsurface is controlled, in part, by their physical–chemical form at the time of release. To facilitate discussion of contaminant distribution we can categorize contaminants as follows.

5.2.2.1 High-Solubility Inorganics (Salts)

Composed principally of ionic solids or salts, this class of contaminants has long been recognized for its capacity to generate very large-scale contamination plumes. Fertilizer-derived nitrates and sodium chloride from road salt storage sites are examples of salts that have been known as groundwater contaminants for many decades. More recently, the perchlorate anion from ammonium perchlorate salts, largely derived from solid rocket propellant, and the anion *N*-nitrosodimethylamine (also known as NDMA) have gained public attention when large-scale aquifer contamination was discovered in several of the western U.S. states. These compounds generally have low natural degradation rates in aerobic aquifers and suffer very little sorption to aquifer solids.

Among the high-solubility inorganics, it is typically the anion that drives environmental concerns. The cations that form these salts are usually naturally occurring elements that are stable and highly soluble. They easily blend into the natural background of soluble cations (e.g., sodium, potassium, and calcium) and usually are not the subject of remedial efforts.

Anions move relatively freely through aquifer matrices because aquifer mineral surfaces tend to be neutral or negatively charged. The halide anions such as chloride and bromide have been used as "conservative" tracer compounds, reflecting their freedom of movement through aquifer matrices. Because chloride is often present in moderately high background concentrations, the bromide anion is favored in tracer work.

Cations are more susceptible to adsorption by aquifer matrices, particularly when clay minerals are present, with their prevalence of negatively charged mineral surfaces.[4] A measure of the cation sorption capacity is the cation exchange capacity, a variable sometimes overlooked in aquifer soil analysis. Cation exchange capacity is discussed in Section 5.2.3.

5.2.2.2 Transition Compounds

Elements from the central portion of the periodic table can assume multiple oxidation states, and compounds of many of these elements, especially the transition metals, exhibit variable solubilities according to solution pH. Compounds of these elements may be observed at high concentrations in groundwater. For metals used in plating processes (e.g., chrome and nickel), compounds of the valence preferred for commercial processes are highly soluble, while compounds of alternative valence states are effectively insoluble under a wide range of natural aquifer chemistries. Most natural soil minerals are also compounds of transition elements, and modification of the aquifer chemistry in reactive zone applications can alter their chemistry to increase or decrease solubilities.

Among the transition elements, which are naturally occurring in aquifer soils, it is usually not their presence but their mobility that drives environmental concerns. The plating metals chromium, nickel, and cadmium are most often handled (and released to aquifers) as soluble salts or acids, and their mobility is somewhat constrained by interactions with the aquifer soil matrix. Transition elements that form oxyanions (e.g., arsenic) may have high mobility in aquifers, especially under reducing conditions.

5.2.2.3 Miscible Organics

These highly soluble organic compounds are fully miscible (infinitely soluble) in water and, consequently, do not form a persistent separate liquid phase when spilled into the subsurface. Many of these compounds are also poorly sorbed on solid-phase organic matter and mineral surfaces in the aquifer matrix. Alcohols, ketones, aldehydes, ethers, and carboxylic acids are examples of miscible or near-miscible organics. Despite their high solubility, these compounds can form a persistent source in an aquifer when they diffuse into nonmigratory water masses, as will be described more completely later.

5.2.2.4 Hydrophobic Organics

These low-solubility organic compounds form NAPLs that may serve as very persistent sources of dissolved-phase contamination in aquifers. Hydrophobic organics occur in a wide range of densities, from very low relative to water (e.g., benzene, $\rho = 0.6$) to very high relative to water (e.g., perchloroethene, $\rho = 1.6$), and their NAPLs may form separated liquid masses in aquifers. The density-driven behaviors of hydrophobic organic liquids observed in free liquid mixtures, for example, the tendency to float or sink through water, are damped by interactions between nonaqueous liquids and the porous medium. Capillary forces attract each liquid to particular matrix pore sizes, thereby limiting the range of motion through a typical aquifer matrix, which is naturally composed of a multitude of pore apertures.

The liquid-phase mobility of hydrophobic organics depends also on the presence of a relatively large mass that can maintain continuity of the liquid body. Experiments reported by Schwille[5] documented the migration of perchloroethene through synthetic porous media under varying conditions of water and solvent saturation. When a large solvent mass was available to form a contiguous separate-phase liquid within the water-saturated porous medium, active spreading was observed. This yields what we refer to as the *drainable fraction* of a NAPL. Capillary tension between liquids and the porous medium generate a retaining force that must be overcome before liquid movement can occur. In a water-saturated aquifer (with water-wetted surfaces in the porous medium), discontinuities form in the nonaqueous liquid masses when active spreading reaches its limit, or when remedial actions remove the drainable fraction. These isolated masses are immobilized by their inability to overcome capillary forces and they form what we refer to as the *residual fraction* of the NAPL mass.

Hydrophobic organics are also generally susceptible to sorption on organic solids and colloids, as well as mineral surfaces in aquifers, so a large portion of the mass of these contaminants may reside outside the aqueous phase under normal physical–chemical conditions.

5.2.3 Sorption and Desorption Processes

Sorption processes transfer contaminants from aqueous phase into nonaqueous binding sites, where they withdraw from participation in aqueous-phase chemistry. Sorption processes are critical to reactive zone designs for several reasons:

- A large fraction of contaminant mass often resides in sorbed phase.
- Sorbed mass is not immediately available to aqueous-phase chemical and biological reaction mechanisms (e.g., Scow and Johnson[6]).
- Sorbed mass can recontaminate groundwater in a reactive zone after a chemical or biological remedy has been applied to remove or destroy dissolved-phase contaminants.
- Desorption from soil organic matter and the associated recontamination of cleaned groundwater controls the elimination of groundwater contamination downgradient from reactive zone barrier systems.

There are three broad categories of sorption occurring in porous media that influence contaminant distribution: (1) ionic adsorption and exchange, especially cation exchange; (2) organic a*b*sorption, also known as partitioning; and (3) organic a*d*sorption. In each of these processes, dissolved-phase molecules or ions migrate into nonaqueous-phase locations, including mineral surfaces (ionic adsorption), mineral crystal lattices (cation exchange), solid-phase soil organic matter (organic absorption and adsorption), and colloidal and liquid-phase organic matter (organic absorption).

5.2.3.1 Adsorption

Adsorption processes occur for both inorganic and organic compounds, and may occur at mineral surfaces or on the surfaces of solid-phase organic matter in the soil (e.g., charcoal). Adsorption processes occur more slowly than absorption processes, and are sometimes referred to as *kinetic* sorption, in contrast to the more rapid partitioning process, which is referred to as *equilibrium* sorption. The adsorption process has also been termed irreversible, but this is a misnomer.

The relationship between aqueous- and solid-phase concentrations is described by equations called isotherms. Two examples are the Langmuir and Freundlich isotherms.[4] Equation (5.6) shows the Freundlich isotherm, which is nonlinear except in the special case in which the exponent, b, equals 1.0. In that case, the distribution follows the Langmuir isotherm, as shown in Equation (5.7).

$$C_{soil} = K_d \cdot C_{aq}^b \tag{5.6}$$

$$C_{soil} = K_d \cdot C_{aq} \tag{5.7}$$

For both isotherms, the concentration units are mg kg^{-1} for soil and mg L^{-1} for aqueous phase. The distribution coefficient, K_d, has units of L kg^{-1}. If the adsorption isotherms follow the Langmuir equation (5.7), the adsorption capacity and binding coefficients can be determined graphically, using Equation (5.8)[4]

$$\frac{C_{aq}}{C_{soil}} = \frac{1}{\beta_1 \beta_2} + \frac{C_{aq}}{\beta_2} \quad \left(\frac{\text{mg L}^{-1}}{\text{mg g}^{-1}}\right) \tag{5.8}$$

where β_1 is a constant that reflects the adsorption binding energy and β_2 is the capacity of the adsorbing matrix. The slope of the plotted line equals $1/\beta_2$, and the binding energy constant equals the slope of the line, divided by the y-intercept.

Two key aspects of adsorption processes are the fact that there is a maximum capacity due to a limited availability of adsorption sites, and that the adsorption process is associated with a binding energy. The adsorption of key contaminants is an exothermic process, and the adsorption process is expected to proceed more quickly than the desorption process. The slower reverse reaction, desorption, leads to the adsorption process being described as irreversible, which is not the case. Adsorption is also a competitive process. Molecules or ions that have been adsorbed can be displaced by others that have a higher binding energy.

5.2.3.2 Absorption

The absorption, or equilibrium partitioning process, occurs rapidly and is probably the dominant sorption process for hydrophobic organics in many aquifers. Aqueous-phase concentrations are related to sorbed-phase concentrations by the partition Equation (5.9), which is essentially the same as the Langmuir equation, above:

$$C_{sorbed} = K_d \cdot C_{aq} \tag{5.9}$$

where C_{sorbed} is the sorbed-phase soil contaminant concentration in mg per kg$_{soil}$, C_{aq} is the groundwater contaminant concentration in mg L^{-1}, and K_d is the distribution coefficient, in L kg^{-1}. The distribution coefficient is calculated from the soil organic carbon fraction, f_{oc}, and the organic carbon partition coefficient, K_{oc}, as follows:

$$K_d = K_{oc} \times f_{oc} \tag{5.10}$$

TABLE 5.1
Values for Soil Organic Carbon/Water Partition Coefficients for Common Chlorinated Alkenes

Compound	K_{oc} (L kg^{-1})
Chlorinated alkenes	
cis-1,2-Dichloroethene	36[a]
1,1,1-Trichloroethane	139
Trichloroethene	94
Perchloroethene	265
Vinyl chloride	19[a]
Aromatics	
Benzene	62
Ethylbenzene	204
Toluene	140
o-Xylene	241
m-Xylene	196
p-Xylene	311
Naphthalene	1191

Note: All values were measured, unless otherwise noted.
[a] Estimated value.
Source: From U.S. EPA,[7] Table 39.

The units for K_{oc} are L per kg$_{org}$, and the units for f_{oc} are kg$_{org}$ per kg$_{soil}$. Combining the results of Equations (5.9) and (5.10),

$$C_{sorbed} = K_{oc} \cdot f_{oc} \cdot C_{aq} \qquad (5.11)$$

Values for the organic carbon partition coefficient, K_{oc}, are available for many compounds, from a number of sources. There is significant variability between data sources for K_{oc} values, which may relate to differences in the nature of the sorbing organic matter, experimental temperatures, or other factors that affect observed partitioning. When available, we utilize organic carbon partition coefficients taken from the U.S. EPA,[7,8] which reports geometric means of observed values from multiple sources. Values for many of the compounds not listed in the U.S. EPA references can be obtained from Montgomery and Welkom.[9] Table 5.1 provides organic carbon partition coefficients for several organic contaminant compounds.

5.2.3.3 Conventional Calculation of Sorbed-Phase and Total Contaminant Mass

It is often necessary to calculate the total mass of contaminant, aqueous-phase plus sorbed-phase, that is expected to reside in any unit volume of an aquifer. The sorbed-phase mass can be calculated from the sorbed-phase concentration and the soil bulk

Building Reactive Zone Strategies

density, and Equations (5.9) and (5.10), as follows:

$$\text{Mass}_{\text{sorbed}} = C_{\text{sorbed}} \times \rho_{\text{bulk}} \quad \left(\frac{\text{mg}}{\text{kg}_{\text{soil}}} \times \frac{\text{kg}_{\text{soil}}}{\text{m}^3_{\text{soil}}}\right) \quad (5.12)$$

We can replace the C_{sorbed} term in Equation (5.12) with an expression based on C_{aq}, derived from Equations (5.9) and (5.10), as follows:

$$\text{Mass}_{\text{sorbed}} = C_{\text{aq}} \times K_{\text{oc}} \times f_{\text{oc}} \times \rho_{\text{bulk}} \quad \left(\frac{\text{mg}}{\text{L}} \times \frac{\text{L}}{\text{kg}_{\text{org}}} \times \frac{\text{kg}_{\text{org}}}{\text{kg}_{\text{soil}}} \times \frac{\text{kg}_{\text{soil}}}{\text{m}^3_{\text{soil}}}\right) \quad (5.13)$$

The aqueous-phase mass can be expressed in terms of the aqueous-phase concentration and the soil pore water content, θ, as follows:

$$\text{Mass}_{\text{aqueous}} = C_{\text{aq}} \times \theta \quad \left(\frac{\text{mg}}{\text{L}} \times \frac{\text{L}}{\text{m}^3_{\text{soil}}}\right) \quad (5.14)$$

The total soil mass is calculated by summing Equations (5.12) and (5.14), expressing the sorbed-phase concentration in terms of the aqueous-phase concentration.

$$\text{Mass}_{\text{total}} = (C_{\text{aq}} \times K_{\text{oc}} \times f_{\text{oc}} \times \rho_{\text{bulk}}) + (C_{\text{aq}} \times \theta) \quad \left(\frac{\text{mg}}{\text{m}^3_{\text{soil}}}\right) \quad (5.15)$$

Dividing Equation (5.13) by Equation (5.15), we obtain the fraction of total aquifer contaminant mass that will reside in aqueous phase, as a function of the organic carbon fraction for the aquifer soil.

$$\frac{\text{Mass}_{\text{sorbed}}}{\text{Mass}_{\text{total}}} = \frac{(C_{\text{aq}} \times K_{\text{oc}} \times f_{\text{oc}} \times \rho_{\text{bulk}})}{(C_{\text{aq}} \times K_{\text{oc}} \times f_{\text{oc}} \times \rho_{\text{bulk}}) + (C_{\text{aq}} \times \theta)} \quad \frac{\text{mg m}^{-3}}{\text{mg m}^{-3}} \quad (5.16)$$

Equation (5.16) has been plotted as a function of soil organic carbon fraction for several common contaminant compounds, and the result is shown in Figure 5.3. This shows that compounds with high K_{oc} values will reside primarily in sorbed phase when the soil organic carbon fraction is high. The partitioning effects observed in Figure 5.3 become very significant as the carbon fraction climbs from 0 to 0.005, corresponding to soil organic matter contents from 0 to 5000 mg kg^{-1}.

The aquifer soil concentrations can be expressed in terms of the relationship between aqueous and sorbed phase concentrations, and is provided in Equation (5.17):[7]

$$C_{\text{total}} = C_{\text{aq}} \times \left[(K_{\text{oc}} \cdot f_{\text{oc}}) + \frac{\theta_W}{\rho_{\text{bulk}}}\right] \quad \left(\frac{\text{mg}}{\text{L}} \times \left[\left(\frac{\text{L}}{\text{kg}_{\text{org}}} \cdot \frac{\text{kg}_{\text{org}}}{\text{kg}_{\text{soil}}}\right) + \frac{\text{L m}^{-3}_{\text{soil}}}{\text{kg}_{\text{soil}} \text{m}^{-3}_{\text{soil}}}\right]\right) \quad (5.17)$$

where θ_W = soil pore water content (L m^{-3}); ρ_{bulk} = soil bulk density (kg$_{\text{soil}}$ m$^{-3}_{\text{soil}}$); K_{oc} = compound-specific organic carbon partition coefficient (L kg$^{-1}_{\text{org}}$); f_{oc} = soil

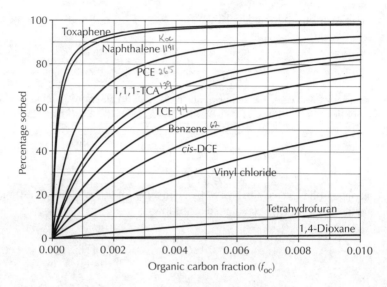

FIGURE 5.3 Relationship between sorbed-phase and total contaminant mass per unit aquifer volume, as a function of soil organic matter content, as measured by the organic carbon fraction, f_{oc}.

organic carbon fraction (kg_{org} per kg_{soil}); and C_{total} = total contaminant mass per kg bulk aquifer matrix. Masses of bulk soil and soil organic matter, kg_{soil} and kg_{org}, respectively, are called out to eliminate ambiguities in units.

The aqueous-phase (groundwater) concentration can be isolated from Equation (5.17), as follows:

$$C_{aq} = \frac{C_{total}}{\left[(K_{oc} \cdot f_{oc}) + \frac{\theta_w}{\rho_{bulk}}\right]} \quad (mg\ L^{-1}) \qquad (5.18)$$

5.2.3.4 Multi-Compartment Sorption Models

The sorption behavior of hydrophobic organic compounds in aquifers suggests that multiple sorption processes occur, leading researchers to suggest what are termed multi-compartment models of the sorption process. These models typically have a rapid-equilibration component that reflects partitioning absorption as expressed in Equation (5.9), and a slower-maturing adsorption component like that embodied in the Freundlich isotherm, Equation (5.6). Luthy et al.[10] provide a comprehensive review of the geosorbents found in aquifers and the nature of their interactions with hydrophobic organic compounds. However, methods that characterize the various geosorbents described by Luthy et al. are not available for routine analysis of aquifer soils. Without that breakdown, the analysis of total organic carbon (TOC) provides a basis for estimating partitioning equilibrium behavior (which is expected to dominate the sorbed mass in most circumstances), and a smaller-scale, lagging desorption can be expected to occur as a result of adsorption processes, the magnitude or duration of which cannot be quantified.

Chen et al.[11] provided an empirically derived two-compartment model that combined a rapid partitioning process with a slower, capacity-limited adsorption process. Termed the dual-equilibrium desorption, or DED model, the sorption behavior is described by the summation of the partitioning and adsorbing processes,

$$q = q^{1st} + q^{2nd} \tag{5.19}$$

where q is the total soil concentration, q^{1st} is the soil concentration predicted by partitioning equilibrium and q^{2nd} is the soil concentration predicted by a linear adsorption isotherm. All values are mg kg^{-1}. The first compartment is calculated according to Equation (5.11), using the DED terminology.

$$q^{1st} = K_{oc}^{1st} \cdot f_{oc} \cdot C_{aq} \tag{5.20}$$

The K_{oc}^{1st} value is the conventional organic carbon partition coefficient. The second compartment is calculated as follows:

$$q^{2nd} = \frac{K_{oc}^{2nd} \cdot f_{oc} \cdot f \cdot q_{max}^{2nd} \cdot C_{aq}}{f \cdot q_{max}^{2nd} + K_{oc}^{2nd} \cdot q_{max}^{2nd} \cdot C_{aq}} \tag{5.21}$$

where

$$K_{oc}^{2nd} = 10^{5.92 \pm 0.16} \tag{5.22}$$

and

$$q_{max}^{2nd} = f_{oc} \cdot \left(\frac{K_{oc}^{1st}}{0.63} \cdot C_{sat}\right)^{0.534} \tag{5.23}*$$

The variable f in Equation (5.21) represents the fraction of the second compartment that is saturated upon exposure. Chen et al.[11] assume the value for f is 1.0 in their calculations. The value for K_{oc}^{2nd} in Equation (5.22) was empirically determined from a population of 41 samples considered by Chen et al.[11] K_{ow} is the octanol-water partition coefficient for the hydrophobic organic and C_{sat} is the aqueous-phase solubility of that compound. The exponent in Equation (5.23) was also empirically determined by Chen et al.[11]

---

*Chen et al.[11] used the value for K_{ow}, the octanol-water partition coefficient in the calculation of q_{max}^{2nd}. We have substituted an estimate of K_{ow} from the relationship $K_{oc} = 0.63 \times K_{ow}$, also given by Chen et al.[11]

Combining terms,

$$q = K_{oc}^{1st} \cdot f_{oc} \cdot C_{aq} + \frac{K_{oc}^{2nd} \cdot f_{oc} \cdot q_{max}^{2nd} \cdot C_{aq}}{q_{max}^{2nd} + K_{oc}^{2nd} \cdot f_{oc} \cdot C_{aq}} \quad (5.24)$$

At low values of C_{aq}, Equation (5.24) reduces to a Langmuir isotherm, and at high values of C_{aq}, the equation reduces to an equilibrium partitioning model. Equation (5.24) provides a "bridge" between the adsorption- and absorption-dominated sectors of the overall sorption process and calls our attention to the bimodal nature of sorption behavior that can be expected for hydrophobic organic compounds in aquifers. Figure 5.4 provides a graphic display of the aqueous-phase–sorbed-phase relationship for perchloroethene predicted by the dual equilibrium desorption model. For perchloroethene, the model suggests that the sorbed fraction significantly exceeds predictions of the "classic" partitioning equilibrium model at concentrations from 0.001 to 10 mg L^{-1}. Because the adsorbed fraction reacts more slowly than the partitioned phase to changes in aqueous-phase concentrations, desorption processes will occur much more slowly under this model, particularly at lower aqueous-phase concentrations.

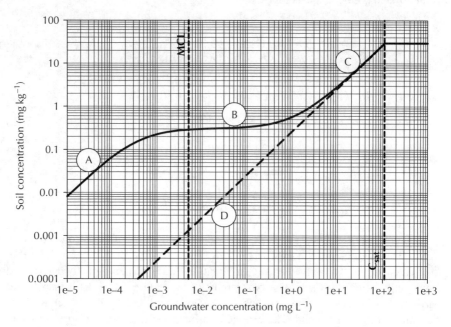

FIGURE 5.4 Dual-equilibrium desorption calculation for perchloroethene, using the model described by Chen et al. (2002). The value for C_{sat} was assumed as 110 mg L^{-1}, the K_{oc} was 265, and the aquifer soil organic carbon fraction was 0.001 (1000 mg kg^{-1} TOC). Curve segment A shows the portion of the DED model dominated by Langmuir isotherm; segment B shows the transition zone between desorption models, and segment C is the zone dominated by partitioning equilibrium. Segment D shows the sorption relationship predicted by "classic" equilibrium partitioning.

One aspect of the DED model that might be controversial is the assertion that K_{oc}^{2nd} can be represented by a single value for a wide range of hydrophobic organic compounds. For comparison, we examined the apparent K_{oc} for a perchloroethene sorption compartment that resisted oxidation in bench trials described in Chapter 3. In that case, the apparent K_{oc} value was 104,000 L kg^{-1} for the oxidant-resistant fraction, compared to an apparent value of 2600 L kg^{-1} for the oxidized fraction. Both values far exceeded the EPA-published value for the K_{oc} of perchloroethene, and the K_{oc} for the oxidant-resistant fraction was 5.5-fold lower than the low end of the range of values suggested by Chen et al.[11] Nonetheless, the dual-equilibrium desorption model appears to provide an order-of-magnitude approximation of field-observed sorption behaviors.

5.2.3.5 Cation Exchange Capacity

Soil minerals and organic matter can bind cations in competitive adsorption. The strength of the binding increases with valence number and mass of the cations. The following are common groundwater cations, shown in order of decreasing exchangeability (increasing binding strength):[4]

$$Na^+ > K^+ > Mg^{2+} > Ca^{2+}$$

If the calcium and sodium concentrations in aqueous phase are equal, and no other cations are present in solution, calcium is likely to displace sodium that may reside on adsorption sites in the aquifer matrix mineral and organic matter. Metal cations, particularly multivalent transition metals such as hexavalent chromium (Cr^{6+}), are likely to be strongly sorbed to cation exchange sites.

Because cation exchange is a competitive process, the cations on any soil can be displaced by flooding the aqueous phase with competing cations. For example, sodium displaces calcium and magnesium ions bound to cation exchange resins in water softeners during the recharge cycle, in which the exchange resin is flooded with a saturated sodium chloride solution. During the operational phase, calcium and magnesium, which are more strongly bound and present at higher concentrations in influent water, displace sodium cations on the exchange resin. If aqueous-phase chromium is added to groundwater, it is expected to displace a portion of the native calcium, magnesium, and other cations adsorbed to the aquifer matrix.

The magnitude of cation exchange capacity depends on the soil particle size distribution (smaller particles, with higher surface area per volume, have a higher exchange capacity than large particles) and the mineral composition (clays have high exchange capacities, sands have lower exchange capacities), both of which control the cation exchange capacity. The cation exchange capacity is expressed as charge equivalents (the sum of exchangeable cation molarities, multiplied by their respective ion charges) per unit soil mass:

$$CEC = \sum([A] \cdot z_A + [B] \cdot z_B + \ldots) \quad (\text{meq } g_{soil}^{-1}) \qquad (5.25)$$

where CEC is the cation exchange capacity, [A], [B], etc. are the molarities, and z_A, z_B, etc. are the charges of all cations in aqueous phase. The cations are measured in a solution of a displacing cation such as ammonium chloride, NH_4Cl, that has been used to flood the soil sample, displacing all the cations other than ammonium, NH_4^+.

Cation exchange capacity is not often measured in aquifer soil samples, but it may be an important determinant in the mobility of contaminant cations, such as transition metals. In source areas, where high-concentration solutions flood an aquifer matrix, transition metal ions will displace native calcium, magnesium, and potassium. When aqueous-phase concentrations of the transition metals decline through groundwater flushing, the competitive replacement of transition metals by native cations can drive a persistent, low-concentration leaching of the transition metals back into the groundwater. During periods of high contaminant metal concentrations in groundwater, the cation exchange process acts as a scavenger. During later periods of low-contaminant metal concentrations, adsorbed material becomes a low-level source.

5.2.4 Contaminant Distribution in Aquifer Matrices

The effects of aquifer matrix structure and physical/chemical contaminant behaviors combine to generate characteristic distribution patterns for each class of contaminants, determining their source persistence and controlling their exposure to reactive zone treatment processes. We have summarized the physical and chemical contaminant characteristics, pore-scale and aquifer-scale hydrogeology, and resulting migration characteristics for the four contaminant classes in a multi-compartment distribution model.

5.2.4.1 A Multi-Phase Conceptual Model for Contaminant Distribution

Dissolved Phases. The assumption that an aquifer is homogeneous and isotropic is a simplification that is useful to support calculations of aquifer hydraulics, but it is unlikely to be valid in real formations. Contaminant distribution and migration patterns are significantly affected by the heterogeneities of real-world formations (e.g., Julian et al.[3] and Rivett et al.[2]). Aquifer water can be divided into two classes.

> *Migratory water.* Flowing (migratory) groundwater is carried in only a portion of the aquifer pore space. The migratory water does not mix fully with the entire aquifer mass, and contaminants dissolved in the migratory water interact with immobile contaminant mass in the static aquifer pore water primarily through diffusion. The migratory water is the contaminant mass transfer carrier, distributing contamination along the aquifer flow path. Groundwater pumping increases the velocity of the migratory water, decreasing its contact time in the contaminant source areas. This causes the migratory water to fall well below equilibrium contaminant concentrations during active pumping, explaining the common observation that monitor well concentrations increase when pumping ceases (longer contact time brings migratory water concentrations closer to equilibrium levels).
>
> *Static water.* A large fraction of the aquifer water mass is essentially static. In a contaminated formation, other contaminant sinks such as nonaqueous-phase liquids and soil organic matter that sorbs hydrophobic organics may be

embedded in the static water. The static water interconnects the various contaminant pools through diffusive migration in the dissolved phase and may, itself, contain a significant contaminant mass. This is the case when the static water comprises a major portion of the total aquifer water mass, and when highly miscible, nonsorbing compounds such as alcohols, ketones, and 1,4-dioxane are spilled into an aquifer. The static water "sink" explains why we sometimes observe long-lasting sources of these compounds — in an isotropic, homogeneous formation, they should simply "wash out." Instead, some sources persist for several years. This can only occur if there is a large pool of static groundwater that bears a large mass of the contaminant, diffusing into migratory groundwater.

Sorbed Phases. Hydrophobic organic compounds partition strongly into soil organic matter typically present in an aquifer matrix. There are many sorption reactions that can occur in an aquifer, and they can be organized into two modes. Recent studies have introduced the concept of "dual equilibrium" sorption to explain the behavior of contaminants in aquifer matrices. The two modes are:

Absorption. The conventional approach to estimation of sorption uses the organic carbon partition coefficient, K_{oc}, for each contaminant. Published values of K_{oc} are used in calculation of a distribution coefficient that expresses the equilibrium partitioning of the contaminant between the aqueous and sorbed phases. As the aqueous-phase concentration increases, the sorbed-phase mass increases steadily until the aqueous-phase concentration reaches its maximum value, C_{sat}. The absorption process reaches equilibrium rapidly, and is often called "equilibrium partitioning."

Adsorption. An aquifer contains mineral surfaces and, in some cases, organic matter that serve as sites of adsorption — an exothermic binding of a contaminant molecule to a binding site. The binding of hydrophobic organic molecules to granular activated carbon, for example, is an adsorption process. Adsorbed contaminants are bound tightly to the sorbing matrix, and the desorption process is extremely slow, relative to absorption equilibration. Consequently, adsorption is often referred to as "irreversible" sorption, although this is not quite the case.

Nonaqueous-Phase Liquids. There are two fractions of NAPLs that may exist in an aquifer matrix.

Drainable NAPL fraction. If the NAPL mass is large, it may occupy sufficient pore space to form a continuous fluid mass. In this case, the NAPL can flow through the aquifer matrix, and thus is "drainable."

Residual NAPL fraction. Smaller masses of NAPL that become disconnected from the main fluid body constitute a residual fraction. This material cannot generate the entry pressures needed to drive its movement through the water-wetted porous or fractured medium, and it becomes stuck in place. Residual NAPL is likely to be a significant portion of the contaminant mass in the source area. It may reside in contact with migratory groundwater, where it dissolves directly into the moving water. It may also reside in the static groundwater, where it is linked to the migratory groundwater through diffusion.

Solids and Colloids. These two phases have received only limited attention in recent years, yet they may be significant persistent sources (in the case of solids) or carriers (in the case of colloids) of migrating groundwater contamination.

Solids. Direct burial of solid-phase transition metal salts or hydrophobic organics that are solid at ambient temperature (e.g., hexachlorobenzene) can form a persistent source of groundwater contamination. It is also possible for spillage of high-concentration plating solutions to be partially neutralized by vadose zone and aquifer soils, reducing the solubility of metals in the solution. The resulting precipitation reactions can deposit soluble salts in the aquifer matrix. The solubility products (K_{sp}) for these solids determines their local aqueous-phase equilibrium concentrations. The dissolution rate from these solids will be a function of the distance between the solid and the nearest migratory groundwater channel, which determines the magnitude of the diffusion gradient that can be established. Long diffusion distances generate lesser gradients and limit the dissolution rates that can be achieved.

Colloids. Colloids are ultra-fine particles that can flow with groundwater or be retained by the porous medium, depending on the soil mineralogy and texture, as well as the pH and ionic strength of the groundwater. Metals and hydrophobic organic compounds can adsorb to colloidal particles and, consequently, may enter the groundwater migration without entering the dissolved phase. Bauman et al.[12] included particles from 0.001 to 10 μm diameter in the colloidal range, and indicated that typical groundwater contains 10^6 to 10^{12} particles per liter in this range. Filtration of groundwater samples by a 0.45-μm filter (a typical break point for analysis of the dissolved-phase constituents) will sort out the larger of the colloidal particles, leaving a significant fraction of the colloidal size range, as defined by Bauman et al.[12]

Due to their small size, colloidal particles are likely to interact electrostatically with soil particles. These interactions can be influenced by groundwater chemistries, altering the mobility of colloidal particles and any contaminants adsorbed to their surfaces. Compere et al.[13] showed that high ionic strength caused retention of clay-derived colloids in fine sand columns. Ryan and Elimelech,[14] in a comprehensive review of colloid behavior in groundwater systems, described the countervailing forces of pH and ionic strength on colloid mobility: increasing pH causes increased colloid mobility, while increasing ionic strength restricts colloidal migration. The pH-induced behavior is caused by surface charge reversal at elevated pH, and the threshold pH of that reversal is likely to be specific to each aquifer matrix. In groundwater that has been driven to very high pH by contact with matrices such as cement kiln dust, the hydroxyl anion concentration may, itself, contribute sufficiently to the ionic strength of the solution as to limit colloidal mobility.

5.2.4.2 Distribution of Hydrophobic Organics

Hydrophobic organic compounds can form persistent nonaqueous-phase liquid mass in aquifers. These are compounds with extremely low water solubility, and with principal fluid characteristics (density, viscosity, and surface tension) so distinct from those of water that their respective fluid masses are immiscible. Many of the hydrophobic organics, such as the chlorinated alkene and alkane solvents, are denser than water. When these compounds are spilled in sufficient mass to reach an aquifer, they are likely to form a dense, nonaqueous-phase liquid mass, or DNAPL. Some of the hydrophobic organics are lighter than water, and form a light, nonaqueous-phase liquid mass, or LNAPL. Neutral-density NAPLs often develop from mixtures of hydrophobic organics, and when the spilled material dissolves organic matter en route to the aquifer, moderating the density of the solvent. In situations where NAPL is present, the aquifer contains two fluids that act (mostly) independently of each other: the wetting fluid (normally water) and the nonwetting fluid (most often the organic, or oil phase).

Dense, nonaqueous-phase liquids can be driven below the aquifer surface when a drainable fraction forms above the aquifer surface, with a thickness sufficient to generate a downforce exceeding the entry pressure of the water-wetted aquifer matrix. LNAPLs can accumulate above an aquifer and then be trapped below the aquifer surface during periods of rising groundwater elevation. <u>Nonaqueous-phase liquid mass that reaches an aquifer matrix will, over time, be partitioned among six distinct pools: migratory and static dissolved phase, adsorbed and absorbed nonaqueous phases, and drainable and residual nonaqueous phases.</u>

Water migrating through a NAPL-contaminated aquifer segment bears only a small fraction of the total contaminant mass in any volume element, yet it is the migratory dissolved-phase mass that constitutes the primary contamination transport mechanism. That disparity contributes to the persistence of hydrophobic organic sources. In areas of the aquifer outside the NAPL-bearing zone, the hydrophobic contaminant mass occupies four of the six pools (all but the nonaqueous-phase liquids).

Consider an aquifer segment that contains DNAPL and has reached equilibrium conditions throughout. Each of the water masses (migratory and static) will be fully saturated by the DNAPL compound(s) in the source area. The sorbent aquifer materials will be at maximum mass in the source area, in equilibrium with the aqueous phase. Downgradient from the source area, the sorbed phase will be in equilibrium with the aqueous phase, and the adsorption sites will be occupied for a significant distance. If dilution or degradation are occurring downgradient from the source area, the steady-state concentration profile along the core of the plume (the longitudinal plume axis), will be determined by the cumulate effects of dilution and degradation. These impacts can be lumped into a first-order attenuation rate constant, and a concentration profile based strictly on the calculated first-order attenuation and time of travel in the aquifer.

Figure 5.5 shows the map view of a conceptual plume, with a persistent source area and downgradient plume distribution areas. If the source area contains sufficient hydrophobic organic mass to form an NAPL, the general cross-sectional profile of the contaminant distributions might appear something like Figure 5.6. The distribution of hydrophobic organics along a plume is determined by the K_{oc} and aqueous-phase solubility of the contaminant, and the organic carbon fraction and distribution of migratory and nonmigratory porosity in the aquifer matrix. The key features to observe in Figure 5.6 are the significance of the sorbed mass and the relatively minor fraction of migratory, relative to total, pore water content of the aquifer matrix. Table 5.2 shows

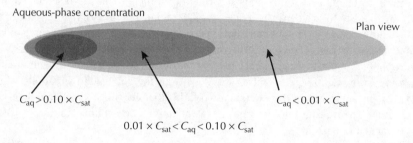

FIGURE 5.5 Plan view of a conceptual contaminant plume, originating from a persistent source zone in which the aqueous-phase concentration exceeds 10% of the aqueous-phase solubility for the contaminant.

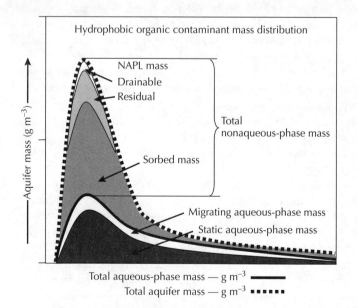

FIGURE 5.6 Conceptual distribution of hydrophobic organics along the core of a plume that has complete saturation of aqueous and sorbent compartments in the source zone, with formation of residual and drainable nonaqueous-phase liquids. Steady-state concentrations in downgradient areas are determined by combined effects of dilution and degradation.

Building Reactive Zone Strategies

TABLE 5.2
Estimates of Maximum Attainable Mass in Each of the Six Contaminant Reservoirs in a Cubic Meter of Aquifer Matrix

Compound	Aqueous-Phase Compartments		Sum of Aqueous Phase	Sorbed-Phase Compartments		Sum of Sorbed and Aqueous Compartments at Saturation	Nonaqueous-Phase Compartment Residual NAPL at 0.5% of Pore Volume	Sum of All Compartments at Saturation
	Static	Migratory		Adsorbed	Absorbed			
Hydrophobic organics								
Perchloroethene	52	8	60	4	93	156	2,430	2,586 (2586)
Trichloroethene	284	46	330	19	181	530	2,190	2,330 (2720)
cis-Dichloroethene	903	147	1050	61	221	1332	1,920	—
1,1,1-Trichloroethane	343	56	399	23	324	746	2,010	2,756
Naphthalene	8	1	9	1	65	65	1,440	1,505
Vinyl chloride	712	116	828	48	92	968	1,365	2,333
Miscible organics								
1,4-Dioxane	258,000	42,000	300,000	17,500	70,000 (700)	318,200	0	318,200
Acetone	258,000	42,000	300,000	17,500	1,006	318,506	0	318,506
Salts								
Ammonium perchlorate	51,600	8,400	60,000	0	0	60,000	0	60,000
Transition Metals								
Nickel plating solution	44,000	7,100	51,000	0	0	51,000	0	51,000

Note: These calculations assume a total porosity of 300 L m^{-3}, a migratory pore fraction of 14% (42 L m^{-3}), and an aquifer matrix f_{oc} of 0.001 (1000 mg kg^{-1} TOC). The k_{oc} and aqueous-phase solubility for each compound were taken from U.S. EPA.[7] For NAPL-forming compounds (hydrophobic organics), the residual NAPL mass was assumed to be 0.5% of the aquifer pore volume. All values are g m^{-3}.

Note: Fig. 5.6 appears to contradict Table 5.2 for sorbed mass of hydrophobic organics.

the distribution of mass, at saturation, among aquifer matrix compartments for selected hydrophobic and miscible organic compounds, salts, and transition metals.

Migratory groundwater passing through a DNAPL-bearing segment of an aquifer may not reach equilibrium concentrations. Consequently, contaminant concentrations observed at monitoring wells may be significantly below the aqueous-phase solubility concentration, C_{SOL}, even though DNAPL is present nearby. To account for this limitation, it is necessary to make a judgment as to the level of contaminant saturation that is likely to signal the presence of DNAPL. An aqueous-phase concentration at or above 10% of C_{SOL} is often used as an indication of DNAPL, and a more conservative indicator is 1% of C_{SOL}. The challenges for remedial system designers are significant any time the aqueous-phase contaminant concentration exceeds 1% of C_{SOL}, due to the large sorbed and static masses that are likely to be present, independent of the possible presence of DNAPL.

To be successful, a remedy applied to any aquifer segment may be required to remove or destroy a significant portion of the nonaqueous-phase, sorbed phase, and static water mass of contaminant. To achieve this, the remedial mechanism must reach beyond the migratory fraction of the groundwater mass, which is the fraction that is available to most injected fluids. Technologies that are severely limited by their lack of access to the nonmigratory mass pools include groundwater pumping and *in situ* oxidation. *In situ* bioremediation through enhanced reductive dechlorination can be managed to enhance access to the nonmigratory contaminant mass, and has been shown to be capable of DNAPL treatment in field applications of PCE and TCE.[15–17]

5.2.4.3 Distribution of Miscible Organics and Soluble Salts

Miscible organic compounds and salts are sufficiently similar in source zone mechanics that they will be treated together for the purpose of describing a conceptual distribution model. The features that distinguish miscible organics and salts from hydrophobic organics and metals are their extremely high aqueous-phase solubility, their lack of separate phase formation in the aquifer matrix, and their low sorption rates on aquifer matrix materials. The mechanism by which these compounds can form a persistent source zone in an aquifer matrix is diffusion into the static pore water fraction of the aquifer matrix. The mass of high-solubility compounds that can be stored is quite large if the migratory pore water fraction is small.

Bromide tracer tests provide a good example of persistent source development for miscible organics and salts. In formations with migratory pore water fractions that are significantly less than the total porosity, there is an immediate dilution of injected tracer when the high-concentration fluid equilibrates with the low-bromide static pore water. Migration of high-concentration solution spreads the center of mass in the downgradient direction, diffusing bromide into static pore water along the flow path. When we track bromide concentrations at a monitoring location downgradient from the injection zone, the bromide concentration vs. time curve often develops a long tail that prevents closure of a mass balance — the area under the curve far exceeds the injected mass, indicating that the monitoring point is measuring bromide in the static water. The tracer study described in Example 5.1 and Figure 5.20 is one such example. When sampling was stopped, the bromide concentration had stabilized at

Building Reactive Zone Strategies

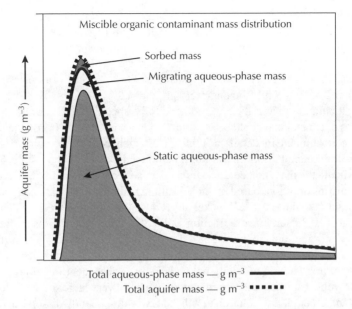

FIGURE 5.7 Conceptual distribution of miscible organics along the core of a plume that has complete saturation of aqueous and sorbent compartments in the source zone. Miscible contaminants cannot form nonaqueous phases; however, their high water solubilities can generate persistent sources in the static water compartment of an aquifer matrix.

20 mg L^{-1}, 3.3% of the peak value at that location. The area under the concentration vs. time curve already showed 150% of the value that would have been observed if all the bromide had been in migrating groundwater, and the area was continuing to increase when sampling stopped.

Figure 5.7 shows the conceptual distribution of a miscible organic or salt, forming a large source mass due to saturation of the static fraction of the aquifer matrix pore volume. This behavior can explain the persistence of miscible organic and salt contamination sources that would otherwise be expected to migrate along with groundwater flows. The center of mass for miscible organic or salt contamination plumes will appear to migrate more slowly than the groundwater velocity when the static aquifer pore fraction is participating in contaminant distribution. In aquifers with very small migratory pore water fractions, miscible organics and salts can conceivably form persistent sources, rooted in the static fraction of the aquifer pore volume.

5.2.4.4 Distribution of Metals

Metals generally match hydrophobic organics for their low solubilities and are subject to significant adsorption in aquifers through the cation exchange process. The source zone mass formation process for metals is very different from what is observed for hydrophobic organics. The solubility of metals varies with pH, ionic strength, and availability of anions or cations, and the chemical habitat in the source zone may initially prevent precipitation of metals (e.g., plating solutions are designed to avoid precipitation), and during that time, the mass of metals stored in the static water fraction can become large. Then, if plating solution stabilizers dissipate or are neutralized, precipitation can occur in the source area, infecting the source zone aquifer matrix with partially soluble metal solids. The solids then can support a long-term leaching of metals into the migratory groundwater.

A large number of variables would have to be specified to determine the maximum mass of a metal that could be accumulated in a source zone. Consequently, metals are not examined in detail in Table 5.2, which compares source zone mass capacities for compounds in the various contaminant classes. It is possible, however, to approximate the mass that could be infused, in aqueous phase, into the static water. A nickel-plating solution, based on nickel sulfate and nickel chloride with boric acid buffer, might contain 170 g L^{-1} nickel at pH 4. If that solution were spilled into an aquifer with a 0.86 static water fraction and a 300 L m^{-3} total porosity, the source zone nickel storage capacity would be 4.4×10^7 mg Ni per m^3 aquifer matrix. If the pH of the solution is neutralized by the aquifer matrix, precipitation occurs, making possible the precipitation of additional nickel from subsequent infusions of plating solution. Cyclical spillage and neutralization can build very large source zone masses of nickel and other plating solution metals that can subsequently dissolve and diffuse into migratory groundwater.

5.2.5 Steady-State and Non-Steady-State Distributions

The concentrations of contaminants entrained in migratory groundwater at steady state reflect a composite of dilution, degradation, and dispersion processes that occur along the flow path. The migrating groundwater passes through the other matrix contaminant compartments, and either gains or loses contaminant molecules through equilibration interactions with those compartments. How closely migrating groundwater reflects the equilibrium concentration at any point along the flow path is controlled by many factors, including the mass transfer mechanism at work and the distance between the equilibrating mass and the migrating groundwater. Figure 5.8 shows a concentration profile along the longitudinal axis of a plume such as that described in Figure 5.5. When the contaminant distribution has reached steady state, the time of travel from the source area to any point downgradient is a function of the linear distance and the groundwater velocity along the plume axis (v_x). The contaminant dilution and attenuation that occurs along the flow path can be expressed in a composite first-order attenuation rate constant, k, and the steady-state concentration for any aquifer volume element can be estimated from a first-order kinetic calculation as shown in Figure 5.8.

For hydrophobic organics and metals, sorption processes may significantly influence concentrations during periods of displacement from steady state. Figure 5.9 describes contaminant inflow and outflow from an aquifer volume element located downgradient from a source zone. During plume expansion, the contaminated groundwater front passes through the volume element. Geosorbents initially contain no contaminant, and the migrating groundwater exceeds local equilibrium concentrations. There will be a net loss of contaminants from the groundwater that continues

Building Reactive Zone Strategies 421

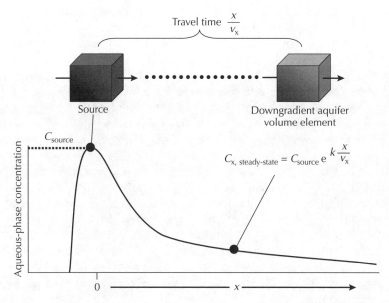

FIGURE 5.8 Development of steady-state contaminant concentrations along a plume axis. Dilution and degradation are represented by a composite first-order attenuation rate constant, k.

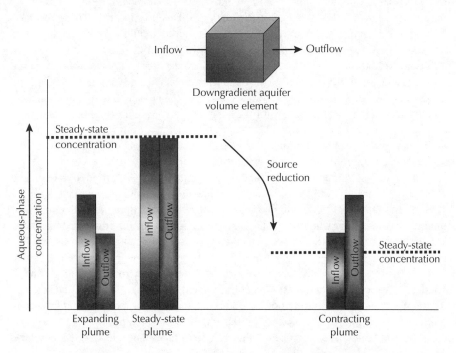

FIGURE 5.9 Displacement from equilibrium contaminant concentrations during plume expansion (left side) following decreases in steady-state concentrations associated with reduction or elimination of contaminant source zones.

until the sites of accumulation reach equilibrium partitioning. During this phase, the outflow from each aquifer volume element is smaller than the inflow. When the geosorbent sites reach equilibrium with the migrating groundwater in an aquifer volume element, steady state is reached and the outflow matches the inflow, as shown in Figure 5.9. If the contaminant source concentration is reduced or eliminated, a new steady-state concentration is established along the entire plume. In each aquifer volume element, the contaminant mass in the geosorbents exceed the newly established equilibrium levels, as shown on the right side of Figure 5.9. At that point, outflows from each volume element exceed inflows until the geosorbent mass reaches the new equilibrium concentration.

During periods of steady state, sorption occurs, but is balanced by desorption and there is no net transfer of contaminants between the flowing groundwater and the static contaminant reservoirs (nonaqueous-phase, sorbed-phase, and dissolved-phase contaminant residing in the static groundwater). The retardation of contaminant flux due to sorption is a non-steady-state phenomenon. That is because sorption is balanced by desorption for the combined population of all contaminant molecules. If we tracked an individual molecule or ion, its migration velocity would be retarded, relative to groundwater velocity, but the entire population behaves as if no sorption effects were occurring.

To examine the propagation of contaminants along a plume axis, we constructed a one-dimensional iterative model that accounts for mass flux and equilibration, through partitioning, with soil organic matter along the flow path. Figure 5.10 shows the model approach. For this exercise, a data grid was constructed that represented 500 1-ft cubes of aquifer matrix, aligned in series along the flow path. A snapshot of the aqueous-phase contaminant concentration was taken at daily intervals, over a 1000-day period. In the data grid, each column represents a daily snapshot, from the source (at the top of the column) to the distance we wish to study (at the bottom of the column). Every day, water moves from each cube to the next cube in line, both in time and distance. That is represented in Figure 5.10 by arrows indicating the water flux, Q. The contaminant mass carried in the water flux was at equilibrium with the cube in which it originated. When that mass arrives in the neighboring cube in the next time interval, a new equilibrium will be established.

The process of mass transfer and re-equilibration was carried out over the 1000-cube length for 1000 days, creating a 1,000,000-cell data matrix that contained 1000 daily snapshots of contaminant mass (hence concentrations) in each cubic foot of aquifer matrix along the flow path. During each mass transfer process, we allowed 1 day's worth of attenuation to occur, using a first-order attenuation rate constant. For convenience, we assumed the volumetric water flux corresponded to $v_x = 1$ ft day^{-1}, and a half-life of 1 year. Hence, after 1 year of travel (365 ft from the source zone at 1 ft day^{-1}), the contaminant mass concentration will be one half of the source concentration. Dispersion effects were not included (i.e., the results represent plug flow conditions) so the equilibration effects on the migrating concentration front would be clear. If significant dispersion occurs, the breadth of the plume's leading edge will be increased.

The first objective of the model was to demonstrate the propagation of a contamination from a constant source. Figure 5.11 shows the progression of a contaminant

Building Reactive Zone Strategies

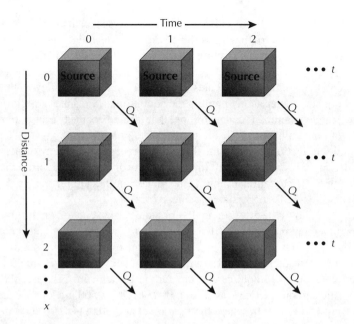

FIGURE 5.10 Iterative process used to calculate the effects of partitioning on contaminant distribution patterns during plume expansion and contraction. Each aquifer volume element is arrayed vertically, from the source zone (top) to the extent of the plume along the x-axis (bottom). The array of aquifer volume elements is repeated to represent each time interval. Q shows the transport of groundwater from volume element x at time t to volume element $x + 1$ and time $t + 1$. The groundwater in each volume element is equilibrated to the total contaminant mass in that volume element, before the next cycle of groundwater interchange between volume elements is initiated.

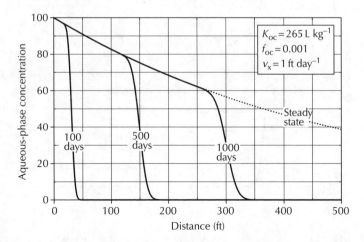

FIGURE 5.11 Iterative calculation of plug-flow plume expansion with a constant source.

front at 100, 500, and 1000 days after initiation of a constant source at $x = 0$ ft. The organic carbon partition coefficient was 265 L kg^{-1} (comparable to perchloroethene), and the aquifer organic carbon fraction was 0.001 (1000 mg kg^{-1} TOC). The contaminant front migrated at approximately one third of the groundwater velocity, due to the diversion of contaminant molecules into partitioning along the flow path. As soon as the soil organic matter at any point in the aquifer matrix reaches equilibrium with the steady-state contaminant concentration, that locale becomes "transparent" to the migration of contaminant. The leading edge of the plume displays a concentration profile similar to that observed in the dispersion process (that is why dispersion was not included in this demonstration). In this case, the dispersion-like profile is created by the diversion of contaminant into sorbed phase in the aquifer matrix. When the sorption sites are filled, steady state is achieved.

Figure 5.12 shows the effect of aquifer organic carbon content on the progression of a contaminant plume expansion. Each curve represents the contaminant concentration profile 1000 days after the source began discharging. At lower aquifer organic carbon fractions, the contaminant front moves at higher velocities. Figure 5.13 shows the effect of variation in aquifer organic carbon content. In this case, the aquifer matrix organic carbon fraction was set at 0.0005 (TOC = 500 mg kg^{-1}), except for a segment from 150 to 250 ft, where the f_{oc} was set at 0.002 (TOC = 2000 mg kg^{-1}). Expansion of the contaminant plume slowed as it passed through the zone of higher organic carbon content, then resumed its pace of expansion after steady state was achieved. Plume expansion after 1000 days can be compared to the 500 mg kg^{-1} plot on Figure 5.12, in which the contaminant front has reached nearly one third farther.

Over time, continuing migration of contaminant mass from the source area brings all the contaminant pools into equilibrium. When equilibrium is achieved for any aquifer segment, there is no further *net* retardation of contaminant flux, and the mass

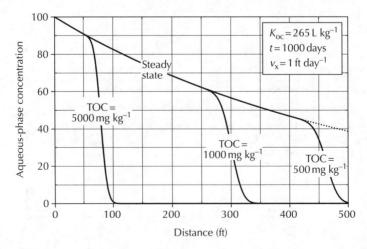

FIGURE 5.12 Effect of organic carbon content on plug-flow plume expansion, with a constant source.

Building Reactive Zone Strategies

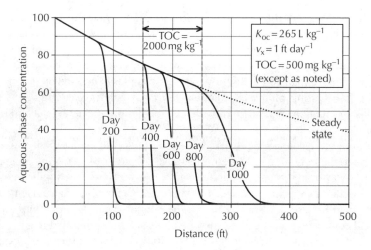

FIGURE 5.13 Effect of variable aquifer organic carbon content on plume expansion.

flux entering the aquifer segment matches the mass flux leaving the aquifer segment (minus degradation losses that may have occurred during transit of the segment). Retardation of contaminant flux continues if we tracked an individual molecule, because each molecule is expected to spend a portion of its transit time through the aquifer matrix stalled in sorbed phase. At steady state, however, when a molecule enters sorbed phase, its entry is balance by the exit of another molecule from sorbed phase, so the loss of the first molecule from the flow path goes unnoticed. That is why there is no *net* retardation after steady state is achieved, but the process continues on a molecular basis. Most importantly, after steady state is reached, the contaminant mass flux proceeds as if there were no retardation of migration.

If a contaminant source is eliminated after the plume reaches steady-state contaminant distribution, migrating groundwater is recontaminated by equilibration with sorbed mass and contaminants dissolved in the static groundwater. Migration of the clean water front is retarded in a manner exactly analogous to the retardation observed during plume expansion.

5.2.6 Propagation of a Clean Water Front

At the downgradient edge of a fully functioning reactive zone, or following complete removal of an aquifer-embedded contaminant source, migratory groundwater is expected to be free of measurable contamination. Although noncontaminant reaction by-product concentrations may be significant at this location, the reactive zone is expected to export only contaminant-free groundwater to downgradient locations. As the migratory water moves through the aquifer matrix, along the plume axis, its concentrations for all dissolved constituents reach equilibrium locally with compounds in the nonmigratory compartments of the aquifer matrix. Assuming the aquifer matrix had reached an equilibrium contaminant distribution prior to reactive zone implementation, the clean groundwater will be undersaturated relative to the local equilibria along the flow path downgradient from the reactive zone. This drives recontamination of the migrating groundwater, and migration of the clean water front is slowed, relative to the velocity of the groundwater. This retardation effect is the reverse of the contaminant migration retardation that occurs during the plume expansion that follows a release, and is expressed on the right half of Figure 5.9. During plume contraction, the mass entering any aquifer volume element is less than the mass departing that element, until the new steady-state concentration is reached.

The iterative calculation process was also applied to plume contraction that occurs due to the migration of a clean water front. Figure 5.14 shows the progression of a clean water front through an aquifer matrix that was initially equilibrated with a constant contaminant source. As clean water reaches an aquifer matrix element that contains sorbed-phase contaminant mass, the contaminant partitions into aqueous phase, and contamination is carried forward along the flow path. In a manner analogous to plume expansion, the aquifer organic carbon content significantly impacts the clean water front progression, as shown in Figure 5.15. Heterogeneous organic matter content in the aquifer matrix can also slow propagation of a clean water front, as described in Figure 5.16.

The movement of clean water, downgradient from reactive zones or other remedial actions that eliminate contaminant sources or establish barriers to contaminant migration, does not proceed at the pace of groundwater flow. The amount of time required for the effects of cleanup to reach any sector of an aquifer is controlled, to a great extent, by the partitioning of contaminants into aquifer matrix organic matter, as well as diffusion into and out of static water segments of the aquifer, along the flow path. These processes are very important design considerations, and will be discussed further in Section 5.4.3.

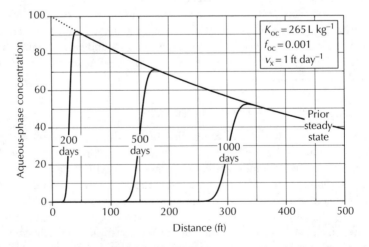

FIGURE 5.14 Progression of a clean water front after the aquifer matrix had reached steady-state distribution. Curves show contamination profiles 100, 500, and 1000 days after groundwater contact with the source zone was eliminated.

Building Reactive Zone Strategies

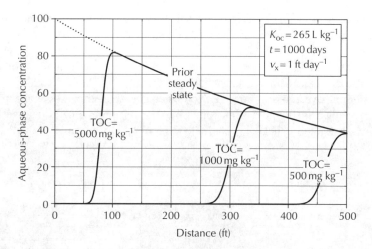

FIGURE 5.15 Effect of organic carbon content on the progression of a clean water front, after the aquifer matrix had reached a steady-state contaminant distribution, with a 1 year contaminant half-life.

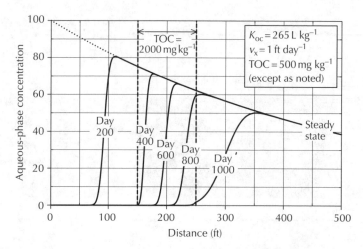

FIGURE 5.16 Effect of variable aquifer matrix organic matter on the rate of migration of a clean water front.

5.3 REACTIVE ZONE STRUCTURE

In situ reactive zones are superimposed on the aquifers they treat and the aquifers, in turn, exert a substantial measure of control on the distribution and migration of reagents injected for the purpose of reactive zone development. Water injections are used as a carrier for reagents that drive the chemical or biological processes for contaminant destruction or precipitation. The flow of migratory groundwater helps to organize reactive zone structure and dilutes injected reagents. The aquifer matrix structure channels injected fluid flows along pathways that mimic contaminant migration and which, therefore, may bypass significant contaminant reservoirs in the static water mass and aquifer organic matter.

In situ reactive zones in flowing aquifers can be divided into several functional zones, as shown in Figure 5.17(A) and (B). The injection zone is defined by the water–reagent mixture that is injected into the formation at the upgradient end of the

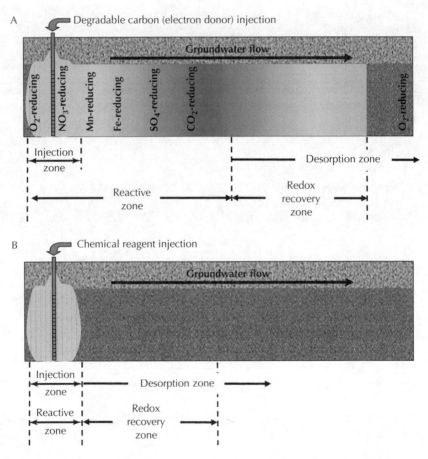

FIGURE 5.17 Conceptual diagram of *in situ* biological (A) and chemical (B) reactive zones. Reactions in the biological reactive zone continue along the groundwater flow path, due to the relatively slow consumption of injected reagents. Reagents injected to create chemical reactive zones are generally consumed within the injection zone. The redox effects of both types of reactive zones are reversed in the redox recovery zone, through equilibration with recharge water and the overlying atmosphere. Desorption from the aquifer matrix and static groundwater recontaminates the clean groundwater leaving the reactive zones of both biological and chemical reactive zones.

Building Reactive Zone Strategies

aquifer segment in which reactions are planned. A reactive zone forms in which microbial communities respond to inputs with metabolic processes that destroy targeted contaminants (Figure 17(A)), or chemical reactions destroy or sequester contaminants through direct interaction (Figure 17(B)). Downgradient from the reactive zone, a desorption zone forms in which the clean groundwater front migrates through the contaminated aquifer. The clean groundwater front migrates along the groundwater flow path as sorbed and static-water contaminants are drawn from the matrix. All reactive zones modify the aquifer chemistry in tangible ways, and a recovery zone forms along the flow path, where the last of the injected reagent is reacted or dissipated. In the recovery zone, the aquifer regains its chemical equilibrium through interaction of matrix minerals, the overlying atmosphere (in unconfined systems) and the inflowing dissolved solids and gases (cf. Champ et al.[18]).

5.3.1 Injection Zone Management

Prediction and control of injected fluid movement through aquifer materials is critically important to the design of IRZ technologies. Chemical reactive zones are especially sensitive to the fluid dynamics because they require direct contact between reagent and target, at prescribed concentrations. Lack of contact prevents reactions; too low a reagent concentration yields incomplete reaction; excessive concentrations may generate unwanted by-products. Biological reactive zones are somewhat less sensitive than chemical reactive zones because the injected reagents have a much longer useful life span in the formation. Nonetheless, the ability to inject fluids in overlapping patterns, with prescribed mixing of reagents and groundwater in the aquifer matrix, is a foundation of all successful reactive zone applications. It is a significant challenge.

Injected reagents merge with the moving groundwater, which participates in the distribution of slow-reacting reagents such as the carbohydrates injected to support reductive dechlorination zones. Groundwater flux can carry away reaction by-products, as well. Figure 5.18 shows the superposition of a series of fluid injections along a reactive zone injection line. Injection locations are marked by crosses on the figure, and the injected fluid volumes were sufficient to spread over a 15-day radius of coverage on the left side of the figure and a 5-day radius of coverage on the right side. Notice that it is very useful to describe groundwater flows in terms of time of travel, rather than simple linear distance. This helps to maintain focus on the fact that injection volumes need to be gauged according to cumulative groundwater flux between injection intervals, in order to maintain design concentrations. In Figure 5.18, longitudinal dispersion (along the flow axis) spreads the injected fluid mass somewhat as it travels from the injection zone. There is essentially no transverse dispersion (perpendicular to the flow axis). The transverse dispersivity of most formations is not sufficient to allow wide spacing of injection points.

Although the progressive migration of injected fluids can be characterized as shown in Figure 5.18, aquifer heterogeneities and the limited fraction of the pore volume that comprise the migratory flow paths limit our predictive capacities regarding injected fluids. To overcome these limitations, tracer studies and pilot-scale injections can be undertaken to gauge the velocities and paths of fluid injections and the fraction

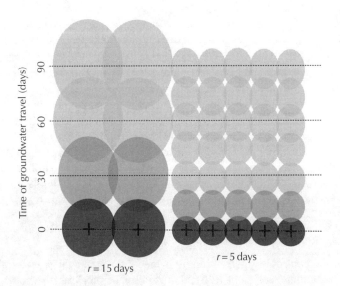

FIGURE 5.18 Superposition of injection events to form a coherent reactive zone. Injection volumes on the left are sufficient to cover a 30-day diameter and are repeated at a 30-day interval. Injection volumes on the right are sufficient to cover a 10-day radius, and must be repeated every 10 days to maintain steady reagent concentrations. Reactive zone tracer tests show very little longitudinal or transverse dispersion and overlap of injection fields is required to avoid gaps in reagent coverage.

5.3.1.2
of aquifer pore volume that participates in groundwater flux. Example 5.1 describes the distribution of a high-concentration bromide tracer after injection into a permeable formation showing how site-specific testing can be used to support the reactive zone design process.

5.3.1.1 Injection Radius

Fluids injected into an aquifer matrix through a vertical injection well travel outward radially if the formation is homogeneous over the aquifer volume, in the vicinity of the injection point, across the depth interval of the injection. Differences between horizontal and vertical hydraulic conductivity determine the relative proportions of horizontal and vertical injected fluid movement. In many aquifers, the vertical hydraulic conductivity is significantly lower than the horizontal hydraulic conductivity, effectively confining injected fluids to the injection screen interval.

Injected fluids travel principally along the migratory flow paths, a concept that was introduced in Section 5.2.1 and Figure 5.2. The migratory fraction of the pore space determines how far fluids travel during an injection event. Equation (5.26) provides a calculation of r_{inj}, the radius occupied by the injected fluid immediately after the injection process is completed.

$$r_{inj} = \sqrt{\frac{V_{inj}}{\pi \cdot h \cdot f_m \cdot N}} \quad (m) \qquad (5.26)$$

Building Reactive Zone Strategies

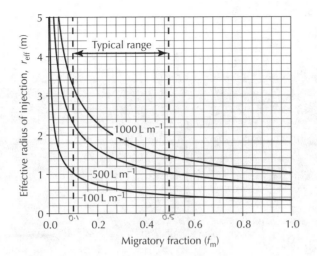

FIGURE 5.19 Effective radius that can be achieved at 0.1, 0.5, and 1.0 m^3 (100, 500, and 1000 L) of injected fluid per meter of injection zone thickness, as a function of the migratory pore fraction. The effective porosity was assumed to be 0.3 for these calculations. A migratory pore fraction of 1.0 would represent plug flow conditions. Values for f_m ranging from 0.1 to 0.5 are most likely.

where h is the injection zone thickness (normally the length of the injection well screen), f_m is the migratory pore fraction, and N is the total drainable pore volume. V_{inj} is the injected volume in cubic meters. Equation (5.26) is only valid when the rate of fluid injection temporarily overwhelms the natural groundwater movement through the injection zone.

If the aquifer matrix migratory fraction, f_m, is 1.0, the injected fluid travels through the entire drainable fraction, a situation we refer to as *plug flow*. As the migratory fraction decreases, the radius covered by an injection event increases. In an aquifer comparable to that described by Julian et al.,[3] one eighth of the pore space is migratory and $f_m = 0.125$. When a 0.1-m^3 (100 L) fluid volume is injected over a 1-m screen length, through a formation in which $f_m = 0.125$, the injected volume will span an effective radius (r_{inj}) of 29 m. Figure 5.19 shows the effective radius obtained by 0.1, 0.5, and 1.0 m^3 (100, 500, and 1000 L) injection volumes, applied over a 1-m-thick injection interval, in a formation with N, the total drainable porosity, equaling 0.3.

5.3.1.2 Injection Zone Tracer Test Example

A tracer test was recently performed using a 1155 L (300 gal) batch of 2000 mg L^{-1} bromide, injected into a 3-m (10-ft) well screen. The first monitoring point was located 3 m downgradient from the injection well, and groundwater samples were collected at 2- to 3-day intervals to observe the passage of bromide. Figure 5.20 shows the time course of bromide concentrations at the monitoring point. A log-normal curve was superimposed on the data points, to represent a possible bromide distribution over time. Tracer concentration curves are expected to be skewed to later times,

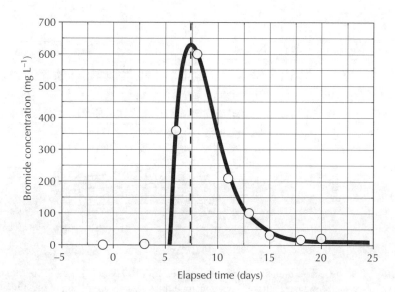

FIGURE 5.20 Interpretation of tracer study data at a monitoring location 3 m downgradient from a 2000 mg L^{-1} bromide injection. Open circles represent observed bromide concentrations, and the solid line indicates a log-normal distribution (hand-fitted) that approximates the expected pattern of leading edge and trailing edge behaviors. The dashed line indicates the estimated arrival time for the center of mass. The injected volume was 385 L m^{-1} of injection zone thickness.

due to the storage of bromide in static water, and subsequent recontamination of migratory water through diffusion from the static water mass. The persistence of miscible contaminant sources is discussed in Section 5.2.4.

Figure 5.20 shows that the center of mass arrived at the monitoring well approximately 7.4 days after injection, traveling at 0.4 m day^{-1}. The leading edge arrived at approximately 5.2 days, from which we can estimate the initial radius of the injected fluid: during 5.2 days, the leading edge traveled 5.2 days × 0.4 m day^{-1} = 2.08 m. Subtracting that distance from the distance between the monitoring point and the injection point, $r_{inj} \approx 0.9$ m. The migratory pore fraction can now be estimated from Equation (5.26). Rearranging,

$$f_m = \frac{V_{inj}}{\pi \cdot h \cdot r^2 \cdot N} \qquad (5.27)$$

The injection volume, V_{inj}, was 0.385 m^3 for each 1 m of injection zone thickness, h. Assuming the total drainable porosity, N, was 0.3,

$$f_m = \frac{0.385}{\pi \cdot 1 \cdot (0.9)^2 \cdot 0.3} = 0.50$$

If we wish to design an injection zone at this site that achieves overlap of injected fluid coverage with wells spaced at a 10-m interval, r_{inj} must be at least 5 m.

Equation 5.26 can be rearranged to estimate the injection volume that will be required to achieve overlap:

$$V_{inj} = r^2 \cdot \pi \cdot h \cdot f_m \cdot N = 11.7 \text{ m}^3 = 11,700 \text{ L}$$

The minimum injection rate is therefore 11,700 L (3116 gal) for every meter of injection zone thickness, when f_m is 0.5 and the desired injection well spacing is 10 m. This can also be estimated from Figure 5.19, for injection volumes up to 1000 L per meter of injection zone thickness.

In another recent tracer test, we pushed a 7700 L (2000 gal) batch of 2000 mg L^{-1} bromide solution into a 1.5-m-long (5-ft) well screen, placed at the top of a semi-confined aquifer. Monitoring wells were placed immediately underneath the injection well to observe any vertical migration and at several radial distances, to determine the initial distribution of the injection and its subsequent migration through the formation. The initial distribution showed a 35 to 50% dilution of injected fluid, over a radius that ranged from 6 to 8 ft from the injection point and bromide concentrations dropped very sharply beyond that distance. Less than 10% of the injected fluid migrated into the underlying 1.5-m interval, although the formation appears unstratified according to well installation logs (consistent with a vertical hydraulic conductivity, K_v, that was approximately $0.1 \times K_h$, the horizontal hydraulic conductivity). In this injection test, dilution of the injected fluid (50%) approximately matched the fluid distribution, which covered double the volume that would have been observed if f_m had been 1.0 (i.e., the observed distribution was twofold greater than plug flow). The observations were consistent with a 50% migratory pore water fraction in the aquifer matrix and dilution of the injected fluid through diffusive equilibration with the static groundwater. Bromide tracking that followed the initial injection showed very little longitudinal or transverse dispersion, in a site that is moderately permeable (5×10^{-4} cm s^{-1}) with a small hydraulic gradient, yielding a slow groundwater flow ($v_x = 0.2$ ft day^{-1}).

In both of these injection tracer tests, the injection process did not push fluids very far from the point of injection and there was approximate agreement between the radius of injection and the dilution of tracer by static water. Continued observations at both sites also showed that transverse dispersion of injection fluids was very limited. This result matched the plume migration pattern observed by Rivett et al.,[2] in which transverse dispersion was small. To achieve complete perfusion of an aquifer cross-section at these sites (and probably many others), overlapping injections will be required to create a coherent perfusion front.

5.3.1.3 Injection Volume Limitations

One of the challenges for reactive zone technologies is to avoid substantial displacement of contaminated groundwater. Ideally, reagents could be injected without displacing any groundwater, but this ideal obviously is not achievable. The impact of groundwater displacement varies among the contaminant classes and aquifer sorption characteristics — contaminants with high sorption potential in aquifers with high geosorbent capacity will not be susceptible to displacement, while low-sorbing contaminant classes in low-geosorbent aquifers can be displaced with relative ease. Injection of fluids into an aquifer necessarily displaces contaminated groundwater, pushing some of the contaminant mass beyond the injection zone. This becomes a problem if a significant contaminant mass is pushed beyond the reach of the treatment process, potentially expanding boundaries of the contaminated formation.

The potential for contaminant displacement varies among the contaminant classes and aquifer matrix characteristics. Near the source area of a hydrophobic organic, only a small fraction of the contamination lies in the migratory fraction of the aquifer pore space. Contaminant displacement under these circumstances will be limited to a very small percentage of the contaminant mass. Conversely, in an aquifer matrix in which a very high percentage of the pore space is migratory and the contaminant source is predominantly aqueous phase (e.g., miscible organics, salts, and metals), injected fluids can displace a significant portion of the contaminant mass. Each reactive zone designer must set limitations on injected volumes that fit the site-specific contaminant–aquifer system.

5.3.1.4 Balancing Injection Volumes and Frequencies

In reactive zone treatments, water is a carrier for reagents that drive the desired reactions, and it is important to separate consideration of the chemical or biological reagent dose to be delivered from the water that must be injected to distribute that dose over the desired aquifer matrix volume. The water injected as a distribution agent and the reagent mass that provides the reactive mechanisms are, ideally, viewed as independent elements of a reactive zone design.

For biological reactive zone treatments, creation of a coherent injection zone requires overlap of injections from adjoining points, but large injection volumes may displace too much contaminated groundwater. This is especially troublesome in low-sorption settings, where a greater portion of the contaminant mass resides in aqueous phase. Decreasing the radius of injection, say from 15 to 5 days, dramatically decreases the potential for contaminant displacement, but forces much smaller well spacing to achieve injection overlap with the lesser volume, and more frequent injections to sustain desired injection zone concentrations. Injection automation can overcome the costs associated with frequent injections, but the injected volume remains a significant fraction of the aquifer flux (as much as 5 to 20%, in some cases). In most cases of hydrophobic organic contamination, the migratory fraction of the aquifer matrix contains only a small fraction of the contaminant mass, and displacement does not cause significant contaminant displacement. However, the reactive zone designer must always consider the issue of displacement and work to minimize or eliminate its impacts.

For chemical reactive zone treatments, injection zone strategies are just as important. The basic model of reagent infusion into an aquifer follows the bromide tracer study (Example 5.1) quite closely. Reagents are injected into the formation at a concentration that exceeds the desired solution strength, and dilution by diffusion from the migratory fraction of the aquifer matrix (which transmits the injected fluids) to the static fraction of the matrix pore water brings the reagent concentration into the desired working range. Pre-design tracer testing or pilot-scale reagent injections can be used to gauge the migratory pore water fraction and the postinjection dilution that will occur.

5.3.2 THE REACTIVE ZONE

The reactive zone is that portion of the overall *in situ* remedial complex in which contaminant destruction or precipitation occurs, and is the namesake of the remedial strategy around which this book is organized. The reactive zone may extend over a long distance, as is the case for enhanced reductive dechlorination systems like those shown in Figure 5.17 (panel A), or they may be very compact, extending over only a few days of groundwater travel, as is commonly the case for chemical reactive zones (Figure 5.17, panel B). The reactive zone is analogous to aboveground reactors that can be constructed to achieve chemical or biological destruction of contaminants, with the potential to encompass very large matrix and groundwater volumes within the reactive system. In the sections that follow, we describe biological reactive zones, using the analogy to bioreactors, then describe chemical reactive zone strategies that can be used in oxidation and precipitation treatments.

5.3.2.1 The *in Situ* Bioreactor Concept

Saturated porous media provide an ideal setup for development of microbial communities that can be managed to perform metabolic or cometabolic destruction of groundwater contaminants. These flow-through systems are analogous to bioreactors used in aboveground wastewater treatment. The most notable difference between aboveground and *in situ* bioreactors is the extended contact time that can be achieved in the latter. Reactive zones that extend for 100 days or more are feasible in anaerobic systems (aerobic reactive zones are limited to shorter zones, due to the rapid consumption of oxygen and excess biomass buildup).

To support an *in situ* anaerobic reactive zone, a mixture of electron donor or acceptor compounds, buffer, and nutrients are injected into the groundwater flowing through the porous medium. The injection mixture from adjoining injection wells must overlap sufficiently to provide for formation of a coherent mixture of influent groundwater and additives that are intended to drive microbial community metabolism in the reactive zone. The volume injected per well, the frequency of injection and the mass of reactant injected at each event are the key operational variables for *in situ* bioreactor development.

If the reactive zone objective is enhanced reductive dechlorination, the injection mixture comprises electron donor compounds. Anaerobic or anoxic oxidation reactors can be developed through injection of alternative electron acceptors such as sulfates or nitrates. Aerobic bioreactors can be established through injection of oxygen or oxygen-releasing compounds such as hydrogen peroxide. Aerobic reactive zones can develop higher-density microbial populations, due to the relatively high-energy yield of aerobic metabolic reactions. This causes difficulty in maintenance of suitable groundwater flow through the reactive zone.

The aquifer solid matrix provides a habitat for microbial populations that process injected reagents through a network of metabolic processes that culminate in destruction or precipitation of targeted contaminants. In some biological reactive zones, the reactions that achieve contaminant destruction do not occur until the downgradient portion of the reactive zone. In these cases, microbial communities in the upgradient regions of the reactive zone lay the groundwork for the contaminant-destroying species, consuming competing electron acceptors such as oxygen or nitrate, or producing secondary electron donors such as hydrogen that are required for dechlorinating bacteria.

The reactive zone extends from the interface between the injection zone and influent groundwater to the point at which injected reagents and their functional by-products are exhausted. At the upgradient interface between injected reagent and influent groundwater, the first microbial communities of the engineered succession take hold. The injected nutrients and electron donors or acceptors, and the reaction products from early stages in the microbial succession, support growth of the later-stage microbial communities along the flow path.

If the reactive zone is designed for enhanced reductive dechlorination, the upgradient interface may be aerobic, and the injected electron donor supports aerobic microbial communities that consume oxygen and nitrates from the influent water. These early-succession communities exhaust the high-energy electron acceptors, opening the downgradient aquifer matrix habitat to bacteria species that rely on less-energy-yielding electron acceptors such as sulfates and chlorinated solvents. The later-succession species cannot compete with oxygen- and nitrate-reducing species in the presence of those electron acceptors.

In situ bioreactor designs are based on groundwater flux into the reactor, the electron acceptor flux carried into the reactor by influent groundwater, and the rate of reaction of injected reagents that will develop as the microbial succession matures. The rate of reagent consumption in the reactive zone must be balanced against the contact time in the reactor required to achieve contaminant reduction objectives. Our experience in reactive zones designed for enhanced reductive dechlorination, for example, indicates that a reactor that is operated to maintain organic carbon availability over a 100-day contact time is sufficient to fully dechlorinate 10 μM influent chlorinated alkenes, delivering 10 μM ethane past monitoring wells located approximately 100 days downgradient from the injection zone. In that and comparable systems, injected carbohydrates and their intermediate metabolites have an approximate 20-day half-life, after the bioreactor system matures. This can provide a basis for calculating carbohydrate dosing in the injection zone that will yield an effective anaerobic bioreactor over the 100-day span along the flow path.

Figure 5.21 recaps the development of an *in situ* bioreactor that occurred at an enhanced reductive dechlorination reactive barrier site in the U.S. midwest. The reactive barrier is located downgradient from a high-concentration source zone. The influent chlorinated alkenes, represented in the figure by 10 μM perchloroethene, were observed at relatively stable levels over an extended pretreatment period. Carbohydrate injections were initiated at time 0 on the graphic, at a loading rate that generated carbon breakthrough at 50 mg L^{-1} dissolved organic carbon, 100 days downgradient from the injection zone.

The concentration of dechlorinated products climbed dramatically, causing a several-fold increase in total alkenes observed in aqueous phase. The surge in total

Building Reactive Zone Strategies

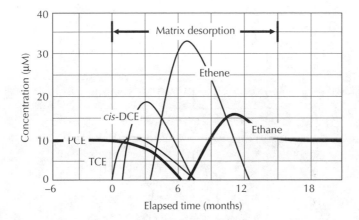

FIGURE 5.21 Development of an *in situ* bioreactor in an enhanced reductive dechlorination reactive zone. A steady 10 μM PCE influent feeds the bioreactor zone, and after 15 months, the *in situ* bioreactor is fully dechlorinating and reducing all ethenes to ethane, which is conservative under reducing conditions. Desorption of chlorinated alkenes from the aquifer matrix causes a significant, temporary increase in the total alkenes in solution that persists until all sorbed-phase mass has been removed from the reactive zone.

alkenes was caused by the desorption and subsequent dechlorination of perchloroethene and trichloroethene from the aquifer matrix. After all the chlorinated alkene mass was desorbed and dechlorinated, the reactive zone began to operate as an *in situ* bioreactor, with a 10 μM chlorinated alkene influent and a 10 μM ethane effluent. The desorption pattern observed here, which results in significant increases in total alkenes in solution, is discussed in Section 5.6.1, which develops a taxonomy of reactive zone concentration patterns.

5.3.2.2 Bioreactor Acclimation — Microbial Population Buildup Phase

Aquifer microbial population densities are generally low prior to startup of an *in situ* bioreactor treatment. The initial electron donor loadings pass through the system with very little consumption, and can be used as a tracer for groundwater flow estimation. If the loading rate is maintained at a steady level, bacterial population densities will expand to utilize all the injected material, and electron donor concentrations measured at a downgradient location fall quickly to levels at, or below, pretreatment levels. If the electron donor loading rate is subsequently increased, another surge of donor will pass through the system, because bacterial population densities are not sufficient to consume all the injected mass. A tracer-like spike of electron donor concentrations accompanies each stepwise increase in donor loading.

Figure 5.22 shows simultaneous dissolved organic carbon concentration profiles that will be observed in the injection zone and at the downgradient extent of a reactive zone. Stepwise carbohydrate loading increases are conducted, each delivering a short-lived spike in dissolved organic carbon arriving at the downgradient location. When

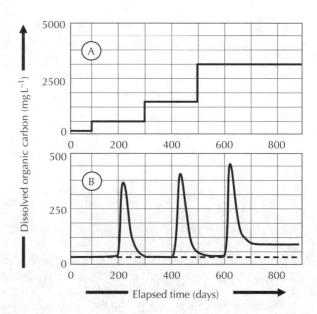

FIGURE 5.22 Dissolved organic carbon breakthrough patterns in reductive dechlorination zones. Graph A shows injection zone carbon concentrations in the migrating groundwater, with stepwise increases in concentration at 200-day intervals, beginning at 100 days elapsed time. Graph B shows dissolved organic carbon concentration patterns that develop at a monitoring location 100 days downgradient from the injection zone. Each injection zone concentration increase generates a tracer-like spike in downgradient concentrations that persists until microbial populations expand to meet the added carbon loading. When additional carbon fails to generate higher microbial population densities, carbon loading breakthrough occurs, as observed after 700 days elapsed time in this conceptual diagram. The dashed line in the lower figure depicts the background aquifer dissolved organic carbon concentration.

the aquifer matrix microbial population reaches carrying capacity (relative to carbon supply), dissolved organic carbon achieves a steady breakthrough from the reactor.

For any position in the reactive zone continuum downgradient from an injection zone, there is a loading rate that induces maximum bacterial population densities. When this density is reached, the surge in donor concentration observed following an increased loading concentration does not dissipate. This is the breakthrough loading rate — bacterial populations are at electron donor saturation. An added donor does not generate any further increases in microbial population levels. Attainment of breakthrough at a specified time of travel from the injection zone serves as a benchmark of reactive zone performance. For example, reactive zones that are designed to achieve electron donor breakthrough 100 days downgradient from the injection zone tend to exhibit rapid, complete dechlorination.

5.3.2.3 Chemical Reactive Zones

The reagents injected to support most chemical reactive zones react very quickly, relative to the pace of carbon source consumption in biological reactive zones. Rapid consumption of injected reagents limits the radial coverage that can be accomplished from a single chemical reactive zone injection point, as shown in Figure 5.17, panel B. With reagent half-lives of minutes (ozone and many hydroxyl-radical-producing formulations) to hours (permanganate and chelated Fenton's reagent systems), the entrainment of reagents in groundwater provides little, if any, benefit in extending the reactive zone along the groundwater flow path. As a result of these constraints, chemical reactive zone injections normally blanket the area to be treated at a high-density spacing. Chemical reactive zone treatments are often deployed in source zones to achieve contaminant source mass reduction, enabling monitored natural attenuation as a remedial strategy for downgradient portions of the plume.

5.3.3 The Desorption Zone

Reactive zones are often designed to cover only a portion of a contaminant plume. In these cases, clean (contaminant-free) groundwater departing the reactive zone is recontaminated through equilibration with nonmigratory contaminant mass in the aquifer matrix. As was shown earlier in Section 5.2.6, the clean water front propagates more slowly than the groundwater velocity. The rate of its propagation is determined by the mass and physical dimensions of the nonmigratory contaminant pools, as well as the strength of binding between contaminants and the sorbing matrix. The desorption zone is defined as the segment of the aquifer matrix, which is rapidly equilibrating with migrating clean water, and its location moves in a downgradient direction as the clean water front progresses.

Predicting the rate of desorption zone migration is one of the more challenging tasks in remedial system design. The prediction is founded on estimates of contaminant sorption capacities in the aquifer matrix (e.g., soil organic matter for hydrophobic organics, and cation exchange capacity for transition metals), along with estimates of the migratory pore water fraction. The desorption zone will migrate most slowly at sites with high sorption capacities and those with low migratory pore water fractions. An ideal site would have a highly conductive, homogeneous porous medium, with a very high migratory pore water fraction and very low sorption capacities. If such sites existed, groundwater pump and treat strategies might be more successful, and there would be less interest in reactive zone technologies.

5.3.4 The Recovery Zone

The functional region of an anaerobic reactive zone extends to the point at which productive reactions are exhausted. At that point, dissolved reduced inorganic chemical species generally remain and react with aquifer minerals and recharging electron acceptors as they migrate in along the flow path. Reduced iron and manganese ions (Fe^{2+} and Mn^{2+}) are two of the major inorganic species exporting reactive reducing equivalents from the reactive zone into the redox recovery zone. Reducing equivalents are also carried by dissolved organic gases such as methane and ethane, both products of anaerobic metabolism in the reactive zone.

The length of the redox recovery zone is determined by several factors, including the groundwater flow velocity, the concentration of reduced inorganic species, the aquifer mineral composition, and the rate of electron acceptor recharge and mixing.

The primary mechanism of redox recovery is electron acceptor (oxidant) recharge, causing a steady decline in the concentration of reduced species along the flow path downgradient from an anaerobic reactive zone. The redox recovery zone extends to the point at which reduced inorganic molecules entrained in the groundwater flux have been exhausted by reaction with electron acceptors.

When an anaerobic reactive zone is first established, minerals in the aquifer matrix may react with reduced inorganic compounds as they depart the reactive zone. This matrix oxidation capacity initially limits the propagation of the reduced groundwater front. The matrix capacity is consumed over time and the redox recovery zone eventually extends to the point at which the cumulative oxidation capacity of recharge along the flow path matches the rate of reducing equivalent discharge from the reactive zone. If the aquifer is closed to recharge, the low-redox condition can be propagated over long distances, as was observed by Champ et al.,[18] in a natural, closed aquifer system.

5.4 REACTIVE ZONE STRATEGIES

Reactive zones can be deployed in several configurations, and the selection of a reactive zone strategy is based on the contaminant distribution, groundwater flow characteristics and long-term project objectives. In many sites, reactive zones can be deployed to achieve site-wide compliance, while in others, a reactive barrier containment strategy may be the most that can be achieved. In all cases, the success of a reactive zone strategy depends on an availability of comprehensive hydrogeology and contaminant distribution analyses and a realistic appraisal of the potential for reactive zone technologies to achieve project treatment objectives.

5.4.1 REACTIVE BARRIERS

A reactive barrier employs one of the reactive zone mechanisms in a system that blocks migration of contaminants in the groundwater. Its containment function may be similar to that of a groundwater pumping system, with cost-effectiveness as the main motivation for deployment. Figure 5.23 shows a cross-sectional view of a reactive barrier that is intercepting groundwater contamination in a zone spanning a 50-ft-long interval of the groundwater flow path. Contaminant concentrations are shown downgradient from the reactive barrier, at several time intervals after the reactive barrier became fully operational. This analysis shows how a reactive barrier can achieve containment and how the propagation of a clean water front can clear downgradient areas of contamination.

Several IRZ technologies can be deployed in a reactive barrier application. Zero-valent iron is used primarily in reactive barrier mode, although it has also been tested in source zones, using nano-scale iron particles. Enhanced reductive dechlorination can be applied in reactive barrier mode, and a key determinant of its feasibility is the intended duration of the barrier operation and the extent of recovery zone migration that will occur. Operation of enhanced reductive dechlorination barriers can be extended over long time spans in open aquifer formations, where recharge of electron acceptors allows more rapid redox recovery. Oxidation systems using Fenton's reagent or permanganate would not typically be considered for reactive barrier

Building Reactive Zone Strategies

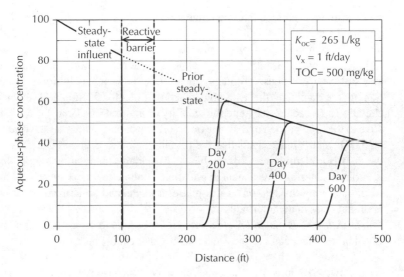

FIGURE 5.23 Propagation of a clean water front downgradient from a reactive barrier.

operation, due to cost disadvantages and for permanganate, limitations due to the buildup of precipitate residues in the aquifer matrix. Ozone can be applied in reactive barrier mode, using an ozone sparge curtain to intercept contaminants.

5.4.2 Source Treatments

The greatest challenge (and potentially their greatest advantage, relative to alternatives) for IRZs is their use in contaminant source areas. Excavation of contaminants below the aquifer surface is usually impractical, and groundwater pumping systems flush clean water through migratory pore spaces, but they remove contamination from the nonaqueous phase and static pore water so slowly that completion times often cannot be projected for these systems. IRZs have a higher potential for success in source area cleanup than other methods, but the challenge is very great.

The nature of the source distribution determines which methods might be successful in gaining contact between contaminants and injected reagents.

Hydrophobic Organics. The majority of contaminant mass in hydrophobic organic source zones usually resides in nonaqueous forms, either sorbed phase or NAPLs. If drainable nonaqueous-phase liquid or solid can be located, it should be removed by pumping or excavation, to the extent possible. The *in situ* remedial effort can then be focused on the residual nonaqueous-phase liquid and sorbed fractions. These materials cannot be accessed directly by microbes or aqueous-phase reagents. There are three approaches that have been considered for source zone treatments of hydrophobic organics, and there is a lack of published outcomes for field applications of these methods.

1. *Aqueous-phase oxidant flood.* One approach that has been suggested is to overwhelm the source zone with oxidants, relying on dissolution of the nonaqueous contaminant mass to bring the contaminant to the reaction. To be

successful, this requires that the reagent be active for a very long period, and that fluid access to the aquifer matrix be maintained over the time needed for the sorbed or nonaqueous-phase mass to dissolve. Permanganate leaves a solid residue, so it is self-limiting at high application rates. Fenton's reagent generates sufficient gas-phase products to also be somewhat self-limiting, in addition to the fact that the available reaction time for Fenton's (even using chelation approaches to extend the viability of injected iron) may be short, relative to the time needed to dissolve or desorb the contaminant mass. An advantage of Fenton's reactions is that their rapid release of heat and gases can stir the matrix, maximizing contact between oxidants and target compounds. A series of intermittent reagent applications may be the most successful approach, allowing dissipation of gas-phase products and solubilization of contaminant mass in the interludes.

2. *Phase-transfer agents and surfactants.* It is possible to break down the interfacial barrier between groundwater and NAPLs, using phase transfer catalysts (see Section 3.4.2) or surfactants. This could improve oxidant access to source material, but solid-phase residue may still cause reagents such as permanganate to be self-limiting. The contaminant-releasing action of surfactants makes it imperative to have a control mechanism in place, downgradient from the source treatment zone, to prevent accelerated migration of contaminants.

3. *Biological reactive zones and partitioning electron donors.* The carbohydrate loadings associated with high-rate reductive dechlorination zones generate biosurfactants, cosolvents, and partitioning carbohydrate compounds that break down the immiscible fluid interface and sorbing matrices, dramatically increasing the exposure of hydrophobic organic compounds to microbial degradation reactions. Extraction of contaminants from the sorbed-phase mass and NAPL generates a characteristic pattern of stepwise increases in alkene molarity as the dechlorination reactions proceed. This pattern is caused by the decreasing organic carbon partition coefficients associated with successive chlorine atom removals. In source areas with NAPLs, monitoring well concentrations for hydrophobic organics often increase to solubility limits, likely due to the cosolvent action of carbohydrate metabolites freeing residual NAPLs from the source zone matrix. These observations are discussed in Section 5.6.1.

Transition Metals. Metal contamination source zones can contain very large masses of metals, especially when undiluted plating solutions remain at the spill point. Plating solutions may contain more than 350 g L^{-1} of dissolved metal (e.g., Ni^{2+}), with pH and redox buffers that resist alteration of the fluid chemistry. If the source zone received multiple aliquots of plating solution, precipitated metal may also be present in the aquifer matrix. There are common elements of source zone control that can precede any of the reactive zone measures.

1. *Pumping.* Because plating solutions are miscible with groundwater (although density flows are possible), pumping withdrawal does not face the

same limitations of pumping for NAPL removal, and a significant mass of contaminant can be removed by pumping from the source area. We have recovered thousands of liters of plating solution by pumping withdrawal from a spill point, removing hundreds of kilograms of dissolved metals from the aquifer matrix.
2. *Neutralization.* Plating solutions are buffered to maintain maximum solubility of metals, and neutralization of the source zone matrix, through injection of mild base, for example, may be necessary to pave the way for subsequent precipitation reactions.

Reactive zone strategies for metals elimination rely on formation of durable precipitates. In these approaches, metals will always be present in the aquifer matrix, but their solubilities will be low enough to sustain acceptable dissolved-phase concentrations, as long as the aquifer chemistry remains at an acceptable pH and reduction–oxidation potential. It will always be possible to resolubilize treated metals through addition of acids or oxidants (reducing agents, in the case of some metals), and the *in situ* treatment strategy must create metal precipitates that are *durable under reasonably foreseeable aquifer chemistry composition*. There are several possible reactive zone mechanisms that can be applied to metals source zones after initial source control measures, above, have been implemented.

1. *Chemical reduction.* Chemical reducing agents such a dithionite and nanoscale zero-valent iron can be injected into the source zone to precipitate redox-sensitive metals such as hexavalent chromium (Cr^{6+}).
2. *Biological reduction.* *In situ* biological reducing zones can be created in a manner identical to that applied for enhanced reductive dechlorination. Injection of carbohydrates or other labile electron donor compounds stimulates the aquifer microbial community and electron acceptor recharge is overwhelmed by microbial consumption, driving the aquifer microbial community metabolism into anaerobic reactions. This floods the aquifer matrix with reduced inorganic species, and lowers the observed oxidation–reduction potential. In particular, sulfide ion availability increases, and under these conditions, metals such as chromium, cadmium, and nickel precipitate rapidly. Section 5.4.2.1 describes the biological nickel precipitation strategy in more detail.

Miscible Organics and Soluble Salts. As with metals, miscible organics can be present as a high-concentration fluid occupying the aquifer matrix, and pumping may recover a very large contaminant mass from the source area. To destroy the remaining mass in source areas and dispersed along an aquifer plume axis, several alternatives are available. Those methods can be applied successfully in concentrated source zone follow-up, as well as reactive barrier lines. Because these compounds have low sorption coefficients, the propagation of clean water fronts in aquifer matrices with high migratory pore fractions can be quite rapid.

1. *Chemical oxidation.* Many of the miscible organics and anions of the soluble salts are susceptible to chemical oxidation treatment. Alcohols, ketones, and

ethers are generally reactive with the hydroxyl radical,[19] including biologically recalcitrant compounds such as 1,4-dioxane. Potential source zone treatments include Fenton's reagent, persulfate/iron, and ozonation (permanganate is not effective against many of the miscible organics — check reactivity before including it in consideration). For reactive barriers, the lower cost of extended operation favors ozone sparging if the aquifer formation is suitably porous and lacking confining strata that would trap injected gases. Example 5.3 shows calculations supporting development of an *in situ* ozone barrier for treatment of NDMA.
2. *Aerobic biological reactive zones.* Aerobic reactive zones can be successful, although monitored natural attenuation is often a viable strategy for aerobically degradable compounds. One notable exception is methyl-*tert*-butyl ether (also known as MTBE), for which aquifer microbial communities require dissolved oxygen at levels exceeding that which is naturally available.
3. *Anaerobic biological reactive zones.* Numerous anions (nitrate, nitrite, sulfate, perchlorate, and others) can be treated by anaerobic biological reactive zones. Reactive barriers can be especially effective for these compounds, and the required carbon loading is much less than for enhanced reductive dechlorination systems.

There are often multiple reactive zone strategies that can be used to solve a specific aquifer contamination problem. The selection of a remedial strategy is based on factors including cost to install and operate, project duration and attainable posttreatment contaminant concentrations. Comparison of the risk profiles for potential remedies normally shows a range of outcomes, with chemical oxidation approaches offering rapid cleanup times and risk of cost exceedances, and biological remedies offering lower cost profiles, but with greater risk of project duration exceedances.

5.4.2.1 Biological Reducing Zone Strategy for Nickel Precipitation

Nickel that is used in plating solutions is supplied by highly soluble salts such as nickel chloride and nickel sulfate, buffered to pH in the range of 4 to 4.5. pH adjustment alone would not reduce the soluble fraction of nickel to acceptable levels. The source zone could, theoretically, be flooded with chloride and sulfate, pushing nickel back into solid phase, but this would only be a short-term effect as chloride and sulfate would wash out of the formation, releasing nickel back into solution. A lasting remedial strategy can be designed that supplies competing anions to react with nickel, forming insoluble nickel compounds.

Nickel can react with anions, such as carbonate and sulfide, forming much less soluble compounds. To develop a reactive zone strategy, we examine candidate nickel salts for their solubility, and determine whether, at equilibrium, their aqueous-phase concentrations meet remedial objectives. The general reaction equations for the dissolution of nickel salts are

$$NiA_{(s)} \rightleftarrows Ni^{2+} + A^{2-} \tag{5.28}$$

where A is a divalent anion, and

$$Ni(B)_{2(s)} \rightleftarrows Ni^{2+} + 2(B^-) \quad (5.29)$$

where B is a monovalent anion. The corresponding solubility product equations are

$$[Ni^{2+}] = \frac{K_{sp}}{[A^{2-}]} \quad (5.30)$$

and

$$[Ni^{2+}] = \frac{K_{sp}}{[B^-]^2} \quad (5.31)$$

The solubility product constants, K_{sp}, for three nickel compounds, nickel carbonate, hydroxide, and sulfide (α, β, and γ forms), are provided in Table 5.3. The solubilities of these compounds are also affected by reactions of the anions when they come into solution. When the anion is a strong base, the solubility of nickel is enhanced through scavenging of the anion as it reaches aqueous phase. Examination of equations (5.30) and (5.31) shows that decreases in anion concentrations yield increases in the dissolved nickel concentration, as was described earlier in Section 3.3.3.3. Recall the equation for the net of nickel dissolution and sulfide acid–base reactions:

$$NiS_{(s)} + 2H^+ \rightleftarrows H_2S + Ni^{2+} \quad K_{net} = K_{sp} \times K_{b1} \times K_{b2} = 3 \times 10^5 \quad (5.32)$$

and the resulting calculation of nickel concentration as a function of pH and sulfide concentration is

$$K_{net} = 3 \times 10^5 = \frac{[H_2S] \cdot [Ni^{2+}]}{[H^+]^2} \quad (5.33)$$

TABLE 5.3
Nickel Solubility Data at 25°C

Compound	Formula	Solubility Product, K_{sp}	Temperature °C	Source
Nickel carbonate	$NiCO_3$	6.6×10^{-9}	25	1
Nickel hydroxide	$Ni(OH)_2$	2.8×10^{-16}	25	1
α-Nickel sulfide	NiS	3×10^{-21}	25	1
β-Nickel sulfide	NiS	1×10^{-26}	25	1
γ-Nickel sulfide	NiS	2×10^{-28}	25	1
Nickel sulfide	NiS	1.4×10^{-24}	18	2

Source: From (1) Kotz and Treichel[21] and (2) Weast.[20]

If nickel sulfide is the only source of sulfide to the solution, Equation (5.33) reduces to the following:

$$K_{net} = 3 \times 10^5 = \frac{[Ni^{2+}]^2}{[H^+]^2} \tag{5.34}$$

and

$$[Ni^{2+}] = \sqrt{3 \times 10^5 \cdot [H^+]^2} \tag{5.35}$$

Equation (5.33) can be used to predict the long-term nickel solubility as a function of pH, on the key assumption that the sulfide ions reaching the aqueous-phase originate from nickel sulfide (a conservative assumption) and that no consumption of sulfide ions occurs other than its redistribution in acid–base reactions (a nonconservative assumption that will be discussed below).

Nickel carbonate gains a boost in solubility due to acid–base reactions of the carbonate anion, in the same manner as does nickel sulfide. The carbonate acid–base reactions were described in Section 3.3.1 and are summarized as

$$H_2CO_3 \rightleftharpoons H^+ + HCO_3^- \quad K_{a1} = 4.5 \times 10^{-7}$$

$$HCO_3^- \rightleftharpoons H^+ + CO_3^{2-} \quad K_{a2} = 4.7 \times 10^{-11}$$

Inversing the carbonate equations and adding the nickel carbonate dissolution,

$$H^+ + HCO_3^- \rightleftharpoons H_2CO_3 \quad K_{b1} = 2.2 \times 10^6$$

$$H^+ + CO_3^{2-} \rightleftharpoons HCO_3^- \quad K_{b2} = 2.1 \times 10^{10}$$

$$NiCO_3 \rightleftharpoons Ni^{2+} + CO_3^{2-} \quad K_{sp} = 6.6 \times 10^{-9}$$

$$NiCO_3 + 2H^+ \rightleftharpoons Ni^{2+} + H_2CO_3 \quad K_{net} = K_{b1} \cdot K_{b2} \cdot K_{sp} = 3.1 \times 10^8$$

The calculation of nickel in solution, as a function of pH, could utilize the form of Equation (5.35), which would assume that the carbonate is supplied entirely by dissolution of nickel carbonate. That assumption is far too conservative, and it is more appropriate to assume that the aquifer alkalinity is the dominant source of carbonate to the aquifer. Therefore, it is appropriate to make the calculation as was done in Equation (5.33), and rely on site-specific data for the aquifer alkalinity to supply the value for H_2CO_3.

$$K_{net} = 3.1 \times 10^8 = \frac{[H_2CO_3] \cdot [Ni^{2+}]}{[H^+]^2} \tag{5.36}$$

The solubility of nickel hydroxide is described by the following reaction equation:

$$Ni(OH)_2 \rightleftharpoons Ni^{2+} + 2OH^- \quad K_{sp} = 2.8 \times 10^{-16} \tag{5.37}$$

Building Reactive Zone Strategies

Reaction Equation (5.37) is linked to aquifer pH, through the hydroxyl anion, and the concentration of nickel in solution due to nickel hydroxide can be calculated by Equation (5.39), which is developed as follows:

$$K_{sp} = 2.8 \times 10^{-16} = [Ni] \cdot [OH^-]^2 \tag{5.38}$$

and

$$[Ni] = \frac{2.8 \times 10^{-16}}{[OH^-]^2} \tag{5.39}$$

The solubilities of each of the candidate nickel compounds can be graphed as a function of pH, as shown in Figure 5.24 (nickel carbonate was graphed at three alkalinities, representing a range that might be encountered in natural aquifers). From these results, we can see that nickel hydroxide would generate very high dissolved-phase nickel concentrations, and is unlikely to control nickel solubility in the presence of carbonate or sulfide anions. Nickel carbonate and sulfide were retained for additional consideration, and a second graph was prepared, focusing on the pH range from 5 to 9, as shown in Figure 5.25. This graph illustrates some of the difficulties in determining the solubility-controlling reaction. There is great variability among values for metals solubility in the published literature. Table 5.3 shows values for solubility of nickel sulfide, obtained from Weast[20] and from Kotz and Treichel.[21] The lower solubility value, obtained at 18°C, more closely matches expected groundwater

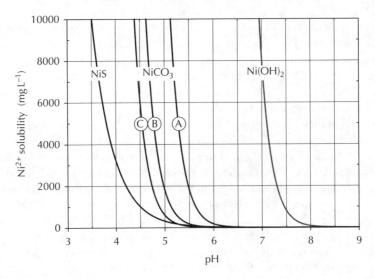

FIGURE 5.24 Predicted dissolved-phase nickel concentration in equilibrium with each of three compounds: nickel sulfide (NiS), nickel carbonate (NiCO$_3$), and nickel hydroxide (Ni(OH)$_2$). Solubility of nickel carbonate was calculated at three alkalinities: 10 mg L^{-1} (A), 100 mg L^{-1} (B), and 300 mg L^{-1} (C), representing low, mid-range, and high aquifer alkalinity values, respectively.

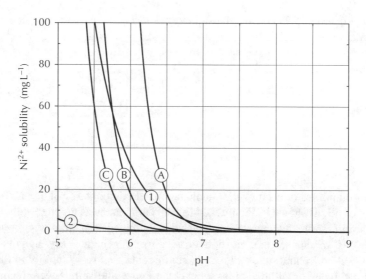

FIGURE 5.25 Predicted dissolved-phase nickel concentration in equilibrium with each of two compounds: nickel sulfide (NiS) and nickel carbonate ($NiCO_3$). Solubility of nickel carbonate was calculated at three alkalinities: 10 mg L^{-1} (A), 100 mg L^{-1} (B), and 300 mg L^{-1} (C), representing low, mid-range, and high aquifer alkalinity values, respectively. Solubility of nickel sulfide was calculated using K_{sp} values from Kotz and Treichel [21] for 25°C (curve 1) and from Weast[20] for 18°C (curve 2).

conditions, but the variation in dissolved Ni^{2+} concentration predicted by the two sources is disconcerting.

From the results of Figure 5.25, we can see that nickel can be precipitated from a high-concentration solution by reaction with carbonate anion, present in most aquifers through the calcium carbonate alkalinity system, or sulfide anions. The sulfide precipitates are likely to generate more insoluble solids at aquifer temperatures typical of the mid to high latitudes, although carbonate precipitates are also likely to form. An IRZ strategy that infuses the aquifer formation with sulfide ions and maintains neutral or high pH through carbonate buffering is likely to meet typical compliance criteria. Because the solubility data are variable, and the predicted solubility limits only marginally meet typical remedial criteria, bench or field trials would be recommended before proceeding to a large-scale IRZ treatment for nickel. The initial results are likely to reflect amorphous nickel sulfate crystal formation, α-NiS, and the maturation of amorphous crystals to form β-NiS is accompanied by a 300,000-fold decrease in solubility (calculated from the 25°C data in Table 5.3). This result suggests that it may be necessary to sustain high-sulfide, high-pH levels over an extended period, to accommodate the crystal maturation process, after which the remaining nickel sulfide precipitate would have a very low solubility.

In Section 2.3.2.3, we showed the results of field application of biological reduction, combined with carbonate buffering, that succeeding in reducing dissolved nickel from 980 to approximately 0.1 mg L^{-1} Ni^{2+}. In the course of that reduction, there were periodic rebounds of the soluble nickel during periods of reduced pH that occurred prior to carbonate buffer addition (refer to Figure 2.49). pH reductions can occur in poorly buffered aquifer matrices, when biological consumption of electron donor compounds, and the associated production of organic acids, delivers acid equivalents to the formation at a rate that exceeds the formation buffering capacity. These pH excursions are easily cured by buffer addition, as was shown in Section 2.3.2.3.

5.4.2.2 *In Situ* Oxidation of *N*-Nitrosodimethylamine

N-nitrosodimethylamine was discovered in groundwater at a site in the western U.S., in a permeable aquifer formation near 50 m below the ground surface. The groundwater flow velocity was approximately 0.3 m day^{-1}, and the aquifer was unconfined. Groundwater concentrations were approximately 100 μg L^{-1} near the site perimeter, and it was necessary to reduce its concentration to 0.02 μg L^{-1} or less to meet regulatory action levels. NDMA is miscible with water and has very low sorption[22] (K_{oc} = 12 L kg^{-1}). The remedial objective was to prevent additional off-site migration in excess of 0.02 μg L^{-1}.

In situ oxidation using ozone was considered as a candidate technology because NDMA is susceptible to oxidation by both the ozone molecule (direct ozonation) and hydroxyl radical that is likely to form in heterogeneous reactions between ozone and aquifer matrix minerals. Dissolved iron levels were relatively low at the site, and initiation of radical chain reactions through the ferrous iron/ferryl ion pathway were not expected to be a significant source of hydroxyl radical (see Section 3.4.3 for additional discussion of ozone chemistry).

The first step in the analysis was to gather second-order kinetic rate constant data for the direct ozonation and hydroxyl radical oxidation reactions. From the NIST database[19]

$$\text{NDMA} + \text{O}_3 \rightarrow \text{product} \quad K_{\text{O}_3} = 1.0 \times 10^1 \text{ M}^{-1} \cdot \text{s}^{-1}$$

and

$$\text{NDMA} + \text{OH}^\bullet \rightarrow \text{product} \quad K_{\text{OH}^\bullet} = 3.3 \times 10^8 \text{ M}^{-1} \cdot \text{s}^{-1}$$

We can assume the ozone generator will produce a 3% ozone atmosphere (30,000 ppmv, from a dry air feed), which yields an aqueous-phase ozone solution of approximately 4.5×10^{-4} M O$_3$ (refer to Figure 3.21(A) and (B) for ozone solubilities). The range of feasible molarities for ozone-driven production of hydroxyl radical is between 1×10^{-13} and 1×10^{-11} M OH$^\bullet$. To calculate reaction rates, we will use the lower range of the estimate for the first calculation of reaction times required to achieve remedial objectives.

The reactive zone approach for *in situ* ozonation is to inject a 3 to 6% ozone-in-air mixture, through an aquifer sparge injection well. Ozone will cross the air–water interface more quickly than oxygen, since its solubility is significantly higher, so ozone is not expected to break out from the aquifer surface when it is injected in air at the base of a 3-m thick aquifer formation. To achieve treatment of NDMA in the aquifer passing through the reactive zone, the contact time in passage through the zone needs to exceed the minimum reaction time for the required concentration reduction. The required reaction time can be estimated by assuming that the ozone and hydroxyl radical concentrations can be maintained at constant levels in the aquifer, allowing calculation of pseudo-first-order reaction rate constants.

For the ozonation of NDMA,

$$k'_{O_3} = k_{O_3} \cdot [O_3] = 1 \times 10^1 \cdot 4.5 \times 10^{-4} \quad (M^{-1} s^{-1} \cdot M)$$

$$k'_{O_3} = 4.5 \times 10^{-3} \quad (s^{-1})$$

and for hydroxyl radical decomposition of NDMA,

$$k'_{OH^{\cdot} \, min} = k_{OH} \cdot [OH^{\cdot}] = 3.3 \times 10^8 \cdot 1 \times 10^{-13} \quad (M^{-1} s^{-1} \cdot M)$$

$$k'_{OH^{\cdot} \, min} = 3.3 \times 10^{-5} \quad (s^{-1})$$

If the rate of hydroxyl radical initiation is at its maximum potential, the molarity of hydroxyl radical would be 100-fold greater, and

$$k'_{OH^{\cdot} \, max} = 3.3 \times 10^{-3} \quad (s^{-1})$$

The composite oxidation rate is the sum of the ozonation and hydroxyl radical rates and ranges as follows:

$$4.5 \times 10^{-3} \leq k'_{comp} \leq 7.8 \times 10^{-3} \quad (s^{-1})$$

We can calculate the time required to achieve the necessary reduction in NDMA concentrations, using pseudo-first-order kinetics, as follows:

$$C = C_0 \cdot e^{-k'_{comp} \cdot t}$$

and

$$\frac{C}{C_0} = e^{-k'_{comp} \cdot t}$$

$$\ln \frac{C}{C_0} = e^{-k'_{comp} \cdot t}$$

The concentration reduction factor, C/C_0, is 1/5000 (from 100 to 0.02 μg L^{-1}), or 2×10^{-4}. Substituting for C/C_0 and k'_{comp}, and solving for t, for the upper and lower reaction times,

$$\frac{\ln(2 \times 10^{-4})}{-4.5 \times 10^{-3}} \leq t \leq \frac{\ln(2 \times 10^{-4})}{-7.8 \times 10^{-3}} \quad (s)$$

and

$$1890 \leq t \leq 189 \quad (s)$$

Building Reactive Zone Strategies

The contact time required to achieve the 5000-fold reduction of NDMA is well within the range of times that can be achieved in a sparge injection process. The sparge air displacement of the aquifer covers a radius similar to the depth of injection, at least up to an injection depth of 8 to 10 m. The groundwater travel time through the sparge zone is slowed by the reduction in hydraulic conductivity that occurs due to the reduction in water saturation. If the injection point is located 5 m below the aquifer surface, the nominal groundwater traverse time across the sparge field is approximately 33 days (10 m radius, at 0.3 m day^{-1}). It is likely that the sparge system could be operated intermittently and achieve satisfactory contaminant reduction.

5.4.3 Reactive Zone Sequencing and Whole-Plume Treatments

It is often impractical to directly apply IRZ treatments across the entire footprint of large-scale contaminant plumes. In sites with sufficient groundwater flow velocities, it is possible to organize a series of reactive zones that combine source treatment with a number of reactive barriers to achieve complete treatment in a reasonable span of time. Figure 5.26 shows an extended plume in plan view, with two reactive barriers (labeled "1") and a source zone treatment (labeled "2") installed. This illustrates how a reactive barrier sequence can be established to take advantage of the propagation of a clean water front that can be accomplished in flowing aquifers with limited sorption.

The reactive barriers are the first elements of the sequence to be installed, to protect against uncontrolled contaminant release from the source zone. Figure 5.27 shows the progression of installations and contaminant reductions from an initial steady-state distribution in the upper panel. The reactive barriers begin generating clean water in the second panel, and when that occurs, the source area treatment can be started. In the lower panel, the source reduction is in progress, while clean water fronts are clearing the aquifer contamination from between the reactive zones and downgradient from the second reactive zone toward the plume margin.

Development of a sequencing treatment can significantly lower the cost of large-scale plume treatment, by limiting the aquifer volume over which chemical or biological

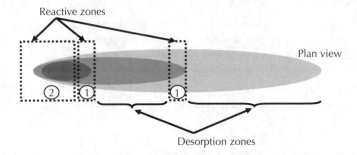

FIGURE 5.26 Reactive zone strategy combining reactive barriers (1) and source zone treatment (2). The desorption zone between the reactive barriers delivers contamination to the downgradient barrier where it is treated. The downgradient barrier is placed in a location where monitored natural attenuation is sufficient to treat the remaining contamination that desorbs and migrates toward the distal end of the plume.

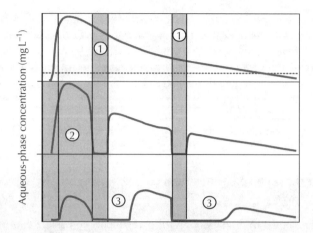

FIGURE 5.27 Contaminant treatment and desorption patterns in a reactive zone barrier sequence and source treatment strategy. Reactive barriers are established first (1), and source treatment (2) is implemented after the reactive barrier is in place to protect against contaminant breakout from the source zone. Desorption (3) clears the contaminated zones downgradient from the reactive barriers.

reaction systems must be sustained. The feasibility of the sequencing strategy is determined primarily by the speed of clean water propagation through the aquifer matrix. Sites with low contaminant sorption and a high migratory pore water fraction make the best candidates for sequencing treatment. Sites with high contaminant sorption rates or low migratory pore water fractions may require very long time frames for clean water propagation. This can be overcome by closer spacing of reactive barriers, and the limiting case for reactive barrier spacing is reached when the downgradient edge of one reactive barrier interlaces with the upgradient edge of the subsequent barrier on the flow path.

5.5 LIMITATIONS OF AQUEOUS-PHASE REACTIONS IN CONTAMINATED POROUS MEDIA

In situ reactive zone technologies provide many attractive advantages relative to alternative approaches to groundwater contamination such as containment through groundwater pumping or monitored natural attenuation. Like any technology, however, there are limitations to the IRZ technologies, including those that derive from the complexities and heterogeneities of the typical aquifer matrix. In the following sections, we examine the clear limitations and a few of the mechanisms that have been considered to overcome these limitations.

5.5.1 ACHIEVING CONTACT BETWEEN REACTANTS AND TARGETS

One of the very obvious and most difficult issues in reactive zone methods is achieving contact between reactants and target contaminants. Injected reagents tend to follow the more permeable, migratory pathways through the aquifer matrix. For reactive barriers that are designed to prevent contaminant migration, the limitation of

Building Reactive Zone Strategies

reagent access to migratory flow pathways is not a problem. However, for any reactive zone system that is intended to achieve complete contaminant removal, especially in source zones, reagent penetration into nonmigratory spaces is critical.

The contaminant mass that has to be treated is likely to be sequestered in sorbed phase, nonaqueous phase, or static groundwater. The contaminant mass distribution is usually the result of a long period of aquifer matrix exposure to high contaminant concentrations and, during that time, NAPL migration and diffusion allow contaminants to migrate into portions of the formation that cannot be reached by injected reagents in the time allowed for remedial activities.

An additional factor that prevents contact between reactants and targets is the self-limiting nature of some reactive zone technologies. Gas-producing reactions, such as Fenton's reagent oxidation (oxygen and carbon dioxide) and permanganate oxidation (carbon dioxide, especially at very high permanganate concentrations), exacerbate the contact issue due to the permeability reductions they generate. Precipitates that form in reactions with permanganate, persulfate, and dithionite can also reduce hydraulic conductivity. These and other effects on aquifer permeability are discussed in Section 5.6.5.

There are no easy solutions to the reagent–contaminant contact issue. Among the actions that can improve contact between reagents and target compounds are

Reduced injection spacing. Decreasing the distance between injection points improves the overlap of injected fluids and allows a reduction in the reagent mass that must be delivered in each point.

Reduce reagent strength, repeat injections. Injection of high-strength reagent sometimes generates aquifer matrix blockage, diverting injected fluids away from their intended path. Reducing injected reagent solution strength can limit the permeability reductions associated with gas production or precipitate formation. In peroxide injections, for example, injection backpressures increase dramatically after a few minutes of high-strength solution injection. This is caused by gas-locking of the formation. Reducing the peroxide solution strength slows the gas formation, and large volumes can be injected, covering a wider radius, without backpressure buildup. Although the injections need to be repeated, the radial penetration of reaction fluid into the formation is much greater.

High pressure, low volume injections. For some reagent systems, it is possible to drive reagents into the formation with specialized equipment that generates formation-penetrating pressures on very low flow volumes. This type of device is pushed vertically through a contaminated formation during the injection process and can infuse fluids into less permeable strata over limited radial distances. Injection point spacing is very close with these systems, and the low flow rates require extended periods of operation and sometimes repeat injections to deliver the necessary reagent volumes. For elimination of source mass, the concentrated effort may be worthwhile.

Achieving contact between reagents and targets is a fundamental requirement of reactive zone systems that need to eliminate the entire contaminant mass in a source zone. There are likely sites at which it is not cost-effective to achieve complete contaminant removal but at which a significant reduction in contaminant migration rates can be achieved through source reduction or reactive barrier application. The test for these applications is whether the reduction in plume profile that can be achieved provides sufficient economic benefit, through elimination of exposures and reduction of long-term site management costs.

5.5.2 Contaminant Shielding by Organic Matter

In aquifer matrices with significant organic carbon fractions (e.g., greater than 1000 mg kg^{-1} TOC), a large fraction of the hydrophobic organic compound mass is likely to reside in sorbed phase, where it is not equally accessible to all reactive zone treatment mechanisms. The best access to this material appears to be gained by enhanced reductive dechlorination reactive zones that generate cosolvents and biosurfactants that solubilize a significant nonaqueous mass. Chemical reactive zones have relatively limited access to sorbed-phase material, and the permanganate ion probably faces the greatest difficulty due to its selectivity.

There are two potential negative outcomes that are driven by organic matter contaminant shielding:

1. *Contaminant release.* If the aquifer organic matter is susceptible to oxidation, a partial treatment has the potential to attack the sorbing matrix more effectively than the target contaminant, releasing untreated contaminant into the formation. This problem can be cured by additional rounds of oxidant injection.
2. *Recalcitrant contaminant mass.* We have observed binding of perchloroethene in oxidant-resistant aquifer organic matter in a case study that was described in Section 3.4.2.5. Permanganate oxidation failed to destroy more than half the soil organic matter and perchloroethene mass was reduced by only 50%. The leaching potential of the remaining perchloroethene was decreased by a factor of 30, which would provide a satisfactory result in many cases. However, the results of that study suggest that mass reduction objectives may be more difficult to achieve than groundwater concentration reductions when there is a significant sorbed contaminant mass fraction.

In either case, the contaminant shielding can be anticipated through field sampling for soil organic matter, bench-scale testing, or pilot-scale field operations. Sorption processes should be considered when setting project objectives (it may be easier to achieve groundwater concentration reductions than mass reductions) and cost estimates (multiple injections may be needed to achieve project objectives).

5.5.3 Phase Transfer Catalysts and Surfactants in Chemical Reactive Zones

Chemical reactive zones face the greatest difficulty in achieving contact between contaminants and injected reagents, which has led to suggestions of phase transfer catalysts and surfactants as aids to those remedial processes.[23–25] Seol and Schwartz[25]

Building Reactive Zone Strategies

suggested the use of phase transfer catalysts for permanganate oxidation of trichloroethene, and Conrad et al.[23] used surfactants to enhance solubilization of trichloroethene NAPL. Gates-Anderson and Siegrist[24] reported no increase in treatment rates for either permanganate or hydrogen-peroxide-based oxidation of trichloroethene with surfactant additions in soil contaminated by perchloroethene, trichloroethene, 1,1,1-trichloroethane, and several polyaromatic hydrocarbons. All three studies were based on laboratory observations.

The potential benefit of surfactant additions is a breakdown of the aqueous-phase–nonaqueous-phase interface, increasing contact between oxidant and target compound. There is also a possibility in field conditions to facilitate increased penetration of hydrophobic organics into aquifer matrices, so the deployment of surfactants should be carefully considered. Phase transfer catalysts, which were discussed in Section 3.4.2, draw permanganate ions into the nonaqueous phase and may have a somewhat reduced risk of enhancing NAPL penetration (assuming the selected phase transfer catalyst is not also an effective surfactant).

5.5.4 Partitioning Electron Donors, Surfactants, and Cosolvents in Biological Reactive Zones

Biological reactive zones can be managed to generate alcohols, ketones, and organic acids that support the dissolution and desorption of hydrophobic organic compounds embedded in aquifer matrices. Many of the intermediate products of microbial carbohydrate consumption can partition into nonaqueous-phase hydrophobic organic mass, drawing those compounds into solution at levels at or above the nominal aqueous-phase solubilities. An example of this is provided in Section 5.6.1.

Compounds that can be injected as electron donors themselves can partition into nonaqueous-phase hydrophobic organics. Alcohols and carbohydrates such as cyclodextrin are examples of possible partitioning electron donor compounds. Alcohols and other cosolvents, when injected, dissolve into nonaqueous-phase hydrophobic organic liquids, reducing their densities and breaking down the interface between the aqueous and nonaqueous phases. Recently, Shirin et al.[26] reported research on substituted cyclodextrins that showed molecules of chlorinated alkenes such as perchloroethene and trichloroethene can be captured as an inclusion in the cyclodextrin molecule, a cyclical ring formed by seven glucose molecules. The maximum solubility enhancement factor observed for trichloroethene was 5.5 and 14 for perchloroethene. These enhancements are within the range of desorption enhancements observed during development of enhanced reductive dechlorination zones using carbohydrates such as sucrose at high loading rates.

5.6 REACTION PRODUCTS AND CONSEQUENCES

Every process that can be applied to aquifer cleanup generates products or otherwise entails consequences that can partially counteract the benefit gained through their application. The intensity at which chemical and biological reactive zone systems must be applied to destroy challenging contaminants *in situ* invites the generation of by-products and consequences. It is very important for the remedial system designer to be aware of the potential secondary effects, design to minimize them, monitor for them, and be prepared to implement counteractive measures, when needed.

5.6.1 Type Curves for Enhanced Reductive Dechlorination

The aqueous-phase hydrophobic organic contaminant concentrations we observe in monitoring wells are an imperfect reflection of the contaminant mass that may be present in the aquifer matrix. As shown earlier in Figure 5.3, the fraction of each hydrophobic organic that resides in nonaqueous phase is dependent on its organic carbon partition coefficient, K_{oc}, and the aquifer matrix organic carbon fraction, f_{oc}. In formations with a high organic carbon fraction, a large portion of the more hydrophobic species mass is obscured from observation due to the partitioning process. These impacts are magnified in the presence of hard carbon species that support adsorption processes.

The sequential removal of chlorine atoms from alkene solvents during enhanced reductive dechlorination reduces the organic carbon partition coefficient with each chlorine removed, unmasking contaminant molecules that were obscured from view in their more highly chlorinated forms. As reductive dechlorination proceeds in an aquifer matrix that harbors sorbed-phase chlorinated alkene mass, the concentration of total alkenes in aqueous phase increases as the fraction of dechlorinated molecules increases. Total alkenes in solution have been observed to increase more than 50-fold during dechlorination from perchloroethene to ethene, demonstrating the significance of the sorbed-phase mass, and the limited perspective on contaminant presence that can be obtained from a groundwater monitoring well result.

The aqueous-phase concentration patterns that develop for chlorinated organics and their degradation products during enhanced reductive dechlorination are characteristic of the contaminant mass distribution in the aquifer matrix. These concentration patterns can be used as a diagnostic tool, assessing the contaminant distribution or the performance of the reactive zone itself. We have developed a six-category taxonomy of reductive dechlorination concentration patterns, which we refer to as the enhanced reductive dechlorination type curves. These curves represent the achievable remedy performance that corresponds to various locations along a contaminant distribution profile such as that given earlier in Figure 5.6. The type curves can be used to determine whether a remedy is performing satisfactorily, based on knowledge of the contaminant distribution at the point of observation, or to draw conclusions regarding the contaminant distribution in an aquifer segment when it is not known.

5.6.2 The Taxonomy of Type Curves

The taxonomy of type curves is based on the number of molecules (molarity) of each contaminant species in solution, over an extended observation period. It is necessary to use molar concentrations, because the loss of mass associated with chlorine atom removal otherwise masks the dramatic increase of alkene molecules that occurs in many sites.

Type 1. Downgradient from active dechlorination reactions, concentrations of all alkenes may be observed to decline at approximately the same rate. This is an indication of passage of a clean water front in the desorption zone, and constitutes a Type 1 curve. Type 1 behavior can also occur during simple displacement of contaminated groundwater by injected solutions, and the observation of Type 1 behavior should always be checked against expected behavior at the observation point. If displacement is suspected, a cross-check can be performed by looking at the concentration of marker compounds associated with injected fluids, and which do not occur in the aquifer. It is often possible to catch the fact that contaminant reductions match dilution due to injected fluids. The simple disappearance of contaminants from a reactive zone should not be trusted without corroborating data, unless the observation was made in the expected desorption zone. Figure 5.28(A) shows the pattern of contaminant decrease that follows Type 1 behavior.

Type 2. When reductive dechlorination occurs in aquifer segments that have very low sorption potential, sequential dechlorination of the alkene mass occurs without an increase in total alkenes in solution. In this case, the molar balance "closes," and is characteristic of behavior that can be expected in the downgradient area of contaminant plumes. If Type 2 behavior is observed in source areas or zones in which significant nonmigratory contaminant mass was expected, it may be the result of combined dechlorination and dilution. Figure 5.28(B) shows the pattern of contaminant decrease that follows Type 2 behavior.

Type 3. In curves of Type 3 and higher, the number of alkene molecules in solution increases as the remedy proceeds. It is the magnitude of the increase that separates Types 3 through 6. In Type 3 behavior, concentrations of the source contaminant (e.g., perchloroethene and trichloroethene) decline during the course of reactive zone development, and significant increases in dechlorinated alkene products is observed. In this case, the remedy achieves extraction of nonaqueous-phase alkene mass that was not observed in groundwater sampling. Although some contaminant displacement might have occurred, extraction of nonaqueous-phase mass and its complete dechlorination dominate the mass balance, which closes to greater than unity. In many cases, the total alkenes in solution reach more than 20 times the initial chlorinated alkene molarity, indicating that less than 5% of the alkenes were observed in aqueous phase in the pretreatment groundwater sampling. Figure 5.28(C) shows a Type 3 contaminant pattern.

Type 4. In curves of Type 4 and higher, significant increases in the contaminant source material are observed during the initial stages of the reductive dechlorination process. This appears to be caused by microbial production of cosolvents and biosurfactants that can mobilize significant quantities of contaminant mass from nonaqueous phase and sorption. It is because of these behaviors, which are indicative of source zone contaminant distribution, that we recommend placement of reactive barriers downgradient from source zones, prior to initiation of source zone treatment (refer to Section 5.4.3). The Type 4 curve is defined by an increase in the original

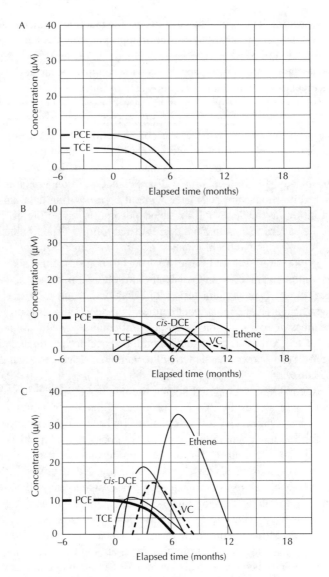

FIGURE 5.28 (A) Type 1 curve–enhanced reductive dechlorination zone concentration pattern occurs in desorption zone (downgradient from reactive zone processes), as a result of clean water migration, or as a result of contaminant displacement by injected fluids. (B) Type 2 curve–enhanced reductive dechlorination zone concentration pattern occurs as a result of enhanced reductive dechlorination in aquifer segments that have very low sorbed-phase contaminant mass. Maximum total alkenes during treatment do not exceed the pretreatment total alkene concentration. (C) Type 3 curve–enhanced reductive dechlorination zone concentration pattern occurs when release of sorbed-phase chlorinated alkene mass generates large increases in aqueous-phase total alkene concentrations, relative to pretreatment levels. It indicates that pretreatment groundwater analysis did not reflect the magnitude of chlorinated alkene contamination.

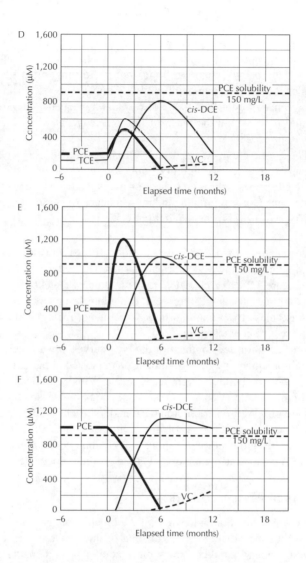

FIGURE 5.28 (continued) (D) Type 4 curve–enhanced reductive dechlorination zone concentration pattern occurs when enhanced reductive dechlorination releases chlorinated alkenes, increasing precursor compounds to levels higher than pretreatment samples. Chlorinated alkene concentrations do not exceed solubility limits. (E) Type 5 curve–enhanced reductive dechlorination zone concentration pattern occurs when nonaqueous-phase chlorinated alkenes are solubilized by an enhanced reductive dechlorination reactive zone. The approximate solubility limit of perchloroethene is indicated. Note that TCE buildup is often negligible when PCE is the primary contaminant. (F) Type 6 curve–enhanced reductive dechlorination zone concentration pattern occurs when pretreatment groundwater sampling captures nonaqueous-phase liquid. Dechlorination reactions achieve reduction to *cis*-DCE before total alkenes begin to decline. Dechlorination of PCE without TCE buildup is commonly observed when PCE is the lone precursor compound. The dashed line indicates the approximate solubility limit of PCE (150 mg L^{-1}).

contaminant molarities, to levels that are below the saturation limit. Type 4 curve behavior is shown in Figure 5.28(D).

Type 5. Type 5 and 6 curves occur in zones that contain nonaqueous-phase liquid contaminant. In the Type 5 curve, pretreatment aqueous-phase concentrations are below saturation limits, but rise during initial treatment to levels at or above the contaminant solubility limits. This is a clear indication of the presence of either drainable or residual nonaqueous-phase contaminant liquid mass within the reach of injected reagents. A Type 5 curve is given in Figure 5.28(E).

Type 6. The Type 6 curve occurs when pretreatment monitoring well concentrations match or exceed solubility limits for one or more of the observed contaminant species. This is an indication of residual nonaqueous-phase liquid contaminant, at a minimum, and may indicate the presence of drainable fluid. A Type 6 curve is provided in Figure 5.28(F).

Each of the type curves provided in Figure 5.28(A)–(F) was drawn based on patterns observed at individual reductive dechlorination sites. The patterns observed in Type 3 and 5 curves (Figure 5.28(C) and (E)) were the first that we picked up as a common thread among multiple sites, indicating significant mobilization of nonaqueous-phase and sorbed-phase mass from aquifers under treatment. Table 5.4 provides a brief summary of curve type characteristics.

In examining data for development of the type curves, it was interesting to note that when perchloroethene was the dominant precursor compound, trichloroethene concentrations did not build to significant levels. Vinyl chloride, also, was not observed at high concentrations in many applications. The small concentration buildups that are observed for trichloroethene and vinyl chloride indicate that their dechlorination rates exceed those of the compound preceding them, that is, trichloroethene is being dechlorinated more quickly than perchloroethene and vinyl chloride is being dechlorinated more quickly than *cis*-dichloroethene.

5.6.3 Results of Partial Oxidation

Oxidation processes often do not destroy all the organic matter in aquifer matrices, either as a result of inadequate oxidant loading or due to the presence of oxidant-resistant organic matter. Partial consumption of aquifer organic matter can generate increases in dissolved organic carbon or release of sorbed-phase contaminant mass.

Oxidation reactions with organic matter proceed in a step-wise manner. Figure 3.14, for example, shows a two-step reaction sequence that leads to complete mineralization of trichloroethene.[27] The oxidation of more complex carbon molecules entail many more reaction steps, and the likelihood of partial degradation is likely to increase when more complex organic matter is present in the aquifer matrix. A consequence of partial oxidation is an increase in the dissolved organic carbon. Droste et al.[28] observed increased reductive dechlorination following a persulfate–permanganate treatment, which they suggested may have resulted from dissolved organic carbon increases related to partial oxidation of organic matter. Other studies[29,30] have shown the increased bioavailability of aquifer organic matter, including

TABLE 5.4
Interpretation of Contaminant Concentration Patterns in Enhanced Reductive Dechlorination Systems

Behavior	ERD Type Curve	Interpretation	Typical Location Where This Is Observed
All alkene concentrations decrease at the same rate	1	No reductive dechlorination is occurring at this location; this may be indicative of desorption zone behavior or may also reflect dilution or normal variance of sampling observations	Redox recovery zone downgradient from an electron donor injection zone, where DOC has returned to very low levels
Dechlorinated alkenes increase over time, but their concentrations do not exceed the precursor alkene concentrations	2	These observations are consistent with reductive dechlorination at locations where there is very little sorbed-phase mass	Downgradient portions of plumes in aquifers that have low soil organic matter content
Dechlorinated alkenes increase dramatically, relative to precursor concentrations; precursor alkene concentrations decrease	3	Indication of reductive dechlorination at a location where there is significant nonaqueous-phase mass; most likely sorbed-phase mass, rather than residual NAPL; dechlorination is running at a rate that matches the desorption rate	Frequently observed in the intermediate zone of plumes and at the edge of source areas
Significant increases in precursor concentrations; dechlorinated compounds may or may not increase	4	This may be an indication of solubilization of residual nonaqueous-phase mass; desorption and/or solubilization is going faster than dechlorination processes	This behavior is observed in source zones
Precursor concentrations increase to levels above the aqueous-phase solubility; degradation products may be present	5	This is a clear indication that NAPL has been solubilized	This behavior is observed due to solubilization of residual NAPL in source zones
Precursor concentrations exceed solubility limits; degradation products may be present	6	This indicates that NAPL was present in the pretreatment sample	This behavior is observed in source zones that may contain drainable NAPL

contaminant mass, as a result of pretreatment with Fenton's reagent or hydrogen peroxide.

When oxidants attack and decompose organic matter that contribute to contaminant sorption in an aquifer, an increase in dissolved-phase contaminant can result. The likelihood of oxidation-generated contaminant increases is increased for target compounds that are marginally reactive with the oxidant. Example 5.4 describes the release of 1,1,1-trichloroethane from sorbed phase, associated with a partially effective aquifer treatment by Fenton's reagent.

5.6.3.1 Release of Sorbed 1,1,1-Trichloroethane Associated with Partial Oxidation by Fenton's Reagent

1,1,1-Trichloroethane is marginally reactive with hydroxyl radical ($k = 1 \times 10^8 \, M^{-1} \, s^{-1}$) compared with other chlorinated alkenes such as perchloroethene ($k = 3 \times 10^9 \, M^{-1} \, s^{-1}$) and trichloroethene ($k = 4 \times 10^9 \, M^{-1} \, s^{-1}$) and has a high organic carbon partition coefficient (139 L kg^{-1}), as shown in Table 5.1. During a chemical oxidation zone treatment using Fenton's reagent, aqueous-phase concentrations of 1,1,1-trichloroethane rebounded to nearly 150% of its pretreatment level, following the first of two oxidant applications, while perchloroethene and trichloroethene were significantly reduced.

Figure 5.29 shows the groundwater concentrations of perchloroethene, trichloroethene, and 1,1,1-trichloroethane in a monitoring well located approximately 2.5 m

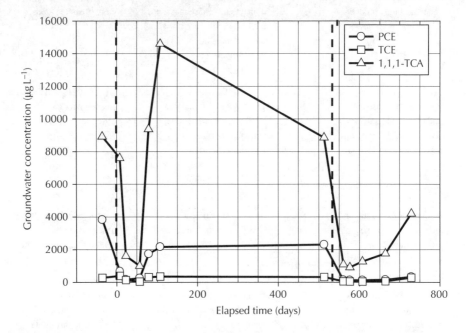

FIGURE 5.29 Temporary increase in concentration of 1,1,1-trichloroethane that followed the application of Fenton's reagent. The nearest oxidant injection point was approximately 3 m upgradient from the monitoring point. Dashed lines show the times at which oxidant was applied.

from the Fenton's reagent injection location. The rebound pattern suggested a release of 1,1,1-trichloroethane from sorbed phase after the first oxidant application, which could have been instigated by the vigorous stirring associated with Fenton's oxidation or with destruction of geosorbent material. Following the second oxidant application, 1,1,1-trichloroethane rebounded again, to 50% of its original pretreatment concentration, while aqueous-phase concentrations of perchloroethene was held to less than 10% of its pretreatment level.

5.6.4 REACTIONS WITH NONTARGET ORGANICS

In most cases, the contaminant mass targeted for removal or destruction is only a fraction of the total organic content of the aquifer matrix. The reagent loadings for IRZ technologies must anticipate interaction with a wide range of naturally occurring organic matter species, as well as an array of contaminants that accompany the compounds of interest, but which are not the focus of remedial efforts. The chemical reactive zone technologies, particularly chemical oxidation, are most susceptible to reactions with nontarget organics.

The oxidant demand generated by the combination of naturally occurring and contaminant organic matter in an aquifer matrix can be viewed as a reductive poise, a concept originally developed by Barcelona and Holm.[31] The organic carbon contribution typically dominates the reductive poise, which can be expressed as follows:

$$\text{Reductive poise} = C_{ox} \cdot \text{TOC} \left(\frac{\text{meq}}{\text{mgTOC}} \cdot \frac{\text{mgTOC}}{\text{kg}_{\text{aquifer soil}}} \right) \quad (5.40)$$

Values for the constant, C_{ox}, are specific to the oxidant applied — stronger oxidants attack larger fraction of the aquifer soil organic matter, and generate higher values for C_{ox}. For dichromate, a strong oxidant, Barcelona and Holm recommended a value of 0.66. Lower values are expected for oxidants such as permanganate, which engage a narrower range of organics.

Nontarget aquifer soil organic matter may also provide a substrate for reactions with oxidants or reducing agents that form by-products, such as acetone or formaldehyde, that are observed during the application of strong oxidants to aquifer soils. Acetone production, for example, was observed in a permanganate bench study at concentrations that exceeded what could have been produced from the target compound (refer to Section 3.4.2.5).

5.6.5 REDOX EFFECTS ON NONTARGET INORGANIC SPECIES

Both chemical and biological reactive zone applications can cause a large swing in aquifer redox levels, an indication that the mobility of electrons among dissolved compounds has either increased or decreased significantly. A decrease in redox indicates that electron mobility has increased, while a redox increase suggests that electron mobility has decreased. Redox-sensitive compounds, especially transition metals, are likely to react to the chemical habitat changes signaled by redox potential changes.

5.6.5.1 Redox Reduction Effects

Enhanced reductive dechlorination and chemical reduction through dithionite and zero-valent iron can lower aquifer redox potentials to strongly negative levels. Under these conditions, metals such as arsenic and barium are likely to react and form soluble ions. In the case of arsenic, solubilization of iron complexes associated with reduction of Fe(III) to Fe(II) may be a major source of mobilization.

Recharge of oxygen and other electron acceptors, along with solid-phase reactions between reduced metal species and the aquifer matrix, gradually eliminates the reduced species from solution and raises the aquifer redox potential in aquifers that are open to the overlying atmosphere.[18] Near the reactive zone, the solid-phase reaction capacity is consumed over time, and long-term operation of a reducing reactive zone is likely to consume the formation's solid-phase electron acceptor capacity. When this capacity has been fully occupied, the distribution of reduced species reaches a steady-state distribution that is determined by their interaction with electron acceptors entering the aquifer through recharge (e.g., oxygen, nitrates, and sulfates).

During enhanced reductive dechlorination and chemical reduction applications, aquifer matrix reduction–oxidation potentials decrease to levels that enable the reduction of iron(III) precipitates from aquifer soils. The resulting iron(II) compounds are highly soluble, and arsenic compounds that were precipitated with iron(III) oxyhydroxy minerals are often released into the aqueous phase. The soluble arsenic compounds reprecipitate in the recovery zone segment of the IRZ complex after reduction–oxidation potentials recover to more aerobic levels.

5.6.5.2 Oxidation Effects

Application of oxidants to an aquifer matrix generates a very rapid increase of redox potentials, to very high levels. Several transition metal species, such as chromium and the actinide metal uranium, can be solubilized in the highly oxidizing chemical habitat. Naturally occurring redox potentials generally do not reach the low levels associated with precipitation of these metals, especially when the aquifer is an open system (open to the atmosphere). These metals may be captured by cation exchange in locations downgradient from the oxidation zone, which is probably the most effective of the potential passive recovery zone mechanisms for these secondary contaminants.

5.6.6 Reaction Product Effects on Hydraulic Conductivity

Reactive zones can alter the hydraulic conductivity of aquifer matrices and, in most of the conceivable mechanisms, the effect is to decrease conductivities. Hydraulic conductivity can be influenced by biological and chemical reactive zones. The most extreme examples occur during chemical oxidation and aerobic biodegradation, and while we have not observed decreases in conductivity during enhanced reductive dechlorination operations, there is a possibility that dissolution and reprecipitation of aquifer minerals could reduce hydraulic conductivity in the recovery zone of an anaerobic reactive barrier.

5.6.6.1 Chemical Oxidation Effects on Aquifer Conductivity

Reactions that occur in conjunction with chemical oxidation can generate very significant reductions in hydraulic conductivity. We routinely observe significant increases in injection backpressures during solution injections for both peroxide-based and permanganate-based oxidation. One or more of the following processes is likely to cause the reductions in hydraulic conductivity that results in injection backpressures and semi-permanent and, sometimes, permanent reductions in hydraulic conductivity.

> *Precipitate formation.* One of the obvious hydraulic conductivity effects of oxidation processes is the formation of manganese dioxide precipitates from reduction of permanganate ion. In sites that require multiple pore volumes of permanganate solution flushing to achieve remedial objectives, the risk of formation blockage is quite high. The manganese dioxide precipitate is very insoluble under typical aquifer conditions, and it is likely to cause a permanent reduction of conductivity.
> *Gas formation.* The largest source of gas among oxidation methods is the oxygen produced by disproportionation of hydrogen peroxide. Carbon dioxide that forms from oxidation of organic molecules adds to the gas volume for both hydrogen peroxide and permanganate oxidation. Gas displaces water from the aquifer pore spaces, reducing the water permeability and hydraulic conductivity.
> *Cation exchange.* When permanganate oxidation is applied to clay-bearing soils, the cation that is used to form the permanganate salt will be present at very high molarities. This can drive cation exchange with aquifer soils, displacing calcium, magnesium, and other polyvalent cations with the monovalent sodium or potassium ions that are used in the formulation of permanganate salts for the oxidation process. In aquifer soils that contain montmorillonite clay, the displacement of divalent cations by monovalent cations, especially sodium, will cause the clay to swell, decreasing hydraulic conductivity. Sites at which hydrofracturing has been applied to clay-bearing aquifer matrices might be especially sensitive to cation-exchange-induced reductions in permeability.
> *Ionic strength effects.* The addition of oxidants to groundwater greatly increases its ionic strength, which can have multiple influences on aquifer matrix chemistries. The most significant effect is likely to be a reduction of gas solubilities in aqueous phase, increasing the magnitude of water displacement from the aquifer matrix. Influences on soil cation exchange capacity are also possible.

5.6.6.2 Chemical Reduction Effects on Aquifer Conductivity

Each of the chemical reduction methods carries the potential to reduce hydraulic conductivity, although in practice this does not appear to be a significant problem.

> *Sulfate precipitation.* The sodium dithionite reduction process generates very high sulfate concentrations as an immediate result of the reduction

reactions. The sulfate can go on to react with dissolved-phase Ca^{2+} to form gypsum ($K_{sp} = 1.95 \times 10^{-4}$ at 10 °C[20]), which can reduce hydraulic conductivity through precipitation and blockage of aquifer pore spaces. In some applications, the sulfate-bearing groundwater is pumped out of the reactive zone to prevent the gypsum-forming precipitation reactions (e.g., Fruchter et al.[32]).

Other mineral precipitation. Mineral precipitation in zero-valent iron reactive walls appears to occur adjacent to the zero-valent iron particle surfaces[33] — good news regarding passivation potential, although this may indicate greater potential for hydraulic conductivity reductions. Blowes et al.[34] indicated that precipitate formation is one of the factors that limit the life of zero-valent reactive barriers. These studies suggest that the focus of precipitation is within the zero-valent reactive barrier rather than downgradient in the aquifer matrix.

5.6.6.3 Biological Reactive Zone Effects on Aquifer Conductivity

Biomass densities. In anaerobic systems, microbial productivity should be fundamentally limited, due to the lower energy yields of anaerobic respiration (approximately one seventh of aerobic respiration of the same electron donor). Aquifer blockage by biomass buildup is a common challenge in aerobic bioreactor systems although, in many cases, it is difficult to separate biomass impacts from carbonates and other precipitates that may form when low-redox groundwater is aerated in support of biological reactive zones.

Precipitate formation. In anaerobic systems, the microbial community may solubilize aquifer minerals that later reprecipitate in the redox recovery zone. Carbonates and ferrous iron compounds are two examples of mineral groups that can enter the aqueous phase in reducing biological reactive zones. During periods of high sulfide generation, iron sulfide compounds (e.g., pyrite) can form in the anaerobic zone. These precipitates all have the potential to reduce hydraulic conductivity over long periods of reactive zone operation, although the reductions would have to be large to be observable in most monitoring well networks.

In aerobic systems, precipitate formation is likely to be a much greater challenge. Groundwater entering an aerobic biological reactive zone is normally anaerobic (hence the need to provide aeration) and is likely to carry dissolved solids that rapidly precipitate upon contact with the aerobic zone. Carbonates and ferrous iron and related compounds are the most likely precipitates. It is usually difficult to separate the effects of precipitation and biomass buildup that can occur in aerobic systems.

REFERENCES

1. Domenico, P.A., An analytical model for multidimensional transport of a decaying contaminant species, *J. Hydrol.*, 91(1–2), 49–58, 1987.

2. Rivett, M.O., S. Feestra, and J.A. Cherry, A controlled field experiment on groundwater contamination by a multicomponent DNAPL: creation of the emplaced source and overview of dissolved plume development, *J. Contam. Hydrol.*, 49, 111–149, 2001.
3. Julian, H.E., J.M. Boggs, C. Zheng, and C.E. Feehley, Numerical simulation of a natural gradient tracer experiment for the natural attenuation study: flow and physical transport, *Ground Water*, 39(4), 534–545, 2001.
4. Fetter, C.W., *Applied Hydrogeology*, Merrill Publishing Company, Columbus, OH, 1988.
5. Schwille, F., *Dense Chlorinated Solvents in Porous and Fractured Media* (English language edition, translated by James F. Pankow), Lewis Publishers, Chelsea, MI, 1988.
6. Scow, K.M. and C.R. Johnson, Effect of sorption on biodegradation of soil pollutants, *Adv. Agron.*, 58, 1–56, 1997.
7. U.S. EPA, Soil Screening Guidance: Technical Background Document, EPA/540/R95/128, May 1996.
8. U.S. EPA, Superfund Chemical Data Matrix, EPA/540/R-96/028, June 1996.
9. Montgomery, J.H. and L.M. Welkom, *Groundwater Chemicals Desk Reference* (2 volumes), Lewis Publishers, Chelsea, MI, 1990.
10. Luthy, R.G., G.A. Aiken, M.L. Brusseau, S.D. Cunningham, P.M. Gschwend, J.J. Pignatello, M. Reinhard, S.J. Traina, W.J. Weber, Jr., and J.C. Westall, Sequestration of hydrophobic organic contaminants by geosorbents, *Environ. Sci. Technol.*, 31(12), 3341–3347, 1997.
11. Chen, W., A.T. Kan, C.J. Newell, E. Moore, and M.B. Tomson, More realistic cleanup standards with dual-equilibrium desorption, *Ground Water*, 40(2), 153–164, 2002.
12. Bauman, T., S. Muller, and R. Niessner, Migration of dissolved heavy metal compounds and PCP in the presence of colloids through a heterogeneous calcareous gravel and a homogeneous quartz sand — pilot scale experiments, *Water Res.*, 36, 1213–1223, 2002.
13. Compere, F., G. Porel, and F. Delay, Transport and retention of clay particles in saturated porous media. Influence of ionic strength and pore velocity, *J. Contam. Hydrol.*, 49, 1–21, 2001.
14. Ryan, J.N. and M. Elimelech, Colloid mobilization and transport in groundwater, *Coll. Surf. A*, 107, 1–56, 1996.
15. ITRC Draft DNAPL Bioremediation Report, as of Feb. 2004. Interstate Technology Regulatory Council.
16. Wymore, R., Northwind, Inc., personal communication.
17. Vidal, A., ARCADIS, personal communication.
18. Champ, D.R., J. Gulens, and R.E. Jackson, Oxidation–reduction sequences in ground water flow systems, *Can. J. Earth Sci.*, 16, 12–23, 1979.
19. Ross, A.B., W.G. Mallard, W.P. Helman, G.V. Buxton, R.E. Huie, and P. Neta, NDRL-NIST Solution Kinetics Database — Version 3. Notre Dame Radiation Laboratory, Notre Dame, IN, and NIST Standard Reference Data, Gaithersburg, MD, 1998.
20. Weast, R.C., *Handbook of Chemistry and Physics*, 50th ed., The Chemical Rubber Company, Cleveland, OH, 1969.
21. Kotz, J.C. and P. Treichel, Jr., *Chemistry and Chemical Reactivity*, 3rd ed., Saunders College Publishing, Fort Worth, TX, 1996.
22. Nyer, E.K., P.L. Palmer, E.P. Carman, G. Boettcher, J.M. Bedessem, F. Lenzo, T.L. Crossman, G.L. Rorech, and D.F. Kidd, *In Situ Treatment Technology*, Lewis Publishers, Boca Raton, FL, 2001.

23. Conrad, S.H., R.J. Glass, and W.J. Peplinski, Bench-scale visualization of DNAPL remediation processes in analog heterogeneous aquifers: surfactant floods and *in situ* oxidation using permanganate, *J. Contam. Hydrol.*, 58, 13–49, 2002.
24. Gates-Anderson, D.D. and R.L. Siegrist, Comparison of potassium permanganate and hydrogen peroxide as chemical oxidants for organically contaminated soils, *J. Environ. Eng.*, 127(4), 337–347, 2001.
25. Seol, Y. and F.W. Schwartz, Phase-transfer catalysis applied to the oxidation of non-aqueous phase trichloroethylene by potassium permanganate, *J. Contam. Hydrol.*, 44, 185–201.
26. Shirin, S., E. Buncel, and G.W. van Loon, The use of β-cyclodextrin to enhance the aqueous solubility of trichloroethylene and perchloroethylene and their removal from soil organic matter: effect of substituents, *Can. J. Chem.*, 81, 45–52, 2003.
27. Yan, Y.E. and F. Schwartz, Kinetics and mechanisms for TCE oxidation by permanganate, *Environ. Sci. Technol.*, 34 (12), 2535–2541, 2000.
28. Droste, E.X., A.M. Lee, P.M. Dinardo, G.E. Hoag, and P.V. Chheda, Observed enhanced reductive dechlorination after *in situ* chemical oxidation pilot test. Proc. Third Internat. Conf. Remediation of Chlorinated and Recalcitrant Compounds Monterey, CA, May 2002, 2002.
29. Chamorro, E., A. Marco, and S. Esplugas, Use of Fenton reagent to improve organic chemical biodegradability, *Water Res.*, 35(4), 1047–1051, 2001.
30. Kao, C.M. and M.J. Wu, Enhanced TCDD degradation by Fenton's reagent preoxidation, *J. Haz. Mat.*, 74(3), 197–211, 2000.
31. Barcelona, M.J. and T.R. Holm, Oxidation–reduction capacities of aquifer solids, *Environ. Sci. Technol.*, 25(9), 1565–1572, 1991.
32. Fruchter, J.S., C.R. Cole, M.D. Williams, V.R. Vermeul, J.E. Amonette, J.E. Szecsody, J.D. Istok, and M.D. Humphrey, Creation of a subsurface permeable treatment zone for aqueous chromate contamination using *in situ* redox manipulation, *Ground Water Monit. R.*, 20(2), 66–77, 2000.
33. Furakawa, Y., J.W. Kim, J. Watkins, and R.T. Wilkin, Formation of ferrihydrite and associated iron corrosion products in permeable reactive barriers of zero-valent iron, *Environ. Sci. Technol.*, 36(24), 5469–5475, 2002.
34. Blowes, D.W., C.J. Ptacek, S.G. Benner, C.W.T. McRae, T.A. Bennett, and R.W. Puls, Treatment of inorganic contaminants using permeable reactive barriers, *J. Contam. Hydrol.*, 45, 123–137, 2000.

Appendix: Physical Properties of Some Common Environmental Contaminants

Compound	Molecular Weight	Henry's Law Constant (atm m³ mol⁻¹)	Vapor Pressure (mm Hg)	Solubility (mg L⁻¹)	Log k_{oc}
A					
Acenaphthene	154.21	7.92×10^{-5} (25°C)	0.00155 (25°C)	3.47 (25°C)	1.25
Acenaphthylene	152.20	2.8×10^{-4}	0.0290 (20°C)	3.93 (25°C)	3.68
Acetaldehyde	44.05	6.61×10^{-5} (25°C)	760 (20.2°C)	Miscible	Unavailable
Acetic acid	60.05	1.23×10^{-3} (25°C)	11.4 (20°C)	Miscible	Unavailable
Acetic anhydride	102.09	3.92×10^{-6} (20°C)	5 (25°C)	12% by wt. (20°C)	Unavailable
Acetone	58.08	3.97×10^{-5} (25°C)	266 (25°C)	Miscible	−0.43
Acetonitrile	41.05	3.46×10^{-6} (25°C)	73 (20°C)	Miscible	0.34
2-Acetylamino-fluorene	223.27	—	—	—	3.20
Acrolein	56.06	4.4×10^{-6} (25°C)	265 (25°C)	200,000 (25°C)	−0.28
Acrylamide	71.08	3.03×10^{3} (20°C)	7×10^{-3} (20°C)	2.155 g l⁻¹ (30°C)	Unavailable
Acrylonitrile	53.06	1.10×10^{-4} (25°C)	110–115 (25°C)	80,000 (25°C)	−1.13
Aldrin	364.92	4.96×10^{-4}	6×10^{-6} (25°C)	0.011 (25°C)	2.61
Allyl alcohol	58.08	5.00×10^{-6} (25°C)	20 (20°C)	Miscible	0.51
Allyl chloride	76.53	1.08×10^{-2} (25°C)	360 (25°C)	—	1.68
Allyl glycidyl ether	114.14	3.83×10^{-6} (20°C)	3.6 (20°C)	141 g L⁻¹	Unavailable
4-Aminobiphenyl	169.23	3.89×10^{-10} (25°C)	6×10^{-5} (20–30°C)	842 (20–30°C)	2.03

Compound	Molecular Weight	Henry's Law Constant (atm m^3 mol^{-1})	Vapor Pressure (mm Hg)	Solubility (mg L^{-1})	Log k_{oc}
2-Aminopyridine	94.12	—	Low	100 wt.% at 20°C	Unavailable
Ammonia	17.04	2.91 × 10^{-4} (20°C)	10 atm (25.7°C)	531 g L^{-1} (20°C)	0.49
n-Amyl acetate	130.19	3.88 × 10^{-4} (25°C)	4.1 (25°C)	1.8 g L^{-1} (20°C)	Unavailable
sec-Amyl acetate	130.19	4.87 × 10^{-4} (20°C)	10 (35.2°C)	0.2 wt.% (20°C)	Unavailable
Aniline	93.13	0.136 (25°C)	0.6 (20°C)	—	1.41
o-Anisidine	123.15	1.25 × 10^{-6} (25°C)	0.1 (30°C)	1.3 wt.% (20°C)	Unavailable
p-Anisidine	123.15	—	—	3.3 (Room temp.)	Unavailable
Anthracene	178.24	6.51 × 10^{-5} (25°C)	1.95 × 10^{-4} (25°C)	0.075 (25°C)	4.41
Antu	202.27	—	≈0 (20°C)	600 (20°C)	

B

Compound	Molecular Weight	Henry's Law Constant (atm m^3 mol^{-1})	Vapor Pressure (mm Hg)	Solubility (mg L^{-1})	Log k_{oc}
Benzene	78.11	0.00548 (25°C)	95.2 (25°C)	1,800 (25°C)	1.92
Benzidine	184.24	3.88 × 10^{-11} (25°C)	0.83 (20°C)	500 (25°C)	1.60
Benzo[a]anthracene	228.30	8.0 × 10^{-6}	1.1 × 10^{-7} (25°C)	0.014 (25°C)	6.14
Benzo[b]fluoranthene	252.32	1.2 × 10^{-5} (20–25°C)	5 × 10^{-7} (20°C)	0.0012 (25°C)	5.74
Benzo[k]fluoranthene	252.32	0.00104	9.59 × 10^{-11} (25°C)	0.00055 (25°C)	6.64
Benzoic acid	122.12	7.02 × 10^{-8}	0.0045 (25°C)	3,400 (25°C)	1.48–2.70
Benzo[ghi]perlene	276.34	1.4 × 10^{-7} (25°C)	1.01 × 10^{-10} (25°C)	0.00026 (25°C)	6.89
Benzo[a]pyrene	252.32	<2.4 × 10^{-6}	5.6 × 10^{-9} (25°C)	0.0038 (25°C)	5.60–6.29
Benzo[e]pyrene	252.32	4.84 × 10^{-7} (25°C)	5.54 × 10^{-9} (25°C)	0.0038 (25°C)	5.6
Benzyl alcohol	108.14	Insufficient vapor pressure data for calculation at 25°C	1 (58°C)	42,900 (25°C)	1.98
Benzyl butyl phthalate	312.37	1.3 × 10^{-6} (25°C)	8.6 × 10^{-6} (20°C)	42.2 (25°C)	1.83–2.54
Benzyl chloride	126.59	3.04 × 10^{-4} (20°C)	1 (22.0°C)	493 (20°C)	2.28
α-BHC	290.83	5.3 × 10^{-6} (20°C)	2.5 × 10^{-5} (20°C)	2.0 (25°C)	3.279
β-BHC	290.83	2.3 × 10^{-7} (20°C)	2.8 × 10^{-7} (20°C)	0.24 (25°C)	3.553
δ-BHC	290.83	2.5 × 10^{-7} (20–25°C)	1.7 × 10^{-5} (20°C)	31.4 (25°C)	3.279

Appendix

Compound	Molecular Weight	Henry's Law Constant (atm m³ mol⁻¹)	Vapor Pressure (mm Hg)	Solubility (mg L⁻¹)	Log k_{oc}
Biphenyl	154.21	4.15×10^{-4} (25°C)	10^{-2} (25°C)	7.5 (25°C)	3.71
Bis(2-chloroethoxy) methane	173.04	3.78×10^{-7}	1 (53°C)	81,000 (25°C)	2.06
Bis(2-chloroethyl) ether	143.01	1.3×10^{-5}	1.55 (25°C)	10,200 (25°C)	1.15
Bis(2-chloroisopropyl) ether	171.07	1.1×10^{-4}	0.85 (20°C)	1,700 (20°C)	1.79
Bis(2-ethylhexyl) phthalate	390.57	1.1×10^{-5} (25°C)	6.2×10^{-8} (25°C)	0.4 (25°C)	5.0
Bromobenzene	157.01	2.4×10^{-3} (25°C)	4.14 (25°C)	409 (25°C)	2.33
Bromochloromethane	129.39	1.44×10^{-3} (24–25°C)	141.07 (24.05°C)	0.129 M (25.0°C)	1.43
Bromodichloromethane	163.83	2.12×10^{-4}	50 (20°C)	4,500 (0°C)	1.79
Bromoform	252.73	5.6×10^{-4}	5.6 (25°C)	3.130 (25°C)	2.45
4-Bromophenyl phenyl ether	249.20	1.0×10^{-4}	0.0015 (20°C)	No data found	4.94
Bromotrifluoromethane	148.91	5.00×10^{-1} (25°C)	149 (20°C)	0.03 wt.% (20°C)	2.44
1,3-Butadiene	54.09	6.3×10^{-2} (25°C)	2,105 (25°C)	735 (20°C)	2.08
n-Butane	58.12	9.30×10^{-1} (25°C)	1,820 (25°C)	61 (20°C)	Unavailable
2-Butanone	72.11	4.66×10^{-5} (25°C)	100 (25.0°C)	25.57 wt.% (25°C)	0.09
1-Butene	56.11	2.5×10^{-1} (25°C)	2.230 (25°C)	222 (25°C)	Unavailable
Butoxyethanol	118.18	2.36×10^{-6}	0.76 (20°C)	Miscible	Unavailable
n-Butyl acetate	116.16	3.3×10^{-4} (25°C)	15 (25°C)	5,000 (25°C)	Unavailable
sec-Butyl acetate	116.16	1.91×10^{-4} (20°C)	10 (20°C)	0.8 wt.% (20°C)	Unavailable
tert-Butyl acetate	116.16	—	—	—	Unavailable
n-Butyl alcohol	74.12	8.81×10^{-6} (25°C)	7.0 (25°C)	74,700 (25°C)	Unavailable
sec-Butyl alcohol	74.12	1.02×10^{-5} (25°C)	13 (20°C)	201,000 (20°C)	Unavailable
tert-Butyl alcohol	74.12	1.20×10^{-5} (25°C)	42 (25°C)	Miscible	Unavailable
n-Butylbenzene	134.22	1.25×10^{-2} (25°C)	1.03 (25°C)	1.26 (25.0°C)	3.40
sec-Butylbenzene	134.22	1.14×10^{-2} (25°C)	1.81 (25°C)	309 (25.0°C)	2.95

Compound	Molecular Weight	Henry's Law Constant (atm m^3 mol^{-1})	Vapor Pressure (mm Hg)	Solubility (mg L^{-1})	Log k_{oc}
tert-Butylbenzene	134.22	1.17×10^{-2} (25°C)	2.14 (25°C)	34 (25.0°C)	2.83
n-Butyl mercaptan	90.18	7.04×10^{-3} (20–22°C)	55.5 (25°C)	590 (22°C)	Unavailable
C					
Camphor	152.24	3.00×10^{-5} (20°C)	0.18 (20°C)	0.12% (20°C)	Unavailable
Carbaryl	201.22	1.27×10^{-5} (20°C)	6.578×10^{-6} (25°C)	0.4105 (25°C)	2.42
Carbofuran	221.26	3.88×10^{-8} (30–33°C)	2×10^{-5} (33°C)	700 (25°C)	2.2
Carbon disulfide	76.13	0.0133	360 (25°C)	2,300 (22°C)	2.38–2.55
Carbon tetrachloride	153.82	0.024 (20°C)	113 (25°C)	1.160 (25°C)	2.35
Chlordane	409.78	4.8×10^{-5}	1×10^{-5} (25°C)	1.85 (25°C)	5.57
cis-Chlordane	409.78	Insufficient vapor pressure data for calculation at 25°C	No data found	0.051 (20–25°C)	6.0
trans-Chlordane	409.78	Insufficient vapor pressure data for calculation at 25°C	No data found	No data found	6.0
Chloroacetaldehyde	78.50	—	100 (20°C)	About 50 wt.%, forms a hemihydrate	Unavailable
α-Chloroacetophenone	154.60	—	0.012 (20°C)	Miscible	Unavailable
4-Chloroaniline	127.57	1.07×10^{-5} (25°C)	0.025 (25°C)	3.9 g l^{-1} (20–25°C)	2.42
Chlorobenzene	112.56	0.00445 (25°C)	11.8 (25°C)	502 (25°C)	1.68
o-Chlorobenzylidene malonitrile	188.61	Not applicable, reacts with water	3.4×10^{-5} (20°C)	Not applicable, reacts with water	Not applicable, reacts with water
p-Chloro-m-cresol	142.59	1.78×10^{-6}	No data found	3,850 (25°C)	2.89
Chloroethane	64.52	0.0085 (25°C)	1,064 (20°C)	5,740 (20°C)	0.51
2-Chloroethyl vinyl ether	106.55	2.5×10^{-4}	26.75 (20°C)	15,000 (20°C)	0.82
Chloroform	119.38	0.0032 (25°C)	198 (25°C)	9,300 (25°C)	1.64
2-Chloronaphthalene	162.62	6.12×10^{-4}	0.017 (25°C)	6.74 (25°C)	3.93
p-Chloronitrobenzene	157.56	$<6.91 \times 10^{-3}$ (20°C)	<1 (20°C)	0.003 wt.% (20°C)	2.68
1-Chloro-1-nitropropane	123.54	1.57×10^{-1} (20–25°C)	5.8 (25°C)	<0.8 wt.% (20°C)	3.34
2-Chlorophenol	128.56	5.6×10^{-7} (25°C)	1.42 (25°C)	28,000 (25°C)	2.56
4-Chlorophenyl phenyl ether	204.66	2.2×10^{-4}	0.0027 (25°C)	3.3 (25°C)	3.6

Compound	Molecular Weight	Henry's Law Constant (atm m^3 mol^{-1})	Vapor Pressure (mm Hg)	Solubility (mg L^{-1})	Log k_{oc}
Chloropicrin	164.38	8.4×10^{-2}	23.8 (25°C)	1.621 g L^{-1} (25°C)	0.82
Chloroprene	88.54	3.20×10^{-2}	200 (20°C)	—	—
Chloropyrifos	350.59	4.16×10^{-6} (25°C)	1.87×10^{-5} (25°C)	2 (25°C)	3.86
Chrysene	228.30	7.26×10^{-20}	6.3×10^{-9} (25°C)	0.006 (25°C)	5.39
Crotonaldehyde	70.09	1.96×10^{-5}	30 (20°C)	18.1 wt.% (20°C)	Unavailable
Cycloheptane	98.19	—	—	30 (25°C)	Unavailable
Cyclohexane	84.16	1.94×10^{-1} (25°C)	95 (20°C)	58.4 (25°C)	Unavailable
Cyclohexanol	100.16	5.74×10^{-6} (25°C)	1 (20°C)	36,000 (20°C)	Unavailable
Cyclohexanone	98.14	1.2×10^{-5} (25°C)	4 (20°C)	23,000 (20°C)	Unavailable
Cyclohexene	82.15	4.6×10^{-2} (25°C)	67 (20°C)	213 (25°C)	Unavailable
Cyclopentadiene	66.10	—	—	0.0103 mol L^{-1} at room temperature	Unavailable
Cyclopentane	70.13	1.86×10^{-1} (25°C)	400 (31.0°C)	164 (25°C)	Unavailable
Cyclopentene	68.12	6.3×10^{-2} (25°C)	—	535 (25°C)	Unavailable

D

Compound	Molecular Weight	Henry's Law Constant (atm m^3 mol^{-1})	Vapor Pressure (mm Hg)	Solubility (mg L^{-1})	Log k_{oc}
2,4-D	221.04	1.95×10^{-2} (20°C)	0.0047 (20°C)	890 ppm (25°C)	1.68
p,p' DDD	320.05	2.16×10^{-5}	1.02×10^{-6} (30°C)	0.160 (24°C)	4.64
p,p' DDE	319.03	2.34×10^{-5}	6.49×10^{-6} (30°C)	0.0013 (25°C)	6
p,p' DDT	354.49	5.2×10^{-5}	1.9×10^{-7} (25°C)	0.0004 (25°C)	6.26
Decahydro-naphthalene	138.25	39.2 (25°C)	1 (22.5°C)	0.889 ppm (25°C)	Unavailable
n-Decane	142.28	1.87×10^{-1} (25°C)	1.35 (25°C)	0.022 (25°C)	Unavailable
Diacetone alcohol	116.16	—	1 (22.0°C)	Miscible	Unavailable
Dibenz[a,h]-anthracene	278.36	7.33×10^{-9}	$\approx 10^{-10}$ (20°C)	0.00249 (25°C)	6.22
Dibenzofuran	168.20	Insufficient vapor pressure data for calculation at 25°C	No data found	10 (25°C)	3.91–4.10

Compound	Molecular Weight	Henry's Law Constant (atm m^3 mol^{-1})	Vapor Pressure (mm Hg)	Solubility (mg L^{-1})	Log k_{oc}
1,4-Dibromobenzene	235.91	5.0×10^{-4} (25°C)	0.161 (25°C)	16.5 (25°C)	3.2
Dibromochloro- methane	208.28	9.9×10^{-4}	76 (20°C)	4,000 (20°C)	1.92
1,2-Dibromo-3- chloropropane	236.36	2.49×10^{-4} (20°C)	0.8 (21°C)	1,000 at room temperature	2.11
Dibromodifluoro- methane	209.82	—	688 (20°C)	—	—
Di-n-butyl phthalate	278.35	6.3×10^{-5}	1.4×10^{-5} (25°C)	400 (25°C)	3.14
1,2-Dichlorobenzene	147.00	0.0024 (25°C)	1.5 (25°C)	145 (25°C)	3.23
1,3-Dichlorobenzene	147.00	0.0047 (25°C)	2.3 (25°C)	143 (25°C)	3.23
1,4-Dichlorobenzene	147.00	0.00445 (25°C)	0.4 (25°C)	74 (25°C)	2.2
3,3'-Dichloro- benzidine	253.13	4.5×10^{-8} (25°C)	1×10^{-5} m L^{-1} (22°C)	3.11 (25°C)	3.3
Dichlorodifluoro- methane	120.91	0.425 (25°C)	4.887 (25°C)	280 (25°C)	2.56
1-3-Dichloro-5,5- dimethylhydantoin	197.03	Not applicable, reacts with water	—	0.21 wt.% (25°C)	Not applicable, reacts with water
1,1-Dichloroethane	98.96	0.00587 (25°C)	234 (25°C)	5,060 (25°C)	1.48
1,2-Dichloroethane	98.96	9.8×10^{-4} (25°C)	87 (25°C)	8,300 (25°C)	1.15
1,1-Dichloroethylene	96.94	0.021	591 (25°C)	5,000 (25°C)	1.81
trans-1,2-Dichloro- ethylene	96.94	0.00674 (25°C)	410 (30°C)	6,300 (25°C)	1.77
Dichlorofluoro- methane	120.91	$\approx 2.42 \times 10^{-2}$ (20–30°C)	760 (8.9°C)	1 wt.% (20°C)	1.57
sym-Dichloromethyl ether	114.96	Not applicable, reacts with water	—	Decomposes	Not applicable, reacts with water
2,4-Dichlorophenol	163.00	6.66×10^{-6}	0.089 (25°C)	4,500 (25°C)	2.94
1,2-Dichlorophenol	112.99	0.00294 (25°C)	50 (25°C)	2,800 (25°C)	1.71
cis-1,3-Dichloro- propylene	110.97	0.00355	43 (25°C)	2,700 (25°C)	1.68
trans-1,3- Dichloropropylene	110.97	0.00355	34 (25°C)	2,800 (25°C)	1.68
Dichlorvos	220.98	5.0×10^{-3}	0.0527 (25°C)	$\approx$1 wt.% (20°C)	9.57
Dieldrin	380.91	2×10^{-7}	1.8×10^{-7} (25°C)	0.20 (25°C)	4.55

Appendix

Compound	Molecular Weight	Henry's Law Constant (atm m³ mol⁻¹)	Vapor Pressure (mm Hg)	Solubility (mg L⁻¹)	Log k_{oc}
Diethylamine	73.14	2.56×10^{-5} (25°C)	195 (20°C)	815,000 (14°C)	Unavailable
2-Diethylaminoethanol	117.19	—	1 (20°C)	Miscible	Unavailable
Diethyl phthalate	222.24	8.46×10^{-7}	0.22 (±0.7) Pa (25°C)	1,000 (25°C)	1.84
1,1-Difluorotetrachloroethane	203.83	—	40 (19.8°C)	—	—
1,2-Difluorotetrachloroethane	203.83	1.07×10^{-1} (20°C)	40 (19.8°C)	0.01 wt.% (20°C)	2.78
Diisobutyl ketone	142.24	6.36×10^{-4} (20°C)	1.7 (20°C)	0.05 wt.% (20°C)	Unavailable
Diisopropylamine	101.19	—	60 (20°C)	Miscible	Unavailable
N,N-Dimethylacetamide	115.18	—	1.3 (25°C)	Miscible	Unavailable
Dimethylamine	45.08	1.77×10^{-5} (25°C)	1,520 (10°C)	Miscible	Unavailable
p-Dimethylaminoazobenzene	225.30	—	—	13.6 (20–30°C)	3
Dimethylaniline	121.18	4.98×10^{-6} (20°C)	1 (29.5°C)	1,105.2 (25°C)	Unavailable
2,2-Dimethylbutane	86.18	1.943 (25°C)	319.1 (25°C)	21.2 (25°C)	Unavailable
2,3-Dimethylbutane	86.18	1.18 (25°C)	234.6 (25°C)	19.1 (25°C)	Unavailable
cis-1,2-Dimethylcyclohexane	112.22	3.54×10^{-1} (25°C)	14.5 (25°C)	6.0 (25°C)	Unavailable
trans-1,4-Dimethylcyclohexane	112.22	8.70×10^{-1} (25°C)	22.65 (25°C)	3.84 ppm (25°C)	Unavailable
Dimethylformamide	73.09	—	3.7 mm (25°C)	Miscible	Unavailable
1,1-Dimethylhydrazine	60.10	2.45×10^{-9} (25°C)	157 (25°C)	Miscible	−0.7
2,3-Dimethylpentane	100.20	1.73 (25°C)	100 (33.3°C)	5.25 (25°C)	Unavailable
2,4-Dimethylpentane	100.20	3.152 (25°C)	98.4 (25°C)	5.50 (25°C)	Unavailable
3,3-Dimethylpentane	100.20	1.84 (25°C)	82.8 (25°C)	5.94 (25°C)	Unavailable
2,4-Dimethylphenol	122.17	6.55×10^{-6} (25°C)	0.098 (25°C)	7,868 (25°C)	2.07
Dimethyl phthalate	194.19	4.2×10^{-7}	0.22 ± 0.7 Pa (25°C)	4,320 (25°C)	1.63
2,2-Dimethylpropane	72.15	2.18 (25°C)	1.287 (25°C)	33.2 (25°C)	Unavailable
2,7-Dimethylquinoline	157.22	—	—	1.795 (25°C)	—
Dimethyl sulfate	126.13	2.96×10^{-6} (20°C)	0.5 (20°C)	2.8 wt.% (20°C)	0.61
1,2-Dinitrobenzene	168.11	$<1.47 \times 10^{-3}$ (20°C)	<1 (20°C)	0.015 wt.% (20°C)	Unavailable
1,3-Dinitrobenzene	168.11	2.75×10^{-7} (35°C)	8.15×10^{-4} (35°C)	0.05 wt.% (20°C)	2.18
1,4-Dinitrobenzene	168.11	4.79×10^{-7} (35°C)	2.25×10^{-4} (35°C)	0.01 wt.% (20°C)	Unavailable

Compound	Molecular Weight	Henry's Law Constant (atm m^3 mol^{-1})	Vapor Pressure (mm Hg)	Solubility (mg L^{-1})	Log k_{oc}
4,6-Dinitro-o-cresol	198.14	1.4×10^{-6}	5.2×10^{-5} (25°C)	250 (25°C)	2.64
2,4-Dinitrophenol	184.11	1.57×10^{-8} (18–20°C)	0.00039 (20°C)	6,000 (25°C)	1.25
2,4-Dinitrotoluene	182.14	8.67×10^{-7}	1.1×10^{-4} (20°C)	270 (22°C)	1.79
2,6-Dinitrotoluene	182.14	2.17×10^{-7}	3.5×10^{-4} (20°C)	≈300	1.79
Di-n-octyl phthalate	390.57	1.41×10^{-12} (25°C)	0.0014 mm (25°C)	3 (25°C)	8.99
Dioxane	88.11	4.88×10^{-6} (25°C)	37 (25°C)	Miscible	0.54
1,2-Diphenyl hydrazine	184.24	4.11×10^{-11} (25°C)	2.6×10^{-5} (25°C)	221 (25°C)	2.82
Diuron	233.11	1.46×10^{-9} (25–30°C)	2×10^{-7} (30°C)	42 (25°C)	2.51
n-Dodecane	174.34	24.2 (25°C)	0.057 (25°C)	0.008 (25°C)	Unavailable

E

Compound	Molecular Weight	Henry's Law Constant (atm m^3 mol^{-1})	Vapor Pressure (mm Hg)	Solubility (mg L^{-1})	Log k_{oc}
α-Endosulfan	406.92	1.01×10^{-4} (25°C)	10^{-5} (25°C)	0.530 (25°C)	3.31
β-Endosulfan	406.92	1.91×10^{-5} (25°C)	10^{-5} (25°C)	0.280 (25°C)	3.37
Endosulfan sulfate	422.92	Insufficient vapor pressure data for calculation	No data found	0.117	3.37
Endrin	380.92	5.0×10^{-7}	7×10^{-7} (25°C)	0.26 (25°C)	3.92
Endrin aldehyde	380.92	3.86×10^{-7} (25°C)	2×10^{-7} (25°C)	0.26 (25°C)	4.43
Epichlorohydrin	92.53	$2.38–2.54 \times 10^{-5}$ (20°C)	13 (20°C)	60,000 (20°C)	1
EPN	323.31	—	0.0003 (100°C)	—	3.12
Ethanolamine	61.08	—	<1 (20°C)	Miscible	Unavailable
2-Ethoxyethanol	90.12	—	4 (20°C)	Miscible	Unavailable
2-Ethoxyethyl acetate	132.18	9.07×10^{-7} (20°C)	2 (20°C)	23 wt.% (20°C)	Unavailable
Ethyl acetate	88.11	1.34×10^{-4} (25°C)	94.5 (25°C)	100 ml l^{-1} (25°C)	Unavailable
Ethyl acrylate	100.12	$1.94–2.59 \times 10^{-3}$ (20°C)	29.5 (20°C)	1.5 wt.% (20°C)	Unavailable
Ethylamine	45.08	1.07×10^{-5} (25°C)	400 (2.0°C)	Miscible	Unavailable
Ethylbenzene	106.17	0.00868 (25°C)	10 (25.9°C)	152 (25°C)	1.98
Ethyl bromide	108.97	7.56×10^{-3} (25°C)	386 (20°C)	0.9 wt.% (20°C)	2.67

Compound	Molecular Weight	Henry's Law Constant (atm m³ mol⁻¹)	Vapor Pressure (mm Hg)	Solubility (mg L⁻¹)	Log k_{oc}
Ethylcyclopentane	98.19	2.10×10^{-2} (25°C)	40 (25.0°C)	245 (25°C)	Unavailable
Ethylene chlorohydrin	80.51	—	8 (25°C)	Miscible	Unavailable
Ethylenediamine	60.10	1.73×10^{-9} (25°C)	10 (21.5°C)	Miscible	Unavailable
Ethylene dibromide	187.86	7.06×10^{-4} (25°C)	11 (25°C)	3,370	1.64
Ethylenimine	43.07	1.33×10^{-7} (25°C)	250 (30°C)	Miscible	0.11
Ethyl ether	74.12	1.28×10^{-3} (25°C)	442 (20°C)	6.05 wt.% (25°C)	Unavailable
Ethyl formate	74.08	2.23×10^{-4} (25°C)	194 (20°C)	118,000 (25°C)	Unavailable
Ethyl mercaptan	62.13	2.74×10^{-3} (25°C)	527.2 (25°C)	1.3 wt.% (20°C)	Unavailable
4-Ethylmorpholine	115.18	—	6.1 (20°C)	Miscible	Unavailable
2-Ethylthiophene	112.19	—	60.9 (60.3°C)	292 (25°C)	Unavailable

F

Compound	Molecular Weight	Henry's Law Constant (atm m³ mol⁻¹)	Vapor Pressure (mm Hg)	Solubility (mg L⁻¹)	Log k_{oc}
Fluoranthene	202.26	0.0169 (25°C)	5.0×10^{-6} (25°C)	0.265 (25°C)	4.62
Fluorene	166.22	2.1×10^{-4}	10 (146°C)	1.98 (25°C)	3.7
Formaldehyde	30.03	3.27×10^{-7}	400 (-33°C)	Miscible	0.56
Formic acid	46.03	1.67×10^{-7} at pH 4	35 (20°C)	Miscible	Unavailable
Furfural	96.09	1.52–3.05×10^{-6} (20°C)	2 (20°C)	8.3 wt.% (20°C)	Unavailable
Furfuryl alcohol	98.10	—	0.4 (20°C)	Miscible	Unavailable

G

Compound	Molecular Weight	Henry's Law Constant (atm m³ mol⁻¹)	Vapor Pressure (mm Hg)	Solubility (mg L⁻¹)	Log k_{oc}
Glycidol	74.08	—	0.9 (25°C)	Miscible	Unavailable

H

Compound	Molecular Weight	Henry's Law Constant (atm m³ mol⁻¹)	Vapor Pressure (mm Hg)	Solubility (mg L⁻¹)	Log k_{oc}
Heptachlor	373.32	0.0023	4×10^{-4} (25°C)	180 ppb (25°C)	4.34
Heptachlor epoxide	389.32	3.2×10^{-5}	2.6×10^{-6} (20°C)	0.350 (25°C)	4.32
n-Heptane	100.20	2.035 (25°C)	45.85 (25°C)	2.24 (25°C)	Unavailable
2-Heptanone	114.19	1.44×10^{-4} (25°C)	2.6 (20°C)	0.43 wt.% (25°C)	Unavailable
3-Heptanone	114.19	4.20×10^{-5} (20°C)	1.4 (25°C)	14,300 (20°C)	Unavailable
cis-2-Heptene	98.19	4.13×10^{-1} (20°C)	48 (25°C)	15 (25°C)	Unavailable
trans-2-Heptene	98.19	4.22×10^{-1} (25°C)	49 (25°C)	15 (25°C)	Unavailable

Compound	Molecular Weight	Henry's Law Constant (atm m^3 mol^{-1})	Vapor Pressure (mm Hg)	Solubility (mg L^{-1})	Log k_{oc}
Hexachlorobenzene	284.78	0.0017	1.089×10^{-5}	0.006 (25°C)	3.59
Hexachloro-butadiene	260.76	0.026	0.15 (20°C)	3.23 (25°C)	3.67
Hexachlorocyclo-pentadiene	272.77	0.016	0.081 (25°C)	1.8 (25°C)	3.63
Hexachloroethane	236.74	0.0025	0.8 (30°C)	27.2 (25°C)	3.34
n-Hexane	86.18	1.184 (25°C)	151.5 (25°C)	9.47 (25°C)	Unavailable
2-Hexanone	100.16	0.00175 (25°C)	3.8 (25°C)	35,000 (25°C)	2.13
1-Hexene	84.16	4.35×10^{-1} (25°C)	186.0 (25°C)	50 (25°C)	Unavailable
sec-Hexyl acetate	144.21	$4.38–5.84 \times 10^{-3}$ (20°C)	4 (20°C)	0.013 wt.% (20°C)	Unavailable
Hydroquinone	110.11	$<2.07 \times 10^{-9}$ (20–25°C)	1 (132.4)	70,000 (25°C)	0.98

I

Compound	Molecular Weight	Henry's Law Constant (atm m^3 mol^{-1})	Vapor Pressure (mm Hg)	Solubility (mg L^{-1})	Log k_{oc}
Indan	118.18	—	—	88.9 (25°C)	2.48
Indeno[1,2,3-cd]pyrene	276.34	2.96×10^{-20} (25°C)	10^{-10} (25°C)	0.062	7.49
Indole	117.15	—	—	3,558 (25°C)	1.69
Indoline	—	—	—	10,800 (25°C)	1.42
1-Iodopropane	169.99	9.09×10^{-3}	43.1 (25°C)	0.1065 wt.% (23.5°C)	2.16
Isoamyl acetate	130.19	5.87×10^{-2} (25°C)	4 (20°C)	0.2 wt.% (20°C)	1.95
Isoamyl alcohol	88.15	8.89×10^{-6} (20°C)	2.3 (20°C)	26,720 (22°C)	Unavailable
Isobutyl acetate	116.16	4.85×10^{-4} (25°C)	20 (25°C)	6,300 (25°C)	Unavailable
Isobutyl alcohol	74.12	9.25×10^{-6} (20°C)	10.0 (20°C)	8.7 wt.% (20°C)	Unavailable
Isobutyl benzene	134.22	1.09×10^{-2} (25°C)	2.06 (25°C)	33.71 (25°C)	3.9
Isophorone	138.21	5.8×10^{-6}	0.38 (20°C)	12,000 (25°C)	1.49
Isopropyl acetate	102.13	2.81×10^{-4} (25°C)	73 (25°C)	18,000 (20°C)	Unavailable
Isopropylamine	59.11	—	478 (20°C)	Miscible	Unavailable
Isopropylbenzene	120.19	1.47×10^{-2} (25°C)	4.6 (25°C)	48.3 (25°C)	3.45
Isopropyl ether	102.18	9.97×10^{-3} (25°C)	150 (25°C)	0.65 wt.% (25°C)	Unavailable

K

Compound	Molecular Weight	Henry's Law Constant (atm m^3 mol^{-1})	Vapor Pressure (mm Hg)	Solubility (mg L^{-1})	Log k_{oc}
Kepone	490.68	3.11×10^{-2} (25°C)	2.25 (25°C)	2.7 (20–25°C)	4.74

L

Compound	Molecular Weight	Henry's Law Constant (atm m^3 mol^{-1})	Vapor Pressure (mm Hg)	Solubility (mg L^{-1})	Log k_{oc}
Lindane	290.83	4.8×10^{-7}	6.7×10^{-5} (25°C)	7.52 (25°C)	3.03

Appendix

Compound	Molecular Weight	Henry's Law Constant (atm m^3 mol^{-1})	Vapor Pressure (mm Hg)	Solubility (mg L^{-1})	Log k_{oc}
M					
Malathion	330.36	4.89×10^{-9} (25°C)	7.95×10^{-6} (25°C)	330 (30°C)	2.46
Maleic anhydride	98.06	Not applicable, reacts with water	5×10^{-5} (20°C)	—	Not applicable, reacts with water
Mesityl oxide	98.14	4.01×10^{-6} (20°C)	8.7 (20°C)	3 wt.% (20°C)	Unavailable
Methoxychlor	345.66	Insufficient vapor pressure data for calculation at 25°C	No data found	0.1 (25°C)	4.9
Methyl acetate	74.08	9.09×10^{-5} (25°C)	235 (25°C)	240,000 (20°C)	Unavailable
Methyl acrylate	86.09	$1.23–1.44 \times 10^{-4}$ (20°C)	70 (20°C)	52,000	Unavailable
Methylal	76.10	1.73×10^{-4} (25°C)	400 (25°C)	33 wt.% (20°C)	Unavailable
Methyl alcohol	32.04	4.66×10^{-6} (25°C)	127.2 (25°C)	Miscible	Unavailable
Methylamine	31.06	1.81×10^{-2} (25°C)	3.1 atm (20°C)	9.590 (25°C)	Unavailable
Methylaniline	107.16	1.19×10^{-5} (25°C)	<1.0 (20°C)	5.624 g L^{-1} (25°C)	Unavailable
2-Methylanthracene	192.96	—	—	0.039 (25°C)	5.12
Methyl bromide	94.94	0.2	1,633 (25°C)	13,000 (25°C)	1.92
2-Methyl-1,3-butadiene	68.12	7.7×10^{-2} (25°C)	550.1 (25°C)	642 (25°C)	Unavailable
2-Methylbutane	72.15	1.35 (25°C)	687.4 (25°C)	49.6 (25°C)	Unavailable
3-Methyl-1-butene	70.13	5.35×10^{-1} (25°C)	902.1 (25°C)	130 (25°C)	Unavailable
Methyl cellosolve	76.10	—	6 (20°C)	Miscible	Unavailable
Methyl cellosolve acetate	118.13	—	7 (20°C)	Miscible	Unavailable
Methyl chloride	50.48	0.010 (25°C)	3,789 (20°C)	7,400 (25°C)	1.4
Methylene chloride	84.93	0.00269 (25°C)	455 (25°C)	13,000 (25°C)	0.94
Methylcyclohexane	98.19	4.35×10^{-1} (25°C)	46.3 (25°C)	16.0 (25°C)	Unavailable
o-Methylcyclohexanone	112.17	—	≈1 (20°C)	—	—
1-Methylcyclohexene	96.17	—	—	52 (25°C)	Unavailable
Methylcyclopentane	84.16	3.62×10^{-1} (25°C)	137.5 (25°C)	41.8 (25°C)	Unavailable
Methyl formate	60.05	2.23×10^{-4} (25°C)	625 (25°C)	30 wt.% (20°C)	Unavailable
3-Methylheptane	114.23	3.70 (25°C)	19.5 (25°C)	0.792 (25°C)	Unavailable
5-Methyl-3-heptanone	128.21	1.30×10^{-4} (20°C)	2 (25°C)	0.26 wt.% (20°C)	Unavailable

Compound	Molecular Weight	Henry's Law Constant (atm m^3 mol^{-1})	Vapor Pressure (mm Hg)	Solubility (mg L^{-1})	Log k_{oc}
2-Methylhexane	100.20	3.42 (25°C)	65.9 (25°C)	2.54 (25°C)	Unavailable
3-Methylhexane	100.20	1.55–1.64 (25°C)	61.6 (25°C)	4.95 (25°C)	Unavailable
Methylhydrazine	46.07	—	49.6 (25°C)	Miscible	Unavailable
Methyl iodide	141.94	5.87×10^{-3} (25°C)	405 (25°C)	2 wt.% (20°C)	1.36
Methyl isocyanate	57.05	3.89×10^{-4} (20°C)	348 (20°C)	6.7 wt.% (20°C)	Unavailable
Methyl mercaptan	48.10	3.01×10^{-3} (25°C)	1,516 (25°C)	23.30 g L^{-1} (20°C)	Unavailable
Methyl methacrylate	100.12	2.46×10^{-4} (20°C)	40 (26°C)	1.5 wt.% (20°C)	Unavailable
2-Methylnaphthalene	142.20	Insufficient vapor pressure data for calculation	No data found	25.4 (25°C)	3.93
4-Methyloctane	128.26	10.27 (25°C)	7 (25°C)	0.115 (25°C)	Unavailable
2-Methylpentane	86.18	1.732 (25°C)	211.8 (25°C)	13.8 (25°C)	Unavailable
3-Methylpentane	86.18	1.693 (25°C)	189.8 (25°C)	17.9 (25°C)	Unavailable
4-Methyl-2-pentanone	100.16	1.49×10^{-5} (25°C)	15 (20°C)	1.91 wt.% (25°C)	0.79
2-Methyl-1-pentene	84.16	2.77×10^{-1} (25°C)	195.4 (25°C)	78 (25°C)	Unavailable
4-Methyl-1-pentene	84.16	6.15×10^{-1} (25°C)	270.8 (25°C)	48 (25°C)	Unavailable
1-Methylphenanthrene	192.26	—	—	269 ppb (25°C)	4.56
2-Methylphenol	108.14	1.23×10^{-6} (25°C)	0.24 (25°C)	25,000 (25°C)	1.34
4-Methylphenol	108.14	7.92×10^{-7} (25°C)	0.108 (25°C)	23,000 (25°C)	1.69
2-Methylpropane	58.12	1.171 (25°C)	10 atm (66.8°C)	48.9 (25°C)	Unavailable
2-Methylpropene	56.11	2.1×10^{-1} (25°C)	2.270 (25°C)	263 (25°C)	Unavailable
α-Methylstyrene	118.18	—	1.9 (20°C)	—	—
Mevinphos	224.16	—	0.003 (20°C)	Miscible	Unavailable
Morpholine	87.12	—	13.4 (25°C)	Miscible	Unavailable

N

Compound	Molecular Weight	Henry's Law Constant (atm m^3 mol^{-1})	Vapor Pressure (mm Hg)	Solubility (mg L^{-1})	Log k_{oc}
Naled	380.79	—	2×10^{-4} (20°C)	—	Not applicable, reacts with water
Naphthalene	128.18	4.6×10^{-4}	0.23 (25°C)	30 (25°C)	2.74
1-Naphthylamine	143.19	1.27×10^{-10} (25°C)	6.5×10^{-5} (20–30°C)	1,700	3.51
2-Naphthylamine	143.19	2.01×10^{-9} (25°C)	2.56×10^{-4} (20–30°C)	586 (20–30°C)	2.11

Appendix

Compound	Molecular Weight	Henry's Law Constant (atm m^3 mol^{-1})	Vapor Pressure (mm Hg)	Solubility (mg L^{-1})	Log k_{oc}
Nitrapyrin	230.90	2.13×10^{-3}	0.0028 (20°C)	40	2.64
2-Nitroaniline	138.13	9.72×10^{-5} (25°C)	8.1 (25°C)	1,260 (25°C)	1.23–1.62
3-Nitroaniline	138.13	Insufficient vapor pressure data for calculation	1 (119.3°C)	890 (25°C)	1.26
4-Nitroaniline	138.13	1.14×10^{-8} (25°C)	0.0015 (20°C)	800 (18.5°C)	1.08
Nitrobenzene	123.11	2.45×10^{-5}	0.28 (25°C)	2,000 (25°C)	2.36
4-Nitrobiphenyl	199.21	—	—	—	—
Nitroethane	75.07	4.66×10^{-5} (25°C)	15.6 (20°C)	45 ml L^{-1} (20°C)	Unavailable
Nitromethane	61.04	2.86×10^{-5}	27.8 (20°C)	22 ml L^{-1} (20°C)	Unavailable
2-Nitrophenol	139.11	3.5×10^{-6}	0.20 (25°C)	2,000 (25°C)	1.57
4-Nitrophenol	139.11	3.0×10^{-5} (20°C)	10^{-4} (20°C)	16,000 (25°C)	2.33
1-Nitropropane	89.09	8.68×10^{-5} (25°C)	7.5 (20°C)	1.4 wt.% (20°C)	Unavailable
2-Nitropropane	89.09	1.23×10^{-4} (25°C)	12.9 (20°C)	1.7 wt.% (20°C)	Unavailable
N-Nitrosodimethylamine	74.09	0.143 (25°C)	8.1 (25°C)	Miscible	1.41
N-Nitrosodiphenylamine	198.22	2.33×10^{-8} (25°C)	No data found	35.1 (25°C)	2.76
N-Nitrosodi-n-propylamine	130.19	Insufficient vapor pressure data for calculation	No data found	9,900 (25°C)	1.01
2-Nitrotoluene	137.14	4.51×10^{-5} (20°C)	0.15 (20°C)	0.06 wt.% (20°C)	Unavailable
3-Nitrotoluene	137.14	5.41×10^{-5} (20°C)	0.25 (25°C)	0.05 wt.% (20°C)	Unavailable
4-Nitrotoluene	137.14	5.0×10^{-5} (25°C)	5.484 (26.0°C)	0.005 wt.% (20°C)	Unavailable
n-Nonane	128.26	5.95 (25°C)	4.3 (25°C)	0.122 (25°C)	Unavailable
O					
Octachloronaphthalene	403.73	—	<1 (20°C)	—	—
n-Octane	114.23	3.225 (25°C)	14.14 (25°C)	0.431 (25°C)	Unavailable
1-Octene	112.22	9.52×10^{-1} (25°C)	17.4 (25°C)	2.7 (25°C)	Unavailable
Oxalic acid	90.04	1.43×10^{-10} (pH 4)	<0.001 (20°C)	9.81 wt.% (25°C)	0.89
P					
Parathion	291.27	8.56×10^{-8} (25°C)	9.8×10^{-6} (25°C)	24 (25°C)	3.68
PCB-1016	257.90	750	4×10^{-4} (25°C)	0.22–0.25	4.7

Compound	Molecular Weight	Henry's Law Constant (atm m³ mol⁻¹)	Vapor Pressure (mm Hg)	Solubility (mg L⁻¹)	Log k_{oc}
PCB-1221	192.00	3.24×10^{-4}	0.0067 (25°C)	1.5 (25°C)	2.44
PCB-1232	221.00	8.64×10^{-4}	0.0046 (25°C)	1.45 (25°C)	2.83
PCB-1242	154–358 with an average value of 261	5.6×10^{-4}	4.06×10^{-4} (25°C)	0.24 (25°C)	3.71
PCB-1248	222–358 with an average value of 288	0.0035	4.94×10^{-4} (25°C)	0.054	5.64
PCB-1254	327 (average)	0.0027	7.71×10^{-5} (25°C)	0.012 (25°C)	5.61
PCB-1260	324–460 with an average value of 370	0.0071	4.05×10^{-5} (25°C)	0.080 (24°C)	6.42
Pentachlorobenzene	250.34	0.0071 (20°C)	6.0×10^{-3} (20–30°C)	2.24×10^{-6} M (25°C)	6.3
Pentachloroethane	202.28	2.45×10^{-3} (25°C)	4.5 (25°C)	7.69 and 500 were reported at 25 and 20°C	3.28
Pentachlorophenol	266.34	3.4×10^{-6}	1.7×10^{-4} (20°C)	20–25 (25°C)	2.96
1,4-Pentadiene	68.12	1.20×10^{-1} (25°C)	734.6 (25°C)	558 (25°C)	Unavailable
n-Pentane	72.15	1.255 (25°C)	512.8 (25°C)	39.5 (25°C)	Unavailable
2-Pentanone	86.13	6.44×10^{-5} (25°C)	16 (25°C)	5.51 wt.% (25°C)	Unavailable
1-Pentene	70.13	4.06×10^{-1} (25°C)	637.7 (25°C)	148 (25°C)	Unavailable
cis-2-Pentene	70.13	2.25×10^{-1} (25°C)	494.6 (25°C)	203 (25°C)	Unavailable
$trans$-2-Pentene	70.13	2.34×10^{-1} (25°C)	505.5 (25°C)	203 (25°C)	Unavailable
Pentycyclopentane	140.28	—	—	0.115 (25°C)	Unavailable
Phenanthrene	178.24	2.56×10^{-5} (25°C)	6.80×10^{-4} (25°C)	1.18 (25°C)	3.72
Phenol	94.11	3.97×10^{-7} (25°C)	0.34 (25°C)	93,000 (25°C)	1.43
p-Phenylenediamine	108.14	—	—	38,000 (24°C)	Unavailable
Phenyl ether	170.21	2.13×10^{-4} (20°C)	0.12 (30°C)	21 (25°C)	Unavailable
Phenylhydrazine	108.14	—	<0.1 (20°C)	—	Unavailable
Phthalic anhydride	148.12	6.29×10^{-9} (20°C)	2×10^{-4} (20°C)	0.62 wt.% (20°C)	1.9
Picric acid	229.11	$<2.15 \times 10^{-5}$ (20°C)	<1 (20°C)	1.4 wt.% (20°C)	—
Pindone	230.25	—	—	18 (25°C)	2.95
Propane	44.10	7.06×10^{-1} (25°C)	8.6 atm (20°C)	62.4 (25°C)	Unavailable

Compound	Molecular Weight	Henry's Law Constant (atm m³ mol⁻¹)	Vapor Pressure (mm Hg)	Solubility (mg L⁻¹)	Log k_{oc}
β-Propiolactone	72.06	7.63×10^{-7} (25°C)	3.4 (25°C)	37 vol.% (25°C)	Unavailable
n-Propyl acetate	102.12	1.99×10^{-4} (25°C)	35 (25°C)	18,900 (20°C)	Unavailable
n-Propyl alcohol	60.10	6.74×10^{-6} (25°C)	20.8 (25°C)	Miscible	Unavailable
n-Propylbenzene	120.19	1.0×10^{-2} (25°C)	3.43 (25°C)	55 (25°C)	2.87
Propylcyclopentane	112.22	8.90×10^{-1} (25°C)	12.3 (25°C)	2.04 (25°C)	Unavailable
Propylene oxide	58.08	8.34×10^{-5} (20°C)	445 (20°C)	41 wt.% (20°C)	Not applicable, reacts with water
n-Propyl nitrate	105.09	—	18 (20°C)	—	Unavailable
Propyne	40.06	1.1×10^{-1} (25°C)	4,310 (25°C)	3,640 (20°C)	Unavailable
Pyrene	202.26	1.87×10^{-5}	6.85×10^{-7} (25°C)	0.148 (25°C)	4.66
Pyridine	79.10	8.88×10^{-6} (25°C)	20 (25°C)	Miscible	Unavailable

Q

Compound	Molecular Weight	Henry's Law Constant (atm m³ mol⁻¹)	Vapor Pressure (mm Hg)	Solubility (mg L⁻¹)	Log k_{oc}
p-Quinone	108.10	9.48×10^{-7} (20°C)	0.1 (20°C)	1.5 wt.% (20°C)	Unavailable

R

Compound	Molecular Weight	Henry's Law Constant (atm m³ mol⁻¹)	Vapor Pressure (mm Hg)	Solubility (mg L⁻¹)	Log k_{oc}
Ronnel	321.57	8.46×10^{-6} (25°C)	8×10^{-4} (25°C)	40 (25°C)	2.76

S

Compound	Molecular Weight	Henry's Law Constant (atm m³ mol⁻¹)	Vapor Pressure (mm Hg)	Solubility (mg L⁻¹)	Log k_{oc}
Styrene	104.15	0.00261	6.45 (25°C)	0.031 wt.% (25°C)	2.87
Strychnine	334.42	—	—	0.02 wt.% (20°C)	2.45
Sulfotepp	322.30	2.88×10^{-6} (20°C)	0.00017 (20°C)	25	2.87

T

Compound	Molecular Weight	Henry's Law Constant (atm m³ mol⁻¹)	Vapor Pressure (mm Hg)	Solubility (mg L⁻¹)	Log k_{oc}
2,4,5-T	255.48	4.87×10^{-8} (25°C)	6.46×10^{-6} (25°C)	278 (25°C)	1.72
TCDD	321.98	5.40×10^{-23} (18–22°C)	7.2×10^{-10} (25°C)	0.0193 ppb (22°C)	6.66
1,2,4,5-Tetrabromo-benzene	393.70	—	—	0.040	4.82
1,1,2,2-Tetrabromo-ethane	345.65	6.40×10^{-5} (20°C)	0.1 (20°C)	0.07 wt.% (20°C)	2.45

Compound	Molecular Weight	Henry's Law Constant (atm m^3 mol^{-1})	Vapor Pressure (mm Hg)	Solubility (mg L^{-1})	Log k_{oc}
1,2,3,4-Tetrachloro-benzene	215.89	6.9 × 10^{-3} (20°C)	2.6 × 10^{-2} (25°C)	5.92 (25°C)	5.4 average value
1,2,3,5-Tetrachloro-benzene	215.89	1.58 × 10^{-3} (25°C)	1 (58.2°C)	5.19 (25°C)	6.0 average value
1,2,4,5-Tetrachloro-benzene	215.89	1.0 × 10^{-2} (20°C)	<0.1 (25°C)	0.465 (25°C)	6.1 average value
1,1,2,2-Tetrachloro-ethane	167.85	4.56 × 10^{-4} (25°C)	6 (25°C)	2,970 (25°C)	2.07
Tetrachloroethylene	165.83	0.0153	20 (25°C)	150 (25°C)	2.42
Tetraethyl-pyrophosphate	290.20	—	1.55 × 10^{-4} (20°C)	Miscible	Not applicable, reacts with water
Tetrahydrofuran	72.11	7.06 × 10^{-5} (25°C)	145 (20°C)	Miscible	Unavailable
1,2,4,5-Tetramethyl-benzene	134.22	2.49 × 10^{-2} (25°C)	0.49 (25°C)	3.48 (25°C)	3.79
Tetranitromethane	196.03	—	13 (25°C)	—	—
Tetryl	287.15	<1.89 × 10^{-3} (20°C)	<1 (20°C)	0.02 wt.% (20°C)	2.37
Thiophene	84.14	2.93 × 10^{-3} (25°C)	79.7 (25°C)	3,015 (25°C)	1.73
Thiram	269.35	—	—	30	—
Toluene	92.14	0.00674 (25°C)	22 (20°C)	490 (25°C)	2.06
2,4-Toluene disocyanate	174.15	—	0.01 (20°C)	Not applicable, reacts with water	Not applicable, reacts with water
o-Toluidine	107.16	1.88 × 10^{-6} (25°C)	0.1 (20°C)	15,0000 (25°C)	2.61
Toxaphene	413.82	0.063	0.2–0.4 (25°C)	0.2–0.4 (25°C)	3.18
1,3,5-Tribromo-benzene	314.80	—	—	2.51 × 10^{-6} (25°C)	4.05
Tributyl phosphate	266.32	—	—	0.1 wt.% (20°C)	2.29
1,2,3-Trichloro-benzene	181.45	8.9 × 10^{-3} (20°C)	1 (40°C)	18.0 (25°C)	3.87
1,2,4-Trichloro-benzene	181.45	0.00232	0.29 (25°C)	31.3 (25°C)	2.7
1,3,5 Trichloro-benzene	181.45	1.9 × 10^{-3} (20°C)	0.58 (25°C)	6.01 (25°C)	5.7 (average)
1,1,1-Trichloroethane	133.40	0.0162 (25°C)	124 (25°C)	950 (25°C)	2.18
1,1,2-Trichloroethane	133.40	9.09 × 10^{-4} (25°C)	19 (20°C)	4,500 (20°C)	1.75
Trichloroethylene	131.39	0.0091	72.6 (25°C)	1,100 (25°C)	1.81
Trichlorofluoro-methane	137.37	1.73 (25°C)	792 (25°C)	1,240 (25°C)	2.2
2,4,5-Trichlorophenol	197.45	1.76 × 10^{-7} (25°C)	0.022 (25°C)	1.2 g L^{-1} (25°C)	2.85

Compound	Molecular Weight	Henry's Law Constant (atm m^3 mol^{-1})	Vapor Pressure (mm Hg)	Solubility (mg L^{-1})	Log k_{oc}
2,4,6-Trichlorophenol	197.45	9.07×10^{-8} (25°C)	0.017 (25°C)	800 (25°C)	3.03
1,2,3-Trichloropropane	147.43	3.18×10^{-4} (25°C)	3.4 (20°C)	—	—
1,1,2-Trichlorotrifluoroethane	187.38	3.33×10^{-1} (20°C)	270 (20°C)	0.02 wt.% (20°C)	2.59
Tri-o-cresyl-phosphate	368.37	—	—	3.1 (25°C)	3.37
Triethylamine	101.19	4.79×10^{-4} (20°C)	54 (20°C)	15,000 (20°C)	Unavailable
Trifluralin	335.29	4.84×10^{-5} (23°C)	1.1×10^{-4} (25°C)	240	3.73
1,2,3-Trimethyl-benzene	120.19	3.18×10^{-3} (25°C)	1.51 (25°C)	75.2 (25°C)	3.34
1,2,4-Trimethyl-benzene	120.19	5.7×10^{-3} (25°C)	2.03 (25°C)	51.9 (25°C)	3.57
1,3,5-Trimethyl-benzene	120.19	3.93×10^{-3} (25°C)	2.42 (25°C)	48.2 (25°C)	3.21
1,1,3-Trimethyl-cyclohexane	126.24	—	—	1.77 (25°C)	Unavailable
1,1,3-Trimethyl-cyclopentane	112.22	1.57 (25°C)	39.7 (25°C)	3.73 (25°C)	Unavailable
2,2,5-Trimethyl-hexane	128.26	2.42 (25°C)	16.5 (25°C)	1.15 (25°C)	Unavailable
2,2,4-Trimethyl-pentane	114.23	3.01 (25°C)	49.3 (25°C)	2.05 (25°C)	Unavailable
2,3,4-Trimethyl-pentane	114.23	2.98 (25°C)	27.0 (25°C)	1.36 (25°C)	Unavailable
2,4,6-Trinitrotoluene	227.13	—	4.26×10^{-3} (54.8°C)	0.013 wt.% (20°C)	2.48
Triphenyl phosphate	326.29	5.88×10^{-2} (20–25°C)	<0.1 (20°C)	0.001 wt.% (20°C)	3.72
V					
Vinyl acetate	86.09	4.81×10^{-4}	115 (25°C)	25,000 (25°C)	0.45
Vinyl chloride	62.50	2.78	2,660 (25°C)	1,100 (25°C)	0.39
W					
Warfarin	308.33	—	—	17 (20°C)	2.96
X					
o-Xylene	106.17	0.00535 (25°C)	6.6 (25°C)	213 (25°C)	2.11
m-Xylene	106.17	0.0063 (25°C)	8.287 (25°C)	173 (25°C)	3.2
p-Xylene	106.17	0.0063 (25°C)	8.763 (25°C)	200 (25°C)	2.31

Sources: Montgomery, J.H. and L.M. Welkom, *Groundwater Chemicals Desk Reference*, Vol. 1, Lewis Publishers, Chelsea, MI, 1990.
Montgomery, J.H., *Groundwater Chemicals Desk Reference*, Vol. 2, Lewis Publishers, Chelsea, MI, 1991.

Index

A

Abiotic pathways, 66, 158–161, 165
Absorption process, 404–406, 413
Acetaldehyde, 198–199, 259, 292
Acetate
    augmentation and, 119–120
    carbohydrates and, 86
    carbon dioxide and, 114
    characteristics of, 375
    cometabolism and, 114–115
    vs. ethanol, 193
    fermentation and, 97, 99–100, 114, 125
    kinetic rate constant for, 247
    methanogenesis and, 97
    nucleophilic substitution and, 260
    perchlorate and, 189-193
    substrate efficiency, 191–193
    uranium and, 207
Acetic acid
    in aquifers, 25
    in carboxyl group, 49
    1,4 dioxane and, 198
    in ERD zones, 326
    half-life of, 259
    nucleophilic substitution and, 259
    perchlorate and, 191
    pesticides and, 307
    solubility of, 49
    TCE and, 72
    titration curve for, 27
    in water, 26
Acetobacterium woodii, 114
Acetogenesis, 58, 88, 96, 97
Acetone
    carboxylation and, 133–134
    dechlorination and, 126
    fermentation and, 96, 114
    NAPL and, 417
    oxidation and, 463
    PDB samplers for, 347
    permanganate and, 279, 284, 463
    reagent injection and, 382
    remediation for, 61
    solubility of, 31
    spargeability of, 197
Acetylene, 159, 230, 299

Achromobacter, 171, 187
Acid–base reactions, 22–27, 251–255
Acidaminobacter hydrogenoformans, 116
Acidity, definition of, 25
Acids, 24–27, 133, 333
Acinetobacter, 187
Active treatment, 317
Activity coefficient, 243–244
Adenosine diphosphate, 105
Adenosine triphosphate, *see* ATP
ADNT, 201
ADP, 105
Adsorption process, 152, 266, 404–405, 413
Advective transport, 54, 360, 401
Aerobic reactions, definition of, 9–10
Aeromonas, 171
Agrobacterium, 171
Air sparging, *see* Sparging
Alabandite, 301
Alcaligens, 76, 187
Alcohols
    acidity of, 48, 49
    aldehydes and, 333
    in aquifers, 403
    fermentation and, 101, 133, 145
    hydrogen bonds and, 47
    hydrolysis of, 22
    metals and, 48
    MMOs and, 76
    molecular structure and, 47
    oxidation and, 47, 71–72, 443–444
    partitioning by, 455
    phenols and, 49
    reactivity of, 47–48
    sMMOs and, 198
    solubility of, 30–31, 47
Aldehydes
    in aquifers, 403
    Fenton's reagent and, 273
    fermentation and, 101
    function group, 47, 48–49
    glutathione and, 72
    hydrolysis of, 22
    oxidation and, 48–49, 71–72
    ozone and, 295
    reduction and, 333
    solubility of, 49

Alicycles, 76
Alkalinity, definition of, 25
Alkanes
　distribution of, 415–418
　function group, 51
　hydrolysis of, 21
　ketones and, 133
　MMOs and, 76
　persulfate and, 296
　reduction and, 96
　solubility of, 76
　in volcanic emissions, 85
Alkenes
　dehydrodehalogenation and, 20
　in elimination reactions, 257
　function group, 51
　hydrolysis of, 21
　ketones and, 133
　MMOs and, 76
　reduction and, 96
　stimulation and, 58, 87–88, 127
　in volcanic emissions, 85
Alkyl aryl ether, 48
Alkylamines, 216
Alkylhalides, 21, 73
Alkynes, 21, 51
Aluminum, 38, 162, 164, 297, 307
Amides, 21, 50, 71
Amines
　biodegradation of, 201–202
　funtional group, 50
　hydrolysis of, 21
　methanogenesis and, 97
　oxidation and, 71
　phase-transfer catalysts, 284–285
Amino acids, 41, 50, 99
Amino-4,6-dinitrotoluene, 201
Ammonia
　in aquifers, 43
　chemolithotrophs and, 76
　covalent bonds in, 35
　formation of, 186
　hydrogen ions and, 24
　hydronium ions and pH, 23
　in mineralization, 5
　nitrogen and, 186
　oxidation of, 76
　pH and, 186
Ammonium
　chloride and, 412
　covalent bonds in, 35–36
　funtional group, 50
　hydrogen ions and, 24
　nitrate and, 186, 326
　oxidation and, 14, 284–285

perchlorate and, 188, 190, 402–403, 417
permanganate and, 284–285
in uranium mining, 207
Amount concentrations, 343–344
Amycolata, 197–198
Anaerobic reactions, definition of, 9–10
Anglesite, 301
Anhydrite, 301
Aniline, 50
Anions, 35, 165, 402
Anoxic conditions, 14–15, 55
Anthraquinone-2,6-disulfonate, 74
Antibiotics, 60
Antimony, 38, 162, 163, 170
AQDS, 74
Aqueous-phase reactions
　at atomic level, 246
　cation classifications, 163
　in chemical reactions, 243–244
　contaminant concentration and, 152
　diffusion rate and, 196
　emulsion and, 145
　limiting factors, 452–455
　metals and, 167, 185
　molecular weight and, 196
　for NAPLs, 441–442
　phase-transfer catalysts and, 284–286
　sulfide accumulation in, 167, 210
Aquifers; *see also* Groundwater
　active treatment of, 317
　augmentation vs. stimulation, 58–60
　biology of, 89–91
　contaminants in
　　CFUs in, 89
　　concentration of, 150
　　distribution of, 412–420, 422–425
　　mass of, 406–408
　definition of, 396
　diffusion in, 70
　drinking water from, 1
　gradients in, 83
　half-life in, 95–96
　isotopic analysis of, 350–351
　matrix, *see* Soil
　natural attenuation by, 87
　passive treatment of, 317
　pε range for, 351–353
　permeability of, 101
　pH range for, 25–26
　　bicarbonate and, 17
　　buffering, 27, 101, 340, 363
　　copper and, 22
　　fermentation and, 101
　　metal precipitation and, 182–183
　　metals and, 22, 167, 341

Index

oxidation of, 267–268
poise, system, 54–55, 333
pore water fraction in, 418–419, 432
reactive barrier in, 440–441
redox potential in, 182–184
steady state distribution in, 420–425
treatment of, 52–55, 88
vitamins in, 90
Archaea, 67, 89, 112
Argon, 33, 38
Aromatic hydrocarbons, 21, 47, 71, 76, 274
Arrhenius equation, 241
Arsenic
  abiotic degradation and, 161
  adsorption of, 266, 306
  in aquifers, 58, 178–182, 306, 403
  carbohydrates and, 180–182
  dissociation of, 306
  electronegativity of, 38
  iron and, 177–182, 185, 305–306, 464
  manganese and, 177, 306
  oxidation of, 182
  oxidation state of, 33, 306
  pH and, 306–307
  precipitation of, 176–182, 207, 306–307, 309
  redox potential and, 306–307
  in soil, 163, 176–178
  solubility of, 58, 167, 306
  specific gravity of, 162
  sulfur and, 177
Ash, 169, 174
Association reaction, 7
Astatine, 38
Atmosphere, hazardous, 328
ATP
  in aquifers, 89
  exergonic reactions and, 104
  in fermentation, 99
  formation of, 104–105
  free energy and, 105–106
  in metabolic cycle, 10, 92, 105
  minerals and, 90
  synthesis of, 68
Autotrophic, definition of, 97
Azo compounds, 16, 202
Azoarcus, 116

B

B vitamins, *see* Vitamin B complex
Bachman Road Residential Well Site studies, 122–123
Bacillus, 89, 171, 187
Bacteria; *see also* specific types of

ATP consumption by, 105
augmentation vs. stimulation, 58–60
in biofilms, 71, 391–393
CAHs and, 73
habitat of
  aquifers, 59, 89
  lab vs. IRZ, 6
  space for, 92
hydraulic conductivity and, 71
solvents and, 9, 83
taxonomy of, 67, 112
Barite, 301
Barium, 38, 163, 167, 301
Bases, definition of, 24
Batch-injection system, 378–381
Benzene; *see also* BTEX; Chlorobenzenes
  in aquifers, 403–404
  density of, 403
  direct oxidation of, 73–74, 198, 201
  fermentation of, 100
  hydrolysis of, 21
  inductive effect and, 47
  iron and, 74
  kinetic rate constant for, 247
  methanogenesis and, 74
  nitrate and, 74–75
  organic carbon partition coefficient for, 406
  perchlorate reduction and, 74
  permanganate and, 274
  persulfate and, 296
  spargeability of, 197
  sulfate and, 74–75
  toxicity of, 74
Benzoate, 79, 85
Beryllium, 38, 163, 164
Bicarbonate; *see also* Carbonates
  in aquifers, 17, 24, 25
  CAHs and, 84, 157, 240, 272
  Fenton's reagent and, 271–273
  hydroxyl and, 17, 240, 249–256, 271–273
  iron oxides and, 178
  kinetic rate constant for, 240, 272
  metals and, 17, 178
  nucleophilic substitution and, 260
  radical reactions and, 249–250
  scavenging by, 246–247, 249–256
Binding energy, 405
Bioattenuation, 87, 315
Bioaugmentation, 58–60, 87–88, 118–123, 135–138, 156–158
Biodegradation, 67–68
Biological oxygen demand, 57–58
Biosparging, *see* Sparging
Biostimulation, 58–60, 87–88, 123–158
Biotic pathways, 66, 165

Biotransformation, 67
Biphenyls, polychlorinated, 1
Bismuth, 38, 170
Bisulfite, 260
Bladder pumps, 345
BOD, 57–58
Boiling point, 40, 41
Bond energy, 12, 231–232
Bonding electrons, 35, 39, 42, 257
Borden aquifer, 270, 400
Boron, 38
Brass, 162
Breakthrough loading rate, 438
Bromide
  function group, 51
  nucleophilic substitution and, 260
  ozone and, 291, 295
  phase-transfer catalysts, 285
  for tracers, 403, 418
Bromine, 20, 38, 47, 230–232
Bronze, 162
BTEX, 68, 73–75, 100; *see also* Benzene; Ethyl benzene; Toluene; Xylene
"bubble strip" method, 348
Butane, 45, 76
Butanol, 45, 114
Butanone, 126, 382
Butyrate, 79, 99, 101, 115, 375

C

C1 compounds, 76–77, 86, 97, 104, 125
Cadmium
  in aquifers, 182, 403
  cation classifications, 164
  persulfate and, 295
  precipitation of, 167–168, 170, 174, 182
  in soil, 163
  solubility of, 167, 174, 263
  sorption capacity of, 411
  specific gravity of, 162
  sulfate reduction and, 443
  sulfide and, 301
Calcite, 301, 363
Calcium
  carbonate and, 25–26, 301
  cation classifications, 164
  chromate and, 300
  electronegativity of, 38
  hydraulic conductivity and, 466
  hydride and, 230
  permanganate and, 465
  solubility of, 43, 301
  sorption capacity of, 411–412

  specific gravity of, 162
  sulfate and, 301, 466
Calories, 11
Carbamates, 21
Carbohydrates
  aerobic metabolism of, 95
  arsenic and, 180–182
  biofilm development from, 390–391
  breakthrough loading rate and, 437–438
  cost of, 373
  emulsifiers and, 145
  half-life of, 436
  metals and, 168–169, 171
  NAPL and, 442, 455
  for reductive dechlorination, 79, 84, 86
  for sulfate reduction, 169
  surfactants and, 145
Carbon
  alkalinity and, 25
  in aquifers, 89–94
  covalent bonds and, 35, 38, 41
  dechlorination and, 60
  in dehydrodehalogenation, 20
  electronegativity of, 38, 257
  hydrogen bonds and, 41
  molecular structure and, 46–47
  nucleophilic substitution and, 256–257
  organic partition coefficient of, 127, 152, 196
  oxidation state of, 230, 330
  van der Waals radius in, 44
Carbon dioxide
  in aquifers, 25, 91–92
  in atmosphere, 25, 91
  at atomic level, 9
  in biodegradation, 67
  carbohydrates and, 86
  carboxylation and, 133
  CT and, 114, 230
  DCE and, 72
  Fenton's reagent and, 269–270, 272–274
  in fermentation, 67, 99–100, 114
  free energy of, 12, 68, 104
  in glucose metabolism, 12, 92
  hydrogen peroxide and, 465
  metals and, 167
  methanogenesis and, 55, 65, 97
  in mineralization, 5, 67
  MMOs and, 76
  molecular structure of, 40
  monitoring of, 57, 347–348
  nucleophilic substitution and, 259
  permanganate and, 57, 277–278, 465
  permeability and, 453
  pH and, 24–25, 27, 68
  polarity of, 40

Index

in soil, 25
solubility of, 24–25
TCE and, 65, 77
VC and, 72
water quality and, 57
Carbon disulfide, 279, 284
Carbon monoxide, 39, 77, 97, 115
Carbon tetrachloride (CT)
   in aquifers, 337
   atomic structure of, 73
   carbon dioxide and, 114, 230
   case study, 119–120, 148
   condensation of, 40
   dithionite and, 299
   half-life of, 21
   hydrogenation and, 297
   iron reduction and, 266, 297, 306
   molecular structure of, 40
   oxidation of, 72–73, 230, 274
   oxidation state of, 230
   polarity of, 40
   reductive dechlorination of, 80, 114
   vitamins for, 308
   zero-valent iron and, 306
Carbon trioxide, 43
Carbonates; *see also* Bicarbonate
   in aquifers, 25
   arsenic and, 178, 179
   equilibrium in aquifers, 251
   Fenton's reagent and, 269–270
   hydraulic conductivity and, 466
   nickel and, 444–448
   precipitation and, 16–17, 167, 466
   radical reactions and, 249–250
   reductive dechlorination and, 84, 124
   scavenging by, 246–247, 249, 253
   in uranium mining, 207
Carbonic acid, 24, 27, 250–253
Carbonyl group, 49, 333
Carboxylic acids
   in aquifers, 403
   biodegradation and, 68
   Fenton's reagent and, 273
   formation of, 49
   hydrogen ions and, 24, 49
   hydrolysis of, 21–22
   oxidation of, 71
   ozone and, 295
   solubility of, 49
Catalysts
   activation energy and, 229
   at atomic level, 246
   enzymes as, 13–14
   free energy and, 12
   kinetics of, 235

in microbial reactions, 67
minerals as, 266
Catechols, 202
Cations
   classification of metal, 163–164
   complex formation, 165
   in ionic bonds, 35
   of salts, 402
   soil and, 465
   sorption capacity of, 402, 411–412
   symmetry in, 164
   toxicity of, 164
Cellulose, 373
Cerussite, 301
Cesium, 38
CF, *see* Chloroform
CFCs, 106
CFUs, 89
Chalcoctite, 301
Chemical oxygen demand (COD), 57–58
Chemical reactions, 227–228
   contaminant classes, 61
   definition of, 17
   ionic strength of, 243–245, 309
   kinetics of, 234–241
   mechanisms for, 228, 245
   parameters for, 105–106
   radical reactions, 248–250
   rate of, 18–19, 234–241
   stability diagrams and, 309
   temperature and, 241–243
   thermodynamics of, 231–234
   water quality and, 22
Chemolithotrophs, 76
Chilean saltpeter, 188
Chitin, 373, 375
Chlorate, 190–191
Chloride
   arsenic and, 178, 179
   electronegativity and, 37–38, 257
   function group, 51
   ionic bonds in, 37–38
   iron oxides and, 178
   nickel and, 444
   nucleophilic substitution and, 257, 260
   perchlorate and, 191
   permanganate and, 285
   phase-transfer catalysts, 285
   TCE and, 77
   for tracers, 403
Chlorine
   in aquifers, 43
   vs. bromine, 20
   carbon and, 46–47
   covalent bonds and, 38, 40

dehalorespiration and, 16
dehydrodehalogenation and, 20
dissociation of, 231–232
electronegativity of, 38, 40, 257
free energy of, 104
hydrogen bonds and, 41
nucleophilic substitution and, 257
oxidation and, 71, 331
oxidation state of, 230
perchlorate and, 189, 191
polarity of, 40
in reductive dechlorination, 65
water and, 20, 43
Chlorite, 190–191
Chlorobenzenes, 47, 61, 71, 279
Chloroethene, *see* Vinyl chloride (VC)
Chlorofluorocarbons, 106
Chloroform
   hydrolysis of, 257
   migration of, 400
   nucleophilic substitution for, 257
   permanganate and, 279–280
   reductive dechlorination of, 80, 119, 148
   toxicity of, 119
Chlorohydroxybenzoate, 98
Chloromethane, 71, 73, 98, 274, 306; *see also* Carbon tetrachloride (CT)
Chlorophenols, 61
Chloropropanes, 259, 261
Chromic acid, 1
Chromium
   in aquifers, 43, 300, 403
   atomic structure of, 171
   case studies, 141
   dithionite and, 19, 298–303
   fluoride and, 301
   hydroxide and, 167–168, 171–174, 301
   iron reduction and, 19, 185, 302–305, 307
   oxidation and, 173–174
   oxidation state of, 33, 171
   permanganate and, 275, 284
   precipitation of, 167–168, 171–174, 309
   in soil, 163
   solubility of, 300–301, 464
   specific gravity of, 162
   sulfate reduction and, 443
Ciliates, 89
Citric acid, 92
Citrobacter, 215
Clays, 89, 210, 402, 411, 465
Clostridium, 89, 96–97, 114, 133
Clostridium bifermentans, 96, 109, 114
Clostridium butyricum, 116
Clostridium limosum, 116
Clostridium rectum, 86

Clostridium sphenoides, 86
Coal tar, 337
Cobalamin, 111, 299, 307–308
Cobalt
   CAHs and, 307–308
   cation classifications, 164
   chloromethane and, 307–308
   hydroxyl and, 290
   persulfate and, 295
   precipitation of, 168, 170
   reductive dechlorination and, 109, 112
   in soil, 163
   specific gravity of, 162
COD, 57–58
Colloidal particles, 185, 214–215, 414–415
Colony-forming units, 89
Combustible gas indicator, 328
Cometabolism
   continuum and, 103
   definition of, 97
   vs. dehalorespiration, 87
   vs. direct oxidation, 75–77
   enzymes and, 111
   methanogenesis and, 97
   MTBE and, 78
   stimulation and, 58, 88
   studies of, 65, 66
   sulfate reduction and, 98
   types of, 110–111
Concentration gradient, 11
Concerted reaction, 256–257
Conductivity, hydraulic
   bacteria and, 71
   in Darcy's law, 397
   definition of, 52–53
   dithionite and, 453
   Fenton's reagent and, 268–269
   horizontal vs. vertical, 430–431
   iron sulfide and, 466
   in IRZs, 321–322, 359, 464–466
   manganese oxides and, 465
   permanganate and, 277, 453, 465
   permeability and, 398
   persulfate and, 453
   precipitation reactions and, 71, 453, 464–466
   pyrite and, 466
   saturation and, 451
   of soil, 71
   viscosity and, 397
   zero-valent iron and, 466
Configuration reaction, 5
Consortiums, 10
Containment curtains, 367
Copper, 22, 76, 162–164, 167–168, 301
Coprecipitation, 261–262

Core electrons, 33
Corn syrup
    characteristics of, 375
    cost of, 373
    delivery of, 381
    entrainment and, 131
    in ERD zones, 79, 326, 374
    fermentation and, 133
    methanogenesis and, 158
    vs. molasses, 131
    perchlorate and, 191, 195
    sulfate in, 366
    utilization of, 84
Corrinoids, 98, 109, 112; see also Cobalt
Covalent bonds, 11, 20, 35–46, 331
Covellite, 301
Creosotes, 61
Cresol, 76, 77
Crotonoate, 115
Cutoff/barrier IRZs, 367
Cyanide, 43, 62, 260
Cyanocobalamin, 111, 299, 307–308
Cyclic ethers, 48, 61
Cyclodextrins, 455
Cyclonite, 200, 203, 204, 206

D

Darcy's flow theory, 383
Darcy's law, 360, 396–398
DCA, 73, 80, 98, 197
DCE
    abiotic degradation of, 161
    augmentation and, 118–123, 135–136, 156–158
    cometabolism and, 76–77, 110–111, 114–115
    dehydrodehalogenation and, 21
    direct oxidation of, 72–73
    enzymes and, 109
    in ERD zones, 80
    fermentation and, 100–101, 114–115
    free energy of, 106–107, 109
    half-life of, 259
    iron and, 19, 149–150
    iron reduction and, 158–159, 305–306
    kinetic rate constant for, 247
    methanogenesis and, 76–77, 97–98, 116, 126
    molecular weight of, 344
    NAPL and, 417
    nucleophilic substitution and, 259
    organic carbon partition coefficient for, 406
    oxidation state of, 331–333
    permanganate and, 279
    reductive dechlorination of
        gradients in, 83
        rate of, 150–156
        thresholds, 155
    spargeability of, 197
    stalling and, 102, 150, 158–159
    stimulation and, 123–158
    sulfate reduction and, 98
    surfactants and, 146
    type curves and, 460
DCM, 80
DCP, 247, 272
DDT, 86
Debye–Hückel equation, 244
Dechlorination
    at atomic level, 9, 15–16
    augmentation vs. stimulation, 58–60
    carbon and, 60
    contaminant classes, 61
    in ERD zones, 78–158
    free energy and, 106
    oxidation vs. reduction, 72
Dechloromonas, 190
Defluorination, 106
Dehalobacter, 58, 88
Dehalobacter restrictus, 98, 114
Dehalococcoides ethenogenes
    augmentation and, 135–137, 156–158
    case studies, 120–121, 123
    cometabolism and, 110, 114, 136
    distribution of, 356
    enzymes and, 109
    fermentation and, 100
    genetic studies of, 114, 138, 356–357
    iron reduction and, 98
    lab vs. IRZ, 105–106, 114
    methanogenesis and, 116, 135–137
    PCR analysis for, 59, 123, 135, 355–356
    stalling and, 102, 150
    stimulation and, 58, 88, 135–138
    sulfate reduction and, 98
Dehalogenation
    of alkylhalides, 73
    augmentation and, 88
    CAHs and, 72–73, 365
    contaminant classes, 61
    EDB and, 85
    enzymes and, 72
    free energy and, 106
    of rivers, 86
    stimulation and, 87
Dehalorespiration
    at atomic level, 16
    CAHs and, 87, 98, 113–115
    carbohydrates and, 16
    chlorine and, 16

cometabolism and, 87
definition of, 16
denitrification and, 365
in ERD zones, 99–100
fermentation and, 16, 99–100
genetic studies of, 114
hydrogen and, 16, 102
iron reduction and, 98, 365
kinetics of, 87
methanogenesis and, 16, 365
rate of, 87
solvents and, 16
substrate and, 87
sulfate reduction and, 98, 365
Dehalospirillum, 58, 88, 114–115
Dehydrodehalogenation, 20–21, 61
δ, definition of, 39
Denaturing Gradient Gel Electrophoresis, 354–355
Denitrification
  acetone and, 133
  alkalinity and, 187
  benzene and, 74
  dehalorespiration and, 365
  direct oxidation and, 73–74
  of groundwater, 187
  hydrogen and, 102
  iron reduction and, 187
  IRZ design and, 15, 340
  manganese and, 187
  metals and, 170
  methanogenesis and, 187
  miscibility and, 444
  MTBE and, 78
  vs. nitrification, 186–187
  oxidation and, 14
  perchlorate and, 190
  pyrite and, 210
  reactive barriers and, 444
  reduction and, 119
  suboxic conditions and, 15
  sulfate reduction and, 187
  uranium and, 207
Denitrobacillus, 187
Dense nonaqueous phase liquid, *see* DNAPL
Desorption, 145, 405, 425–429, 439
Desulfitobacterium, 115, 121
Desulfitobacterium chlororespirans, 98, 115
Desulfitobacterium frappieri, 98
Desulfomonile, 58, 88
Desulfomonile tiedjei, 97, 98, 115
Desulfotomaculum, 98
Desulfovibrio, 98, 171
Desulfovibrio desulfuricans, 170
Desulfuromonas, 58, 88, 115

Desulfuromonas chloroethenica, 115
DHE, *see Dehalococcoides ethenogenes*
Dialkyl ether, 48
Diaryl ether, 48
Diatomic molecules, 40
Dibromoethane, 85
Dichlorobenzene, 61
Dichlorodiphenyltrichloroethane, 86
Dichloroethane, *see* DCA
Dichloroethene, *see* DCE
Dichloromethane, 80
Dichlorophenoxyacetic acid, 270
Dichloropropene, 116, 256
Dichromate, 463
Diethylene dioxide, *see* 1,4 dioxane (Dioxacyclohexane)
Diffusion, 70, 91–92, 196, 249, 381, 453
Dilution factor, 387
Dimethylamine, 215
Dimethylbenzimidazole, 308
Dimethylhydrazine, 215
1,4 dioxane (Dioxacyclohexane)
  in aquifers, 60, 195
  biodegradation of, 197
  carbon partition coefficient of, 196
  chemical oxidation of, 195, 444
  cometabolism and, 197, 198–199
  description of, 195
  direct oxidation of, 69, 197–198, 292
  enzymes and, 109, 198
  hydroxyl reactivity of, 292–293, 444
  iron and, 294
  kinetic rate constant for, 247
  molecular structure of, 48
  NAPL and, 417
  ozonation and, 196, 290, 292–294
  permanganate and, 292
  persulfate and, 297
  phytoremediation of, 196–197, 292
  reduction of, 199–200
  solubility of, 196
  spargeability of, 196–197, 294
  THF and, 197–199, 292
Dioxygenases, 69
Dipoles, 39–41, 44–45
Dispersion, 45, 360, 385, 399, 429, 433
Disproportionation, 92, 102, 267, 274, 465
Dissociation equation, 262–263
Dissolved organic carbon (DOC)
  breakthrough loading rate and, 437–438
  chlorinated ethene and, 131
  in flow tests, 94
  half-life of, 95
  for microbial reactions, 382
  oxidation and, 339, 460–462

VC levels and, 128
Dissolved oxygen, 348, 444
Dissolved-phase reactions, 246
Distribution coefficient, 405–406
Disulfides, 16, 51
Dithionate ion, 199–200
Dithionite
   at atomic level, 246
   chromium and, 19, 298–303
   dechlorination and, 297–299
   hydraulic conductivity and, 453
   hydrolysis of, 298
   iron and, 298–300, 302–303
   metals and, 297–303, 443
   mineral activation by, 266
   precipitation of, 465–466
   redox potential and, 464
DMA, 215
DNAPL
   in aquifers, 337–338, 415–418
   definition of, 337
   in ERD zones, 139–141
   geologic layers and, 142
   in IRZs, 321, 322
   modeling of, 337
   monitoring of, 344
   surfactants and, 145
   VOC concentrations and, 145
DOC, *see* Dissolved organic carbon
1-dodecanol, 30–31
Dolomite, 363
Domenico's fate and transport equation, 399–400
Dover Air Force Base, 118–119
Dual-equilibrium desorption model, 409–411
Dual-porosity model, 400–401

E

EDB, 85
Electric dipole, 39
Electrolytes, 29
Electronegativity, 37–41, 257
Electrophiles, 18, 48, 68
Electrostatic forces, 40
Elementary reactions, 5
Elimination reaction, 5, 20–21, 61, 257–261, 299
Emulsifiers, 145
Encapsulation, 261–262
Endergonic reactions, 103–104, 107, 233
Endocrine disruptors, 60
Endothermic reaction, 13
Enhanced reductive dechlorination, *see* ERD zones
Enterobacter, 115, 171

Enterobacter agglomerans, 96
Enthalpy, 12–13
Entropy, 12–13
Enzymes
   as catalysts, 13–14
   definition of, 13, 107
   kinetics of, 235
   molecular weight of, 109
   oxidation and, 69, 72
   reduction and, 98, 103, 107–112
   synthesis of, 109
   van der Waals interactions and, 44
Epoxides, 21, 48, 72
Equilibrium
   in chemical reactions, 6, 251
   concentration, 30
   definition of, 6
   dissolution and, 262–263
   in Gibbs law, 12
   in microbial reactions, 12–13
   modeling of, 422–425
   partitioning, 404–406, 413
   pε, relationship to, 352–353
   pH, relationship to, 26
   in redox reactions, 6, 352–353
   reversibility and, 50
   solubility and, 262–263
ERD zones, 78–85
   abiotic degradation in, 158–161
   case studies, 118–158, 436–437
   microbiology of, 85–87, 102–103
   modes of, 87–88
   pH range for, 364–365
   reactive barriers and, 440
   redox potential in, 158, 464
   type curves for, 456–461
Escherichia, 171
Escherichia coli, 86
Esters, 41, 61, 112
Esthers, 51, 71
Ethane
   at atomic level, 9
   condensation of, 40
   covalent bonds and, 36
   dechlorination and, 65
   hydrolysis of, 261
   in IRZs, 347
   nucleophilic substitution and, 257, 261
   oxidation of, 71, 274
   oxidation state of, 331
   in recovery zone, 439
   reduction of, 97–98
   spargeability of, 197
Ethanol, 31, 99, 114–115, 193, 257, 259
Ethenes; *see also* Chloroethene

abiotic degradation of, 161
at atomic level, 9
augmentation vs. stimulation, 58–60
biodegradation of, 67
case studies, 150–155
cometabolism and, 110, 111, 114
dehydrodehalogenation and, 21
direct oxidation of, 73
EDB and, 85
enzymes and, 109
fermentation and, 100
free energy of, 104, 106
half-life of, 259
iron reduction and, 305–306
in IRZs, 347
methanogenesis and, 97, 116, 117
molecular weight of, 344
nucleophilic substitution and, 257, 259
oxidation state of, 331–333
ozonation of, 286
sulfate reduction and, 97
VC and, 83
Ethers, 22, 47–49, 76, 112, 403, 443–444
Ethyl acetate ester, 20
Ethyl benzene, 100, 197, 406; *see also* BTEX
Ethylene, 104
Ethylene glycol, 292
Ethylene oxide, 48
Ethyne, 37
Eucarya, 67, 89, 112
Evaporite, 188
Exergonic reactions, 103–104, 107, 233
Exothermic reactions, 12–13, 405
Explosives, 186, 188, 200–206

F

Facultative organisms, 96, 98
FADH, 109
Fate and transport equation, 399–400
Fatty acids, 84, 97, 121, 125–126
Fenton's reagent
at atomic level, 246
disproportionation and, 267, 274
goethite and, 271
handling of, 326–328
hydroxyl and, 268–273
iron and, 182, 267–272, 274
mechanisms for, 266–267
miscibility and, 444
monitoring of, 57, 273–274
for NAPLs, 442
organic matter and, 460–463
oxidation reactions, 19, 227, 266–274

permeability and, 453
vs. persulfate, 296–297
pH and, 267–272, 327
phosphate and, 270
radical reactions and, 248–249, 267–273
for reactive barrier, 440–441
Fermentation, 16, 67, 92, 96–102, 114–115, 125–126, 133–134, 145
Fertilizers, 186, 326–327, 402–403
Fire, 189
First-order reactions, 235
Flagellates, 89
Flavins, 109
Flour, 373
Fluoride, 257, 260
Fluorine, 37–38, 41, 43, 47, 230, 231–232, 247
Formaldehyde, 76, 86, 295, 463
Formate, 97, 99, 114–115, 207
Formic acid, 76, 77, 86, 247, 278
Fracture gradient, 385
Free energy, 11–14, 50, 68, 103–109, 164, 232–234
Freon, 1, 197
Freundlich isotherm, 405
Fructose, 92, 104, 131
Fuel, jet, 77, 274, 337
Fulvic acids, 31, 171
Fumarate, 115
Fumigants, 85, 256, 258–260
Functional groups, 46–48
Fungi, 89
Funnel and gate systems, 2, 315
Furan, 48

G

Galena, 301
Gallic acid, 270
Gallium, 38
Gasoline, 66, 77, 337
Germanium, 38
Gibbs law, 12, 103; *see also* Free energy
Gluconobacter, 187
Glucose, 11–13, 84, 91–92, 99, 103–104, 115
Glutathione, 72
Glycerols, 31, 114
Glycolic acid, 278
Glycolipids, 145
Glycolysis, 92, 109
Glyoxylic acid, 77, 247, 278
Goethite, 213, 266, 271, 301
Gold, 162
Gradient, hydraulic, 52–53
Gravity, 16, 397

Index

Gravity-feed injections, 381
"green rust" minerals, 158–159, 307
Greenockite, 301
Greigite, 159
Groundwater; *see also* Aquifers
    boiling of, 327
    clean water front and, 425–429, 439–440, 443, 451, 457
    dispersivity of, 385
    flow characteristics
        bacteria and, 118
        matrix porosity and, 397–398
        modeling of, 333–339
        reagents and, 53–54, 382
        reduction and, 96
        velocity, 320–322, 340, 359–360, 397–398
    odor of, 57–58
    oxygen transfer to, 69–71
    sampling of, 342–351
    taste of, 57–58
    temperature range in, 105–106
Gypsum, 466

H

Half-life, 7, 21, 95–96, 237–238
Halogens, 20–21, 37–38, 46–47, 68, 257
Hanksite, 188
Heat capacity, 40
Helium, 33, 38
Hematin, 111
Hematite, 210, 213
Heme group, 299
Henderson-Hasselbach equation, 26
Henry's law, 196–197
Heptane, 44, 45
Heterogeneous reactions, 5
Heterotrophs, 76
Hexachlorobenzene, 40, 414
Hexachlorocyclohexanes, 86
Hexachloroethane, 21
Hexachloroethane (HCA), 72
Hexahydro-1,3,5-trinitro-1,3,5-triazine, 200, 203, 204, 206
Hexane, 45
Hexoses, 191
High melting explosive, 200, 203, 206
High-solubility inorganics, 402–403
HMX, 200, 203, 206
Homogeneous reactions, 5
Hot-spot IRZs, 367
Humic substances, 31, 74, 171, 214, 347
Hydraulic fracturing, 385

Hydrochloric acid, 23–25
Hydrogen
    in aquifers, 79, 297
    at atomic level, 9
    bonds, 40–45
    carbohydrates and, 86
    characteristics of, 375
    covalent bonds and, 35, 39–41
    in dehalorespiration, 16, 102
    in dehydrodehalogenation, 20
    denitrification and, 102
    1,4 dioxane and, 199
    electronegativity of, 38–41, 257
    explosive reduction and, 203
    fermentation and, 98–102, 114
    fluoride and, 39
    genetic studies of, 114–115
    interspecies transfer of, 99
    ions, *see* Hydronium
    iron reduction and, 102, 297, 304, 306
    methanogenesis and, 93, 97, 101–102
    in mineralization, 5
    molecular structure and, 46
    monitoring of, 348
    nucleophilic substitution and, 257
    in oxidation, 18
    oxidation state of, 230
    polarity of, 40
    release rate, 84
    sulfate reduction and, 102
    supply rate, 93
    van der Waals radius in, 44
Hydrogen peroxide
    direct oxidation of, 70–71, 376
    disproportionation and, 267, 274, 465
    Fenton's reagent and, 267–271, 274, 376
    gas-locking and, 453
    goethite and, 271
    handling of, 326–328
    iron and, 267–271, 274, 376
    kinetic rate constant for, 247
    organic matter and, 460–462
    oxidation potential of, 376
    oxidation state of, 230
    ozonation and, 286–289, 291
    vs. persulfate, 296–297
    radical reactions and, 248–250
    surfactants and, 455
    in uranium mining, 207
Hydrogen Release Compound, 373
Hydrogen sulfide
    arsenic and, 181
    covalent bonds and, 36
    dissociation of, 169–170, 263–264
    free energy of, 104

in IRZs, 347
metals and, 169–170, 263–264, 301
solubility of, 170, 263–264
ventilation of, 328
water quality and, 57
Hydrogenotrophic bacteria, 93, 100, 102
Hydrolysis, 19–21, 256–258
Hydronium, 22–24, 103, 243, 250, 257
Hydrophiles, 41
Hydrophobic bonds, 44–45
Hydroquinone, 109
Hydroxides, 16–17, 20, 165–168, 445–447
Hydroxyl group
bicarbonate and, 17, 240, 250–255, 271–272
cobalt and, 290
diffusion rate and, 249
Fenton's reagent and, 268–273
free energy of, 104
half-life of, 240
hydrogen bonds and, 45
hydronium and, 22–23
kinetic rate constant for, 246–247
miscibility and, 443–444
nucleophilic substitution and, 256–260
ozone and, 286–291, 449
vs. permanganate, 274
persulfate and, 296
pH and, 243, 268–272
radical reactions and, 246–250
solubility and, 41, 45
Hyphomicrobium, 73
Hypochlorite, 19

I

Imines, 50
Immobilization, 52, 66, 185
In situ reactive zones (IRZs)
advantages of, 4–5
closure of, 391–392
configuration of, 367–370
cost of, 358, 367, 391
definition of, 54
design of, 339–340, 343, 357–372
development of, 2–4
documentation of, 330
effectiveness of, 3, 54–55, 317, 343
ERD zones, *see* ERD zones
vs. extraction, 315
geochemistry of, 361–363, 435–440
hydrogeology of, 52–54, 333–337, 358–359, 382, 396–402
limiting factors, 452–455
mapping of, 330
modeling of, 330, 333–339
performance measurement, 341–357, 390–391
pH range for, 321–322, 363–365
pilot test for, 386–387
plumes in, 338–339, 358
vs. pump and treat systems, 2, 4, 156, 315, 439
pumps for, 345
regulations for, 55–57
safety of, 322, 326–328
sequential treatments in, 451–452
site selection for, 320–330, 385–388
size of, 180
vs. soil vapor extraction, 2, 4, 315
source zones, 337–338
vs. sparging, 2, 4, 315
stability diagrams, 309
strategies for, 395–396, 440–452
structure of, 427–429
total electron acceptor flux, 303
tracer studies, 329, 431–433
type curves for, 456–461
vapor migration in, 326–328
wells in, *see* Wells
Indigo carmine method, 348
Indium, 38
Inductive effect, 47
Interspecies hydrogen transfer, 99
Iodine, 38, 47, 188, 230–232, 260
Ion-dipole forces, 40
Ion-ion forces, 40
Ionic bonds, 35, 37–39, 42–43, 331
Ionic strength, 243–245, 309, 465
Iron
abiotic degradation and, 158–161
adsorption sites for, 266
anoxic conditions and, 15, 55, 304–305
in aquifers, 43, 58, 74, 185
arsenic and, 177–182, 185, 305–306, 464
at atomic level, 9, 246
BTEX and, 74
cation classifications, 164
chelation of, 253, 269–270
chromium and, 174, 302–305
contaminant classes, 62
DCE and, 19, 149–150
dehalorespiration and, 98, 365
denitrification and, 187
1,4 dioxane and, 200
dithionite and, 298–300, 302–303
entrainment and, 297, 303
in ERD zones, 464
Fenton's reagent and, 182, 267–272, 274
free energy of, 68, 104
goethite and, 266

humic substances and, 74
hydrogen and, 102, 297, 304
IRZ design and, 340
kinetic rate constant for, 247
metals and, 171, 185
methanogenesis and, 68
MTBE and, 78
NDMA and, 216, 449
oxic conditions and, 304
oxidation and, 74, 117, 440
oxidation state of, 230
ozone and, 290
perchlorate and, 189
persulfate and, 296, 444
petroleum hydrocarbons and, 65
pH and, 185, 302
precipitation of, 167–168, 170, 466
in recovery zone, 439–440
in redox reactions, 14–15
in soil, 185
solubility of, 58, 301
sulfate and, 68, 174
sulfide and
    abiotic degradation and, 158–161
    arsenic and, 180–181
    clays and, 210
    dissociation constant for, 169–170
    dithionite and, 302
    formation of, 170, 209–210
    hydraulic conductivity and, 466
    pH and, 302
    regenerable reactions and, 266
    solubility of, 301–302
    techicium and, 213
    uranium and, 207–212
surface area concentration, 303, 305
system poise and, 55
techicium and, 213
uranium and, 210–211
VC buildup and, 149–150, 158
zero-valent, *see* Zero-valent iron

J

Jarosite, 210
Joules, 11

K

Kaolinite, 89
Kelly Air Force Base, 120–121
Kerosene, 337

Ketones
    in aquifers, 403
    carboxylation and, 133–134
    Fenton's reagent and, 273
    fermentation and, 96, 101, 133–134, 145
    hydrolysis of, 22
    oxidation and, 71, 443–444
    ozone and, 295
    partitioning by, 455
    reduction and, 96–97, 126, 333
    remediation for, 61
    solubility of, 31, 49
Kinetic rate constant, 241, 246, 251, 301
Kinetics
    adsorption processes, 404
    classification of reactions, 234–241
    definition of, 6, 10–11
    of dehalorespiration, 87
    of enzymes, 103, 107–110, 133, 150–151
    feasibility and, 239
    oxidation and, 68
    steady-state approximation, 7–8
    vs. thermodynamics, 6, 234, 239
Krypton, 38, 163

L

Lactate
    augmentation and, 118–119, 122
    bioavailability of, 84
    characteristics of, 375
    cost of, 373
    dehalorespiration and, 16
    delivery of, 381
    in ERD zones, 79, 326, 374
    fermentation of, 99, 114–115
    perchlorate and, 191
    sulfide precipitation and, 170
    uranium and, 207
    well construction and, 378
Lactic acid, 27
Langmuir isotherm, 405
Lanthanum, 215
Largo Department of Energy Pinellas site, 118
Lateral stress gradient, 385
Lattice energy, 43
Lead
    in aquifers, 43
    carbonate and, 301
    cation classifications, 164
    electronegativity of, 38
    precipitation of, 167–168, 170, 174
    in soil, 163
    solubility of, 167, 174, 263

specific gravity of, 162
sulfur and, 301
Lemon juice, 23
Lewis acid/base ligands, 163–164
Ligands, 35, 163–165
Light nonaqueous-phase liquids, 337, 415
Lime, 23, 62, 167, 327
Limestone, 25, 363
Lindane, 86
Lipids, 99
Lithium, 38
Lithotrophic bacteria, 93
Liver disease, 215
LNAPL, 337, 415; *see also* NAPL
London forces, 45
Low-flow pumps, 345

M

Mackinawite, 158, 302
Magnesite, 301
Magnesium
  carbonate and, 301
  cation classifications, 164
  electronegativity of, 38
  in fertilizer, 326–327
  oxidation and, 71
  permanganate and, 465
  sorption capacity of, 411–412
  specific gravity of, 162
  water and, 43
Magnetite, 159, 305
Manganese; *see also* Permanganate
  in aquifers, 43, 58
  arsenic and, 177, 180
  cation classifications, 164
  chromium and, 173–174
  denitrification and, 187
  free energy, pH, and, 68
  hydraulic conductivity and, 465
  MTBE and, 78
  oxidation state of, 33, 230, 277
  perchlorate and, 189
  permanganate and, 57, 277, 285
  phase-transfer catalysts, 285
  precipitation of, 168, 170, 277
  in recovery zone, 439–440
  in redox reactions, 14–15
  solubility of, 58
  suboxic conditions and, 15
  sulfide and, 301
  system poise and, 55
  water quality and, 57
Marcasite, 159

Mass concentrations, 343–344
Matrix demand, 276–277, 363
MC, 71, 73, 114, 148, 285
MEK, 31, 133–134, 197
Melting point, 41
Membranes, 11
Mercury, 162–164, 167–168
Metabolism, definition of, 102
Metals; *see also* specific types of
  in acid-base reactions, 25
  adsorption sites for, 266
  aging effects, 166
  alcohols and, 48
  in aquifers
    BTEX and, 74
    equilibrium and, 340
    leaching of, 419–420
    pH range for, 22, 167
    precipitation of, 4, 16–17
  burial of, 414
  carbohydrates and, 141
  carbonates and, 16–17
  as catalyst, 297
  categorization of, 162
  cation classifications, 163–164
  dithionite and, 19, 297–303
  electronegativity of, 38
  Fenton's reagent, 267, 270–271
  gravity and, 16
  hydrogen sulfide and, 263–264
  hydroxides, 16–17
  ionic bonds, 43
  in IRZs, 4, 321
  lime and, 62, 167
  molecular weight and, 199
  neutralization of, 443
  oxidation state of, 230
  ozonation, 290
  permanganate, 275, 284
  persulfate and, 295–297
  pH and, 16–17, 165–167, 170
  plating operations
    microbial remediation for, 66, 141
    mobility of, 403
    NAPL and, 417
    neutralization of, 414, 443
    precipitation and, 420
    pump withdrawl and, 442–443
  precipitation of, *see* Precipitation reactions
  pump withdrawl of, 442–443
  radio nuclides and, 207–212, 215
  soil and, 16, 164, 171, 178
  solubility of, 57–58, 262–265, 419, 443
  sorption capacity of, 411
  specific gravity of, 161–162

Index   501

sulfate reduction and, 58, 168–171, 443
sulfides and, 52, 98, 164–170, 263–264
toxicity of, 163
water quality and, 57–58, 170
Methane; *see also* Chloromethane
 at atomic level, 9
 in biodegradation, 67
 condensation of, 40
 covalent bonds in, 35–36
 free energy of, 104
 hydrolysis of, 21
 in IRZs, 347
 oxidation of, 246, 296
 oxidation state of, 231
 permanganate and, 284
 persulfate and, 296
 in recovery zone, 439
 reduction of, 231
 vapor migration in, 327–328
 water quality and, 57
Methane monooxygenase, 75–77, 109, 111, 198
Methanobacterium thermoautotrophicum, 97–98
Methanogenesis
 anoxic conditions and, 15, 55
 in aquifers, 58, 89, 96
 BTEX, 74
 BTEX and, 65
 in continuum, 59, 68, 83
 dechlorination and, 95
 definition of, 97
 dehalorespiration and, 16, 365
 denitrification and, 187
 direct oxidation and, 73
 fermentation and, 99, 101
 free energy, 107
 hydrogen in, 93, 97, 101–102
 iron reduction and, 68
 lab vs. IRZ, 142
 MTBE and, 78
 oxidation and, 76–77, 111
 pH range for, 101
 reduction and, 97–98, 116–117, 155
 stimulation and, 58, 88, 126, 135, 142
 sulfate reduction and, 68, 83, 93
Methanol
 bacteria and, 115
 bioavailability of, 84
 CAHs and, 373
 case studies, 120
 dehalorespiration and, 16
 for denitrification, 187
 formation of, 75–76
 hydrogen bonds in, 41
 kinetic rate constant for, 247
 methanogenesis and, 97

 perchlorate and, 191
 solubility of, 30–31
Methanosarcina, 89, 115
Methanosarcina mazei, 97, 115
Methanosariuna thermoplula, 97–98, 115
Methanothrix soehngenii, 98
Methyl alcohol, 48, 86
Methyl bromide, 20, 242–243, 256–261
Methyl cellulose, 373
Methyl ethyl ketone, 31, 133–134, 197
Methyl isobutyl ketone, 134
Methyl orange, 26
Methyl tertiary butyl ethers, *see* MTBE
Methylamine, 41, 115
Methylene chloride, 71, 73, 114, 148, 285
Methylococcus capsulatus, 76
Methylomonas, 76–77
Methylosinus trichosporium, 76–77
Metolachlor, 307
MIBK, 134
Microbial reactions, 65–66
 assessment of, 353–356
 categorization of, 9–10
 concentration gradients in, 11
 consortiums in, 10
 contaminant classes, 61
 definition of, 8
 equilibrium in, 12–13
 pressure and, 11
 rate of, 12–13
 temperature and, 11–12, 112
 terms for, 67
 types of, 66–68
Microbubbles, 71
Micrococcus, 171, 187
Micropurge procedures, 344–345
Millerite, 301
Mineralization, 5, 67, 69
Minerals, 90, 303
Miscible, definition of, 30
Mixtures, definition of, 28
MMOs, 75–77, 109, 111, 198
Molasses
 augmentation and, 135–137
 bioavailability of, 84
 case studies, 124, 127, 128, 131, 138
 characteristics of, 375
 composition of, 169
 vs. corn syrup, 131
 cost of, 373
 dehalorespiration and, 16
 delivery of, 381
 in ERD zones, 79, 326, 373–374
 explosives and, 203
 half-life of, 95–96

handling of, 326
metals and, 141
sulfur in, 159, 365, 366
well construction and, 378
vs. whey, 131
Molozonide, 286
Monitored natural attenuation (MNA), 315–317
Monooxygenases, 69
Montmorillonite, 89, 299
MTBE
in aquifers, 1
bromide and, 291
cometabolism and, 77–78, 111
denitrification and, 78
direct oxidation of, 77–78
dissolved oxygen and, 444
enzymes and, 109, 111, 198
iron reduction and, 78
kinetic rate constant for, 247
manganese reduction and, 78
methanogenesis and, 78
MMOs and, 109
molecular structure of, 48
PDB samplers for, 347
permanganate and, 274, 280
peroxone and, 291
propanes and, 78
spargeability of, 197
sulfate reduction and, 78
Mycobacterium, 76
Mycobacterium vaccae, 197–198

N

NAD, 69, 109
Naphthalene, 247, 286, 406, 417
NAPL; *see also* DNAPL; LNAPL
in aquifers, 403–404, 413–418
carbohydrates and, 442, 455
drainable fraction of, 404, 413
entrainment and, 320, 453
in ERD zones, 139–141, 143
Fenton's reagent for, 442
modeling of, 337
permanganate and, 274, 442
phase-transfer catalysts and, 284–286, 442, 455
residual fraction of, 404, 413–414
in soil, 337
solubility and, 31
surfactants and, 145, 442, 455
type curves for, 456–461
types of, 337
VOC concentrations and, 145

Natroalunite, 210
Natural attenuation, 87, 315–316, 329, 338, 386
NDMA
in aquifers, 60, 215–216, 402–403
formation of, 215–216
hydroxyl and, 449–450
iron and, 449
kinetic rate constant for, 247
miscibility of, 449
odor of, 215
oxidation of, 449–451
ozonation of, 293, 444, 449–450
remediation for, 216
sorption capacity of, 449
Neon, 33, 38
Neptunium, 62
Nernst equation, 233, 352
Niacin, 109; *see also* Nicotinic acid
Nickel
in aquifers, 43, 175–176, 403, 420
carbonates and, 444–448
cation classifications, 164
chloride and, 444, 444–445
dissociation of, 174
hydroxides and, 445–447
in IRZs, 4
persulfate and, 295
pH range for, 444–449
plating solutions, 417, 420, 444
precipitation of
chemical, 263–264, 309
microbial, 167–168, 170, 174–176, 444–448
regulatory limit, 175
remediation for, 62
in soil, 163
solubility of, 174, 263–264, 445–447
specific gravity of, 162
sulfur and, 301, 443–448
Nicotinamide adenine dinucleotide, 69, 109
Nicotinic acid, 109
Nitrate
in aquifers, 90–91, 119, 402–403
BTEX and, 74–75
chromium and, 305
complex formation, 165
dehalogenation and, 115
in fertilizer, 186, 326–327
miscibility and, 444
nucleophilic substitution and, 260
perchlorate and, 189, 190
poise setting for, 189
pyrite and, 210
reactive barriers for, 444
reduction of, *see* Denitrification

Index

solubility of, 186
sources of, 186
sulfate and, 74–75
uranium and, 207
Nitric acid, 25, 186
Nitric oxide, 39
Nitriles, 21, 50
Nitrilotriacetic acid, 270
Nitrite, 190, 444
Nitrobenzene, 247
Nitrogen
    ammonia and, 186
    in aquifers, 43, 90
    at atomic level, 9
    in cobalamin, 308
    condensation of, 40
    covalent bonds and, 35, 38–41
    electronegativity of, 38
    fertilizers and, 186
    free energy of, 68, 104
    funtional group, 50
    hydrogen bonds and, 41
    in mineralization, 5
    molecular structure and, 46
    oxidation and, 68
    oxidation state of, 330
    pH and, 68
    polarity of, 39–40
    reduction of, *see* Denitrification
    sources of, 186
    system poise and, 55
    van der Waals radius in, 44
Nitrosodimethylamine, *see* NDMA
Nitrosomonas, 76
Nonaqueous phase liquids, *see* NAPL
NTA, 270
Nuclear reactions, 11
Nucleic acids, 37
Nucleophiles, 19–20, 48, 241, 255–261

O

Obligate organisms, 97, 98, 99, 112–113
Octahydro-1,3,5,7-tetranitro-1,3,5,7-tetrazocine, 200, 203, 206
Oils, 30
Orbitals, 33
Organic carbon partition coefficient, 405–406, 456
Organohalogens, 47
Orpiment, 180–181
Orthophosphate, 327
Oxalic acid, 247, 278
Oxic conditions, 14–15, 55

Oxidation
    at atomic level, 6, 330
        in chemical reactions, 18–19, 229–230, 245–247
        in microbial reactions, 8–10, 14–15, 68–69
    cost of, 69
    definition of, 18, 68
    methods for, 69–71, 78
    testing for, 8
    vapor migration in, 326–328
    water table and, 161
Oxidation-reduction reactions, *see* Redox reactions
Oxidation states, 229–230, 330–333
Oxyanions, 62
Oxygen
    aerobic vs. anaerobic reactions and, 10
    in aquifers, 90–92
    in atmosphere, 25
    at atomic level, 9
    chromium and, 173–174
    condensation of, 40
    covalent bonds and, 36–41
    dehalogenation and, 115
    diffusion of, 70, 91–92
    disproportionation and, 465
    electronegativity of, 38–40
    Fenton's reagent and, 57, 267–268, 272
    free energy, pH, and, 68
    functional group, 47–49
    handling of, 326–327
    hydrogen bonds and, 41
    in microbial reactions, 68–73
    molecular structure and, 46
    monitoring of, 348
    MTBE and, 77–78
    nitrate and, 74–75
    oxidation state of, 230
    vs. ozone, 449
    permeability and, 453
    polarity of, 39–40
    in redox reactions, 69
    in soil, 25
    system poise and, 55
    van der Waals radius in, 44
    ventilation of, 326–328
    water quality and, 57
Oxygenases, 72
Ozone
    aldehydes and, 295
    at atomic level, 246
    bromide and, 291, 295
    carboxylic acid and, 295
    1,4 dioxane and, 196, 290, 292

formaldehyde and, 295
half-life of, 290–291
handling of, 327
hydrogen peroxide and, 286–289, 291
hydroxyl and, 286–291, 449
iron and, 290
ketones and, 295
kinetic rate constant for, 246–247
miscibility and, 444
monitoring of, 348
oxidation potential of, 376
vs. oxygen, 449
perchlorate and, 189
radical reactions and, 248–250
for reactive barrier, 441
solubility of, 286–289, 449
structure of, 286

P

PAHs, 21, 61, 338
Palladium, 297, 303
Partitioning, 404, 405–406, 413
Passive Diffusion Bag, 345–347
Passive treatment, 317
PCBs, 308
PCE
    abiotic degradation of, 161
    in aquifers, 337, 403–404
    atomic structure of, 73
    augmentation and, 120–123, 135–136, 156–158
    cometabolism and, 76–77, 110–111, 114–115
    cyclodextrin and, 455
    dehalorespiration and, 356
    dehydrodehalogenation and, 21
    density of, 403
    direct oxidation of, 72–73
    dithionite and, 299
    dual-equilibrium desorption model for, 410–411
    enzymes and, 109, 111
    in ERD zones, 80, 418
    faculative organisms and, 96
    Fenton's reagent and, 270, 462–463
    fermentation and, 100–101, 115
    free energy of, 106–107
    genetic studies of, 116
    hydrogen and, 101
    hydroxyl groups and, 286, 462
    iron and
        oxidation of, 270
        reduction of, 65, 158–159, 305–306
    kinetic rate constant for, 247

    methanogenesis of, 73, 76–77, 116–117
    migration of, 400, 404
    molecular weight of, 344
    montmorillonite and, 299
    NAPL and, 417
    organic carbon partition coefficient for, 406
    oxidation state of, 231, 331–333
    ozonation of, 286
    permanganate and, 280–284
    rate of reductive dechlornation, 150–155
    smectite and, 299
    solubility of, 31
    spargeability of, 197
    stalling by, 158–159
    stimulation and, 123–158
    sulfate reduction and, 98
    syntrophism and, 97
    thresholds for, 155
    TOC and, 454
    type curves and, 460
    vitamins for, 308
PCR analysis
    cost of, 354
    for *Dehalococcoides*, 59, 123, 135, 355–356
Pentachlorophenol, 61, 86
Pentanediol, 31
Pentanol, 31
Pentyltriphenylphosphonium bromide, 285
Peptides, 41
Perchlorate
    in aquifers, 74, 190–193, 402–403
    benzene and, 74
    contamination by, 60, 188, 190
    formation of, 188–190
    in IRZs, 4, 17, 191
    miscibility and, 444
    oxidation state of, 230
    poise setting for, 189, 191
    reactive barriers for, 444
    reduction of, 190–195
    regulatory limit, 191
    in soil, 194–195
    uses for, 188
Perchloroethene, *see* PCE
Periodic table, 34
Peristaltic pumps, 345
Permanganate
    acetone and, 279, 284, 463
    at atomic level, 246
    carbon dioxide and, 57, 277–278
    carbon disulfide and, 279, 284
    chloroform and, 279–280
    chromium and, 275, 284
    dechlorination and, 279–280
    handling of, 327

Index

hydraulic conductivity and, 277, 453, 465
  vs. hydroxyl, 274
  ionic strength of, 244
  kinetic rate constant for, 246–247
  matrix demand and, 276–277
  metals and, 275
  methane and, 284
  miscibility and, 444
  for NAPLs, 274, 442
  oxidation and, 19, 227, 274 286, 327
  oxidation potential of, 376
  oxidation state of, 230
  permeability and, 453, 465
  pH and, 365
  phase-transfer catalysts, 284–286, 455
  poise and, 463
  precipitation of, 275, 277
  for reactive barrier, 440–441
  solubility of, 274–276
  surfactants and, 455
  TOC and, 454
Permeability, 397
Permitivity, 29
Peroxides, 230; *see also* Hydrogen peroxide
Peroxone process, 291
Persulfate, 241, 295, 326, 376, 444, 453
Pertechnate, 212
Pertechnetate, 213
Pesticides, 215, 307
Phase-transfer agents, 284–286, 442, 454–455
Phenanthrene, 247
Phenol group, 22
Phenolphthalein, 26
Phenols, 49, 76–77, 201, 247, 272, 274
Phosphate, 119, 178–179, 215, 260–261, 270
Phospholipid Fatty Acid technique, 354–355
Phosphoric acid, 21, 27, 35
Phosphorus, 5, 35, 38, 90
Photosynthesis, 11
Phthalates, 347
Picolinic acid, 270
Pinellas culture, 119
Playa crust, 188
Plug flow, 54, 399, 431
Plume-wide IRZs, 367
Plutonium, 214–215
PMOs, 198
Poise, 54–55, 180, 361, 463
Polarity, 29–30, 33, 38–40, 163
Polarizability, 39
Polonium, 38
Polychlorinated biphenyls, 308
Polycyclic aromatic hydrocarbons (PAHs), 21, 61, 338

Polymerase chain reaction analysis, *see* PCR analysis
Polynuclear aromatic hydrocarbons, 4
Polysaccharides, 92
Poplar trees, 196, 198
Porosity dilation wave, 384
Potash, 167, 188
Potassium
  alcohols and, 48
  bromide and, 39
  cation classifications, 164
  chromate and, 300
  electronegativity of, 38
  nitrate and, 188
  oxidation and, 19
  perchlorate and, 188
  sorption capacity of, 411–412
  specific gravity of, 162
  water and, 43
Potassium permanganate
  carbon dioxide and, 57
  color of, 376
  handling of, 327
  hydraulic conductivity and, 465
  manganese dioxide and, 57
  matrix demand in, 277
  molecular weight of, 244, 274
  oxidation, 19, 274–282
  petroleum hydrocarbons and, 376
  solubility of, 274–275, 277
Potential energy, 10–12
Precipitation reactions, 52, 62
  abiotic pathway, 66
  of carbonates, 167
  chemical reagents and, 17, 228, 261–266, 297, 309, 443
  crystal formation, 164–165, 265
  definition of, 164
  encapsulation in, 261–262
  hydraulic conductivity and, 71, 453, 464–466
  leaching after, 420
  microbial reagents and, 16–17, 161–185, 443
  modes of, 261–262
  pH and, 17
  plating solutions and, 420
  of radionuclides, 207, 211–213, 215
  rate of, 265
  reversibility of, 52, 262
  sequence of, 265–266
  solubility of, 164, 166–171, 262–265
  sulfate reduction and, 98
  water quality, 57–58
  zero-valent iron and, 297, 443
Pressure injection, 381
Prokaryotes, 67

Propane monooxygenase, 198
Propanes, 4, 76, 78, 296; *see also* Chloropropanes
Propanol, 198
Propellant, rocket, 188, 215
Propene, 116
Propionate, 99, 101
Proteins, 37
Protons, 14, 22
Pseudo-first-order reactions, 235, 237
Pseudomonas, 73, 76, 89, 171, 187, 199
Pseudomonas cepacia, 77
Pseudomonas fluorescens, 73
Pseudomonas putida, 77
Pseudomonas stutzeri, 119–120
PTPP, 285
Pump and treat systems, 2, 4, 156, 315, 418, 439
Pumps, 345
Pyrite, 159, 181, 210–211, 301, 466
Pyrochroite, 301
Pyrophosphate, 270
Pyrrhotite, 159
Pyruvate, 92, 99, 114–115

Q

Quantitative structure/activity relationships (QSARs), 21
Quartz, 178
Quinones, 74

R

Radical chain reaction, 248–250
Radionuclides, 62, 206–216, 235
Radon, 33, 38
Random walk effect, 399
RDX, 200, 203, 204, 206
Reaction mechanism, 5–8, 12–13
Reactions, definition of, 5
Reactive barriers, 440–441, 444, 451–452
Reagents
   automatic feed system for, 390
   carbohydrates, *see* Carbohydrates
   chemical vs. microbial systems, 229
   cost of, 372–373
   definition of, 3
   delivery of, 326, 366–367, 378–385, 429–435
   efficiency of, 383
   Fenton's, *see* Fenton's reagent
   in heterogeneous reactions, 266
   injection radius, 430–434
   ionic strength of, 244
   kinetics of, 240–241
   PPT for, 383–385
   in radical chain reactions, 249–250
   reaction order and, 235, 237
   toxicity of, 326
   well location and, 377–378, 432–434
Realgar, 180–181
Redox reactions
   definition of, 6, 54
   vs. E_H, 351–353
   vs. fermentation, 102
   pε and, 351–353
   sequence of, 14–15, 92
   standard state potential, 351–352
Reduction
   at atomic level, 6, 330
      in chemical reactions, 18–19, 229–230, 245–247
      microbial, 8–10, 14–15, 68–69, 73
   case studies, 118–158
   conditions for, 9–10, 14–15
   contaminant classes, 61
   definition of, 6, 18, 68, 79
   in ERD zones, *see* ERD zones
   genetic studies of, 114–115, 138
   metabolism in, 87, 92, 97
   microbiology of, 85–87, 102–103
   modes of, 87–88
   monitoring of, 343–344
   rate of, 72–73, 150–155
   sequence of, 150–155
   testing for, 8
   types of, 66–68
Reversible processes, 50
Rhodococcus, 197–198
Riboflavin, 109
Rivers, 25, 86, 90–91
Royal demolition explosive, 200, 203, 204, 206
Rubidium, 38

S

Safe Drinking Water Act, 56
Scavengers, 246–247, 249–250, 253–255
Schoolcraft case study, 119
Seawater, 23, 244
Second-order reactions, 235–238
Selenium, 38, 62, 162–163, 207
Shiprock site, 207
Siderite, 210
Silicate, 178, 178–179
Silicon, 38
Silver, 162–164, 167–168
Smectite, 299
sMMOs, 76, 109, 111, 198

Index

Smythite, 159
Sodium
   bicarbonate and, 327
   boron and, 19
   bromide and, 385, 387
   carbonate and, 176
   cation classifications, 164
   chloride and, 39–43, 331, 402–403
   electronegativity and, 37–38, 257
   hydroxide and, 119
   oxidation and, 19
   perchlorate and, 188
   sorption capacity of, 411
   specific gravity of, 162
   sulfur and, 19, 167–169
   water and, 43
Sodium permanganate
   dechlorination and, 279–280
   handling of, 327
   hydraulic conductivity and, 465
   molecular weight of, 276
   oxidation, 19, 276–282
   solubility of, 276, 376
Sodium pyrophosphate decahydrate, 270
Soil
   adhesion of, 123
   bacteria and, 65, 85, 92, 114
   CAHs and, 73, 76, 114
   cation exchange capacity, 411–412
   contaminants in
      CFUs in, 89
      distribution of, 412–420, 422–425
      mass of, 406–408
   genetic studies of, 89, 356–357
   hydraulic conductivity of, 71
   hydronium ions and pH, 23
   matrix demand in, 276–277, 363
   for microbubbles, 71
   minerals in, 159
   modeling of, 400–401
   NAPLs in, 337
   natural attenuation in, 87
   oxidation capacity of, 440
   oxygen in, 25
   PCR analysis of, 123
   permeability of, 398
   pH range for, 25
   porosity of, 398
   redox potential, 182–184
   sorptive capacity of, 361
   well construction and, 377–378
Soil vapor extraction, 2, 4, 196, 315
Solid-phase reactions, 246
Soluble methane monooxygenase, 76, 109, 111, 198

Solutions, 28–29
Solvents; *see also* specific types of
   in aquifers, 1, 337
   at atomic level, 9
   bacteria and, 83
   dehalorespiration and, 16
   IRZs and, 4, 28–29
   miscibility of, 30
   molecular structure of, 30–31
   polarity and, 29–30
   in redox reactions, 29
   saturation concentration, 30
Sorption process, 404–412
Source zones, 337–338
Sparging
   definition of, 70
   development of, 2
   effectiveness of, 70
   vs. Fenton's reagent, 268, 272
   Henry's law and, 196–197
   IRZs and, 2, 4
   methods for, 70, 286–287
   vs. microbubbles, 71
   molecular weight and, 196
   vs. pump and treat systems, 2, 315
Specific heat, 41
Sphaelerite, 301
Spirillum, 187
Spirochaetes, 116
sPMOs, 198
Spores, 89, 113
Sporomusa ovata, 115
Starch, 11
Steady-state distribution, 7–8, 420–425
Steel, 307
Stoichiometry, 98
Strontium, 38
Suboxic conditions, 14–15, 55
Substitution reaction, 5, 255–261
Substrate, definition of, 108
Succession, definition of, 90
Succinate, 99, 115
Sucrose, 11, 92, 169, 187, 378, 455
Sugar, 9, 11, 92, 99, 104, 114, 169
Sulfate; *see also* Persulfate; Sulfuric acid
   abiotic degradation and, 158–159
   acetone and, 133
   anoxic conditions and, 15, 55
   in aquifers, 58, 95–96, 210
   arsenic and, 178–179
   bacteria for, 113, 115
   BTEX and, 65, 74–75
   carbohydrates and, 169
   chromium and, 174, 305
   clays and, 210

cometabolism and, 98
definition of, 98
dehalorespiration and, 98, 365
fermentation and, 101–102, 133–134, 145
free energy and, 74, 107
hydrogen and, 93, 102
iron and, 68, 174, 178, 305
IRZ design and, 15, 68, 83, 340
metals and, 58, 168–171, 443
methanogenesis and, 68, 93, 116, 365
miscibility and, 444
MTBE and, 78
nickel and, 444
nitrate and, 74–75
nucleophilic substitution and, 260
perchlorate and, 190
petroleum hydrocarbons and, 65, 74–75
pH range for, 101
precipitation of, 465–466
radical reactions and, 295–296
reactive barriers for, 444
reductive dechlorination and, 97–98
spores and, 113
stimulation and, 58–59, 87–88
uranium and, 207–210
Sulfides
metals and, 16–17, 52, 62, 167–169
nickel and, 444–448
oxidation of, 14–15, 274
perchlorate and, 189
reduction by, 19
scavenging by, 263–264
solubility of, 263–264
techicium and, 213
uranium and, 207
water quality and, 57–58
Sulfites, 19, 298
Sulfonic acid, 21, 49
Sulfur
in aquifers, 43
arsenic and, 177–178
covalent bonds and, 36
dithionite and, 298
electronegativity of, 38
in ERD zones, 365–366
funtional group, 50
hydrogen bonds and, 41
molecular structure and, 46
oxidation state of, 330
reduction of, 115
uranium and, 210–211
van der Waals radius in, 44
Sulfuric acid; see also Sulfate
in aquifers, 25, 43
at atomic level, 9

covalent bonds and, 36
dithionite and, 298
free energy of, 68, 104
hydrolysis of, 21
oxidation and, 68
pH and, 68
system poise and, 55
Sulfurous acid, 298
Sunlight, 11, 216
Surfactants, 145–146, 442, 454–455
SW3810 method, 348
Syntrophism, definition of, 95
Systems, definition of, 6

T

TAT, 202, 206
TBA
cometabolism and, 111
enzymes and, 111
kinetic rate constant for, 247
methanogenesis and, 78
persulfate and, 297
potassium and, 48
spargeability of, 197
TCA
case studies, 138–139
dehydrodehalogenation and, 20–21
direct oxidation and, 73
dithionite and, 299
in ERD zones, 80
Fenton's reagent and, 462–463
hydrolysis of, 256
hydroxyl groups and, 462
kinetic rate constant for, 247
nucleophilic substitution and, 257–258
organic carbon partition coefficient for, 406, 462
oxidation state of, 231
persulfate and, 296
spargeability of, 197
TCE
abiotic degradation, 161
in aquifers, 337
augmentation and, 118–123, 135–136, 156–158
bicarbonate and, 240, 253–255, 272
carbon dioxide and, 65, 77
cometabolism and, 75–77, 110–111, 114–115
cyclodextrin and, 455
direct oxidation of, 72–73
dithionite and, 299
enzymes and, 75–77, 109, 111
in ERD zones, 80, 418

Fenton's reagent and, 271, 462–463
fermentation and, 100
free energy of, 107, 109
hydroxyl groups and, 280, 286, 462
iron reduction and, 65, 305–306
kinetic rate constant for, 247, 280
methanogenesis and, 75–77, 116
migration of, 400
mineralization of, 460–462
molecular weight of, 344
NAPL and, 417
organic carbon partition coefficient for, 406
oxidation state of, 331–333
ozonation and, 286
permanganate and, 279, 280, 285
persulfate and, 296
phase-transfer catalysts for, 285, 455
rate of reductive dechlorination, 150–155
regulatory limit, 255
scavenging and, 253–255
solubility of, 31
spargeability of, 197
stimulation and, 123–158
surfactants and, 455
syntrophism and, 97
thresholds for, 155
type curves and, 460
vitamins for, 308
water quality and, 57–58, 77
TDS, 57, 185
TEA, 285
Techicium, 212–214
Tellurium, 38
Temperature
 ambient, 13
 chemical reactions and, 241–243
 dissociation and, 22
 free energy and, 12–14
 for glucose oxidation, 13
 of groundwater, 105–106
 kinetic rate constant and, 241
 microbial reactions and, 11–12, 112
 persulfate and, 241
Tertiary butyl alcohol, see TBA
Tetra-n-ethylammonium bromide, 285
Tetrachloroethane, 21
Tetrachloroethene, see PCE
Tetrachloroethylene, 104
Tetrachloromethane, 98
Tetrachlorophenol, 61
Tetrahydrofuran, 61, 197–199, 292, 297
Thermodynamics
 equilibrium and, 352–353
 feasibility and, 239
 free energy and, 233–234

vs. kinetics, 6, 234, 239, 309
porosity and, 384
precipitation, control of, 261
reversibility and, 6
saturation and, 384
stability diagrams and, 309
THF, 61, 197–199, 292, 297
Thio complexes, 168
Thiobacillus, 187
Thiols, 51, 242–243, 260, 298
Thyroid, 188
Tin, 38, 163
Titanium, 162
Titanium citrate, 308
Titration curve, 27
TNT, 200–202, 206
Toluene; see also BTEX
 biodegradation of, 201–202
 cometabolism and, 76–77
 direct oxidation of, 201
 fermentation and, 100
 organic carbon partition coefficient for, 406
 persulfate and, 296
 spargeability of, 197
Total dissolved solids (TDS), 57, 185
Total electron acceptor flux rate, 303–304
Total mass of contaminant, 406–408
Total organic carbon (TOC)
 in aquifers, 101, 376
 ERD zones and, 454, 460–463
 partitioning equilibrium and, 408–409
 VC buildup and, 158
Transformation, definition of, 66–67
Transport mechanism, 5
Transverse dispersion, 360
Triaminotoluene, 202, 206
Trichloroacetonitrile, 299
Trichloroethane, see TCA
Trichloroethene, see TCE
Trichloroethlyene, 104
Trichloromethane, 98, 400; see also Chloroform
Trimethylamine, 115
Troilite, 301
Type curves, 456–461

U

Ultraviolet photolysis, 216
Underground Injection Control (UIC) permit, 56
Unimolecular reaction, 6
Universal gas constant, 241
Uranium, 62, 207–212, 464
Ureas, 50, 186
Urethanes, 50

V

Vadose zone
  1,4 dioxane in, 196
  Fenton's reagent and, 268, 273–274
  gas in, 57, 91–92
  plating solutions in, 414
  water in, 55, 196
Valence electrons, 33
van der Waals interactions, 44, 45
Vanadium, 163, 207
van't Hoff equation, 103, 233
Vapor pressure, 40
Vaporization, 41
VC, see Vinyl chloride (VC)
Vegetable oil, 191, 326, 373, 375
Vermiculite, 89
Vinegar, 23
Vinyl chloride (VC)
  abiotic degradation, 161
  augmentation and, 59–60, 122–123, 135–136, 156–158
  biodegradation of, 67
  carbon dioxide and, 72
  cometabolism and, 76–77, 110–111, 114
  dehydrodehalogenation and, 21
  direct oxidation and, 73
  direct oxidation of, 71–73
  dithionite and, 297
  drinking water standards for, 122
  enzymes and, 76, 109–110
  in ERD zones, 80
  fermentation and, 100, 114
  free energy and, 104, 106–107
  half-life of, 259
  inhibition of, 156
  iron and, 149–150, 158–159, 305
  IRZ gradient, 83
  metals and, 117
  methanogenesis and, 76–77, 116, 126
  molecular weight of, 344
  NAPL and, 417
  nucleophilic substitution and, 259
  organic carbon partition coefficient for, 406
  oxidation state of, 331–333
  oxidation vs. reduction, 65, 72
  rate of reduction, 150–156
  spargeability of, 197
  stalling by, 102, 150, 158–159
  stimulation and, 123–158
  thresholds for, 155
  TOC and, 158
  type curves and, 460
  water quality and, 57
Viscosity, 40, 385, 397

Vitamin B complex
  in aquifers, 90
  at atomic level, 246
  cobalamin, 111, 299, 307–308
  corrinoids and, 109
  niacin, 109
  in redox reactions, 109, 307–308
  riboflavin, 109
Volatility, 41
Volume, 6

W

Water; see also Aquifers; Groundwater; Rivers
  acidity of, 25
  activity of, 243
  alkalinity of, 25–26
  in biodegradation, 67
  buffering capacity of, 25
  contaminants in
    CFUs in, 89
    clean water front and, 425–429, 439
    concentration of, 150
    distribution of, 412–420, 422–425
  density of, 397–398
  dipole moment of, 39–40
  dispersion forces in, 45
  dissociation of, 22, 243
  for drinking, 1
  electrostatic forces and, 40
  energy of hydration, 43
  for extraction, 2
  in fermentation, 99–100
  free energy of, 104
  in glucose oxidation, 12
  hydrogen bonds in, 41–42
  hydrophobic bonds and, 44–45
  ionic bonds and, 43
  ionization constants for, 289
  kinetic rate constant for, 247
  liquid-gas phase change, 42
  metals in, 170
  in mineralization, 5
  molecular structure of, 37, 41–42
  nucleophiles and, 19–20, 256–260
  pH range of, 14, 22, 24–27
  polarity of, 39–41
  quality of, 16, 22, 56–58
  solubility in, 28–31
  TCE and, 77
  VC and, 72
  viscosity of, 385, 397
Wells
  biofilm development in, 71, 391–393

construction of, 377–378
cost factors for, 357
design of, 360
injection points in, 366–367, 377
location of
    accessibility, 320–326
    Bachman Road Residential Site, 122
    for biostimulation, 124, 128, 138
    Dover Air Force Base, 118
    entrainment and, 453
    groundwater flow and, 340, 370–372, 378
    hydrogeology, 52–53, 333–337, 358–359
    IRZ gradient, 83, 124
    for pilot tests, 384–386
    performance evaluation, 372
    reagent injection and, 432–434
    Schoolcraft, 119–120
monitoring of, 342–351, 366
types of, 361, 366, 377–378
Whey
    characteristics of, 375
    cost of, 373
    delivery of, 381
    in ERD zones, 79, 131–133, 326, 374
    vs. molasses, 131
    perchlorate and, 191
    well construction and, 378
Witherite, 301
Wood, 79, 196–198, 297, 373

X

Xanthobacter autotrophicus, 73

Xenon, 38
Xylene, 61, 100, 196–197, 406; *see also* BTEX

Y

Yeast, 375

Z

Zero-order reactions, 235
Zero-valent iron
    hydraulic conductivity and, 466
    magnetite and, 305
    PPT and, 385
    precipitation and, 466
    reactivity of, 297, 303–307
    in recovery zone, 440
    redox potential and, 464
    total electron acceptor flux rate and, 303–304
Zinc
    cation classifications, 164
    half-cell reactions for, 352
    in IRZs, 4
    persulfate and, 295
    precipitation of, 167–168, 170
    remediation for, 62
    in soil, 163
    specific gravity of, 162
Zinc hydroxide, 301
Zinc sulfide, 301
Zwitterion, 286